Chemicals from Microalgae

To my parents for their unshaken confidence;
to Fruma, Ofer and Ayal for their love and
support

Chemicals from Microalgae

Edited by

ZVI COHEN

Ben Gurion University of the Negev, Israel

CRC Press
Taylor & Francis Group
Boca Raton   London   New York

CRC Press is an imprint of the
Taylor & Francis Group, an **informa** business

CRC Press
Taylor & Francis Group
6000 Broken Sound Parkway NW, Suite 300
Boca Raton, FL 33487-2742

First issued in paperback 2019

© 1999 by Taylor & Francis Group, LLC
CRC Press is an imprint of Taylor & Francis Group, an Informa business

No claim to original U.S. Government works

ISBN-13: 978-0-7484-0515-2 (hbk)
ISBN-13: 978-0-367-39971-9 (pbk)

Every effort has been made to ensure that the advice and information in this book is true and accurate at the time of going to press. However, neither the publisher nor the author can accept any legal responsibility or liability for any errors or omissions that may be made. In the case of drug administration, any medical procedure or the use of technical equipment mentioned within this book, you are strongly advised to consult the manufacturer's guidelines.

British Library Cataloguing-in-Publication Data
A catalogue record for this book is available from the British Library.

Library of Congress Cataloging-Publication-Data are available

Visit the Taylor & Francis Web site at
http://www.taylorandfrancis.com

and the CRC Press Web site at
http://www.crcpress.com

Contents

Preface

Almost twenty years ago I received a letter from Professor Amos Richmond, who was at that time the director of the Blaustein Institute for Desert Research and the head of the algae unit, offering me the chance to study the production of valuable chemicals from microalgae. In my reply I asked him what valuable chemicals are to be found in microalgae and he said that there are over 30 000 different species of microalgae which contain a variety of unique chemicals, some of which must be of commercial value. I strongly believe that this statement still holds today. We have expanded our knowledge on the contents of various algae; we know much more about the physiological role of some of these chemicals and about the role physiology has on their content. However we have just started to scratch the surface of this relatively untapped resource. As close as it may seem, we still do not produce any microalgal-derived chemical on a large scale. In contrast, the production of agar, carrageenan and alginic acid from macroalgae is a well established industry.

The study of microalgae has seen a major influx of interest in recent years. Perhaps the most important advance in this field is the development of closed systems for cultivation of algal biomass. While open ponds could be utilized for growing microalgae of special niches, closed systems will allow the cultivation of almost any microalgae, enabling the production of valuable chemicals such as astaxanthin from *Haematococcus*. The major obstacle for large scale production of many valuable chemicals is now the cost of production of microalgal biomass which is still too high.

Several volumes were published concerning various aspects of algal biotechnology. More and more studies described the potential of microalgae to produce pigments, fatty acids, polysaccharides and other compounds. However, to date no effort has been made to put together a volume dedicated to these studies. I feel that the level of know-how has developed so as to justify the publication of this volume not as a summary but as an introduction. This book is intended for both established researchers in these fields and students who make their first strides, as well as a wider audience of people interested in the various aspects of microalgae.

The first few chapters of this book deal with the production of polyunsaturated fatty acids (PUFAs), mostly eicosapentaenoic acid (EPA) and arachidonic acid. Subsequent chapters describe research concerning the identification of carotenoids and xanthophylls as well as production of β-carotene from *Dunaliela* and astaxanthine from *Haematococcus*. One chapter describes algal phycobilins. The next group of chapters deals with a variety of compounds: polymers such as polysaccharides and polyhydroxyalkanoates. One chapter describes the variety of compounds of potential pharmaceutical value. The last two chapters are of a general nature and deal with outdoor cultivation of microalgae and economic evaluation of their products.

The contributors to this volume are the foremost authorities in their respective areas and I am very grateful to each and every one of them for joining this venture and for their excellent work. I would like to thank a dear friend, Professor Yair M. Heimer, whose critique and clear thought was, for me, the equivalent of a lighthouse in a stormy night at sea. Lastly, I would like to thank my coworkers, Dr Inna Khozin and Ms Shoshana Didi for their trusted work of many years and Ms Ilana Saller who took care of typing and all the administrative work, without which this volume would never have seen the light of day.

ZVI COHEN
Sede Boker

Foreword

The range of chemicals, agrichemicals and pharmaceuticals which can be isolated in commercially significant quantities from microalgae is impressive. This book brings together articles from the key active groups from around the world. Since there is rapidly increasing interest in 'clean', 'sustainable' and 'organic' technologies to produce products for human consumption, we can expect more and more catalytic use of biological systems to fulfil these requirements. Microalgae have distinct advantages in this respect, especially the cyanobacteria, since they require minimal inputs of light, water and nutrients and do not use arable land. Indeed, in the new trend toward using photobioreactors instead of open land (ponds), especially for the production of very high quality/specification chemicals, we are seeing more closed cycle production systems with little or no effluents – highly desirable in modern society.

What comes across clearly in this book is the dedication of the groups in bringing their technologies to fruition. We see that it has taken many years from biochemical/physiological observations to practical products. There is no doubt that these pioneers have laid the foundations for future basic and practical work which will enhance the production of microalgal chemicals.

The editor and authors are to be congratulated on what I am sure will become a landmark book in the microalgal field. We look forward to seeing how these technologies will be applied in the future to the production of increasing quantities and varieties of clean chemicals from microalgae.

D.O. HALL
King's College, London

Contributors

Professor Diego López Alonso
Universidad de Almería
Facultad de Ciencias Experimentales
Departamento de Biologia Aplicada
Unidad de Genetica
La Cañada de San Urbano
04120 Almería
Spain

Professor Shoshana (Malis) Arad
The Institutes for Applied Biosciences
Ben-Gurion University of the Negev
PO Box 653
Beer-Sheva 84105
Israel

Professor Ami Ben-Amotz
The National Institute of Oceanography
Israel Oceanographic and Limnological Research
Tel-Shikmona
PO Box 8030
Haifa 31080
Israel

Professor Michael A. Borowitzka
Algal Research Laboratory
School of Biological and Environmental Sciences
Murdoch University
Murdoch, WA 6150
Australia

Dr F. García Camacho
Departamento de Ingeniería Química
Universidad de Almería
E-04071 Almería
Spain

Dr Clara I. Segura del Castillo
Universidad de Almería
Facultad de Ciencias Experimentales
Departamento de Biologia Aplicada
Unidad de Genetica
La Cañada de San Urbano
04120 Almería
Spain

Professor Zvi Cohen
Microalgal Biotechnology Laboratory
Jacob Blaustein Institute for Desert Research
Ben Gurion University of the Negev
Sede Boker Campus
84990 Israel

Dr Einar Skarstad Egeland
Organic Chemistry Laboratories
Norwegian University of Science and Technology
N-7034 Trondheim
Norway

Dr F.G. Acién Fernández
Departamento de Ingeniería Química
Universidad de Almería
E-04071 Almería
Spain

Dr A. Giménez Giménez
Departamento de Ingeniería Química
Universidad de Almería
E-04071 Almería
Spain

Professor Alexander N. Glazer
Department of Microbiology and Immunology
University of California, Berkeley
Berkeley
California
USA

Professor Emilio Molina Grima
Departamento de Ingeniería Química
Universidad de Almería
E-04071 Almería
Spain

Dr Claude Largeau
Université Pierre et Marie Curie
Ecole Nationale Superieure de Chimie de Paris
11, Rue Pierre et Marie Curie
75231 Paris Cedex 05
France

Professor Yuan-Kun Lee
Department of Microbiology
National University of Singapore
10 Kent Ridge Crescent
Singapore 119260

Professor Synnøve Liaaen-Jensen
Organic Chemistry Laboratories
Norwegian University of Science and Technology
N-7034 Trondheim
Norway

Dr A. Robles Medina
Departamento de Ingeniería Química
Universidad de Almería
E-04071 Almería
Spain

Dr P. Metzger
Université Pierre et Marie Curie
Ecole Nationale Superieure de Chimie de Paris
11, Rue Pierre et Marie Curie
75231 Paris Cedex 05
France

Dr Roberto de Philippis
Dipartimento di Scienze e Tecnologie
Alimentarie Microbiologiche
Universita'degli Studi
and Centro di Studio dei Microrganismi Autotrofi, CNR
Piazzale delle Cascine 27
I-50144 Firenze
Italy

Professor Amos Richmond
Microalgal Biotechnology Laboratory
Jacob Blaustein Institute for Desert Research
Ben Gurion University of the Negev
Sede Boker Campus
84990 Israel

Dr Assaf Sukenik
Israel Oceanographic and Limnological Research
National Institute of Oceanography
PO Box 8030, Haifa 31080
Israel

Professor Massimo Vincenzini
Dipartimento di Scienze e Tecnologie
Alimentarie Microbiologiche
Universita'degli Studi
and Centro di Studio dei Microrganismi Autotrofi, CNR
Piazzale delle Cascine 27
I-50144 Firenze
Italy

Dao-Hai Zhang
Department of Microbiology
National University of Singapore
10 Kent Ridge Crescent
Singapore 119260

Porphyridium cruentum

ZVI COHEN

The unicellular red alga *Porphyridium cruentum* is a primitive member of the Rhodophyta, order of Porphyridiales. It can be found in sea water and in humid soils. The cells, which can be solitary or massed together in colonies, are spherical, their diameter ranging from 4 to 9 μm. The wall-less cells contain a single large chloroplast and are surrounded by a sheath consisting of a water soluble sulphated polysaccharide (Ramus, 1972; see also Chapter 12). For the taxonomy, morphology and growth conditions of *P. cruentum*, see Vonshak (1987).

Anyone envisaging the large-scale production of *Porphyridium*, should consider the possibility of obtaining useful products from the biomass. The presence in the biomass of sulphated polysaccharides (see Chapter 12) and phycoerythrin, a red proteinaceous pigment (see Chapter 11), can add substantially to the economic feasibility of large scale cultivation of *Porphyridium*. Maintenance of a monoalgal culture may be facilitated by cultivation on high salinity medium which provides a unique ecological niche. Harvesting is less problematic than in certain other species as *Porphyridium* has a tendency towards autoflocculation, especially at low pH (Vonshak *et al.*, 1985).

Fatty acid composition

P. cruentum is unique among those microalgae that produce polyunsaturated fatty acids (PUFA) in its ability to produce both arachidonic acid (AA, 20:4ω6) and eicosapentaenoic acid (EPA, 20:5ω3) in significant quantities. At first glance, this property might seem a liability since the two fatty acids are of antagonistic properties, a circumstance that normally makes separation costly. Fortunately, they are concentrated in different classes of lipids. EPA is abundant in the chloroplastic lipids monogalactosyldiacylglycerol (MGDG), digalactosyldiacylglycerol (DGDG), sulphoquinovosyldiacylglycerol (SQDG) and phosphatidylglycerol (PG) where, under optimal growth conditions, it constitutes 35–62 per cent of the fatty acids whereas that of AA does not exceed 6 per cent (Table 1.1). In the cytoplasmic lipids on the other hand, AA is the major PUFA and its proportion in the phospholipids

Table 1.1 Lipid and fatty acid composition of wild type (WT) and HZ3 mutant of *Porphyridium cruentum*.

Strain	Lipid class	% of total lipids	Fatty acid composition (% of total fatty acids)[a]									
			16:0	16:1	18:0	18:1 ω9	18:2 ω6	18:3 ω6	20:2 ω6	20:3 ω6	20:4 ω6	20:5 ω3
WT	Total	100	26	8	t	t	5	1	t	1	18	41
HZ3	lipids	100	31	5	t	t	4	1	t	1	23	27
WT	MGDG	40	26	1	t	t	4	t	–	t	6	63
HZ3		33	33	2	t	t	7	t	t	t	14	42
WT	DGDG	22	46	t	t	t	5	t	–	t	2	45
HZ3		22	46	t	t	t	5	t	t	t	6	41
WT	SQDG	14	50	2	1	1	2	–	2	t	4	37
HZ3		14	53	t	1	1	2	–	3	t	8	29
WT	PC	10	27	1	1	1	4	3	t	2	57	5
HZ3		8	18	1	1	1	2	3	t	2	68	3
WT	PE	2	35	3	4	3	3	3	–	t	34	13
HZ3		4	43	3	2	2	2	6	–	1	34	9
WT	PG	9	24	37	t	t	1	–	–	–	3	35
HZ3		7	23	36	t	1	2	–	t	–	10	28
WT	PI	1	54	3	1	2	30	1	–	1	6	3
HZ3		2	53	1	3	2	25	2	–	1	13	1
WT	PA	5	16	2	3	2	3	2	–	11	63	7
HZ3		3	17	1	2	1	2	3	–	2	68	4
WT	TAG	2	21	1	2	1	21	1	–	1	33	17
HZ3		10	22	1	2	22	24	2	–	1	40	5

Cultures were cultivated at 25°C under a regime of daily dilution.
[a]Figures are rounded to the nearest whole number.
t = trace levels (less than 1%).

phosphatidyl choline (PC) and phosphatidyl ethanolamine (PE) has been shown to be 57 and 34 per cent, respectively (Table 1.1). In contrast to most other known microalgae, the triacylglycerols (TAG) of *P. cruentum* are rich in PUFA and contain 33 per cent AA as well as 17 per cent EPA (see below). Furthermore, by choosing the appropriate strain and growth conditions it is possible to determine which PUFA will be preferentially accumulated.

Triacylglycerols

The ability of cells to accumulate PUFA is intrinsically limited in most algae since these fatty acids are components of polar, membranal lipids whose content cannot be increased beyond a certain limit. However, certain algal species can be induced to synthesize and accumulate extremely high proportions of TAG (Shifrin and Chisholm, 1981). Unfortunately, while PUFA-rich TAG of higher plant oils are quite common, algal TAG are generally characterized by saturated and monounsaturated fatty acids thought to serve as storage material (Piorreck *et al.*, 1984;

Henderson and Sargent, 1989). This characteristic seems to be common to most algae studied for their potential to produce C_{20} PUFA. The TAG of the eustigmatophyte *Nannochloropsis,* an EPA producer (Seto *et al.,* 1984), contain mainly 14:0, 16:0 and 16:1 (Sukenik *et al.,* 1993). Similar results have been reported for other EPA-producing algae, namely the eustigmatophyte *Monodus subterraneus* (Chapter 2) and the diatom *Phaeodactylum tricornutum* (Chapter 4). Likewise, the TAG of the docosahexaenoic acid (DHA)-rich cryptomonad *Chroomonas salina* (Henderson and McKinlay, 1992) are almost entirely made of C_{18} fatty acids while the proportion of C_{20} fatty acids is less than 3 per cent. In contrast, Cohen *et al.* (1988a, 1995) have shown that TAG of *P. cruentum,* which as noted produces both AA and EPA, are unusually rich in AA. The presence there of AA relates to the unique role that TAG fulfil as a source of eukaryotic MGDG in this alga as will be discussed later. TAG containing high levels of AA appear to be a general feature of many, if not all, rhodophytes. Thus, the proportion of AA in TAG was reported to be 36 per cent in *Chondrus crispus,* 49 per cent in *Polysyphonia lanosa* (Pettitt *et al.,* 1989), and 40 to 64 per cent in various *Gracilaria* sp. (Araki *et al.,* 1990).

As TAG are considered to be the preferred chemical form for addition of AA to milk formulae for the newborn, separation of the AA-rich TAG of *P. cruentum* could be advantageous for the production of milk formulae additives. It could also provide AA-free EPA for pharmaceutical purposes for which highly purified EPA are required (see Chapter 6).

Biosynthesis of AA and EPA

In higher plants and in algae, enzymes synthesizing saturated fatty acids as far as stearate (18:0) are all soluble and located in plastids (Stumpf, 1984). Synthesis of oleic acid (18:1), resulting from the subsequent desaturation of 18:0-ACP (acyl carrying protein), is also carried out by a soluble plastid enzyme system which has been well characterized (Jaworski, 1987). The PUFA 18:2 and 18:3 are formed by sequential desaturation of 18:1, which is esterified to a specific glycerolipid (rather than by desaturation of thioesters) while palmitic acid (16:0) may be desaturated when it is esterified at the *sn*-2 position of MGDG or PG (Browse *et al.,* 1990).

Two major pathways have been proposed for lipid-linked fatty acid desaturation of 18:1 to 18:3ω3 in higher plants and microalgae (Frentzen, 1986; Heemskerk and Wintermans, 1987; Harwood, 1988). In the prokaryotic scheme, MGDG is first synthesized as the 18:1/16:0 species within the chloroplast, and is subsequently desaturated at both positions *in situ* to form 18:3/16:3 MGDG. In the eukaryotic pathway, 18:1-CoA (coenzyme A), synthesized in the chloroplast, is exported across the chloroplast envelope into the endoplasmic reticulum, where it is employed in the synthesis of 18:1/18:1 PC. After desaturation to 18:2/18:2 PC, the phospholipid is either further desaturated to 18:2/18:3ω3 PC, or transported back to the chloroplast for galactosylation of the diacylglycerol moieties in the chloroplast envelope, yielding 18:2/18:2 MGDG, which is finally desaturated to 18:3ω3/18:3ω3 MGDG. These sequences result in two distinctive types of chloroplastic lipids: prokaryotic lipids possess C_{16} fatty acids at the *sn*-2 position and both C_{16} and C_{18} at the *sn*-1 position of glycerolipids, while eukaryotic lipids possess C_{18} fatty acids at the *sn*-2 and both C_{16} and C_{18} acids at the *sn*-1 position. Separation of MGDG and DGDG from *P. cruentum* by reverse-phase HPLC to their constituent molecular species, has shown that the major prokaryotic molecular species of both MGDG and DGDG were 20:5/16:0, 20:4ω6/16:0 and 18:2/16:0 (Shiran *et al.*, 1996). The eukaryotic species 20:5/20:5, 20:4ω6/20:5, 20:5/20:4ω6 and 20:4ω6/20:4ω6 were quite abundant in MGDG, whereas DGDG was predominantly prokaryotic and had very low proportions of these eukaryotic species.

While the PUFA presented in leaf lipids and in many algae contain 16 or 18 carbon atoms, some algae – *P. cruentum* among them – are unique in having PUFA with longer carbon chains such as AA and EPA. A thorough understanding of the biosynthetic pathways leading to the production of these valuable fatty acids is essential for the development of genetic engineering approaches aimed at creating improved algal strains with enhanced PUFA levels. Unfortunately, little is known about the metabolic pathways of PUFA in algae, and what is known appears to be species specific. In plants, there is only one way, at the fatty acid level, by which 18:1 can be converted to 18:3ω3. In algae, the added number of carbon atoms and double bonds is consistent with the existence of at least 12 different theoretical routes, at the fatty acid level for the conversion of 18:2 to 20:5 (Figure 1.1). Indeed, several pathways have been suggested for PUFA biosynthesis in various algae (Nichols and Appleby, 1969). In *P. cruentum*, 18:2 is converted to AA according to Sequence 1.1.

$$18:2\omega 6 \rightarrow 18:3\omega 6 \rightarrow 20:3\omega 6 \rightarrow 20:4\omega 6 \tag{1.1}$$

Feeding with fatty acid precursors

Except for 20:5ω3, the fatty acid profile of *P. cruentum* is composed of ω6 fatty acids only, in keeping with Sequence 1.1 (Table 1.1). However, other fatty acids could be produced at transitory levels implying the existence of additional pathways. A possible approach to increasing the endogenous content of a putative metabolite is feeding algal cells with exogenously supplied fatty acid intermediates (Howling *et al.*, 1968; Gurr, 1971; Higashi and Murata, 1993; Phan Quoc *et al.*, 1993). The presence of 18:3ω3 and 20:4ω3 was detected upon cultivation of *P. cruentum* on a medium supplemented with 18:2ω6, indicating a capability of the algal cells for ω3 desaturation (Shiran *et al.*, 1996). However, no fatty acids of the ω3 series (except for EPA) were identified when the alga was grown in the presence of 18:3ω6 or 20:3ω6. It was inferrred that in contrast to 18:2, neither 18:3ω6 nor 20:3ω6 is a likely substrate for ω3 desaturation. In order to determine whether the appearance of ω3 pathway intermediates upon supplementation with 18:2ω6 indeed indicated the presence of an ω3 pathway, the alga was fed with 18:3ω3. All the fatty acids appearing in Sequence 1.2 below were detected, suggesting that 18:3ω3 could be further converted to 20:4ω3 via 18:4ω3, but also via 20:3ω3 (Sequence 1.2):

$$\begin{array}{ccc} 18:3\omega 3 & \xrightarrow{\Delta 6D} & 18:4\omega 3 \\ \scriptstyle 9,12,15 & & \scriptstyle 6,9,12,15 \\ \downarrow & & \downarrow \\ 20:3\omega 3 & \xrightarrow{\Delta 8D} & 20:4\omega 3 \\ \scriptstyle 11,14,17 & & \scriptstyle 8,11,14,17 \end{array} \tag{1.2}$$

Supplementation with 18:4ω3 led to the appearance of 20:4ω3. However, not even traces of of 20:4ω3 could be detected upon addition of 20:3ω3, suggesting that

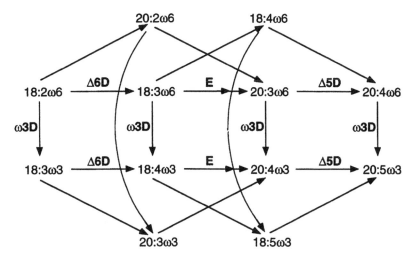

Figure 1.1 Theoretically possible pathways of EPA biosynthesis in *P. cruentum*. Δ6D represents Δ6 desaturase; E, the elongation system; Δ5D, Δ5 desaturase and ω3D, ω3 desaturase.

18:4ω3, rather than 20:3ω3 is the precursor for 20:4ω3 that in the postulated ω3 pathway (Sequence 1.3). Similarly, incorporated 20:2ω6 failed to enhance the level of any of the fatty acids implicated in the ω6 or ω3 pathways, indicating that 20:2ω6 is not an intermediate of EPA biosynthesis.

$$18:2\omega6 \rightarrow 18:3\omega3 \rightarrow 18:4\omega3 \rightarrow 20:4\omega3 \rightarrow 20:5\omega3 \qquad (1.3)$$

Like the ω6 intermediates, the ω3 pathway intermediates, 18:4ω3 and 20:4ω3 were primarily concentrated in the phospholipid and neutral lipid fractions. This finding suggests that the Δ6 desaturation, elongation, and Δ5 desaturation steps of both pathways are cytoplasmic and that they utilize similar enzymatic activities and substrates.

Radiolabelling studies

To elucidate further the various pathways of EPA biosynthesis at the fatty acid and lipid level we pulse-labelled the cultures with [1-^{14}C]linoleic acid (Khozin *et al.*, 1997). Among the lipids, PC and TAG exhibited the highest level of radioactivity directly after the pulse. Thereafter, however, their radioactivity decreased rapidly, while that of the chloroplastic lipids increased. These changes in the labelling pattern suggested a flux of fatty acids from cytoplasmic to chloroplastic lipids – mainly from PC and TAG to the galactolipids. This phenomenon is a characteristic of eukaryotic pathways (Figure 1.2).

The labelling kinetics of individual fatty acids supported Sequence 1.1 implicating PC as the major lipid substrate for the desaturation of 18:2 to 18:3ω6 and 20:3ω6 to 20:4ω6 (Figure 1.3A), and MGDG and DGDG as the main substrates for the ω3 desaturation of 20:4ω6 to 20:5ω3 (Figure 1.3B). The flow of radioactivity in the

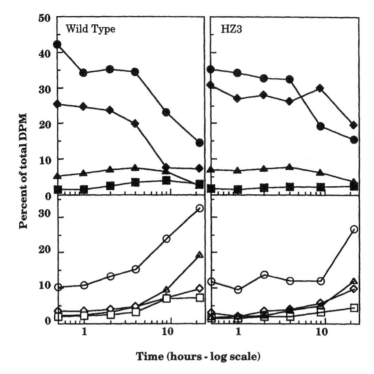

Figure 1.2 Redistribution of total radioactivity in lipids of wild type and HZ3 mutant of *P. cruentum* after labelling with 25 μCi of [1-^{14}C]linoleic acid. Lipids were separated by 2-D TLC.

molecular species of PC from 20:4/18:2* and 16:0/18:2* to 20:4*/20:4* and 16:0/20:4*, respectively, enabled us to further elaborate the ω6 pathway (Sequence 1.4).

$$20:4\omega6/18:2^* \rightarrow 20:4\omega6/18:3\omega6^* \rightarrow 20:4\omega6/20:3\omega6^* \rightarrow 20:4\omega6^*/20:4\omega6^* \quad (1.4)$$
$$\text{(16:0)} \qquad\qquad \text{(16:0)} \qquad\qquad \text{(16:0)} \qquad\qquad \text{(16:0)}$$

In MGDG, both the prokaryotic 20:4ω6*/16:0 and the eukaryotic 20:4ω6*/20:4ω6* were gradually desaturated to 20:5ω3*/16:0 and 20:5ω3*/20:5ω3*, respectively. No molecular species that might be considered precursors for the prokaryotic or eukaryotic C$_{20}$-containing species was found, suggesting, first, that the latter are imported from the cytoplasm, and, second, that only Δ17 (ω3) desaturation takes place in the chloroplast (Sequences 1.5 and 1.6).

$$20:4\omega6/16:0 \rightarrow 20:5\omega3/16:0 \quad (1.5)$$

$$20:4\omega6/20:4\omega6 \rightarrow 20:4\omega6/20:5\omega3 \rightarrow 20:5\omega3/20:5\omega3 \quad (1.6)$$

Following the incorporation of radiolabelled arachidonic acid, EPA was detected in galactolipids but not in PC. It appears, therefore, that while AA is the major product of fatty acid biosynthesis in PC, its further desaturation to EPA in this lipid is either insignificant or non-existent. However, since EPA was observed in PC following a label with [^{14}C]18:2, it follows that 18:2-PC can be converted to EPA in a sequence differing from Sequence 1.2 that does not include AA-PC. In contrast, in both *Nannochloropsis* (Schneider and Roessler, 1994) and *P. tricornutum* (Arao and

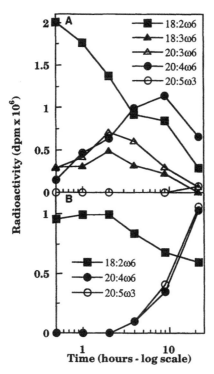

Figure 1.3 Redistribution of total radioactivity in fatty acids of PC (A) and MGDG (B) after labelling (see Figure 1.1) with [1-^{14}C]linoleic acid. Lipids were separated by TLC and fatty acids were determined by radio-HPLC.

Yamada, 1994; Arao *et al.*, 1994) AA is desaturated to EPA by cytoplasmic ω3 (Δ17) desaturases. Thus, PUFA biosynthesis in *P. cruentum* is mostly cytoplasmic; only the ω3 desaturation of AA appears to be chloroplastic.

Labelling with 18:3ω3 produced the ω3 fatty acids depicted in Sequence 1.3. However, in none of the labelling experiments was it possible to obtain direct evidence for the first, distinctive step of the ω3 pathway, the desaturation of 18:2 to 18:3ω3. It seems that an ω3 desaturase can convert 18:2- to 18:3ω3-phospholipid or else that the chloroplastic ω3 desaturase that converts AA to EPA can also desaturate 18:2. The occurrence of 13 per cent EPA in PE could indicate involvement of this lipid in the ω3 pathway (Table 1.1; see also Chapter 2).

The current understanding of the biosynthetic routes of EPA in *P. cruentum* is outlined in Figure 1.4. It appears that EPA is biosynthesized by two pathways originating at the level of 18:2, namely the ω6 and the ω3 pathways. In the ω6 pathway, 18:2-PC is sequentially converted, in the cytoplasm, to 20:4ω6/20:4ω6 PC. In the ω3 pathway, 18:2-PC could be desaturated by a Δ15 (ω3) desaturase to 18:3ω3-PC, which is further converted, by the same enzymatic sequence as utilized in the ω6 pathway, to 20:5ω3-PC. The products of both pathways are exported as their diacylglycerol (DAG) moieties to the chloroplast, to be galactosylated into the respective MGDG molecular species. The AA in the molecular species that evolves from the ω6 pathway is further desaturated to EPA by a D17 (ω3) desaturase.

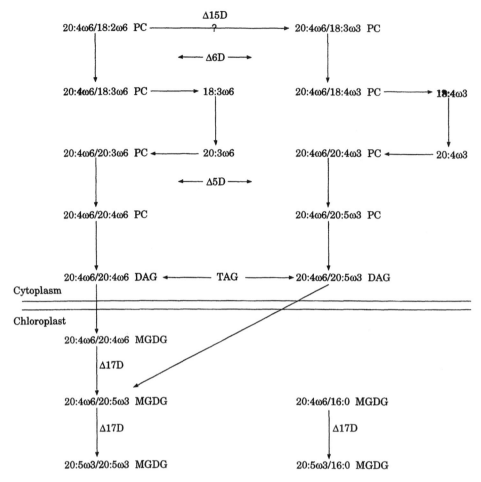

Figure 1.4 Suggested outline of EPA biosynthesis in *P. cruentum.*

The complete map however is still far from clear, and several intriguing questions remain unanswered. While both C_{18} and C_{20} fatty acids can be incorporated into the *sn*-2 position of PC, the former are excluded from the *sn*-1 position. This finding brings up the question of the origin of C_{20} fatty acids residing in the *sn*-1 position. Both the eukaryotic and the prokaryotic molecular species of galactolipids require a source of AA. As in higher plants, eukaryotic species of MGDG import 20:4ω6/ 20:4ω6 from PC and probably also from TAG (see below). However, in contrast to higher plants, where the fatty acids to be incorporated into the *sn*-1 position of prokaryotic species are synthesized in the chloroplast, the *P. cruentum* chloroplast is unable to produce AA for this position. Presumably, the AA for the *sn*-1 position of prokaryotic lipids – and of eukaryotic MGDG as well– originates in the cytoplasm. However, the source AA for prokaryotic lipids is still unknown. Thus, the prokaryotic lipids of *P. cruentum* are structurally, but not biosynthetically, similar to the prokaryotic species of higher plants.

Selection of mutants deficient in EPA biosynthesis

The biosynthesis of plant PUFA was elucidated primarily by studying mutants of
Arabidopsis thaliana deficient in various stages of this process (Browse and
Somerville, 1991). Schneider *et al.* (1995) recently isolated a mutant of
Nannochloropsis deficient in EPA production which was shown to lack the ability
for cytoplasmic $\Delta 17$ desaturation of AA to EPA. However, these mutants were
obtained by random and tedious selection. In a more focused approach, Wada
and Murata (1989) selected desaturase-deficient mutants of the cyanobacterium
Synechocystis based on their chill sensitivity. Wada *et al.* (1992) subsequently proved
that PUFA are necessary for growth and tolerance to photoinhibition at low temperatures.

A search for chill-sensitive mutants of *P. cruentum* was launched on the assumption that EPA in this microalga fulfils a role similar to that of 18:3ω3 in cyanobacteria and *Arabidopsis* (Cohen *et al.*, 1988a). By comparing the growth of putative
mutants on agar plates at 15 and 25°C with that of the wild type we succeeded in
selecting a series of chill-sensitive lines. In one of these, the HZ3 mutant growth was
severely inhibited at 15°C (Figure 1.5) although some inhibition could also be
observed at the optimal temperature of 25°C.

The lipid composition of the HZ3 mutant as compared with the wild type showed
a threefold increase in the proportion of TAG and a corresponding decrease in that
of PC (Table 1.1). Concomitantly, the proportion of EPA decreased to 27 per cent,
(versus 41 per cent in the wild type), while the proportions of 16:0, 18:2 and 20:4ω6
increased. The lipid most affected was MGDG, where the proportion of EPA
decreased from 63 to 42 per cent. However, the other chloroplastic lipids were not
affected. This difference between the lipids is presumably due to the fact that
MGDG is partly eukaryotic, whereas the other chloroplastic lipids are almost

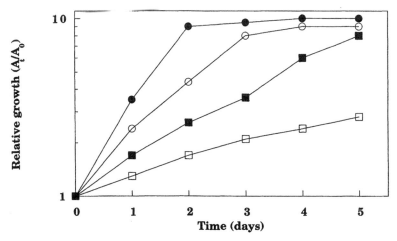

Figure 1.5 Growth (expressed as chlorophyll concentration relative to day 0) of wild type
(filled symbols) and HZ3 mutant (hollow symbols) of *P. cruentum* cultivated at 25°C (circles)
and 15°C (squares).

entirely prokaryotic. Indeed, the proportion of the dominant eukaryotic molecular species of MGDG, 20:5/20:5, decreased whereas that of the prokaryotic species 18:2/16:0, 20:4/16:0 and 20:5/16:0 increased (Figure 1.6). However, the molecular species composition of DGDG was not significantly changed, indicating that the mutation had affected the eukaryotic pathway.

On application of a $[^{14}C]18:2$ pulse, the radioactivity of TAG remained relatively stable initially in the mutant. By contrast, wild-type TAG and both the wild-type and the mutant PC, lost their label rapidly in favour of the chloroplastic lipids (Figure 1.2). As expected, mutant eukaryotic MGDG showed lesser and slower labelling than that of the wild type, indicating a deficiency in the eukaryotic pathway (Khozin *et al.*, 1997).

In leaf lipids of higher plants, TAGs share a common DAG pool with phospholipids, primarily with PC. However, the conversion of DAGs to TAGs is generally considered to be unidirectional (Browse and Somerville, 1991). The findings of Khozin *et al.* (1997) clearly show that, in addition to PC, there is a significant contribution of TAGs to the synthesis of chloroplastic lipids in *P. cruentum*. In contrast to most microalgae, whose TAGs contain short chain saturated and mono-unsaturated fatty acids, those of *P. cruentum* contain both 20:4 and 20:5 fatty acids.

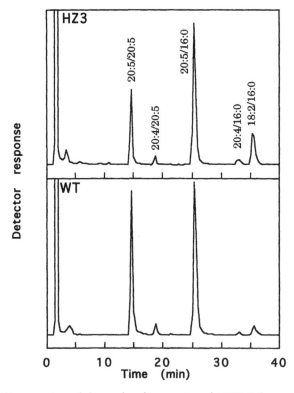

Figure 1.6 HPLC separation of the molecular species of MGDG from wild type and HZ3 mutant of *P. cruentum*. Separation was performed by reverse phase chromatography, detection accomplished by a light-scattering detector as described by Adlerstein *et al.* (1997).

The ability of this organism to utilize its TAGs may explain the unique fatty acid composition of these lipids.

P. cruentum is mainly found in shallow marshes, an environment in which temperature fluctuations are rapid and more pronounced than in deeper bodies of water. The increase in the proportion of EPA in MGDG and especially in that of the eukaryotic component of MGDG, 20:5/20:5, may reflect the organism's attempt to cope with stress inflicted by sudden drops in temperature. However, *de novo* synthesis of EPA is apparently not fast enough to accommodate the increased demand for EPA. Arguably, TAG could be utilized as a buffering capacity for 20:4- and 20:5-containing DAG, which could be relatively rapidly mobilized for the production of eukaryotic molecular species of MGDG. The HZ3 mutant is deficient in the eukaryotic pathway, and compensation by enhanced production of prokaryotic species is inherently limited.

Environmental factors affecting PUFA content

The fatty acid composition and content of algae can be modulated by changing growth conditions. When growth is retarded in response to any limiting factor, such as light restriction, nutrient depletion, or non-optimal pH, temperature or salinity, lipid and carbohydrate synthesis may be enhanced at the expense of protein synthesis (Cohen *et al.*, 1995). Generally, the newly synthesized lipid will serve as reserve material and thus will consist mainly of triglycerides (TAG). Since the fatty acids that constitute algal TAG are quite different from those of the polar lipids, EPA and other membrane associated fatty acids will be diluted. The highest levels of TAG are generally found at the stationary phase (Iwamoto *et al.*, 1955; Klyachko-Gurevich *et al.*, 1967).

Environmental conditions affect fatty acid composition and hence the production of PUFA by the various microalgae. However, optimal conditions for PUFA productivity may vary from species to species. For example, in strains where the EPA content peaks at suboptimal growth temperatures, EPA production rate (mg $EPAe^{-1} \cdot d^{-1}$) could be severely reduced by elevated temperatures. Indeed, Sukenik *et al.* (1989) have shown that the cellular content of total fatty acids and EPA in *Nannochloropsis* decreases as growth temperature increases, the highest proportion of EPA being attained at a temperature slightly below the optimum for growth.

Temperature

Growth temperature is one of the most important factors affecting PUFA productivities. It can affect biomass productivity, lipid and fatty acid content and the degree of unsaturation of the fatty acids, all of which must be considered simultaneously in evaluating the potential of temperature modulation as a means of controlling PUFA production.

The effects of temperature on the total fatty acid content of microalgae cannot be generalized, as only a small number of species have been studied and the results are not consistent between species. In most of the PUFA-producing microalgae studied so far, fatty acid content increases with temperature. The lipid content of *Ochromonas danica* for instance, increased from 39 to 53 per cent as the temperature was raised from 15 to 30°C (Aaronson, 1973). A similar trend was observed also in *P. tricornutum* and *Monodus subterraneus* (Iwamoto and Sato, 1986). On the other hand, in *Nannochloropsis* the fatty acid content was higher at 18°C than at 32°C (Sukenik *et al.*, 1989).

An increase in the degree of fatty acid unsaturation in response to a decrease in growth temperature is an almost universal phenomenon, and it has been observed in blue–green algae (Holton *et al.*, 1964; Kleinschmidt and McMahon, 1970), bacteria (Kates and Hagen, 1964; Duran, 1970), eukaryotic algae (Patterson, 1970; Ahern *et al.*, 1983; Seto *et al.*, 1984), yeast (Brown and Rose, 1969), and fungi (Sumner *et al.*, 1969; Hansson and Dostalek, 1988; Shimizu *et al.*, 1988), as well as in higher plants. While the phenomenon is widely documented, little is known about its origin. The need to maintain membrane fluidity was implicated as early as 1962 (Marr and Ingraham, 1962). However, in many PUFA-rich microalgae, the level of unsaturation remains high even at elevated temperatures. In *P. cruentum*, the proportion of EPA decreases sharply under cultivation at supra-optimal temperatures, whereas that of the highly unsaturated arachidonic acid increases (Cohen *et al.*, 1988a).

The proportion of EPA in both MGDG and DGDG increases with decreasing growth temperature in dense cultures. EPA proportions in MGDG of 10, 37 and 51 per cent have been obtained in cultures cultivated at 30, 25 and 20°C, respectively (Table 1.2); reducing the growth temperature from 30 to 25°C in low-density cultures produces a similar increase in the proportion of EPA from 37 to 48 per cent. However, a further drop in temperature to 20°C results in a reduction, rather than an increase, in the proportion of EPA (44 per cent in MGDG).

Comparison of the molecular species compositions of MGDG at different temperatures shows that the proportion of eukaryotic components in low density cultures is inversely related to growth temperature (Adlerstein *et al.*, 1997): 20–22 per cent of the molecular species were eukaryotic at 30°C, as against 58 per cent at 20°C, the most significant increase being noted in the proportion of 20:5/20:5, the ultimate eukaryotic species.

In *P. cruentum* the level of $\Delta 17$ ($\omega 3$) desaturation in prokaryotic and eukaryotic molecular species of both MGDG (Table 1.3) and DGDG decreases with increasing

Table 1.2 Fatty acid composition of MGDG in *P. cruentum* cultivated at different temperatures and biomass densities.

T (°C)	Cell density	Fatty acid composition (% of total fatty acids)										
		16:0	16:1 $\omega 11$	16:1	18:0	18:1 $\omega 9$	18:1 $\omega 7$	18:2	20:2 $\omega 6$	20:3 $\omega 6$	20:4 $\omega 6$	20:5 $\omega 3$
20	High[a]	29	t	1	1	1	t	10	t	t	7	51
	Low[b]	33	t	1	1	1	t	16	t	t	5	44
25	High	26	t	1	t	t	t	5	t	t	29	37
	Low	38	t	t	1	1	t	6	1	t	5	48
30	High	29	5	t	7	18	3	11	1	1	17	10
	Low	17	1	1	4	8	3	8	2	t	19	37

[a]High cell density, cultures diluted daily to 8–16 mg chl/ml; [b]low cell density, cultures diluted daily to 2–6 mg chl/ml; t, trace levels.

Table 1.3 Molecular species composition of MGDG of *P. cruentum* cultivated at different temperatures and biomass concentrations

Molecular species (% of total)	Cell density	20°C		25°C		30°C	
		High	Low	High	Low	High	Low
Eukaryotic species							
20:5/20:5		40	46	19	26	6	12
20:4/20:5		13	11	17	2	9	7
20:4/20:4		5	1	9	t	8	2
Total eukaryotic		58	58	46	29	22	21
Prokaryotic species							
20:5/16:0		28	20	26	54	23	54
20:4/16:0		5	2	20	2	46	16
18:2/16:0		10	20	8	15	10	10
Total prokaryotic		42	42	54	71	78	80

High, cultures diluted daily to 8–16 mg chl/ml; low, cultures diluted daily to 2–6 mg chl/ml; t, trace levels.

Pairs of fatty acids separated by slash designate the components in the *sn*-1 and *sn*-2 positions, respectively, of the molecular species (except for 20:4/20:5, which is apparently a mixture of both isomers). Molecular species composition was determined using HPLC equipped with a light scattering detector. Figures are rounded to the nearest whole number.

temperature. The same holds for the molecular species of DGDG. At 30°C the prokaryotic precursor for this desaturation, 20:4ω6/16:0, constituted about 46 per cent of MGDG in dense cultures and 16 per cent in dilute cultures, with similar values in DGDG. At 20°C AA-containing molecular species accounted for less than 6 per cent of either MGDG or DGDG, while the proportion of EPA-containing species was high compared with cultures at higher temperature.

The content of eukaryotic MGDG, which consists predominantly of 20:5/20:5, is inversely proportional to growth temperature (Table 1.3). This finding suggests that the 20:5/20:5 molecular species fulfils a unique role in this organism, possibly in promoting accommodation to changes resulting from variations in ambient temperature (e.g., changes in membrane fluidity). In higher plants, the difference between the prokaryotic (18:3/16:3) and eukaryotic (18:3/18:3) molecular species end products is quite subtle and is expressed in a slightly shorter acyl group at the *sn*-2 position of the prokaryotic species; shifts between prokaryotic and eukaryotic molecular species are thus of lesser consequence. In *P. cruentum*, however, the eukaryotic species end product (20:5/20:5) contains five more double bonds (and four more carbon atoms) than the prokaryotic one (20:5/16:0). Indeed, the HZ3 mutant of *P. cruentum* which is characterized by retardation of growth at 15°C has been shown to be lacking in its ability to produce eukaryotic molecular species of MGDG (Khozin *et al.*, 1997).

Just as eukaryotic molecular species may play a role in chill resistance, prokaryotic species may contribute to heat tolerance. The HZ3 mutant which produces less of the former must compensate for this deficiency by producing prokaryotic molecular species. A high proportion of the latter, although detrimental at low temperatures, could be advantageous at higher temperatures.

For optimal EPA productivity, the growth temperature of *P. cruentum* must not be allowed to exceed 25°C, as this is the temperature that yields the highest growth rate while maintaining a high proportion and content of EPA. For optimal AA production, on the other hand, the culture should be maintained at about 30°C to assure a maximum proportion of AA, minimum proportion of EPA and maximum fatty acid content.

Light intensity and biomass concentration

The total fatty acid content of *P. cruentum* (percentage of ash-free dry weight) increases with increasing biomass concentration, apparently due to reduced penetration of light. Light limitation also affects the fatty acid composition, and especially the proportion of C_{20} PUFA and the ratio of ω3 to ω6 fatty acids. When cells were cultivated in a chemostat at 25°C under high light intensity, EPA and AA were the major fatty acids, constituting 40 and 30 per cent, respectively, of total fatty acids (Table 1.4) (Cohen *et al.*, 1988a; Lee and Tan, 1988). Reducing the light intensity for a fixed rate of dilution resulted in a decrease in the proportion of both EPA (to 16 per cent) and AA (to 22 per cent) together with an increase in 16:0 (from 8 to 25 per cent). Changing the dilution rate from high (0.3 d^{-1}) to low (0.08 d^{-1}) led to a gradual decrease in the proportion of EPA (from 45 to 36 per cent) and increase in AA (from 27 to 33 per cent). Thus, the relative content of AA and EPA is affected by both growth rate and light intensity (Lee and Tan, 1988). Similar results have been reported by Cohen *et al.* (1988a), who have shown that reducing the growth rate by other means, such as increasing the salinity or changing the pH of the medium, increases the proportion of AA at the expense of EPA. The fatty acid composition of stationary phase cultures resembles that of cultures with a high biomass density. Characteristically, the proportion of AA is at its highest in such cultures whereas that of EPA is minimal (Cohen *et al.*, 1988a). This finding may explain earlier findings of high levels of AA and little, if any, EPA in *Porphyridium* (Nichols and Appleby, 1969; Kost *et al.*, 1984; Lee *et al.*, 1989). A few reports even state that AA is the only C_{20} PUFA present (Ahern *et al.*, 1983); however, failure to separate additional PUFA could be due to poor choice of analytical conditions.

Studies dealing with other EPA-producing microalgae suggest an opposite response to light intensity. For example, Sukenik *et al.* (1989) (see also Chapter 3) have shown that under growth-limiting light conditions, *Nannochloropsis* sp. contains high proportions of EPA; however, 16:0 and 16:1 predominate as the culture becomes light-saturated, indicating that EPA content is inversely related to irradiance level. Included in this group are the diatoms *Cyclotella*, *Nitzchia closterium* (Orcutt and Patterson, 1974; Yongmanitchai and Ward, 1989), *P. tricornutum* (Iwamoto and Sato, 1986), *Chaetoceros calcitrans*, and *Thalassiosira pseudonana* as well as the prymnesiophyte *Isochrysis galbana* (Harrison *et al.*, 1990).

The differences in response to environmental conditions between *P. cruentum* and almost any other alga that produces C_{20} PUFA could be related to differences in their biosynthetic pathways. To elucidate this point, Adlerstein *et al.* (1997) studied the effect of biomass density on the molecular species composition of the major galactolipids MGDG and DGDG of *P. cruentum*, the main depots of EPA in most algae.

Our data show that diluted cultures of *P. cruentum* contain a larger proportion of EPA-containing species and a lower proportion of AA-containing species than in denser cultures (Table 1.3). This situation holds in both prokaryotic (20:5/16:0)

Table 1.4 Effect of light intensity on the composition of major fatty acids in *P. cruentum*.

Light intensity (w m^{-2})	Fatty acid composition (% of total fatty acids)					
	16:0	18:0	18:1	18:2	20:4	20:5
15	8	4	3	15	30	40
10	25	14	8	16	22	16
5	20	16	26	7	10	20

Cultures were cultivated in a chemostat at a constant dilution rate (adapted from Lee and Tan, 1988).

and eukaryotic (20:5/20:5) species (Table 1.3). For example, the ratio of 20:5ω3/16:0 to 20:4ω6/16:0 in MGDG and DGDG increased, at 25°C, from about 1 and 2 in the denser cultures to 23 and 57 in the diluted cultures, respectively. Likewise, the ratio of the eukaryotic species 20:5/20:5 to 20:4/20:4 in MGDG increased from 2 to 131.

At the optimal growth temperature, light-induced enhancement in the eukaryotic (20:5/20:5) and prokaryotic (20:5/16:0) EPA-containing molecular species occurs at the expense of their respective precursors, 20:4/20:4 and 20:4/16:0 (Table 1.4). It appears therefore that when light intensity is limited, the cells accumulate AA in the various molecular species of their galactolipids. When the light per cell increases, either by dilution of the culture or by an increase in the light intensity, a relatively rapid desaturation of the accumulated AA can take place (Khozin *et al.*, 1997).

P. cruentum is apparently unique in that its EPA is primarily produced by Δ17 (ω3) desaturation of AA in the chloroplast. In other algae this desaturation is either entirely or predominantly cytoplasmic (Arao *et al.*, 1994; Schneider *et al.*, 1995). We therefore venture to speculate that light intensity has a specific effect on chloroplastic Δ17 desaturation in *P. cruentum*, leading to a response quite different from that observed in organisms that predominantly desaturate AA to EPA via a cytoplasmic desaturase.

Nitrogen starvation

Imposing nitrogen limitation when light is in excess results in cessation of growth. Since photosynthetic fixation of carbon continues, the cellular ratio of C/N is thereby increased (Mayzaud *et al.*, 1989) and energy is channelled into production of non-nitrogenous reserve materials such as TAG, which serve as a sink for photo-synthetically fixed carbon. Nitrogen starvation has indeed been reported to bring about an increased accumulation of neutral lipids. For example, Shifrin and Chisholm (1981) reported a 130–320 per cent increase in oil content under nitro-gen-deficient conditions in 15 chlorophycean strains. Likewise, increasing light inten-sity during the N-starvation of *Navicula saprophilla* (Kyle *et al.*, 1989) resulted in a higher lipid content.

While nitrogen starvation generally leads to a sharp reduction in both proportion and content of PUFA (Cohen, 1986), several exceptions have been reported. The proportion of EPA in *Navicula saprophilla* (Kyle *et al.*, 1989) exhibited little change following nitrogen deprivation while the lipid content nearly doubled, resulting in a substantial increase in EPA content. In an exponential culture of the AA-rich strain

Table 1.5 Effect of nitrogen starvation on the composition of the major fatty acids in *P. cruentum.*

Lipid	Nitrogen	Fatty acid composition (% of total fatty acids)						
		16:0	18:0	18:1	18:2	20:3	20:4	20:5
Total	+	35	1	1	11	1	21	29
	−	28	2	3	21	3	30	10
TAG	+	26	11	4	17	2	20	8
	−	23	5	5	27	3	31	4
MGDG	+	34	1	1	11	1	5	43
	−	37	6	7	27	1	3	17
DGDG	+	40	1	1	4	−	6	35
	−	46	5	4	7	−	4	26
PC	+	29	4	1	4	−	46	17
	−	15	2	2	6	7	61	3

Cultures were batch cultivated for four days in the presence (+) or absence (−) of nitrogen at 28°C.

P. cruentum 113.80, four days of N-starvation led to an increase in total fatty acid content of 60 per cent, primarily due to enhanced accumulation of TAG which increased from 21 to 61 per cent of lipids (Cohen, 1990). Concomitantly, the content of AA common to the TAG tripled, from 1 to 3 per cent, while that of EPA decreased from 2 to 1 per cent. The content of AA increased in the neutral lipids and phospholipids, whereas that of EPA remained stable at 3–4 per cent in these lipids (Table 1.5). Surprisingly, the proportion of AA in the galactolipids did not increase. By contrast, when growth is retarded by other means such as light, temperature or pH limitations, the proportion of AA increases in all lipids (Cohen *et al.*, 1988a).

The negative correlation between N-starvation and PUFA may be related to the fact that in most algae the PUFA are concentrated in the polar lipids while the TAG consist mainly of saturated and monounsaturated short-chained fatty acids. Under nitrogen starvation the synthesis of TAG increases at the expense of the polar lipids and the PUFA they contain. However, in certain algae, in addition to serving as reserve material, the TAG have a role that requires the presence of PUFA. In such algae, presumably, conditions which enhance the production of TAG, e.g., N-starvation, will also enhance the content of their characteristic PUFA.

Increasing the concentration of CO_2 resulted in a reduction in the level of unsaturation in *Chlorella vulgaris*, but did not significantly affect fatty acid composition in *P. cruentum* (Tsuzuki *et al.*, 1990).

Strains

Evaluation of seven strains of *P. cruentum* for their ability to produce AA under conditions conducive to maximal AA production i.e., elevated temperature (30°C) in the stationary phase, revealed that several strains had an AA proportion of over 40 per cent already in the exponential phase. The most promising strains in this respect were judged to be 1380-1e and 113.80 (Table 1.6). In the stationary phase, both strains had an AA content of 2.5 per cent of dry weight and an EPA proportion

Table 1.6 Fatty acid composition of *P. cruentum* strains grown at 30°C.

Strain	Growth phase	Fatty acid composition (% of total fatty acids)		Fatty acid content (% of AFDW)		R
		20:4 ω6	20:5 ω3	AA	EPA	
1380-1a	E	40	7	2.0	0.3	6.1
	S	35	10	2.2	0.6	3.7
1380-1b	E	39	14	1.5	0.5	2.8
	S	42	7	2.0	0.3	6.4
1380-1c	E	41	12	1.8	0.5	3.4
	S	37	16	2.0	0.8	2.3
1380-1d	E	33	19	1.8	1.0	1.7
	S	38	9	1.8	0.4	4.5
1380-1e	E	41	13	1.8	0.6	3.2
	S	38	4	2.5	0.3	9.4
1380-1f	E	41	12	1.9	0.6	3.3
	S	40	5	2.2	0.3	7.8
113.80	E	39	7	1.6	0.3	5.2
	S	32	3	2.5	0.2	11.5

AFDW, ash-free dry weight; R, AA/EPA; E, exponential phase, 4–5 mg chlorophyll l^{-1}; S, stationary phase, 27–30 mg chlorophyll l^{-1}.

of 4 per cent or less. The low EPA content is of considerable significance, since separation of EPA from AA could be quite expensive (see p. 32).

If one is to rate the strains of *P. cruentum* in terms of their potential to produce EPA, three factors should be primarily considered: growth rate under optimal conditions; EPA content (percentage of ash-free dry weight); and ratios of AA to EPA (R value) at the exponential and at the stationary phase. Growth rate and EPA content determine the EPA production rate. The R value in the exponential phase reflects the extent of AA 'contamination' and the degree of difficulty in EPA purification. The lower the R value, the easier the separation from AA. The differences in the corresponding R values for the exponential and stationary phases in the various strains could be used as indicators of the expected variability in EPA contents under outdoor conditions, as a result of light and cell concentration changes. On this basis, strain 1380-1d was judged to be the most promising of the seven mentioned above from the standpoint of EPA production and relative purity (Cohen, 1990). It had the highest content of EPA (2.4 per cent) and a low R value (0.33) in the exponential phase, which increased to only 0.90 in the stationary phase (Table 1.7).

Outdoor cultivation

P. cruentum cultures were grown outdoors in Israel for over a year and fluctuations in fatty acid composition and content were assessed (Cohen *et al.*, 1988b). While the dependence of the fatty acid composition on cell concentration was similar to the pattern found indoors (Cohen *et al.*, 1988a), the temperature dependence was different. More especially, the proportion of EPA was higher at suboptimal tempera-

Table 1.7 Fatty acid composition of *P. cruentum* strains grown at 25°C.

Strain	Growth phase	Fatty acid composition (% of total fatty acids)		Fatty acid content (% of AFDW)		
		20:4ω6	20:5ω3	AA	EPA	R
1380-1a	E	22	35	0.8	1.9	0.6
	S	34	19	1.9	1.1	1.7
1380-1b	E	15	44	0.9	2.1	0.4
	S	29	23	1.6	1.2	1.3
1380-1c	E	17	40	0.9	2.1	0.4
	S	24	28	1.3	1.5	0.9
1380-1d	E	15	44	0.8	2.4	0.3
	S	25	27	1.2	1.4	0.9
1380-1e	E	18	39	1.0	2.1	0.5
	S	34	16	1.9	0.9	2.1
1380-1f	E	18	40	0.9	2.1	0.4
	S	33	18	1.7	0.9	1.8
113.80	E	18	42	0.8	1.9	0.4
	S	27	25	1.4	1.3	1.1

AFDW, ash-free dry weight; R, AA/EPA; E, exponential phase, 4–5 mg chlorophyll l^{-1}; S, stationary phase, 27–30 mg chlorophyll l^{-1}.

tures, than at the optimal growth temperature (25°C). This apparent discrepancy could be due to the temperature regime outdoors, which differed greatly from the constant temperature regime maintained in the laboratory, as well as to factors such as day length and periodicity. The latter, while are not fully understood, could affect the fatty acid composition.

Maintaining *Porphyridium* in outdoor culture at low cell concentrations resulted in a fatty acid mixture rich in EPA and correspondingly poor in AA. A higher content of AA was obtained when cell concentrations were kept high in the summer. The proportion of EPA peaked in December at 50 per cent (of fatty acids), corresponding to a content of 2.5 per cent (of dry weight), while the concentration of AA peaked in September at 30 per cent of fatty acids for a content of 1.4 per cent. However, the productivity of either EPA or AA was extremely low, due to very low biomass productivity. Presumably, cultivation in an enclosed bioreactor would result in much higher productivities. Indeed, Iqbal *et al.* (1993) cultivated *P. cruentum* outdoors in a flat-sided bioreactor and reached a growth rate of 1.1 d^{-1} at a light intensity of 8000 lx.

EPA overproduction

Since the manipulation of environmental conditions is intrinsically limited, selection of EPA-overproducing strains of *P. cruentum* was attempted. The content of specific cell components can in principle be enhanced by inducing resistance to inhibitors of specific steps in biosynthetic pathways. At the time, no inhibitors of either the elongation or any of the desaturation steps were known, but several inhibitors for ω3 desaturation were available. These inhibitors were herbicides of the substituted pyridazinone family, primarily SAN 9785, whose effect on chloroplast desaturase linoleic acid level is a direct inhibition of the enzyme activity (Norman *et al.*, 1991). Over 90 per cent of the EPA in *P. cruentum* is located in chloroplastic galactolipids (Table 1.1), which are regarded as structural components of the photosynthetic apparatus (Quinn and Williams, 1983). Thus, it was reasonable to assume that inhibitors having an effect on desaturation reactions involved in PUFA biosynthesis would also affect growth (i.e., by impairment of photosynthetic efficiency).

Cultivation of *P. cruentum* in the presence of SAN 9785 resulted in growth inhibition as well as a reduction in the level of EPA and an increase in AA, its precursor (Cohen *et al.*, 1993a). It was inferred that the growth inhibition stemmed from inhibition of the ω3 desaturation of AA to EPA. Resistance to SAN 9785 was built up by continuous cultivation in the presence of gradually increasing concentrations of the herbicide. The resistant culture tolerated a concentration of 0.4 mM SAN 9785 while the wild type collapsed after a few days. When transferred back to inhibitor-free medium, the growth rates of the resistant lines returned to control levels. The resistance was maintained even after 50 generations in an inhibitor-free medium, apparently indicating an adaptive change of a genetic nature. The resistant lines also differed in fatty acid composition. While the proportion of EPA in the galactolipids of SAN 9785-treated wild type cells was reduced from 50 to 29 per cent, that of the resistant culture was only slightly affected (5 per cent reduction) (Cohen *et al.*, 1993b).

Resistance could have developed due to the emergence of genetic changes leading to overproduction of EPA. However, it could also be the result of decreased uptake or enhanced metabolism of the inhibitor, or reduced affinity of the target enzyme. Overproduction seems to be responsible for the resistance in at least some lines since elevated EPA levels were sustained for many generations of cultivation in herbicide-free medium.

Following agar plating, six colonies were selected at random. The EPA proportion of two of these lines (Table 1.8) was found to be slightly but significantly higher than in the wild type (41 versus 36 per cent) (Cohen *et al.*, 1992). The EPA content was even further improved from 1.8 to 2.3 per cent. These findings suggest that development of resistance to specific inhibitors could indeed result in overproduction of EPA or any other PUFA. Even higher levels of overproduction could be attained by scanning large batches of mutagenized cultures consisting of at least several hundred colonies. However, further exploitation of this approach may be hampered by the low specificity of the herbicide, and more specific inhibitors, such as transition stage analogues, are being sought.

Table 1.8 Fatty acid composition of wild-type (WT) and SAN 9785-resistant (SRP-6) cultures of *P. cruentum*.

Culture	Fatty acid composition (% of total fatty acids)									
	16:0	16:1	18:0	18:1	18:2	18:3	20:2	20:3	20:4	20:5
WT	33	4	1	–	6	1	1	1	17	36
SRP-6	29	5	1	1	5	1	1	–	15	41

EPA purification

Fish oil is an adequate source of ω3 dietary supplements, but its use for production of pharmaceutical-grade EPA is unlikely to be economic (see Chapter 6) due to the high cost of separation from other PUFA such as AA and DHA. In most algae, EPA is characteristically contained in the galactolipids whereas AA is mainly concentrated in the TAG and phospholipids. Both classes of lipids should in principle be much simpler to separate than the PUFA. This concept was tested on biomass of *P. cruentum* (Cohen and Cohen, 1990) grown under conditions conducive to high EPA content. Under these conditions, EPA constituted 41 per cent of total fatty acids and AA 15 per cent. Separation of the galactolipids was accomplished by washing the lipid extract on a thin pad of silica gel. The acetone fraction eluted glycolipids with a somewhat higher EPA content and reduced proportion of AA (5 per cent; Table 1.9). By transmethylation, the fatty acids of these lipids were converted to their methyl esters, which were subsequently treated with urea. This treatment resulted in the formation of urea inclusion complexes of the saturated and monounsaturated fatty acids, the removal of which served to increase the EPA content to 82 per cent. In a trial run, the remaining 7 per cent of AA were removed by reverse-phase chromatography, and EPA was recovered at a purity of 97 per cent.

Feeding with external fatty acids

Another strategy that could increase the content of PUFA is feeding with its precursors. Okumura *et al.* (1986) found that addition of 18:1, 18:2 or 18:3 to the cultivation medium of *Euglena gracilis* enhanced the production of PUFA and especially that of AA and EPA. Similarly, the EPA content of fungi was doubled after supplementation with linseed oil (rich in α-linolenic acid) (Shimizu *et al.* 1989). In *P. cruentum*, Ahern *et al.* (1983) claimed that feeding with linoleic acid at 18°C enhanced the content of AA by a threefold factor. As mentioned earlier, the analytical conditions utilized in that study did not enable separation of AA from EPA. However, we have not been able to reproduce these results.

Table 1.9 Purification of methyl esters of EPA from *P. cruentum*.

Lipid treated	Purification	Fatty acid composition (% of total fatty acids)								
		16:0	16:1	18:0	18:1	18:2	18:3	20:3	20:4	20:5
Total	None	31	5	–	1	5	1	2	15	41
Glycolipids	Silica gel	35	–	1	1	10	–	–	5	48
PUFA-ME[a]	Urea addition	3	–	–	1	7	–	–	7	82
PUFA-ME[b]	RP chrom[c]	–	–	–	–	1	–	–	1	97

[a]Polyunsaturated fatty acid methyl esters of glycolipids, [b]PUFA-ME after urea treatment; [c]reverse-phase chromatography.

References

AARONSON, S., 1973, Effect of incubation temperature on the macromolecular and lipid content of the phytoflagellate *Ochromonas danica*, *J. Phycol.*, **9**, 111–113.

ADLERSTEIN, D., KHOZIN, I., BIGOGNO, C. and COHEN, Z., 1997, Effect of environmental conditions on the molecular species composition of galactolipids in the alga *Porphyridium cruentum*, *J. Phycol.* **33**, 975–979.

AHERN, T.J., KATOH, J.S. and SADA, E., 1983, Arachidonic acid production by the red alga *Porphyridium cruentum*, *Biotechnol. Bioeng.*, **25**, 1057–1070.

ARAKI, S., SAKURAI, T., OOHUSA, T., *et al.*, 1990, Content of arachidonic and eicosapentaenoic acids in polar lipid from *Gracilaria* (Gracilariales, Rhodophyta), *Hydrobiologia*, **204/205**, 513–519.

ARAO, T. and YAMADA, M. 1994, Biosynthesis of polyunsaturated fatty acids in the marine diatom, *Phaeodactylum tricornutum*, *Phytochemistry*, **35**, 1177–1181.

ARAO, T., SAKAKI, T. and YAMADA, M., 1994, Biosynthesis of polyunsaturated lipids in the diatom, *Phaeodactylum tricornutum*, *Phytochemistry*, **36**, 629–635.

BROWN, C.M. and ROSE, A.H., 1969, Fatty acid composition of *Candida utilis* as affected by growth temperature and dissolved oxygen tension, *J. Bacteriol.*, **99**, 371–376.

BROWSE, J. and SOMERVILLE, C.R., 1991, Glycerolipid synthesis: biochemistry and regulation, *Ann. Rev. Plant Mol. Biol.*, **42**, 467–506.

BROWSE, J., MIQUEL, M. and SOMERVILLE, C., 1990, Genetic approaches to understanding plant lipid metabolism, in Quinn, P.J. and Harwood, J.L. (Eds), *Plant Lipid Biochemistry, Structure and Utilization*, pp. 431–438, London: Portland Press Ltd.

COHEN, Z., 1986, Products from microalgae, in Richmond, A. (Ed.), *Handbook for Microalgal Mass Culture*, pp. 421–454, Boca Raton, FL: CRC Press.

COHEN, Z., 1990, The production potential of eicosapentaenoic and arachidonic acids by the red alga *Porphyridium cruentum*, *J. Am. Oil Chem. Soc.*, **67**, 916–920.

COHEN, Z. and COHEN, S., 1990, Preparation of eicosapentaenoic acid (EPA) concentration from *Porphyridium cruentum*, *J. Am. Oil Chem. Soc.*, **68**, 16–19.

COHEN, Z. and HEIMER, Y.M., 1990, Δ6 desaturase inhibition: a novel mode of action of Norflurazon, *Plant Physiol.*, **93**, 347–349.

COHEN, Z., DIDI, S. and HEIMER, Y.M., 1992, Overproduction of γ-linolenic and eicosapentaenoic acids by algae, *Plant Physiol.*, **98**, 569–572.

COHEN, Z., VONSHAK, A. and RICHMOND, A., 1988a, Effect of environmental conditions on fatty acid composition of the red alga *Porphyridium cruentum*: correlation to growth rate, *J. Phycol.*, **24**, 328–332.

COHEN, Z., VONSHAK, A., BOUSSIBA, S. and RICHMOND, A., 1988b, The effect of temperature and cell concentration on fatty acid composition of outdoor cultures of *Porphyridium cruentum*, in Stadler, T., Mollion, J., Verdus, M.-C., Karamanos, Y., Morvan, H. and Christiaen, D. (Eds), *Algal Biotechnology*, pp. 421–430, London: Elsevier Applied Science.

COHEN, Z., REUNGJITCHACHAWALI, M., SIANGDUNG, W. and TANTICHAROEN, M., 1993a, Production and partial purification of γ-linolenic acid and some pigments from *Spirulina platensis*, *J. Appl. Phycol.*, **5**, 109–115.

COHEN, Z., NORMAN, H.A. and HEIMER, Y.M., 1993b, Potential use of substituted pyridazinones for selecting polyunsaturated fatty acid overproducing cell lines of algae, *Phytochemistry*, **32**, 259–264.

COHEN, Z., NORMAN, H.A. and HEIMER, Y.M., 1995, Microalgae as a source of omega-3 fatty acids, in Simopoulos, A. (Ed.) *Plants in Human Nutrition, World Review of Nutrition and Dietetics*, Vol. 77, pp. 1–32, Basel: Karger.

DURAN, H.H., 1970, Fatty acid composition of lipid extracts of a thermophilic *Bacillus* species, *J. Bacteriol.*, **101**, 145–151.

FRENTZEN, M., 1986, Biosynthesis and desaturation of the different diacylglycerol moieties in higher plants, *J. Plant Physiol.*, **124**, 193–209.

GURR, M.I., 1971, The biosynthesis of polyunsaturated fatty acids in plants, *Lipids*, **6**, 266–273.

HANSSON, L. and DOSTALEK, M., 1988, Effect of culture conditions on mycelial growth and production of γ-linolenic acid by the fungus *Mortierella ramanniana*, *Appl. Microbiol. Biotechnol.*, **28**, 240–246.

HARRISON, P.J., THOMPSON, P.A. and CALDERWOOD, G.S., 1990, Effects of nutrient and light limitation on the biochemical composition of phytoplankton, *J. Appl. Phycol.*, **2**, 45–56.

HARWOOD, J.L., 1988, Fatty acid metabolism, *Ann. Rev. Plant Physiol. Plant Mol. Biol.*, **39**, 101–138.

HEEMSKERK, J.W.M. and WINTERMANS, J.F.G.M., 1987, Role of the chloroplast in leaf acyl lipid synthesis, *Physiol. Plant*, **70**, 558–568.

HENDERSON, R.J. and McKINLEY, E.E., 1992, Radiolabelling studies of lipids in the marine cryptomonad *Chroomonas salina* in relation to fatty acid desaturation, *Plant Cell Physiol.*, **33**, 395–406.

HENDERSON, R.J. and SARGENT, J.R., 1989, Lipid composition and biosynthesis in aging cultures of the marine cryptomonad, *Chroomonas salina*, *Phytochemistry*, **28**, 1355–1362.

HIGASHI, S. and MURATA, N., 1993, An *in vivo* study of substrate specificities of acyl-lipid desaturases and acyltransferases in lipid synthesis in *Synechocystis* PCC6803, *Plant Physiol.*, **102**, 1275–1278.

HOLTON, R.W., BLACKER, H.H. and ONURE, M., 1964, Effect of growth temperature on the fatty acid composition of a blue-green alga, *Phytochemistry*, **3**, 595–602.

HOWLING, D., MORRIS, J.L. and JAMES, A.T., 1968, The influence of chain length on the dehydrogenation of saturated fatty acids, *Biochim, Biophys. Acta*, **152**, 224–226.

IQBAL, S., YASMIN, A., KHAN, M.R. and SUFI, N.A., 1993, Fatty acid composition of apricot kernel oil, *Sarhad J. Agric.*, **9**, 11–16.

IWAMOTO, H. and SATO, S., 1986, EPA production by freshwater unicellular algae, *J. Am. Oil Chem. Soc.*, **63**, 434.

IWAMOTO, H., YONEKAWA, G. and ASAI, T., 1955, Fat synthesis in unicellular algae. I. Culture conditions for fat accumulation in *Chlorella* cells, *Bull. Agric. Chem. Soc. Jpn.*, **19**, 240–246.

JAWORSKI, J.G., 1987, Biosynthesis of monoenoic and polyenoic fatty acids, in Stumpf, P.K. (Ed.), *The Biochemistry of Plants, Vol. 9, Lipids: Structure and Function*, pp. 159–174, NY: Academic Press.

KATES, M. and HAGEN, P.O., 1964, Influence of temperature on fatty acid composition of psychrophilic and mesophilic *Serratia* species, *Can. J. Biochem.*, **42**, 481–488.

KHOZIN, I., ADLERSTEIN, D., BIGOGNO, C., HEIMER, Y.M. and COHEN, Z., 1997, Elucidation of the biosynthesis of eicosapentaenoic acid in the microalga *Porphyridium cruentum*. II. Studies with radiolabeled precursors, *Plant Physiol.*, **114**, 223–230.

KLEINSCHMIDT, M.G. and McMAHON, V.A., 1970, Effect of growth temperature on the lipid composition of *Cyanidium caldarium*. II. Glycolipid and phospholipid components, *Plant Physiol.*, **46**, 290–293.

KLYACHKO-GUREVICH, G.L., ZHUKOVA, T.S., VLADIMIROVA, M.G., *et al.*, 1967, Comparative characteristics of the growth and direction of biosynthesis of various strains of *Chlorella* under nitrogen deficiency conditions. III. Synthesis of fatty acids, *Sov. Plant Physiol.*, **16**, 205–209.

KOST, H.P., SENSER, M. and WANNER, G., 1984, Effect of nitrate and sulphate starvation on *Porphyridium cruentum* cells, *Z. Pflanzenphysiol.*, **113S**, 231–249.

KYLE, D.J., BEHRENS, P.W., BINGHAM, S., *et al.*, 1989, Microalgae as a source of EPA-containing oils, in Applewhite, T.H. (Ed.), *Biotechnology for the Fats and Oils Industry*, pp. 117–121, Champaign: Amer. Oil Chem. Soc.

LEE., Y.K. and TAN, H.M., 1988, Effect of temperature, light intensity and dilution rate on the cellular composition of red alga *Porphyridium cruentum* in light-limited chemostat cultures, *Mircen J.*, **4**, 231–237.

LEE, Y.K., TAN, H.M. and LOW, C.S., 1989, Effect of salinity of medium on cellular fatty acid composition of marine alga *Porphyridium cruentum* (Rhodophyceae), *J. Appl. Phycol.*, **1**, 19–23.

MARR, A.G. and INGRAHAM, L.J., 1962, Effect of temperature on the composition of fatty acids in *E. coli*, *J. Bacteriol.*, **84**, 1260–1268.

MAYZAUD, P., CHANUT, J.P. and ACKMAN, R.G., 1989, Seasonal changes of the biochemical composition of marine particulate matter with special reference to fatty acids and sterols, *Mar. Ecol. Prog. Ser.*, **56**, 189–204.

NICHOLS, B.W. and APPLEBY, R.S., 1969, The distribution of arachidonic acid in algae, *Phytochemistry*, **8**, 1907–1915.

NORMAN, H.A., PILLAI, P. and ST JOHN, J.B., 1991, *In vitro* desaturation of monogalactosyldiacylglycerol and phosphatidylcholine molecular species by chloroplast homogenates, *Phytochemistry*, **30**, 2217–2222.

OKUMURA, M., II, S., FUJII, R. *et al.*, 1986, Tokyo Koho, JP61:177,990.

ORCUTT, D.M. and PATTERSON, G.W., 1974, Effect of light intensity upon lipid composition of *Nitzchia closterium* (*Cylindrotheca fusiformis*), *Lipids*, **9**, 1000–1003.

PATTERSON, G.W., 1970, Effect of culture temperature on fatty acid composition of *Chlorella sorokiniana*, *Lipids*, **5**, 597–600.

PETTITT, T.R., JONES, A.L. and Harwood, J.L., 1989, Lipids of the marine red algae, *Chondrus crispus* and *polysiphonia lanosa*, *Phytochemistry*, **28**, 399–405.

PHAM QUOC, K., DUBACQ, J.P., JUSTIN, A.M., DEMANDRE, C. and MAZLIAK, P., 1993, Biosynthesis of eukaryotic lipid molecular species by the cyanobacterium *Spirulina platensis*, *Biochim. Biophys. Acta*, **1168**, 94–99.

PIORRECK, M., BAASCH, K.H. and POHL, P., 1984, Preparatory experiments for the axenic mass culture of microalgae. Part 1. Biomass production, total protein, chlorophylls, lipids and fatty acids of freshwater green and blue-green algae under different nitrogen regimes, *Phytochemistry*, **23**, 207–216.

QUINN, P.J. and WILLIAMS, W.P., 1983, The structural role of lipids in photosynthetic membranes, *Biochim. Biophys. Acta*, **737**, 223–266.

RAMUS, F., 1972, The production of extracellular polysaccharides by the unicellular alga *Porphyridium aerugineum*, *J. Phycol.*, **8**, 97–111.

SCHNEIDER, J.C. and ROESSLER, P., 1994, Radiolabelling studies of lipids and fatty acids in *Nannochloropsis* (Eustigmatophyceae), an oleaginous marine alga, *J. Phycol.*, **30**, 594–598.

SCHNEIDER, J.C., LIVNE, A., SUKENIK, A. and ROESSLER, P.G., 1995, A mutant of *Nannochloropsis* deficient in eicosapentaenoic acid production, *Phytochemistry*, **40**, 807–814.

SETO, A., WANG, H.L. and HESSELTINE, C.W., 1984, Culture conditions affect eicosapentaenoic acid content of *Chlorella minutissima*, *J. Am. Oil Chem. Soc.*, **61**, 892–894.

SHIFRIN, N.S. and CHISHOLM, S.W., 1981, Phytoplankton lipids: interspecific differences and effects of nitrate, silicate, and light-dark cycles, *J. Phycol.*, **17**, 374–384.

SHIMIZU, S., SHINMEN, Y., KAWASHIMA, H., *et al.*, 1988, Fungal mycelia as a novel source of eicosapentaenoic acid production at low temperature, *Biochem. Biophys. Res. Commun.*, **150**, 335–341.

SHIMIZU, S., AKIMOTO, K., KAWASHIMA, H., *et al.*, 1989, Microbial conversion of an oil containing alpha-linolenic acid to an oil containing eicosapentaenoic acid, *J. Am. Oil Chem. Soc.*, **66**, 342–347.

SHIRAN, D., KHOZIN, I., HEIMER, Y.M. and COHEN, Z., 1996, Biosynthesis of eicosapentaenoic acid in the microalga *Porphyridium cruentum*. I. The use of externally supplied fatty acids, *Lipids*, **31**, 1277–1282.

STUMPF, P.K., 1984, Plant fatty acid biosynthesis – present status, future prospects, in Numa, S. (Ed.), *Fatty Acid Metabolism and its Regulation*, pp. 312–317, Amsterdam: Elsevier/ North Holland.

SUKENIK, A., CARMELI, Y. and BERNER, Y., 1989, Regulation of fatty acid composition by growth irradiance level in the eustigmatophyte *Nannochloropsis* sp., *J. Phycol.*, **25**, 686–692.

SUKENIK, A., YAMAGUCHI, Y. and LIVNE, A., 1993, Alterations in lipid molecular species of the marine eustigmatophyte *Nannochloropsis* sp., *J. Phycol.*, **29**, 620–626.

SUMNER, J.L., MORGAN, E.D. and EVANS, H.C., 1969, The effect of growth temperature on the fatty acid composition of fungi in the order Mucorales, *Can. J. Microbiol.*, **15**, 515–520.

TSUZUKI, M., OHNUMA, E., SATO, N., *et al.*, 1990, Effects of carbon dioxide concentration during growth on fatty acid composition in microalgae, *Plant Physiol.*, **93**, 851–856.

VONSHAK, A., 1987, *Porphyridium*, in Borowitzka, M. and Borowitzka, L. (Eds), *Micro-algal Biotechnology*, pp. 122–134, Cambridge: Cambridge University Press.

VONSHAK, A., COHEN, Z. and RICHMOND, A., 1985, The feasibility of mass culture of *Porphyridium*, *Biomass*, **8**, 13–25.

WADA, H. and MURATA, N., 1989, *Synechocystis* PCC 6803 mutants defective in desaturation of fatty acids, *Plant Cell Physiol.*, **30**, 971–978.

WADA, H., GOMBOS, Z., SAKAMOTO, T. and MURATA, N., 1992, Genetic manipulation of the extent of desaturation of fatty acids in membrane lipids in the cyanobacterium *Synechocystis* PCC 6803, *Plant Cell Physiol,*, **33**, 535–540.

YONGMANITCHAI, W. and WARD, O.P., 1989, Omega-3 fatty acids: alternative source of production, *Process Biochemistry*, August, 117–125.

2

Monodus subterraneus

ZVI COHEN

Fatty acid composition of individual lipids

The fresh water eustigmatophyte *Monodus subterraneus* is one of the most promising of the microalgal sources of eicosapentaenoic acid (20:5ω3, EPA). EPA is generally the major fatty acid. Depending on cultivation conditions the proportion of EPA in this species ranges from 20 to 37 per cent of total fatty acids (Table 2.1). Other major fatty acids are, in decreasing order of occurrence, 16:1ω7, 16:0 and 18:1ω9.

In *M. subterraneus*, as in most other microalgae containing this fatty acid, EPA is primarily located in the galactolipid fraction. However, it also constitutes a significant proportion of the phospholipids and to a smaller extent of the triacylglycerols (TAG) (Table 2.1). The other three major fatty acids contribute significantly to all lipid groups but are concentrated mainly in the TAG. In contrast, most of the fatty acids of the ω6 family (18:2ω6, 18:3ω6 and 20:4ω6) reside in the phospholipids (Cohen, 1994).

The major molecular species of MGDG is the eukaryotic 20:5/20:5 (57 per cent). The main prokaryotic-like species are 20:5/14:0 (13 per cent), 20:5/16:1 (14 per cent), 20:5/16:0 (6 per cent) and 20:4ω6/16:0 (9 per cent) (Khozin and Cohen, 1996).

Effects of nutritional and environmental conditions

Temperature

Iwamoto and Sato (1986) found that the EPA content (percentage of dry weight) of *M. subterraneus* was inversely related to temperature, the highest proportion (35 per cent) being obtained at 20°C (Table 2.2). However, EPA productivity (mg EPA $l^{-1} d^{-1}$) was maximal at 25°C, the optimal temperature for growth. Furthermore, temperature-related differences in *M. subterraneus* decreased with diminishing light intensity.

Table 2.1 Fatty acid composition of lipid classes of *M. subterraneus*.

Lipid class	% of total lipids													

Lipid class	% of total lipids	14:0	16:0	16:1 ω11	16:1 ω7	18:0	18:1 ω9	18:1 ω7	18:2	18:3 ω6	18:3 ω3	20:3 ω6	20:4 ω6	20:5 ω3
Total		6.0	13.7	3.2	0.4	1.1	4.5	0.9	1.3	0.9	0.3	0.5	6.0	39.1
MGDG	31.9	5.3	14.2	22.1	0.4	0.3	3.2	0.7	0.7	0.7	0.3	0.4	4.2	46.3
DGDG	21.0	2.5	26.0	46.0	–	0.4	6.4	–	0.8	0.2	1.0	–	0.7	15.3
SQDG	17.8	4.0	48.8	38.5	0.7	0.6	4.3	–	0.2	0.2	–	–	0.3	0.3
DGTS	14.2	4.1	29.5	30.5	0.7	1.8	4.2	–	0.9	0.7	0.5	–	7.6	18.5
PC	5.0	1.9	41.9	17.2	1.2	4.7	9.9	0.1	7.4	6.7	0.1	–	2.9	4.4
PG	3.2	0.9	24.2	7.2	26.8[a]	2.0	3.4	2.2	1.0	0.2	0.1	–	0.4	30.5
PE	1.8	1.5	12.4	7.1	0.3	2.5	5.0	4.3	2.0	1.3	0.9	2.5	23.8	35.6
TAG	5.0	2.6	14.5	34.5	1.1	3.1	17.1	–	1.3	0.9	1.0	0.7	3.4	18.7

[a] 16:1ω13.

Effect of light intensity

Both proportion of EPA and total fatty acid content increase as light intensity decreases. Reducing the light intensity from 170 to 90 μE m^{-2}s^{-1} (where E = elongation) for example, enhanced the proportion of EPA from 30 to 36 per cent and fatty acid content from 10 to 12 per cent (Cohen, 1994), inducing a corresponding increase in EPA content from 3.1 to 4.4 per cent of dry weight. Similarly, EPA content rose proportionally when cell density was increased from 0.2 to 3.4 g l^{-1}: while the EPA content of the dilute culture (200–300 mg l^{-1}) was only 2.5 per cent (Figure 2.1), increasing the biomass density to over 2000 mg l^{-1} resulted

Table 2.2 Effect of temperature and light intensity on EPA content and yield in *M. subterraneus* (adapted from Iwamoto and Sato, 1986).

Temp. (°C)	Light (Klux)	Biomass yield (g l^{-1})	Lipid content (% dry wt)	EPA (% TFA[b])	EPA content (% dry wt)[a]	EPA prod. (mg l^{-1} d^{-1})
20	5	1.28	25.6	35.2	4.5	5.8
	10	2.66	25.8	31.6	4.1	10.8
	20	3.20	26.8	31.2	4.2	13.4
	40	3.28	28.1	31.9	4.5	14.7
25	5	1.70	34.2	33.3	5.7	9.7
	10	3.02	33.9	30.8	5.2	15.8
	20	4.56	35.4	28.3	5.0	22.8
	40	5.68	37.4	26.3	4.9	27.9
30	5	1.05	29.1	26.4	3.8	4.0
	10	2.74	33.7	20.2	3.4	9.3
	20	3.90	38.7	18.7	3.6	14.1
	40	5.30	36.5	15.2	2.8	14.7

[a] Total fatty acids.
[b] Assuming an average fatty acid content in lipids of 50%.

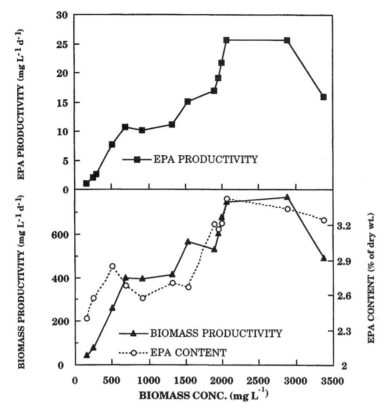

Figure 2.1 Changes in biomass productivity, EPA content and EPA productivity during growth of *M. subterraneus* in batch mode (Cohen, 1994).

in an increase in EPA content to 3.5 per cent. Furthermore, the low biomass productivity of the dilute culture (less than 100 mg l^{-1} d^{-1}) resulted in a very low EPA productivity, 2–3 mg l^{-1} d^{-1}, whereas the high biomass productivity (700 mg l^{-1} d^{-1}) of the dense culture promoted EPA productivity to a peak of 26 mg l^{-1} d^{-1}. The highest EPA content was obtained towards the end of the exponential phase of growth (Figure 2.1), for although growth slows down at this stage, biomass productivity is at its highest and EPA content is close to its maximum. It follows that in *M. subterraneus* the highest EPA contents and productivites will be obtained by semi-continuous cultivation at relatively high biomass concentrations.

Nitrogen starvation

Several studies aimed at achieving maximal cell contents and overall productivities of PUFA in microalgae have focused primarily on nutrient stress (e.g., Suen *et al.*, 1987; Reitan *et al.*, 1994). Nitrogen starvation has been reported to enhance fatty acid content in many species of algae (Shifrin and Chisholm, 1981). Generally, however, nitrogen deprivation is accompanied by a sharp decrease in the proportion of PUFA (Ben-Amotz *et al.*, 1985). Furthermore, unfavourable culture conditions may be expected to reduce overall biomass productivity, as well as introducing instability in continuous cultures and promoting contamination (Richmond *et al.*, 1990). A Japanese patent (Meiji Milk Products, 1985) has claimed that when *M. subterraneus* was cultivated under a high CO_2 concentration (5 per cent in air) for several days in complete medium and for another 12 days in nitrogen-free medium, fatty acid content increased to 36 per cent and EPA content to 11 per cent. However, in experiments conducted in the author's laboratory (Cohen, 1994), increasing the CO_2 concentration did not affect fatty acid content but reduced the proportion of EPA to 20 per cent. Subsequent nitrogen starvation of stationary phase cultures did indeed increase the fatty acid content to an unprecedented value of 17 per cent, but the proportion of EPA plummeted to 11 per cent, corresponding to an EPA content of only 1.9 per cent.

Salinity

Iwamoto and Sato (1986) reported that increasing the NaCl concentration range from 0 to 3 per cent resulted in an increase in lipid content from 34 to 46 per cent. Unfortunately, the proportion of EPA dropped from 34 to 15 per cent. In the marine eustigmatophyte *Nannochloropsis*, on the other hand, cultivation in a range of salinities (1–3.5 per cent) revealed a small increase in the proportion of EPA at low salinity (Renaud and Parry, 1994). EPA content, however, may increase with salt content, as suggested by the observed increase in lipid content with rising salinity.

Outdoor cultivation

M. subterraneus was cultivated outdoors in a flat plate bioreactor (Figure 2.2) with a narrow light path and intensive stirring to facilitate high cell concentrations (Hu *et al.*, 1997). The effect of cell density on the fatty acid composition of this alga is shown in Table 2.3. At the optimum cell density for growth (ca. 4 g l^{-1}), the proportion

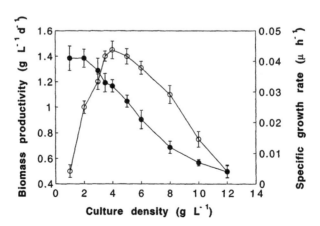

Figure 2.2 Specific growth rate (filled circles) and biomass production (open circles) as affected by culture concentration of *M. subterraneus* grown outdoors in a flat plate bioreactor. Bars indicate ±SD (Hu *et al.*, 1997).

Table 2.3 Effect of cell density on fatty acid content and composition of *Monodus subterraneus* (adapted from Hu *et al.*, 1997)

Cell density (g l^{-1})	Fatty acid composition (% TFA)													Fatty acid content (% of dry wt)	
	14:0	16:0	16:1 ω11[a]	16:1 ω7[a]	16:1 ω5[a]	18:0	18:1 ω9	18:1 ω7	18:2 ω6	18:3 ω6	20:3 ω6	20:4 ω6	EPA	EPA	TFA
0.5	2.4	24.1	4.3	23.1	1.3	1.3	9.0	1.5	3.3	1.5	0.7	5.2	21.2	2.1	9.6
0.8	2.5	24.3	4.4	23.7	1.2	1.3	8.9	1.5	3.1	1.7	0.7	5.0	20.9	2.1	10.2
1.5	2.3	23.3	4.3	23.3	1.3	1.2	8.9	1.5	3.2	1.7	0.7	5.4	22.2	2.3	10.6
2.0	2.1	23.4	2.3	25.8	1.4	1.0	6.2	0.6	2.2	0.7	0.6	5.4	26.9	3.4	12.7
4.0	2.3	20.2	2.0	26.9	1.6	0.6	4.5	–	2.0	1.0	0.3	4.7	32.1	3.8	11.8
5.0	1.9	22.9	4.2	25.9	1.5	0.7	5.1	1.0	2.6	0.5	0.3	5.0	29.4	3.4	11.6
6.0	5.2	20.9	4.2	24.9	1.1	0.6	7.0	1.0	3.0	3.0	0.3	5.4	27.8	3.2	11.7
7.0	3.9	22.5	3.1	28.0	–	0.7	7.2	1.2	3.4	3.4	0.2	5.3	25.6	2.7	10.4
10.0	7.1	23.0	–	30.5	–	0.8	8.9	–	2.4	2.4	–	4.7	20.9	2.4	10.5
12.0	2.6	28.4	–	29.4	–	0.9	7.8	1.1	2.7	2.7	–	5.2	20.5	1.9	9.3

Cultures were semi-continuous, harvested daily in the late afternoon. The data shown are mean values of at least two independent samples, each analysed in duplicate. Maximal difference between duplicates was less than 10%.
TFA, total fatty acids.
[a]Tentative assignment.

of EPA reached its highest level (32 per cent), whereas the proportions of most other fatty acids were at their lowest. In keeping with the results of the indoor experiments, lower than optimal cell densities resulted in a significant decrease in both total fatty acid content and EPA proportion. EPA is primarily concentrated in the galactolipids, which are photosynthetic membrane components (Nichols and Wood, 1968; Cohen *et al.*, 1988; Sukenik *et al.*, 1989). There is evidence that shade adaptation involves an increase in the cellular content of the photosynthetic membranal complex as well as in its surface area (Sukenik *et al.*, 1987, 1989; Berner *et al.*, 1989). The increase in the share of EPA in total fatty acids in response to a high cell density and associated light limitation might conceivably reflect such a photoadaptative process. Nevertheless, when cell densities increased well above the optimum (ca. above 7 $g l^{-1}$), the proportion of EPA gradually declined (Table 2.3).

EPA productivity in outdoor cultures as affected by cell density is shown in Figure 2.3. In mid-summer, maximal EPA productivity of 59 mg $l^{-1} d^{-1}$ (3.8 per cent dry wt) was achieved in cultures maintained at the optimal cell density (1.6 g dry wt $l^{-1} d^{-1}$). This value represents the highest EPA productivity reported for any microalga (Table 2.4).

It appears then, that manipulating cell density is preferable to imposing stress as a means of maximizing EPA content and productivity, insofar as it allows all factors affecting growth to function optimally. In algal mass culture the light regime of the single cell, which is controlled mainly by cell density, should be the sole environmental limitation for photoautotrophic production (Goldman, 1979; Richmond, 1988).

Evaluation of EPA-producing species

The emerging pharmaceutical importance of EPA has triggered the study and development of several microalgal sources of this PUFA, each having its own advantages. To identify the most promising of these sources, one must decide upon a set of

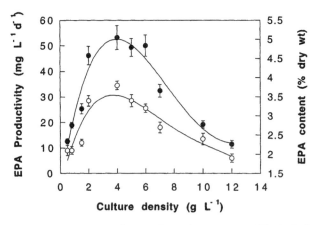

Figure 2.3 EPA content (open circles) and production rate (filled circles) as affected by biomass density in *M. subterraneus* cultures in mid-summer. Bars indicate ±SD (Hu *et al.*, 1997).

Table 2.4 Comparison of production rate and cell-content of eicosapentaenoic acid (EPA) in various species of microalgae (adapted from Hu *et al.*, 1997)

Algal species	Cultivation vessel	Cultivation mode	Growth condition	EPA content (% dry wt)	EPA productivity (mg l^{-1} day^{-1})	References
Nannochloropsis sp.	Glass column, 5 cm ID	B	CL	[a]	3.6	Sukenik *et al.* (1991)
Phaeodactylum tricornutum	Glass tank, 5.6 l	C	LD	3.3	25.1	Yongmanitchai and Ward (1992)
P. tricornutum	Tubular reactor, 50 l	C	Outdoors	1.9	47.8	Molina Grima *et al.* (1994b)
Monodus subterraneus	Erlenmeyer, 0.1 l	SC	CL	3.4	25.7	Cohen (1994)
M. subterraneus	Flat plate reactor, 14 l	SC	Outdoors	3.8	58.9	Hu *et al.* (1997)

ID, internal diameter; B = batch; C = continuous; SC = semi-continuous; CL = continuous illumination; LD = light–dark cycle; outdoors, natural conditions.

[a] EPA content, 0.22 pg per cell.

31

criteria. The ultimate criterion will obviously be the cost of producing the EPA. However, the latter cannot be accurately estimated in the absence of data relating to large-scale cultivation. Until such data become available, the parameters discussed below could prove helpful in a comparative evaluation.

The cost of production of EPA is directly related to the cost of production of the biomass, which in turn depends on the type of facility used for cultivation, i.e., open raceways versus a closed bioreactor.

The productivity of EPA (P_{EPA}, mg l^{-1} d^{-1}), as expressed in Equation 2.1, is a function of four parameters: the proportion of EPA ($\%_{EPA}$); the fatty acid content (C_{FA}); the specific growth rate μ and the biomass concentration (X) at harvest. The product of $\%_{EPA}$ times C_{FA} gives the EPA content (C_{EPA}), while the product $\mu \cdot X$ yields the biomass productivity (P_B) (Equations 2.2 and 2.3, respectively):

$$P_{EPA} = C_{EPA} \cdot P_B \qquad\qquad (2.1)$$

where

$$C_{EPA} = \%_{EPA} \cdot C_{FA} \qquad\qquad (2.2)$$

and

$$P_B = \mu \cdot X \qquad\qquad (2.3)$$

The highest proportion of EPA (44 per cent) was reported in *Nannochloropsis oculata* (Sukenik *et al.*, 1989). However, this value was reached under cultivation at 20°C, at which temperature both the fatty acid content and the growth rate were significantly depressed. As a result, both C_{EPA} and P_B were low, as reflected in a marginal EPA productivity of 3.7 mg l^{-1} d^{-1}. At 25°C, the proportion of EPA was lower but the growth rate was much higher and consequently EPA productivity rose to 5.5 mg l^{-1} d^{-1}. Similar findings have been reported for *Phaeodactylum tricornutum* (Yongmanitchai and Ward, 1991): although EPA content decreased from 5 to 3.3 per cent as the culture temperature was raised from 20 to 25°C, the concomitant increase in the growth rate enhanced EPA productivity to 19 mg l^{-1} d^{-1}. *Porphyridium cruentum* is unique in that its EPA proportion was maximal under conditions leading to the highest growth rate (Cohen *et al.*, 1988). Maximum biomass productivity was obtained at a biomass concentration higher than that responsible for the highest growth rate. However, increasing biomass concentration caused EPA proportion and content to decline, cancelling the benefits of the increase in biomass productivity. In *M. subterraneus*, both EPA proportion and fatty acid content were high, if not maximal, at a relatively high biomass concentration. The resulting combination of high biomass productivity and high EPA content made it possible to obtain EPA productivities as high as 26 mg l^{-1} d^{-1} indoors and 59 mg l^{-1} d^{-1} outdoors.

Production of high purity EPA

The major source of commercial EPA is oil of marine cold water fish. While the quality of this source is adequate for current health food uses, pharmaceutical applications require highly purified EPA (over 90 per cent). Unfortunately, fish oils are mixtures of different PUFA. The separation of EPA from such mixtures requires preparative HPLC, a very expensive technique which is economically prohibitive on a large scale. While the cost of fish oil is only a few dollars per kilogram, 70 per cent pure EPA sells at $100–200/kg and high purity EPA (> 90 per cent) at about $3000/ kg. A recent cost-sensitive analysis of EPA purification showed preparative HPLC to be the single most expensive component of the final product, affecting the cost of the purified product much more than the oil extract itself (Molina Grima *et al.*, 1996). It was further demonstrated that separation of EPA from the microalga *Phaeodactylum tricornutum* would be more economical than from fish, the cheaper source of oil, primarily due to the high cost of HPLC purification. The higher initial proportion of EPA and lower contents of other PUFA could be expected to reduce the need for HPLC and perhaps even eliminate it. Thus, although the cost of production of algal oil is currently much higher than that of fish oil, significant reduction in purification costs could more than offset this shortcoming.

The richest reported algal sources of EPA, *P. tricornutum*, *N. oculata* and *M. subterraneus* have EPA contents of around 4 per cent, not high enough to compete with fish oil as a source of EPA. However, unlike fish oils which are composed of TAG, algal oils also contain glycolipids and phospholipids. Fortunately, in many algae most of the cellular EPA resides in the chloroplast, predominantly in the glycolipids, while the remaining PUFA are often concentrated in the phospholipids and sometimes in the TAG. The separation of glycolipids from other lipids is a fairly simple matter. It can be done by stepwise washing of the lipid mixture off a thin pad of silica gel, a procedure which increases the concentration of EPA in the oil with minimal loss of EPA at a relatively low cost. The fatty acid composition of the glycolipid fraction in leading microalgal sources is shown in Table 2.5. The separation of the glycolipids fraction results in an increase in the proportion of EPA from 31 per cent in total lipids to 35 per cent in *P. tricornutum*, from 41 to 53 per cent in *Porphyridium cruentum*, from 37 to 73 per cent in *N. oculata* (Seto *et al.*, 1992) and from 37 to 49 per cent in *M. subterraneus*. Still, the most crucial predictor for the economic viability of the process is the proportion of other PUFA in the glycolipids. While separation of EPA from saturated and monounsaturated fatty acids by urea complex formation is rather simple (Cohen and Cohen, 1990), other PUFA can only be removed by preparative HPLC. Thus, an ideal algal source should have the lowest possible proportions of other PUFA in the glycolipid fraction. We have estimated the maximum EPA proportion (MEP) in the oil following a complete removal of saturated and monounsaturated fatty acids. For fish oil this value would be only 42 per cent, since EPA is not the major PUFA (Table 2.5). In *P. tricornutum*, EPA occurs in abundance in MGDG (40 per cent of fatty acids) but also in PC and TAG (32–33 per cent) (Molina Grima *et al.*, 1994a). Therefore, separation of the glycolipids would result in a significant loss of EPA. The presence of other PUFA, mainly 16:3 and 16:4, reduces the MEP value to 60 per cent, indicating that highly purified EPA cannot be obtained from this source except by invoking HPLC. In *P. cruentum* the theoretical MEP is 76 per cent, although a value as high as 82 per cent has been obtained (Cohen and Cohen, 1990). The difference is due to the fact that, although 18:2 is a PUFA, it can, to a certain extent, form urea complexes and be partially removed. Still, the presence of the antagonistic 20:4 may necessitate the use of preparative HPLC. The most promising sources to emerge from this analysis are *Nannochloropsis* and *M. subterraneus* with MEP values of 92 and 96 per cent, respectively. *M. subterraneus*, however, appears to be superior to *Nannochloropsis* due to its higher EPA content, much higher biomass, and EPA productivity under indoor as well as outdoor conditions. These figures clearly indicate that highly

Table 2.5 Fatty acid composition of the galactolipid fraction of EPA-rich microalgae as compared with fish oil.

Source	Fatty acid composition (% of total fatty acid)										MEP[a]	EPA content (% dry wt)	EPA prod[b]
	14:0	16:0	16:1	16:3	16:4	18:1	18:2	20:4	20:5	22:6			
Monodus[1]	0.7	19.0	23.9	–	–	2.3	0.4	1.5	49.3	–	96.3	3.8	60.9[2]
Nannochloropsis[3]	7.7	4.3	4.5	–	–	0.8	1.1	5.2	72.5	–	92.0	3–4	2–3
Porphyridium[4]	–	32.4	1.0	–	–	0.5	4.7	7.1	53.1	–	81.7	2.0	25.4
Phaeodactylum[5]	2.5	8.8	15.3	16.9	3.1	0.3	1.3	1.0	34.6	0.6	60.2	3.0	47.8
Cod liver oil[6]	4.2	10.6	7.8	–	–	21.6	1.5	0.5	9.4	11.0	42.0	–	–

[a]MEP, maximal EPA proportion, achievable utilizing urea treatment only. EPA productivity – mg EPA l^{-1} d^{-1}. [b]Productivity – mg EPA l^{-1} d^{-1}.
[1]Cohen, 1994; [2]Hu et al., 1997; [3]Seto et al., 1992; [4]Cohen, unpublished data; [5]Molina Grima et al., 1994; [6]Robles Medina et al., 1995.

purified EPA can be obtained using simple, inexpensive, methodologies that do not involve preparative HPLC.

Biosynthesis

A detailed understanding of the pathways responsible for the biosynthesis of EPA is essential not only for future genetic manipulations but also as a basis for modifying growth conditions so as to obtain higher EPA productivity and higher proportion of EPA in the glycolipids.

One of the most useful tools for the elucidation of biosynthetic pathways is exposure to specific inhibitors. Unfortunately, the arsenal of such inhibitors is quite limited. The substituted pyridazinone SAN 9785 (BASF 13-338, 4-chloro-5(dimethylamino)-2-phenyl-3(2H) pyridazinone) (Murphy *et al.*, 1985; Norman and St John, 1986, 1987) has been shown to inhibit chloroplastic, but not cytoplasmic, desaturation of 18:2 to 18:3ω3 in both prokaryotic and eukaryotic pathways in higher plants and algae (Murphy *et al.*, 1985; see also Chapter 1). However, SAN 9785 affects the lipids of *M. subterraneus* and of *P. cruentum* very differently (Khozin and Cohen, 1996). In the presence of the inhibitor, the proportion of EPA in the MGDG of *P. cruentum* decreased from 72 to 53 per cent and the share of eukaryotic molecular species was reduced to 29 per cent of total MGDG, primarily due to a decrease in the proportion of 20:5/20:5 from 46 to 27 per cent. Concurrently, the main prokaryotic species, 20:5/16:0, increased from 35 to 58 per cent. In *M. subterraneus*, in a similar situation, the fatty acid composition of the biomass surprisingly displayed an increase in the proportion of EPA, from 33 to 43 per cent (Table 2.6). In MGDG, the main depot of EPA in this organism, the proportion of EPA increased even more steeply, from 54 to 81 per cent, while that of most other fatty acids and especially that of 14:0, 16:0 and 16:1 decreased (Table 2.6). Molecular species analysis of MGDG revealed that, under the influence of SAN 9785, the eukaryotic-like 20:5/20:5 becomes the predominant molecular species, comprising 80 per cent of this lipid class (Table 2.7).

In *P. cruentum*, 18:2-PC is sequentially converted by the ω6 pathway to 20:4ω6-containing molecular species of PC, which serves as the major source of 20:4ω6 for both prokaryotic and eukaryotic molecular species of MGDG (Khozin *et al.*, 1996) (see Chapter 1). Chloroplastic EPA is primarily produced by ω3 desaturation of the eukaryotic 20:4ω6/20:4ω6 and prokaryotic 20:4ω6/16:0 molecular species of MGDG. In *M. subterraneus*, on the other hand, where 20:5/20:5 is the major molecular species of MGDG, we could not detect even traces of its presumed precursors 20:4ω6/20:4ω6 or 20:4ω6/20:5, while 20:4ω6/16:0 was quite abundant. A possible explanation is that, while 20:4/16:0 can serve as a possible substrate for chloroplastic ω3 desaturation, the 20:5/20:5 molecular species of MGDG is mostly obtained by import of the respective DAG from the cytoplasm, rather than by chloroplastic desaturation of the 20:4ω6/20:4ω6 species (Figure 2.4). Chloroplastic desaturation which predominantly involves the prokaryotic species 20:4ω6/16:0 and 20:4ω6/14:0, appears to be inhibited by SAN 9785, while cytoplasmic desaturation, which utilizes phospholipid-bound fatty acids as substrates, is not affected. Under these conditions, EPA would mostly be produced by cytoplasmic desaturation, a mechanism which increases the eukaryotic to prokaryotic ratio and hence the proportion of EPA in MGDG and in total lipids.

Table 2.6 Effect of SAN 9785 on the fatty acid composition of the major lipids of *M. subterraneus.*

Lipid class	SAN 9785	14:0	16:0	16:1 ω11	16:1 ω7	18:0	18:1 ω9	18:1 ω7	18:2	18:3 ω6	18:3 ω3	20:3 ω6	20:4 ω6	20:5 ω3
								Fatty acid composition (% of total fatty acids)						
Total lipids	–	3.3	20.6	3.3	26.1	0.7	5.9	0.8	0.9	1.2	0.5	0.5	3.7	32.6
Total lipids	+	3.1	16.2	2.7	23.3	0.9	2.1	0.5	2.4	0.3	0.5	0.2	3.2	42.7
MGDG	–	5.4	13.5	1.5	13.4	1.3	4.3	0.8	1.8	0.6	0.3	0.9	2.2	54.2
MGDG	+	2.7	4.3	t	5.7	0.2	0.8	0.2	0.4	0.2	0.7	0.2	3.3	81.3
PC	–	1.0	33.1	–	13.2	2.2	7.4	0.9	7.3	15.4	–	0.4	4.4	13.9
PC	+	2.1	21.3	–	21.1	2.9	12.2	0.4	18.0	0.2	–	t	5.7	14.6

t, trace level.

Table 2.7 Effect of SAN 9785 on the molecular species composition of MGDG in *M. subterraneus.*

SAN 9785	Molecular species composition (%)					
	20:5/20:5	20:5/14:0	20:5/16:1	X	20:5/16:0	20:4/16:0
−	57.2	12.7	14.1	1.3	5.9	8.9
+	80.1	5.4	5.8	3.0	4.4	1.5

X, unidentified; 20:5 and 20:4 stand for 20:5ω3 and 20:4ω6, respectively.

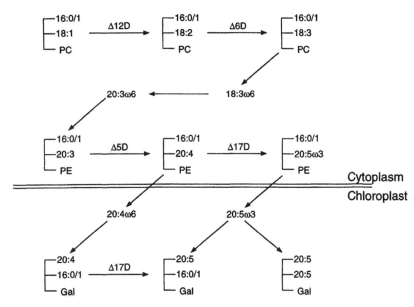

Figure 2.4 Likely pathways of EPA biosynthesis in *M. subterraneus.*

Elevated light intensity promotes chloroplastic ω3 desaturation. Thus, under high light more 20:4/14:0 would be desaturated to 20:5/14:0. Light limitation on the other hand, inhibits chloroplastic but not cytoplasmic desaturation. Since the latter produces only the eukaryotic molecular species of MGDG (20:5/20:5), it is obvious that low light will be associated with a greater level of 20:5/20:5 but a lower level of prokaryotic molecular species such as 20:5/14:0. A eukaryotic molecular species contains two EPA moieties per molecule versus only one in prokaryotic molecular species. Consequently, the proportion of EPA in MGDG will increase at the expense of 14:0, 16:0 and 16:1.

Strains that predominantly synthesize MGDG via the eukaryotic pathway then, are advantageous for EPA production. The MGDG of such strains will mostly consist of the 20:5/20:5 molecular species, which contains twice as much EPA as the 20:5/16:0 species produced by the prokaryotic pathway.

These pathways could be common in other algae. Thus, EPA production in the marine microalgae *Chroomonas salina* and *Nannochloropsis oculata* was not inhibited by SAN 9785 (Henderson *et al.*, 1990), indicating a lack of chloroplastic ω3 activity.

Indeed, Schneider and Roessler (1994) have recently demonstrated that EPA bio-synthesis in the latter alga is entirely cytoplasmic, and they suggest that 20:4ω6 PE may be the precursor of 20:5 PE. In view of the resemblance between *N. oculata* and *M. subterraneus* it is likely that a similar mechanism operates in the latter as well. Indeed, the only phospholipid in *Monodus* with an EPA content high enough to contain the 20:5/20:5 molecular species is PE.

The level of 20:4ω6 in MGDG could thus serve as an indicator of the type of EPA biosynthetic pathway operating in a given alga. The chloroplasts of algae such as *P. cruentum*, which have a high abundance of 20:4, are capable of ω3 desaturase activity, whereas in algae such as *N. oculata* and *P. tricornutum*, whose galactolipids are devoid of 20:4, ω3 desaturation is exclusively cytoplasmic. The low level of 20:4 in *M. subterraneus* could point to a capacity for both chloroplastic and cytoplasmic ω3 desaturation activities. An analogous distinction can be seen in higher plants. In the leaves of pea, an 18:3 plant, 18:3ω3 is synthesized exclusively via the eukaryotic pathway, mostly in the endoplasmic reticulum. By contrast, in the leaves of spinach, a 16:3 plant, both the prokaryotic and the eukaryotic pathways are operative, but 18:3ω3 is mostly produced in the chloroplast (Mazliak, 1994).

Future directions

The drastic shift from prokaryotic to eukaryotic MGDG induced by exposure to the herbicide SAN 9785 suggests a possible approach to manipulation of lipid composition. For example, reducing the proportion of the prokaryotic components in the galactolipids and increasing that of the eukaryotic component (20:5/20:5), it should be possible to enhance the proportion of EPA in the galactolipids at the expense of C_{14} and C_{16} fatty acids. Pham Quoc *et al.* (1993) recently reported that addition of oleic acid to the growth medium of the cyanobacterium *Spirulina platensis* caused the normally prokaryotic lipids to become partly eukaryotic, producing the 18:3ω6/ 18:3ω6 molecular species of MGDG. It can be deduced that the externally added fatty acid overcame the natural preference for C_{16} fatty acids at the sn-2 position of glycolipids and was incorporated instead. By analogy, addition of oleic or linoleic acids to the growth medium of *M. subterraneus*, may be expected to lead to their displacing the C_{16} fatty acids residing in the sn-1 position of the phospholipids in this alga to form C_{18}/C_{18} molecular species. Eventually, this could result in the preferential production of EPA-rich eukaryotic glycolipids.

The question of the separation of EPA-rich glycolipids from algal oils was discussed earlier. Efficient separation could be facilitated by the adoption of sophisticated extraction methods. For instance one could envisage the development of a super critical extraction method for preferential separation of galactolipids from other lipids.

Further elucidation of the effect of environmental conditions on the interrelation between prokaryotic and eukaryotic lipids could also be useful. Adlerstein *et al.* (1997) have recently shown that in *P. cruentum* this interrelation is strongly affected by temperature and light intensity (see also Chapter 1).

Finally, a thorough understanding of the underlying pathways of EPA biosynthesis is a prerequisite for the utilization of genetic engineering tools to retailor the genes responsible for lipid biosynthesis so as to obtain higher productivities of eukaryotic lipids.

References

ADLERSTEIN, D., KHOZIN, I., BIGOGNO, C. and COHEN, Z., 1997, Effect of environmental conditions on the molecular species composition of galactolipids in the alga *Porphyridium cruentum*, *J. Phycol.*, **33**, 975–979.

BEN-AMOTZ, A., TORNABENE, T.G. and THOMAS, W.H., 1985, Chemical profile of selected species of microalgae with emphasis on lipids, *J. Phycol.*, **21**, 72–81.

BERNER, T., WYMAN, K., DUBINSKY, Z. and FALKOWSKI, P.G., 1989, Photoadaptation and the 'package' effect in *Dunaliella tertiolecta* (Chlorophyceae), *J. Phycol.*, **25**, 70–78.

COHEN, Z., 1994, Production of eicosapentaenoic acid by the alga *Monodus subterraneus*, *J. Am. Oil Chem. Soc.*, **71**, 941–946.

COHEN, Z. and COHEN, S., 1990, Preparation of eicosapentaenoic acid (EPA) concentration from *Porphyridium cruentum*, *J. Am. Oil Chem. Soc.*, **68**, 16–19.

COHEN, Z., VONSHAK, A. and RICHMOND, A., 1988, Effect of environmental conditions on fatty acid composition of the red alga *Porphyridium cruentum*: correlation to growth rate, *J. Phycol.*, **24**, 328–332.

GOLDMAN, J.C., 1979, Outdoor algal mass cultures II. Photosynthetic yield limitations, *Water Res.*, **13**, 119–136.

HENDERSON, R.J., HODGSON, P. and HARWOOD, J.L., 1990, Differential effects of the substituted pyridazinone herbicide Sandoz 9785 on lipid composition and biosynthesis in non-photosynthetic marine microalgae. 1. Lipid composition and synthesis, *J. Exp. Bot.*, **41**, 729–736.

HU, Q., HU, Z-y., COHEN, Z. and Richmond, A., 1997, Enhancement of eicosapentaenoic acid (EPA) and γ-linolenic acid (GLA) production by manipulating cell density in outdoor cultures of *Monodus subterraneus* (Eustigmatophyte) and *Spirulina platensis* (Cyanobacterium), *Eur. J. Phycol.*, **32**, 81–86.

IWAMOTO, H. and SATO, S., 1986, EPA production by freshwater unicellular algae, *J. Am. Oil Chem. Soc.*, **63**, 434.

KHOZIN, I. and COHEN, Z., 1996, Differential response of microalgae to the substituted pyridazinone Sandoz 9785 reveal different pathways in the biosynthesis of eicosapentaenoic acid (EPA), *Phytochemistry*, **42**, 1025–1029.

KHOZIN, I., ADLERSTEIN, D., BIGOGNO, C. and COHEN, Z., 1996, Elucidation of the biosynthesis of eicosapentaenoic acid (EPA) in the microalga *Porphyridium cruentum*, in Williams, J.P., Khan, M.U. and Lem, N.W. (Eds), *Physiology, Biochemistry, and Molecular Biology of Plant Lipids*, pp. 93–95, Dordrecht, The Netherlands: Kluwer Academic Publishers.

MAZLIAK, P., 1994, Desaturation processes in fatty acid and acyl lipid biosynthesis, *J. Plant Physiol.*, **143**, 399–406.

MEIJI MILK PRODUCTS, 1985, Japanese Pat. JP 60 87,798.

MOLINA GRIMA, E., GARCIA CAMACHO, F., SÁNCHEZ PÉREZ, J. A., ACIÉN FERNÁNDEZ, F.G., FERNÁNDEZ SEVILLA, J.M. and VALDÉS SANZ, F., 1994a, Effect of dilution rate on eicosapentaenoic acid productivity of *Phaeodactylum tricornutum* UTEX 640 in outdoor chemostat culture, *Biotechnol. Lett.*, **16**, 1035–1040.

MOLINA GRIMA, E., SÁNCHEZ PÉREZ, J. A., GARCIA CAMACHO, F., FERNÁNDEZ-SEVILLA, J.M., ACIÉN FERNÁNDEZ, F.G. and URDA CARDONA, J., 1994b, Biomass and eicosapentaenoic acid productivities from an outdoor batch culture of *Phaeodactylum tricornutum* UTEX 640 in an airlift tubular photobioreactor, *Appl. Microbiol. Biotechnol.*, **20**, 279–290.

MOLINA GRIMA, E., ROBLES, A., GIMENEZ, A., *et al.*, 1996, Gram scale purification of EPA from *Phaeodactylum tricornutum*, *J. Appl. Phycol.*, **8**, 359–367.

MURPHY, D.J., HARWOOD, J.L., LEE, K.A., ROBERTO, F., STUMPF, P.K. and ST JOHN, J.B., 1985, Differential responses of a range of photosynthetic tissues to a substituted pyridazinone, Sandoz 9785. Specific effects on fatty acid desaturation, *Phytochemistry*, **24**, 1923–1929.

NICHOLS, B.W. and WOOD, B.J.B., 1968, The occurrence and biosynthesis of gamma-linolenic acid in a blue-green alga, *Spirulina platensis*, *Lipids*, **3**, 46–50.

NORMAN, H.A. and ST JOHN, J.B., 1986, Metabolism of unsaturated monogalactosyldiacylglycerol molecular species in *Arabidopsis thaliana* reveals different sites and substrates for linolenic acid synthesis, *Plant Physiol.*, **81**, 731–736.

NORMAN, H.A. and ST JOHN, J.B., 1987, Differential effects of a substituted pyridazinone, BASF 13-338 on pathways of monogalactosyldiacylglycerol synthesis in *Arabidopsis*, *Plant Physiol.*, **85**, 684–688.

PHAM QUOC, K., DUBACQ, J.P., JUSTIN, A.M., DEMANDRE, C. and MAZLIAK, P., 1993, Biosynthesis of eukaryotic lipid molecular species by the cyanobacterium *Spirulina platensis*, *Biochim. Biophys. Acta*, **1168**, 94–99.

REITAN, K.I., RAINUZZO, J.R. and OLSEN, Y., 1994, Effect of nutrient limitation on fatty acid and lipid content of marine microalgae, *J. Phycol.*, **30**, 972–979.

RENAUD, S.M. and PARRY, D.L., 1994, Microalgae for use in tropical aquaculture, II: effect of salinity on growth, gross chemical composition and fatty acid composition of three species of marine microalgae, *J. Appl. Phycol.*, **6**, 347–356.

RICHMOND, A., 1988, A prerequisite for industrial microalgaculture: efficient utilization of solar irradiance, in *Algal Biotechnology*, Stadler, T., *et al.* (Eds), pp. 237–244, London: Elsevier Applied Science Publishers.

RICHMOND, A., LICHTENBERG, E., STAHL, B. and VONSHAK, A, 1990, Quantitative assessment of the major limitations on productivity of *Spirulina platensis* in open raceways, *J. Appl. Phycol.*, **2**, 195–206.

ROBLES MEDINA, A., GIMÉNÉZ GIMÉNÉZ A., GARCIA CAMACHO, F., SÁNCHEZ PÉREZ, J.A., MOLINA GRIMA, E. and CONTRERAS GÓMEZ, A., 1995, Concentration and purification of Stearidonic, eicosapentaenoic, and docosahexaenoic acids from cod liver oil and the marine microalga *Isochrysis galbana*, *J. Am. Oil Chem. Soc.*, **72**, 575–583.

SCHNEIDER, J.C. and ROESSLER, P., 1994, Radiolabeling studies of lipids and fatty acids in *Nannochloropsis* (Eustigmatophyceae), an oleaginous marine alga, *J. Phycol.*, **30**, 594–598.

SETO, A., KUMASAKA, K., HOSAKA, M., *et al.*, 1992, Production of eicosapentaenoic acid by a marine microalga and its commercial utilization for aquaculture, in Kyle, D.J. and Ratledge, C. (Eds), *Single Cell Oil*, pp. 219–234, Champaign: Am. Oil Chem. Soc.

SHIFRIN, N.S. and CHISHOLM, S.W., 1981, Phytoplankton lipids: interspecific differences and effects of nitrate, silicate, and light-dark cycles, *J. Phycol.*, **17**, 374–384.

SUEN, Y., HUBBARD, J.S., HOLZER, G. and TORNABENE, T.G., 1987, Total lipid production of the green alga *Nannochloropsis* sp. QH under different nitrogen regimes, *J. Phycol.*, **23**, 289–296.

SUKENIK, A., FALKOWSKI, P.G. and BENNETT, J., 1987, Potential enhancement of photosynthetic energy conversion in algal mass culture, *Biotechnol. Bioeng.*, **30**, 970–977.

SUKENIK, A., CARMELI, Y. and BERNER, Y., 1989, Regulation of fatty acid composition by growth irradiance level in the eustigmatophyte *Nannochloropsis* sp., *J. Phycol.*, **25**, 686–692.

SUKENIK, A., LEVY, R.S., LEVY, Y., FALKOWSKI, P.G. and DUBINSKY, Z., 1991, Optimal algae biomass production in an outdoor pond: a simulation model, *J. Appl. Phycol.*, **3**, 197–209.

YONGMANITCHAI, W. and WARD, O.P., 1991, Growth of and omega-3 fatty acid production by *Phaeodactylum tricornutum* under different culture conditions, *Appl. Environ. Microbiol.*, **57**, 419–425.

YONGMANITCHAI, W. and WARD, O.P., 1992, Growth and eicosapentaenoic acid production by *Phaeodactylum tricornutum* in batch and continuous culture systems, *J. Am. Chem. Soc.*, **69**, 581 590.

3

Production of eicosapentaenoic acid by the marine eustigmatophyte *Nannochloropsis*

A. SUKENIK

Introduction

Very long chain polyunsaturated fatty acids (VLC-PUFA) that belong to the omega-3 group, are essential fatty acids for the development of marine organisms and important in a balanced human diet. Medical studies showed that these fatty acids in the human diet may reduce morbidity risks of coronary heart diseases (Simopoulos, 1986). Recent studies indicated that omega-3 VLC-PUFA, mainly docosahexaenoic acids (DHA), are essential for the development of the brain, the retina and the nervous system in human (Innis, 1992). DHA is an important fatty acid found in human breast milk, but is lacking in commercially available infant formulas (Kyle *et al.*, 1992). The current major source for omega-3 VLC-PUFA is fish oil that contains both eicosapentaenoic acid (EPA) and DHA. Although fish oil is an abundant source of both EPA and DHA, there are some significant problems associated with the use of fish oil as a food supplement (Boswell *et al.*, 1992). Unicellular algae that synthesize EPA and DHA would be a useful source for these essential fatty acids. Microalgae are a large and diverse group of photosynthetic microorganisms. Microalgae are known as the primary producers and the foundation of essential fatty acids in the food web of marine life. The cultivation of microalgae for EPA production has been developed and is mainly applied in the aquaculture industry, although attempts were made to use this source of essential fatty acids as human diet supplements and animal feed.

In this chapter we review the genus *Nannochloropsis* and describe its biochemical characteristics, with special emphasis on lipid and fatty acid production. Recent attempts to elucidate the biosynthetic pathway of EPA in this genus are described herein. Physiological and biochemical responses of *Nannochloropsis* sp. to various environmental conditions are discussed and optimized production of EPA is demonstrated. *Nannochloropsis* biomass is examined and evaluated as a reliable enriched source of EPA and current applications and potentially commercial implementations are reviewed.

Nannochloropsis and other eustigmatophytes

The genus *Nannochloropsis* was first described by Hibbered (1981), who removed it from the class Chlorophyceae and placed it in the class Eustigmatophyceae in the family Monodopsidaceae. The genus *Nannochloropsis* is difficult to identify because the cells are small and indistinguishable from other chlorophytes by light microscopy observations. Furthermore *Nannochloropsis* cells are difficult to fix for electron microscopy (Hibbered, 1981). Four species of *Nannochloropsis* were described: *N. oculata* and *N. salina* (Hibbered, 1981), *N. gaditana* (Lubian, 1982) and recently Karlson *et al.* (1996) described a new species: *N. granulata* sp. nov. Many other isolates of *Nannochloropsis* were deposited in various culture collections (Anderson *et al.,* 1991) and specific mutants were selected and characterized (Schneider *et al.,* 1995).

Species in the genus *Nannochloropsis* are characterized by various unique biochemical and ultrastructure features such as the absence of chlorophyll *b* or *c* and the composition of the cellular xanthophyll pigments (Whittle and Casselton, 1975; Volkman *et al.,* 1993), relatively high content of EPA (Maruyama *et al.,* 1986) and the presence of specific sterols (Patterson *et al.,* 1994; Gladu and Patterson, 1995). The unique ultrastructure features of *Nannochloropsis* are: lamellate vesicles present in the cytoplasm; and the connection of the chloroplast envelope with the nuclear envelope (Figure 3.1) (Santos and Leedale, 1995). Molecular techniques that determine the 18S rRNA gene sequences were recently applied to distinguish among species and to assist in establishing sounder taxonomic criteria (Karlson *et al.,* 1996). The unique biochemical and ultrastructural features of *Nannochloropsis* were used to reclassify various algal species that were previously identified as species of the genus *Chlorella* (Maruyama *et al.,* 1986; Gladu and Patterson, 1995) or *Nannochloris* (Anita *et al.,* 1975).

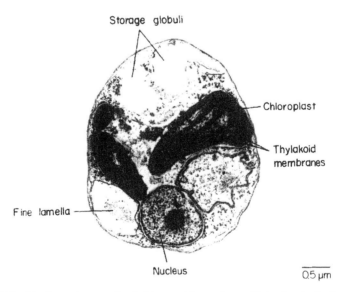

Figure 3.1 Electronmicrograph of *Nannochloropsis* sp. Note the typical ultrastructure features of the Eustigmatophyceae: a lamellate vesicle present in the cytoplasm and the connection of the chloroplast envelope with the nuclear envelope. Other organelles labelled are chloroplast, nucleus, thylakoid membranes and lipid bodies.

Chemical composition

The chemical composition of *Nannochloropsis* was studied as a taxonomic indicator (Menzel and Wild, 1989; Mourente *et al.*, 1990; Karlson *et al.*, 1996) and to assess the value of this genus as primary producers in the artificial food chain required in many marine fish hatcheries (Volkman *et al.*, 1989; Seto *et al.*, 1992).

Pigments

The absence of chlorophyll *b* or *c* is a common phenomenon to the class Eustigmatophyceae (Whittle and Casselton, 1975). In addition to β-carotene, the major xanthophyll pigments found in all *Nannochloropsis* species are violaxanthin and a vaucherxanthin-like pigment (Anita and Cheng, 1982; Karlson *et al.*, 1996). Zeaxanthin and anthraxanthin were detected as minor xanthophylls only after cells were exposed to a high irradiance level (Figure 3.2). Epoxidation of violaxanthin and the formation of anthraxanthin and zeaxanthin were suggested as components of the xanthophyll cycle in *Nannochloropsis* (L.M. Lubian, personal communication).

Sugars

The sugar composition of the polysaccharides of *Nannochloropsis* is quite similar to that of diatoms, except that it does not contain arabinose. The dominant sugar is glucose, which composes 68 per cent of the polysaccharides. Other typical sugars are fucose, galactose, manose, rhamnose, ribose and xylose, each consisting of less than 10 per cent of the polysaccharides (Brown, 1991).

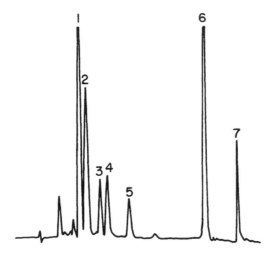

Figure 3.2 High performance liquid chromatogram of *Nannochloropsis* pigments. Note the absence of chlorophyll *b* or *c*. The major xanthophyll pigments present are violaxanthin (1) and a vaucherxanthin-like pigment (2). Anthraxanthin (3) and zeaxanthin (5) were detected as minor xanthophylls only after cells were exposed to a high irradiance level. Other detected pigments are chlorophyll *a* (6), β-carotene (7) and an unidentified pigment (4).

Amino acids

Nannochloropsis, like many other unicellular algae, is poor in methionine, trypto-
phan, cystine, histidine and hydroxyproline, whereas aspartate and glutamate were
found in relatively high concentrations (Brown, 1991). *Nannochloropsis oculata* was
reported to contain high levels of proline (Brown, 1991) and a similar elevated
proportion of proline was also found in several unidentified *Nannochloropsis* species
(James *et al.,* 1989; Sukenik *et al.,* 1993a).

Lipids and sterols

The glycerolipids monogalactosyl diacylglycerol (MGDG) and digalactosyl diacyl-
glycerol (DGDG) represent about 45 per cent of the polar lipid classes in
Nannochloropsis (Menzel and Wild, 1989; Schneider and Roessler, 1994). Among
the phospholipids, phosphatidylcholine (PC) was found in a high proportion.
Phosphatidylglycerol (PG), phosphatidylethanolamine (PE) and phosphatidylinosi-
tol (PI) exist in *Nannochloropsis,* but in relatively small quantities (Schneider and
Roessler, 1994). Sulphoquinovosyl diacylglycerol (SQDG) was also reported in
Nannochloropsis (Menzel and Wild, 1989). The presence of diacylglycerol trimethyl-
homoserine (DGTS) in *Nannochloropsis* was reported by Schneider and Roessler
(1994). Neutral lipids are essentially made up of triacylglycerol (TG), but hydro-
carbon-wax and sterol esters were reported as well (Suen *et al.,* 1987; Emdadi and
Berland, 1989). Cholesterol is the principal sterol in all *Nannochloropsis* species
studied to date. Other minor sterols are fucosterol and isofucosterol (Patterson *et
al.,* 1994).

Fatty acids

Polyunsaturated fatty acids produced by *Nannochloropsis* are the major cellular
chemical component that attracted scientists' attention to this organism. The
major fatty acids in *Nannochloropsis* are: 14:0, 16:0, 16:1, 20:4ω6 and 20:5ω5
(Maruyama *et al.,* 1986; Sukenik *et al.,* 1989; Schneider and Roessler 1994). Fatty
acid distibution in major lipid classes is shown in Table 3.1. Triacylglycerol is characterized by a high proportion of saturated and monounsaturated short chain fatty
acids, 14:0, 16:0 and 16:1. The galactolipids and specifically MGDG are enriched
with the VLC-PUFA 20:5ω3. A high proportion of EPA is also found in DGTS and
PG, whereas PE is enriched with arachidonic acid (AA – 20:4ω6) and contains a
relatively low amount of EPA. PC represents a unique fatty acid composition with a
relatively high proportion of 18:1ω9 and 18:2ω6.

    The superb quality of *Nannochloropsis* as a feedstuff in the aquaculture industry is
attributed to its high EPA content, which makes it a preferred nutritional feed for
rotifers in an artificial food chain for many fish hatchlings. The content of EPA and
other cellular fatty acids of *Nannochloropsis* can be modulated by varying growth
conditions (see below).

Table 3.1 Fatty acid composition of *Nannochloropsis* sp. glycerolipids.

	Fatty acid composition in lipid class (mol %)									
	14:0	16:0	16:1 ω7	18:0	18:1 ω9	18:2 ω6	18:3 ω3	18:3 ω6	20:4 ω6	20:5 ω3
Total lipids	6.9	19.9	27.4	–	1.7	3.5	0.7	–	4.2	34.9
TAG	10.8	39.6	45.0	1.7	1.6	0.8	–	–	–	–
MGDG	20.0	11.8	13.0	tr	tr	0.6	–	–	0.7	53.2
DGDG	8.0	30.7	40.7	tr	tr	1.0	–	–	tr	18.5
SQDG	5.5	52.0	41.3	0.6	tr	–	–	–	–	tr
PG	1.7	51.8	14.3	0.7	tr	0.6	–	–	–	30.5
PC	2.6	14.1	46.7	0.6	6.6	17.4	3.4	tr	2.1	6.1
PE	2.7	8.6	28.3	0.7	1.7	2.4	1.0	2.2	25.9	26.5
DGTS	3.7	6.9	23.5	0.8	0.9	1.3	–	–	5.0	57.8
PI	4.4	28.6	36.1	1.1	3.9	4.9	–	–	14.1	6.8

Relative fatty acid distribution was determined by gas chromatography of FAME, prepared from isolated lipid classes. Values for peaks less than 0.5 per cent are designated by tr (trace) (adapted from Schneider *et al.*, 1995).

Biosynthesis of lipid and polyunsaturated fatty acids

Although many of the biosynthetic steps leading to the formation of lipids are similar in many respects for higher plants and algae, many algae contain unique lipids such as DGTS and diacylglycerol hydroxymethyltrimethyl-β-alanine (DGTA) (Vogel and Eichnberger, 1992). In addition, highly elongated and polyunsaturated fatty acids, such as 20:4ω6, 20:5ω3 and 22:6ω3, are frequently found in various algae genera, but are rare in higher plants.

Fatty acid and lipid biosynthesis was intensively studied in higher plants, leading to a widely accepted model of the biosynthetic pathways (Roughan and Slack, 1982; Browse and Somerville, 1991). This serves as a preliminary model for understanding lipid synthesis in algae, which was further developed for PUFA synthesis in *Nannochloropsis* (Figure 3.3). Labelling studies in plants with exogenously applied acetate suggested that PC was an intermediate in the synthesis of PUFAs which eventually accumulate in MGDG (reviewed by Heinz, 1993). However, analysis of the change of radioactivity in lipids and fatty acids of *Nannochloropsis,* after pulse-chase labelling with [^{14}C]-acetate, suggested that PE and PC are both intermediates in the synthesis of 20:5ω3 (Schneider and Roessler, 1994). From these experiments and the fatty acid composition of lipid classes in *Nannochloropsis*, it may be proposed that C18 fatty acids are desaturated to 18:3ω6 while attached to PC. The 18:3ω6 is elongated by an unknown mechanism that does not occur in higher plants and is then desaturated from 20:3ω6 to 20:5ω3 (EPA) (Figure 3.3), probably while attached to PE. Recent support for the proposal that PE is an important intermediate in 20:5ω3 synthesis comes from the discovery of a novel enzyme activity that incorporates unsaturated C20 fatty acids into PE at both positions of the glycerol backbone (Schneider and Roessler, 1995).

The model (Figure 3.3) stipulates that the introduction of a double bond at the ω3 position occurs using 20:4ω6 (arachidonic acid, AA) as a substrate, instead of

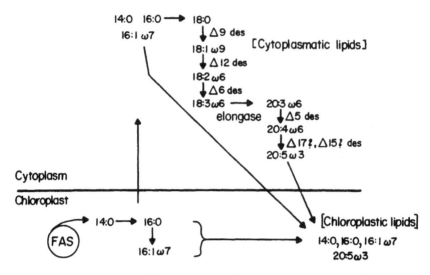

Figure 3.3 A schematic model for the major pathway of polyunsaturated fatty acid synthesis in *Nannochloropsis*. FAS = fatty acid synthetase. Fatty acids 14:0, 16:0 and 16:1 are produced by chloroplastic enzymes and either exported to the cytoplasm or used for synthesis of chloroplast lipids. The exported fatty acids are incorporated into lipids and desaturated and elongated by specific enzymes. The final product of these steps, 20:5ω3, is imported into the chloroplast for synthesis of chloroplast lipids that contain this fatty acid. This scheme proposes that desaturation reactions that lead to the formation of 20:5ω3 occur only in the cytoplasm and the final desaturation step is performed by a putative Δ17 desaturase or a non-specific Δ15 desaturase.

18:2ω6, as occurs in higher plants. Strong evidence to support this hypothesis comes from the substrate–product relationship between the EPA and AA (Schneider and Roessler, 1994). Additional support for this model comes from a recently isolated mutant of *Nannochloropsis* which is completely devoid of EPA. The elimination of 20:5ω3 in the mutant strain led to unexpected changes in the fatty acid composition of lipids: i.e., 20:4ω6 accumulated to the greatest extent in extra-chloroplastic lipids and to a lesser extent in chloroplastic lipids. Molecular species of MGDG containing 20:4ω6 were much reduced compared to the level of species containing 20:5ω3 in the wild type strain, concomitant with an increase in species containing 16:0 and 16:1ω7 (Schneider *et al.*, 1995). These changes in the fatty acid composition of lipids indicate that the ω3 desaturase is located in the microsomes. Thus, all 20:5ω3 fatty acids which occur in chloroplast lipids would need to be imported from the cytosol. The end product of the PUFA pathway in the mutant, 20:4ω6, was found at a reduced level in chloroplastic lipids. Thus it may be postulated that the enzymes that synthesize chloroplast lipids from imported fatty acids exhibit a high selectivity for EPA, relative to intermediates such as AA. The complete absence of 20:5ω3, which is partly compensated by AA in this mutant, may suggest that there is a single genetic locus for the ω3 desaturase; however, a complete genetic analysis is required to prove this point. Evidence that 20:4ω6 is a substrate for an ω3 desaturation has been presented for two other algae (Cohen *et al.*, 1992; Arao and Yamada, 1989, 1994) and some fungi (Gellerman and Schlenk, 1979; Shimizu *et al.*, 1988).

Modulation of fatty acid composition

The chemical composition of unicellular algae can be manipulated by varying growth conditions (Spoehr and Milner, 1949; Shifrin and Chisholm, 1981). Under optimal growth conditions, cells mainly synthesize proteins that are required for growth. However, as growth rate is retarded by a limiting factor, such as nutrient availability or by unoptimal temperature, cellular composition is changed due to accumulation of lipids, carbohydrates or other metabolites. Although total lipid content may increase under specific growth conditions, the cellular content of a specific lipid class can be reduced concomitant with disproportional variations in fatty acid composition. The effect of various growth conditions on lipid content and fatty acid composition has been studied in *Nannochloropsis* (Teshima *et al.*, 1983, 1991; James *et al.*, 1989; Sukenik *et al.*, 1989, 1993a, 1993b; Sukenik and Carmeli, 1990; Renaud *et al.*, 1991; Seto *et al.*, 1992), and integrated into operational conclusions that were aimed at increasing the yield of EPA from this alga (Sukenik, 1991).

Modulation of lipid content and fatty acid composition has been achieved in higher plants and cyanobacteria by genetic manipulation of relevant genes (Murata *et al.*, 1992; Falcone *et al.*, 1994). A similar approach has been suggested by Roessler (1990), but could not be materialized until a reliable transformation system was developed for unicellular algae (Dunahay *et al.*, 1995). Consequently, a significant breakthrough is expected in our ability to genetically control lipid and fatty acid production by unicellular algae.

Light intensity

Modulation of lipid content and composition by light intensity has been demonstrated in *Nannochloropsis* (Sukenik *et al.*, 1989, 1993b). *Nannochloropsis* grown in saturating light conditions was characterized by a high content of lipids and fatty acids, as compared to cells grown in light limiting conditions (Sukenik *et al.*, 1989). At subsaturating light conditions, *Nannochloropsis* preferentially synthesizes galactolipids, which are enriched with EPA (Sukenik *et al.*, 1993b). Under these conditions, EPA accumulated up to 40 per cent of total cellular fatty acids. Similar results were reported by Renaud *et al.* (1991), who found a significant decrease in the relative abundance of EPA when cultures of *Nannochloropsis oculata* were grown at high photon densities. Seto *et al.* (1992), however, reported that high light intensity imposed a net increase in growth rate with almost no effect on the relative distribution of EPA.

It is important to note that experimental evaluation of the effect of growth irradiance level on cellular chemical composition should be performed with optically thin cultures that are continuously diluted to keep cell density constant. These constraints are obviously not applicable to large scale cultures which are commonly maintained as batch cultures (Renaud *et al.*, 1991).

Light–dark cycles

The increase in cellular lipid content imposed by light saturating conditions was associated with an increase in the rate of triacylglycerol synthesis and a decrease in the relative synthesis rate of the major galactolipid component, MGDG. These changes in the relative synthesis rates of various lipid classes coincided with variations in the percentage of EPA and AA, as compared to the relative abundance of palmitic acid (C16:0). It was shown that in *Nannochloropsis*, storage triacylglycerols are synthesized in the light and are metabolized rapidly in the dark. Polar lipids such as MGDG are synthesized in the light as well, but turn over very slowly in the dark (Sukenik and Carmeli, 1990).

A repetitive light–dark cycle may cause the processes of the algal cell cycle to occur almost simultaneously in the whole population. In such a culture, cell division occurs during the dark period, while many metabolic processes, such as pigment, lipid and fatty acid synthesis, including accumulation of polyunsaturated fatty acids, take place during the light period. In *Nannochloropsis* the cellular content of EPA increased during the day and decreased in the night, mainly due to cell division (Figure 3.4). The total mass of EPA per unit culture increased during the light period, but remained fairly stable during the dark period. Moreover, the consumption of neutral storage lipids for maintenance during the dark period caused an increase in the relative proportion of EPA at the end of the diurnal cycle (Sukenik and Carmeli, 1990).

Temperature

Various *Nannochloropsis* strains showed optimal growth at temperatures between 20 and 25°C (James *et al.*, 1989; Sukenik, 1991; Seto *et al.*, 1992; Sukenik *et al.*, 1993a). Cultures of *Nannochloropsis* sp. failed to grow at temperatures below 10°C and above 38°C (Sukenik, 1991). The total ω3 PUFA content in *Nannochloropsis* strain

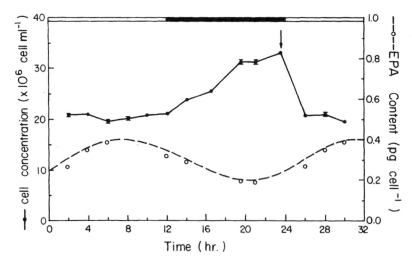

Figure 3.4 Variations in cell number and relative distribution in *Nannochloropsis* sp. cultures grown in a light–dark cycle of 12:12 h. Cultures were maintained in a light–dark regime with a dilution at the end of the dark period for several days before the analyses were performed. The dark period is presented by the blackened horizontal bar. The arrow indicates the time of culture dilution (adapted from Sukenik *et al.*, 1991).

MFD-2 increased with increasing temperature up to 25°C. A declining trend in total ω3 PUFA was observed above 25°C (James *et al.*, 1989). Seto *et al.* (1992) reported that by increasing the culture temperature from 20 to 25°C, the EPA content was reduced but the growth rate was unaffected. Slightly different results were reported by Teshima *et al.* (1983, 1991) who have shown a maximal value of EPA content at 25°C. The productivity of EPA slightly increased as growth temperature increased towards the optimal temperature for growth. Above that temperature, EPA productivity declined, evidently due to a low synthesis of fatty acids and a reduction in growth rate (Sukenik, 1991).

Nutrients

Nutrient limitation severely reduced cellular growth rate in *Nannochloropsis*. Cellular chlorophyll *a* content decreased in response to the reduction of nitrate availability. Under strict nitrogen limitation, the percentage of 16:0 fatty acid was fairly high at the expense of EPA. Under these conditions, photosynthetically fixed carbon cannot be used for synthesis of proteins and is directed into short chain saturated fatty acids to form storage lipids. Optimal concentration of nitrate imposed an increase in the relative abundance of EPA and a reduction in the proportion of saturated fatty acids (Sukenik *et al.*, 1993a).

Salinity

Nannochloropsis oculata has a relatively wide salinity tolerance between 10 and 35 ppt, with maximal growth rate at 25 and 30 ppt (Renaud and Parry, 1994). The highest total amount of fatty acids was found in cultures grown at 35 ppt. At these salinities, 16:0 fatty acid was at its highest abundance, whereas EPA percentage was at its minimal value. The optimal salinity for the production of a maximum amount of lipids and essential fatty acids, such as EPA, was between 25 and 30 ppt for *Nannochloropsis oculata* (Renaud and Parry, 1994).

Mass production

The eustigmatophyte *Nannochloropsis* was recognized as an essential component of the live food chain required in the cultivation of various marine fish larvae. In Japan, *Nannochloropsis* is cultivated in large scale outdoor tanks all year round to feed rotifers and to establish a reliable artificial food chain for the production of fish larvae (Okauchi, 1991).

Cultivation of *Nannochloropsis* sp. in large-scale outdoor shallow ponds was described by Sukenik *et al.* (1993a) at facilities in Eilat, Israel. Shallow raceway-type ponds of different areas, from a few square metres to 3000 m^2, were used (Figure 3.5). Cultures were maintained at 20 cm depth, mixed and forced to move along the raceway channel by a paddle wheel. CO_2 was provided via a solenoid valve activated by a pH controller, which maintained the culture pH at 7.8. Large scale ponds were inoculated from cultures grown in smaller size ponds. Cultures were diluted daily by replacing 10 to 50 per cent of the pond volume with fresh medium.

Figure 3.5 Large scale outdoor ponds used for mass culture of *Nannochloropsis* in Israel. (**A**) Two ponds of 60 m^2 each used in the National Center for Mariculture, Israel Oceanographic and Limnological Research, Eilat. (**B**) A 3000 m^2 pond in the Nature Beta Technology algae production plant in Eilat.

This operational scheme was maintained for several weeks up to 2 months, until the culture collapsed, usually due to the appearance of algal predators or contaminating algal species (Sukenik *et al.*, 1993a). Daily production rate of *Nannochloropsis* biomass exceeded 7 g d.w. m^{-2} in winter and 22 g d.w. m^{-2} in summer (Zmora and Sukenik, unpublished data).

Outdoor mass scale production of *Nannochloropsis* sp. all year round demonstrated the effect of seasonal variations in temperature and light availability on gross chemical composition and fatty acid distribution. During winter, when daily water temperatures varied between 8 and 16°C and the daily total solar radiation was lower than 25 mol quanta m^{-2}, the highest EPA percentage and cellular content were measured (Sukenik *et al.*, 1993a). It should be noted that outdoor cultures cultivated in shallow raceway ponds are exposed to daily fluctuations in temperature and light intensity, in addition to seasonal variations in the daily period of dark. Furthermore, light intensity is attenuated with the culture depth at a rate which is exponentially related to the algal concentration. Therefore, the actual light intensity to which algal cells are exposed is much lower than the measured light intensity at the culture surface (Sukenik *et al.*, 1991).

Based on the biological quality of *Nannochloropsis* grown under various environmental conditions in the laboratory and in different seasons under outdoor conditions, Sukenik *et al.* (1991) recommended that maximization of biomass and a relatively high level of EPA is achieved under nutrient-sufficient conditions and daily irradiance levels lower than 40 mol quanta m^{-2}. In addition, it is important to maintain the culture at an areal density around 65 g m^{-2}.

Nannochloropsis oculata is extensively cultured all year round in Japan in large outdoor tanks filled with enriched seawater. Chemical fertilizers are used to support the massive growth of the algae (Okauchi, 1991). Outdoor cultures suffer from time to time from a sudden decrease in cell density, mainly during rainy seasons. This phenomenon was partly attributed to predation by the protozoa *Paraphysomonas* sp. (Zmora, unpublished data). Chlorination and application of the antibiotic terramycin were found effective, but did not completely prevent the problem. Another problem observed in outdoor cultures of *Nannochloropsis* is a contamination by blue–green algae, diatoms and benthic microalgae (Okauchi, 1991). Recent experimental data indicated that *Nannochloropsis* has a reduced sensitivity to the herbicide DCMU, a phenomenon that may be used to selectively eliminate contaminating algal species from outdoor cultures (Gonen-Zurgil *et al.*, 1996).

Commercial applications and feeding experiments

The cultivation of *Nannochloropsis* for EPA production is mainly applied in the aquaculture industry, although experiments were carried out to demonstrate the potential use of this source of essential fatty acids as supplements in human diet and animal feed.

The steady growth of the world fresh fish market dictates an increase in the quantities of rotifers and brine shrimp required as feed for most juvenile fish. Thus, cultivation of microalgae has become an important element in the food chain of fish hatcheries. Aprroximately 50 per cent of the culture area in a typical aquaculture facility in Japan is used for microalgae cultivation (Seto *et al.*, 1992). *Nannochloropsis* has been used in aquaculture as an essential component of the artificial food chain for fish larvae for more than two decades. The main reason that *Nannochloropsis* was chosen as a preferable microalga in many marine hatcheries is the substantial quantities of EPA found in this genus.

Cultivation of *Nannochloropsis* for aquaculture requires highly controlled culture techniques and provides high quality food only for rotifers, but not for brine shrimp. An attempt to overcome limitations of algae production in the hatchery site by providing rotifers with frozen, high quality *Nannochloropsis* biomass which is produced and processed in a central algal production site, was suggested by several studies (Snell, 1991; Lubzens *et al.*, 1995). Seto *et al.* (1992) found that treatment of *Nannochloropsis* biomass with digestive enzymes and removal of much of the algal cell improved the effectiveness of this alga as a food source for brine shrimp.

Although *Nannochloropsis* was recognized as a potentially important direct source of dietary ω3 fatty acids in human diet, feeding experiments were performed only with animals (Markovits *et al.*, 1992; Seto *et al.*, 1992; Sukenik *et al.*, 1994). The effect of the addition of EPA containing *Nannochloropsis* biomass on reducing blood pressure in spontaneous hypersensitive rats was examined (Seto *et al.*, 1992). The bioavailability of ω3 fatty acids and specifically EPA from algal sources was evaluated in rats that were fed with diets supplemented with *Nannochloropsis* biomass (Sukenik *et al.*, 1994). Addition of *Nannochloropsis* to the rat diet resulted in an increase in the ω3/ω6 ratio in liver and blood. This enhancement was accompanied by a reduction in the relative abundance of AA.

The role of ω3 VLC-PUFA originating from *Nannochloropsis* biomass on fatty acid composition of the developing brain was evaluated in rats (Mokady and Sukenik, 1995). It was demonstrated that feeding pregnant and lactating rat dams a diet enriched with *Nannochloropsis* biomass as the major source for dietary ω3 PUFA resulted in a higher level of DHA in the brain lipids of the pups (Mokady and Sukenik, 1995). Furthermore, the dietary ω3 lipids of *Nannochloropsis* are efficiently transferred to the developing foetus and pups to support the requirements for brain development.

An additional possible application of *Nannochloropsis* biomass could be in the enrichment of eggs with ω3 polyunsaturated fatty acids. Such modified eggs may provide an attractive source of these essential and healthful lipids. In a recent study we have shown that diets containing dried *Nannochloropsis* biomass or extracted lipids caused significant changes in the yolk colour and its visual score increased proportionally to the amount of algae biomass provided with the diet (Figure 3.6a). A significant increase in the relative abundance of DHA was found in yolks from laying-hens that were fed a diet containing 1 per cent *Nannochloropsis*. Increasing the

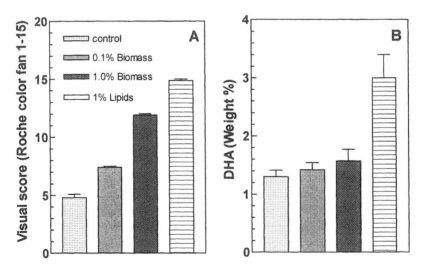

Figure 3.6 Modification of table egg quality by providing *Nannochloropsis* biomass to laying hens. Diets containing 1 per cent (w/w) dried *Nannochloropsis* biomass or 1 per cent (w/w) of extracted lipids from *Nannochloropsis* caused a significant increase in the visual score of the yolk colour (**A**) and in the relative abundance of DHA (**B**).

amount of lipids originating from *Nannochloropsis* in the diet further increased the proportion of DHA in the egg yolk (Figure 3.6b).

Acknowledgements

Funding for this work was provided by the US–Israel Binational Agriculture Research and Development Fund, IS 1836-90C, US–Israel Binational Science Foundation, 86-00376, by the MAGNET program, Israel Ministry of Commerce and Industry and by the Strauss/Hanauer Memorial Fund Inc.

References

ANDERSON, R.A., JACOBSON, D.M. and SEXTON, J.P., 1991, Catalog of strains. Provasoli–Guillard Center for Culture of Marine Phytoplankton. West Boothbay Harbor, Maine, USA, p. 98.

ANITA, N.G. and CHENG, J.Y., 1982, The keto-carotenoids of two marine coccoid members of the Eustigmatophyceae, *British Phycological Journal*, **17**, 39–50.

ANITA, N.G., BISALPUTRA, T., CHENG, J.Y. and KALLEY, J.P., 1975, Pigment and cytological evidence for reclassification of *Nannochloris oculata* and *Moallantus salina* in the Eustigmatophyceae, *Journal of Phycology*, **11**, 339–343.

ARAO, T. and YAMADA, M., 1989, Positional distibution of fatty acids in galactolipids of algae, *Phytochemistry*, **28**, 805–807.

ARAO, T. and YAMADA, M., 1994, Biosynthesis of polyunsaturated fatty acids in the marine diatom *Phaeodactylum tricornutum*, *Phytochemistry*, **35**, 1177–1181.

BOSWELL, K.D.B., GLADUE, R.M., PRIMA, B. and KYLE, D.J., 1992, SCO production by fermentative microalgae, in *Industrial Applications of Single Cell Oils*, edited by D.J. Kyle and C. Ratledge, Champaign, Illinois: American Oil Chemists' Society, pp. 274–286.

BROWN, M.R. 1991, The amino-acid and sugar composition of 16 species of microalgae used in mariculture, *Journal of Experimental Marine Biology and Ecology*, **145**, 79–99.

BROWSE, J. and SOMERVILLE, C., 1991, Glycerolipid synthesis: biochemistry and regulation, *Annual Review of Plant Physiology and Plant Molecular Biology*, **42**, 467–506.

COHEN, Z., DIDI, S. and HEIMER, Y.M., 1992, Overproduction of gamma-linolenic and eicosapentaenoic acids by algae, *Plant Physiology*, **98**, 569–572.

DUNAHAY, T.G., JARVIS, E.E. and ROESSLER, P.G., 1995, Genetic transformation of the diatoms *Cyclotella cryptica* and *Navicula saprophila*, *Journal of Phycology*, **31**, 1004–1012.

EMDADI, D. and BERLAND, B., 1989, Variations in lipid class composition during batch growth of *Nannochloropsis salina* and *Pavlova lutheri*, *Marine Chemistry*, **26**, 215–225.

FALCONE, D.L., GIBSON, S., LEMIEUX, B. and SOMERVILLE, C., 1994, Identification of a gene that complements an *Arabidopsis* mutant deficient in the chloroplast ω6 desaturase activity, *Plant Physiology*, **106**, 1453–1459.

GELLERMAN, J.L. and SCHLENK, H., 1979, Methyl-directed desaturation of arachidonic to eicosapentaenoic acid in the fungus, *Saprolegnia parasitica*, *Biochim. Biophys. Acta*, **573**, 23–30.

GLADU, P.K. and PATTERSON, G.W., 1995, Some biochemical characteristics of UTEX 2341, a marine eustigmatophyte identified previously as *Chlorella minutisima* (Chlorophyceae), *Journal of Phycology*, **31**, 774–777.

GONEN-ZURGIL, Y., CARMELI-SCHWARTZ, Y. and SUKENIK, A., 1996, The selective effect of the herbicide DCMU on unicellular algae – a potential tool to maintain a monoalgal culture of *Nannochloropsis*, *Journal of Applied Phycology*, **8**, 415–419.

HEINZ, E., 1993, Biosynthesis of polyunsaturated fatty acids, in *Lipid Metabolism in Plants*, edited by T.S. Moore, Jr., Ann Arbor: CRC Press, pp. 39–89.

HIBBERED, D.J., 1981, Notes on the taxonomy and nomenclature of the algal classes Eustigmatophyceae and Tribophyceae (synonym Xantophyceae), *Botanical Journal of the Linnean Society*, **82**, 93–119.

INNIS, S.M., 1992, n-3 fatty acid requirements of the newborn, *Lipids*, **27**, 879–885.

JAMES, C.M., AL-GINTY, S. and SALMAN, A.E., 1989, Growth and ω3 fatty acid and amino acid composition of microalgae under different temperature regimes, *Aquaculture*, **77**, 337–357.

KARLSON, B., POTTER, D., KUYLENSTIERNA, M. and ANDERSON, R.A., 1996, Ultrastructure, pigment composition and 18s rRNA gene sequence for *Nannochloropsis granulata* sp. nov. (Monodopsidaceae, Eustigmatophyceae), a marine ultraplankton isolated from the Skagerrak, North East Atlantic Ocean, *Phycologia*, **35**, 253–262

KYLE, D.J., BOSSWELL, K.D.B., GLADUE, R.M. and REAB, S.E., 1992, Designer oil from micro-algae as nutritional supplements, in *Biotechnology and Nutrition*, edited by D.D. Bills and S. Kung, Boston: Butterworth-Heinemann, pp. 451–468.

LUBIAN, L.M., 1982, *Nannochloropsis gaditana* sp. nov. una nueva Eustigmatophyceae marina, *Lazaroa*, **4**, 287–293.

LUBZENS, E., GIBSON, O., ZMORA, O. and SUKENIK, A., 1995, Potential advantages of frozen algae (*Nannochloropsis* sp.) for rotifer (*Brachionus plicatilis*) culture, *Aquaculture*, **133**, 295–309.

MARKOVITS, A., CONJEROS, R., LOPEZ, L. and LUTZ, M., 1992, Evaluation of microalga *Nannochloropsis* sp. as a potential dietary supplement: chemical, nutritional and short term toxicological evaluation, *Nutrition Research*, **12**, 1273–1284.

MARUYAMA, I., NAKAMURA, T., MATSUBAYASHI, T., ANDO, Y. and MAEDA, T., 1986, Identification of the algae known as 'marine *Chlorella*' as a member of the Eustigmatophyceae, *Journal of Phycology*, **34**, 319–325.

MENZEL, K. and WILD, A., 1989, Fatty acid composition in the lipids of some marine Chlorococcales and Eustigmatales, *Zeitschrift für Naturforschung*, **44c**, 743–748.

MOKADY, S. and SUKENIK, A., 1995, A marine unicellular alga in diets of pregnant and lactating rats as a source of ω3 fatty acids for the developing brain of their progeny, *Journal of Science and Food Agriculture*, **68**, 133–139.

MOURENTE, G., LUBIAN, L.M. and ODRIOZOLA, J.M., 1990, Total fatty acid composition as a taxonomic index of some marine microalgae used in marine aquaculture, *Hydrobiology*, **203**, 147–154.

MURATA, N., ISHIZAKI-NISHIZAKI, O., HIGASHI, S., HAYASHI, H., TASAKA,Y. and NISHIDA, I., 1992, Genetically engineered alteration in the chilling sensitivity of plants, *Nature*, **356**, 710–713.

OKAUCHI, M., 1991, The status of phytoplankton production in Japan, in *Rotifers and Microalgae Culture Systems*, edited by W. Fulks and K.L. Main, Proceedings of a US–Asia workshop. Honolulu, HI, pp. 537–577.

PATTERSON, G.W., TEITSA-TZARDIS, E., WIKFORS, G.H., GHOSH, P., SMITH, B.C. and GLADU, P.K., 1994, Sterols of eustigmatophytes, *Lipids*, **29**, 661–664.

RENAUD, S.M. and PARRY, D.L., 1994, Microalgae for use in tropical aquaculture. II. Effect of salinity on growth, gross chemical composition and fatty acid composition of three species of marine microalgae, *Journal of Applied Phycology*, **6**, 347–356.

RENAUD, S.M., PARRY, D.L., THINH, L-V., KUO, C., PADOVAN, A. and SAMMY, N., 1991, Effect of light intensity on the proximate biochemical and fatty acid composition of *Isochrysis* sp. and *Nannochloropsis oculata* for use in tropical aquaculture, *Journal of Applied Phycology*, **3**, 43–45,

ROESSLER, P.G., 1990, Environmental control of glycerolipid metabolism in microalgae: com-mercial applications and future research directions, *Journal of Phycology*, **26**, 393–399.

ROUGHAN, P.G. and SLACK, C.R., 1982, Cellular organization of glycerolipid metabolism, *Annual Review of Plant Physiology*, **33**, 97–132.

SANTOS, L.M.A. and LEEDALE, G.F., 1995, Some notes on the ultrastucture of small azoosporic members of the algal class Eustigmatopyceae, *Nova Hedwigia*, **60**, 215–225.

SCHNEIDER, J.C. and ROESSLER, P., 1994, Radiolabeling studies of lipids and fatty acids in *Nannochloropsis* (Eustigmatophyceae), an oleaginous marine alga, *Journal of Phycology*, **30**, 594–598.

SCHNEIDER, J.C. and ROESSLER, P.G., 1995, Novel acyltransferase activity in an oleaginous alga, in *Plant Lipid Metabolism*, edited by J.C. Kader and P. Mazliak, Dordrecht, Netherlands: Kluwer Academic Publishers, pp. 105–107.

SCHNEIDER, J.C., LIVNE, A., SUKENIK, A. and ROESSLER, P.G., 1995, A mutant of *Nannochloropsis* deficient in eicosapentaenoic acid production, *Photochemistry*, **40**, 807–814.

SETO, A., KUMASAKA, K., HOSAKA, M., KOJIMA, E., KASHIWAKURA, M. and KATO, T., 1992, Production of eicosapentaenoic acid by a marine microalgae and its commercial utilization for aquaculture, in *Industrial Applications of Single Cell Oils,* edited by D.J. Kyle and C. Ratledge, Champaign, Illinois: American Oil Chemists' Society, pp. 219–234.

SHIFRIN, N.S. and CHISHOLM, S.W., 1981, Phytoplankton lipids: interspecific differences and effects of nitrate, silicate and light-dark cycles, *Journal of Phycology*, **17**, 364–384.

SHIMIZU, S., KAWASHIMA, H., SHINMEN, Y., AKIMOTO, K. and YAMADA, H., 1988, Production of eicosapentaenoic acid by *Mortierella* fungi, *Journal of the American Oil Chemists' Society*, **65**, 1455–1459.

SIMOPOULOS, A.P., 1986, Summary of the NATO advanced research workshop on dietary omega 3 and omega 6 fatty acids. Biological effects and nutritional essentiality, *Journal of Nutrition*, **119**, 521–528.

SNELL, T.W., 1991, Improving the design of mass culture systems for the rotifer *Brachionus plicatilis*, in *Rotifer and Microalgae Culture Systems*, Proceedings of the US–Asia Workshop, Honolulu, HI, 1991. The Oceanic Institute, pp. 61–71.

SPOEHR, H.A. and MILNER, H.W., 1949, The chemical composition of *Chlorella*: effect of environmental conditions, *Plant Physiology*, **24**, 120–149.

SUEN, Y., HUBBARD, J.S., HOLZER, G. and TORNABENE, T.G., 1987, Total lipid production of the green alga *Nannochloropsis* sp. QII under different nitrogen regimes, *Journal of Phycology*, **23**, 289–296.

SUKENIK, A., 1991, Ecophysiological considerations in optimization of eicosapentaenoic acid production by *Nannochloropsis* sp. (Eustigmatophyceae), *Bioresource Technology*, **35**, 263–269.

SUKENIK, A. and CARMELI, Y. 1990, Lipid synthesis and fatty acid composition in *Nannochloropsis* sp. (Eustigmatophyceae) grown in a light-dark cycle, *Journal of Phycology*, **26**, 463–469.

SUKENIK, A., CARMELI, Y. and BERNER, T., 1989, Regulation of fatty acid composition by irradiance level in the eustigmatophyte *Nannochloropsis* sp., *Journal of Phycology*, **25**, 686–692.

SUKENIK, A., LEVI, R.S., LEVI, Y., FALKOWSKI, P.G. and DUBINSKY, Z., 1991, Optimizing algal biomass production in an outdoor pond: a simulation model, *Journal of Applied Phycology*, **3**, 191–201.

SUKENIK, A., ZMORA, O. and CARMELI, Y., 1993a, Biochemical quality of marine unicellular algae with special emphasis on lipid composition. II, *Nannochloropsis* sp., *Aquaculture*, **117**, 313–326.

SUKENIK, A., YAMAGUCHI, Y. and LIVNE, A., 1993b, Alterations in lipid molecular species of the marine Eustigmatophyte *Nannochloropsis* sp., *Journal of Phycology*, 29, 620-626.

SUKENIK, A., Takahashi, H. and Mokady, S., 1994, Dietary lipids from marine unicellular algae enhance the amount of liver and blood ω3 fatty acids in rats, *Annals of Nutrition and Metabolism*, **38**, 85-96.

TESHIMA, S., YAMASAKI, S., KANAZAWA, A. and HIRATA, H., 1983, Effects of water temperature and salinity on eicosapentaenoic acid level of marine *Chlorella*, *Bulletin of the Japanese Society of Scientific Fisheries*, **49**, 805.

TESHIMA, S., YAMASAKI, S., KANAZAWA, A., KOSHIO, S., MUKAI, H. and HIRATA, H., 1991, Fatty acid composition of Malaysian marine *Chlorella*, *Nippon Suisan Gakkaishi*, **57**, 1985.

VOGEL, G. and EICHNBERGER, W., 1992, Betaine lipids in lower plants, biosynthesis of DGTS and DGTA in *Ochromonas danica* (Chrysophyceae) and the possible role of DGTS in lipid metabolism, *Plant Cell Physiology*, **33**, 427–436.

VOLKMAN, J.K., JEFFREY, S.W., NICHOLS, P.D., ROGERS, G.I. and GARLAND, C.D., 1989, Fatty acid and lipid composition of 10 species of microalgae used in mariculture, *Journal of Experimental Marine Biology and Ecology*, **128**, 219–240.

VOLKMAN, J.K., BROWN, M.R., DUNSTON, G.A. and JEFFREY, S.W., 1993, The biochemical composition of marine microalgae from the class Eustigmatophyceae, *Journal of Phycology*, **29**, 69–78.

WHITTLE, S. and CASSELTON, P., 1975, The chloroplast pigments of the algal class Eustigmatophyceae and Xanthophyceae. I. Eustigmatophyceae, *British Phycological Journal*, **10**, 179–191.

4

Production of EPA from *Phaeodactylum tricornutum*

E. MOLINA GRIMA, F. GARCíA CAMACHO AND
F.G. ACIÉN FERNÁNDEZ

Introduction

In recent years, wide interest has focused on the production and pharmaceutical applications of eicosapentaenoic acid (EPA). However, until now, the studies released have been limited to determining the EPA content of microorganisms and the influence of different culture parameters under laboratory conditions, those carried out under outdoor conditions being very scarce. Likewise, much research has been done on the influence of culture parameters on biomass productivity, but always under laboratory conditions and by modifying single parameters. Under outdoor conditions, most published papers refer to the production of *Spirulina*, *Dunaliella* and *Chlorella* in open systems with no control of culture conditions, and very little has been published concerning the outdoor production of such very sensitive strains as *Monodus*, *Porphyridium*, *Isochrysis* or *Phaeodactylum*, for the production of fine chemicals.

For the industrial production of EPA from microalgae, four main tasks should be undertaken: first, select/obtain the appropriate strain of microalga; second, design culture systems in which the biomass productivity and control of culture conditions are optimized at minimum cost; third, optimize the culture conditions for the improvement of EPA productivity; and fourth, optimize the fatty acid extraction and EPA purification processes. Previous chapters have been devoted to the first and fourth of these. In this chapter we will deal with the second and third: design and operation of tubular reactors for biomass production, and optimization of culture conditions for EPA productivity enhancement.

Production of microalgal biomass in photobioreactors is faced with the double challenge of both making use of the incident high photon flux densities on the culture surface more efficient and improving the utilization of the main nutrient (CO_2). System productivity in continuous operating mode is obtained by multiplying the steady state biomass concentration by the optimized dilution rate. Both are related to the average irradiance inside the photobioreactor, which is a function of the irradiance on the reactor surface, operational variables (mixing and dilution rate) and pigment content.

The effect of light has two aspects. First, light availability inside the culture is determined by the incident solar irradiance, which is a function of the geographical location and day of the year (Incropera and Thomas, 1978), the design and position of the reactor, which determines the system capacity for collecting incident energy (Lee and Low, 1992), and the biomass concentration and pigment content in the culture, due to self-shading (Molina Grima *et al.*, 1994b, 1996b). Second, the light regime, that is, the irradiance and time at which cells are exposed inside the culture (Phillips and Myers, 1954; Grobbelaar, 1994). Thus, the effect of the light can be modified by manipulating the light path (design of the reactor), the incident irradiance (location of the reactor), mixing and biomass concentration (operating conditions).

Finally, the influence of culture conditions such as light availability, nutrient saturation, pH, culture temperature, fluidodynamic conditions, etc., on not only the biomass productivity, but also the EPA content of the biomass, must be determined in order to establish the optimal culture conditions for EPA production.

To understand the interrelationships between the various parameters that influence outdoor cultures, this chapter summarizes work carried out with *Phaeodactylum tricornutum* in tubular airlift photobioreactors over a three-year period: first, to improve the biomass productivity; and second, to optimize EPA productivity.

Criteria for selecting *Phaeodactylum tricornutum* UTEX 640 as a potential source of EPA

In order to select an appropriate strain for EPA production in outdoor photobioreactors, three specific parameters of the strain must be compared: the optimal culture conditions, the EPA content and EPA proportion, and the specific growth rate.

The optimal culture conditions of the strain must be similar to the environmental conditions in the factory in order to diminish the cost associated with controlling culture conditions. Thus, the environmental conditions are a function of the geographic location and climate conditions, which determine temperature, daily solar cycles and seasonal variations. For this reason, most of the factories in the world are located in temperate or warm climates where temperature and solar irradiance are more favourable.

The EPA content of the biomass (as percentage of dry weight) influences the total EPA productivity of the system, whereas the EPA proportion (as percentage of total fatty acids) describes the purity of the fatty acid fraction. The first parameter allows a determination of the maximum EPA yield of the system. However, the real EPA yield is determined by the EPA proportion, which expresses the fatty acid composition of the microalgal oils extracted, and determines the cost of downstream purification of EPA, which represents a substantial part (ca. 80 per cent) of the final cost of the product (Molina Grima *et al.*, 1996a).

Finally, the maximum specific growth rate is a parameter characteristic of the selected strain, that determines the maximum biomass productivity under optimal culture conditions. In this sense, strains for which maximum biomass productivity and EPA content are reached under the same culture conditions should be selected, as this increases system performance.

Of the more than 30 000 defined species of microalgae, only a limited number were analyzed to determine their lipid composition and fatty acid profiles (Table 4.1).

However, as a source of polyunsaturated fatty acids, microalgae must compete with other microorganisms such as fungi and bacteria. Any comparison must not only be done in terms of EPA content, but also of EPA productivity (Table 4.2).

Only one bacterial strain has been referenced as an EPA source, the dry weight content being low, 10.4 mg g^{-1} of dry matter, although due to the high growth rate of the bacteria versus microalgae, the EPA productivity is high, 64.5 mg l^{-1} d^{-1}. Nevertheless, the low dry weight content increases the cost of EPA extraction and purification. With regard to the fungi, the EPA content on dry weight is similar to that of microalgae but the fungi have problems related with toxin accumulation. In

Table 4.1 Fatty acid composition of arachidonic (AA), eicosapentaenoic (EPA) and docosahexaenoic (DHA) acid profiles in highly valuable microalgal biomass.

Strain	20:4n6 (AA)	20:5n3 (EPA)	22:6n3 (DHA)	Reference
Diatoms				
Asterionella japonica	20.0	–	–	Wood (1974)
Bidulphia sinensis	24.2	–	–	Volkman *et al.* (1989)
Chaetoceros septentrionale	21.0	–	–	Wood (1974)
Lauderia borealis	1.3	30.3	–	Pohl (1982)
Navicula biskanteri	–	26.7	–	DeMort *et al.* (1972)
Navicula laevis (heterotrof.)	–	23.2	–	Koon and Johns (1996)
Navicula laevis	–	16.7		Koon and Johns (1996)
Navicula incerta	–	25.2		Koon and Johns (1996)
Stauroneis amphioxys	–	26.1	1.9	Gillan *et al.* (1981)
Navicula pelliculosa	–	26.0	–	Wood (1974)
Bidulphia aurtia	–	26.0	–	Cobelas and Lechazo (1989)
Nitzschia alba	–	29.0	–	Cobelas and Lechazo (1989)
Nitzschia closterium	–	45.0	–	Cobelas and Lechazo (1989)
Phaeodactylum tricornutum	1.3	30.5	2.5	Molina Grima *et al.* (1994a)
Phaeodactylum tricornutum	–	25.8	–	Kates and Volcani (1966)
Skeletonema costatum	–	30.0	–	Wood (1974)
Chrysophyceae				
Pseudopedinella sp.	1.0	27.0	–	Wood (1974)
Cricosphaera elongata	2.0	28.0	–	Ziboh *et al.* (1986)
Eustigmatophyceae				
Monodus subterraneus	5.1	39.2	–	Cohen (unpublished)
Nannochloropsis	3.4	44.7	–	Seto *et al.* (1984)
Prymnesiophyceae				
Rodela violacea 115.79	18.9	32.9	–	Cohen (unpublished)
Porphyr. cruentum 1380.1d	14.7	44.1	–	Cohen (1990)
Prasinophyceae				
Pavlova salina	–	28.2	10.9	Volkman *et al.* (1989)
Dinophyceae				
Cochlodinium heteroloblatum	–	11.4	28.0	Joseph (1975)
Crypthecodinium cohnii	–	–	30.0	Pohl (1982)
Gonyaulax catenella	–	11.2	33.9	Joseph (1975)
Gyrodinium cohnii	–	–	30.0	Wood (1974)
Prorocentrum minimum	–	5.3	24.5	Joseph (1975)

Table 4.2 Comparison of EPA content (mg g^{-1} of dry matter) and productivity (mg l^{-1} d^{-1}) in some promising microalgae and fungi.

Species	Content (%)	Prod. (mg l^{-1} d^{-1})	Reference
Microalgae			
Chlorella minutissima[a,b]	31.8	3.0	Seto *et al.* (1984)
Chlorella minutissima	26.0	nr	Borowitzka (1988b)
Isochrysis galbana ALII4[c]	46.5	23.2	Molina Grima *et al.* (1995)
Phaeodactylum tricornutum[b] *WT*	33.0	19.0	Yongmanitchai and Ward (1991)
Phaeodactylum tricornutum[d] *WT*	18.7	47.8	Molina Grima *et al.* (1994a)
Phaeodactylum tricornutum[a] *WT*	7.0	nr	Veloso *et al.* (1991)
Phaeodactylum tricornutum[a] *WT*	17.3	3.5	López Alonso *et al.* (1996)
Phaeodactylum tricornutum I14[a]	31.5	4.0	López Alonso *et al.* (1996)
Phaeodactylum tricornutum I1242[a]	38.6	5.0	López Alonso *et al.* (1996)
Porphyridium cruentum	24.0	nr	Cohen and Heimer (1992)
Porphyridium cruentum	12.2	0.8	Ohta *et al.* (1992)
Monodus subterraneus[b]	34.4	25.7	Cohen (1994)
Fungi			
Mortierella alpina	29.0	nr	Ratledge (1989)
Mortierella alpina 1S-4[e]	67.0	118.0	Jareonkitmongkol *et al.* (1993)
Pythium irregulare	24.9	nr	O'Brien *et al.* (1993)
Bacteria			
SCRC-2738	10.4	64.5	Suzuki *et al.* (1991)

[a] Batch cultures.   nr = not referenced
[b] Air supplemented with carbon dioxide.
[c] Batch cultures indoor at 20°C.
[d] Continuous cultures 20°C outdoor.
[e] At 12°C and linseed oil added to the medium.

conclusion, microalgae seem to be the most suitable source of EPA, *Phaeodactylum*, *Isochrysis* and *Monodus* being the most interesting.

In this sense, the most notable characteristic of the microalga *Phaeodactylum tricornutum* UTEX 640 is its ability to produce a high proportion of EPA, which reduces the cost of downstream processing. In addition, although its EPA content is not very high, its productivity is one of the highest due to its high maximum specific growth rate, of between 0.06 and 0.09 h^{-1} (Siron *et al.*, 1989; Kaixian and Borowitzka, 1992). Moreover, this strain can grow to high cell densities (Yongmanitchai and Ward, 1991; Molina Grima *et al.*, 1994a, 1997) which reduces the harvesting cost, and its optimum growth temperature of between 18°C (Mann and Myers, 1968) and 22°C (Kaixian and Borowitzka, 1992) is similar to the annual average temperature of the temperate climate, because the cost of cooling outdoor cultures is not high.

This microorganism has several additional advantages: although it is a freshwater species, it can be grown in a seawater medium and on slant agar, facilitating maintenance of stock culture. In addition, its comparatively large cell size, 3 μm thick and 8 to 20 μm long for oval and fusiform cells (Lewin *et al.*, 1958), facilitates cell harvesting procedures. Figure 4.1 shows two electronic microscopic photographs

(A)

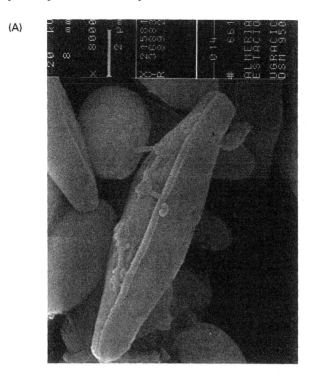

(B)

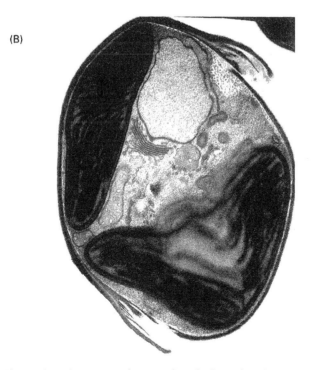

Figure 4.1 Electronic microscopy photographs of *Phaeodactylum tricornutum* UTEX 640, showing (A) oval and fusiforme characteristic morphotypes, and (B) internal structure of the cell. Molina *et al.*, 1997

of *Phaeodactylum tricornutum* UTEX 640. Figure 4.1A shows the two basic morphotypes of this microalga, oval and fusiform; Figure 4.1B shows the internal organization of the cell, with a large Golgi apparatus.

Phaeodactylum tricornutum UTEX 640 biomass production in horizontal tubular air-lift photobioreactors

Culture system description

Tubular photobioreactors were pioneered in the 1950s (Cook, 1950; Little, 1953; Tamiya *et al.*, 1953), but were ignored because of the lower cost of operation of the open systems for production of classic microalgae like *Spirulina*. These ponds were later improved with temperature control systems, supply of appropriate nutrients, optimization of pond depth, CO_2 injection systems, etc. However, in spite of these improvements, productivity remained fairly low; moreover only the production of certain microalga species was possible. Therefore, closed reactors soon began to appear for large-scale production of the so-called 'sensitive' strains with higher productivities (Richmond, 1990). Thus, in recent years, research has again been focused on this type of closed photobioreactors for three reasons:

(1)   Open systems cannot produce the so-called 'valuable strains', because of easy contamination and poor control of culture conditions.
(2)   The price and quality of the materials necessary to build closed systems is similar to open systems.
(3)   The possibility of producing fine chemicals, with high added value, has enabled private business to invest in them.

In this sense, very different designs for closed reactors as outdoor open ponds (Terry and Raymond, 1985), vertical alveolar panels (Tredici and Materassi, 1992), flat plate reactors (Ratchford and Fallowfield, 1992; Hu *et al.*, 1996), tubular reactors (Pirt *et al.*, 1983; Gudin and Therpenier, 1986; Lee, 1986; Richmond *et al.*, 1993; Torzillo *et al.*, 1993; Watanabe and Hall, 1996), etc., have been proposed. Among them, the tubular photobioreactor, in which the circulation of fluids is induced by bubbling air in the riser section of the reactor, allows better control of the culture variables, enabling higher productivities, and reducing power consumption and cellular stress (Gudin and Chaumont, 1991; Molina Grima *et al.*, 1994a, 1994c, 1997). Autotrophic microalgal production under outdoor conditions in these bioreactors is the subject of an important world-wide research trend. Studies have determined the effect of culture conditions, nutrient supply, biomass concentration, temperature control, incident radiation, etc., on the biomass productivity and biochemical composition (Lee and Low, 1992; Richmond *et al.*, 1993; Chrismadha and Borowitzka, 1994; Molina Grima *et al.*, 1994a, 1994c). Nevertheless, light availability, which is one of the most important factors affecting cell growth, is difficult to control in outdoor cultures, due to the variation in solar radiation during the day and its non-homogeneous distribution in a cylindrical geometry.

  Two horizontal tubular photobioreactors (HTPB) type air-lift, like those pioneered by Gudin (1989), HTPB0.06 (external diameter = 0.06 m, total volume = 0.220 m³) and HTPB0.03 (external diameter = 0.03 m, total volume = 0.050 m³), located in Almería Spain (36°48′N, 2°54′W), were used at pilot-plant scale (Figure 4.2). Both photobioreactors can be divided into an air-lift pump and an external loop. The air-lift pump, composed of a riser, degasser and down-comer, provides circulation of the culture and gas removal. The external loop, con-sisting of transparent Plexiglas tubes connected by joints of the same material, acts as a solar receiver. The tubes, positioned horizontally over the ground, are connected to the airlift system. In both photobioreactors, a constant operating speed of 0.30 m s^{-1} was maintained to prevent algal cells from settling and to ensure a favourable light regime in the system. Therefore, due to their different tube diameters and lengths (Table 4.3), the pressure drop is higher in HTPB0.03 than HTPB0.06, because the

Table 4.3 Dimensions of the horizontal tubular photobioreactors used.

Dimensions	HTPB0.06	HTPB0.03
Outer tube diameter (m)	0.060	0.030
Inner tube diameter (m)	0.050	0.024
External loop length (m)	98.800	80.800
Area occupied by the solar receiver (m^2)	21.400	8.300
Area covered by the cross-section of tubes (m^2)	4.940	1.940
Outer curved surface of tubes (m^2)	15.500	6.090
External loop volume (m^3)	0.174	0.036
Reactor volume (m^3)	0.200	0.050
Degasser height (m)	2.050	2.600
Air flow rate ($m^3 h^{-1}$)	1.800	0.480

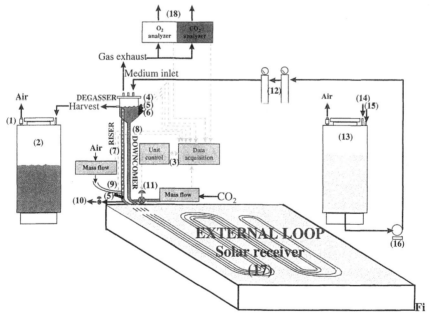

Figure 4.2 Scheme of outdoor culture system. 1 Air filters, 2 harvest tank, 3 control unit, 4 temperature sensor, 5 dissolved O_2 probe, 6 pH probe, 7 riser, 8 downcomer, 9 air injection, 10 sampler, 11 CO_2 injection, 12 medium sterile filters, 13 fresh medium, 14 nutrients inlet, 15 sea water inlet, 16 pump, 17 thermostatic water pool, 18 gas composition analyser. Molina Grima *et al.*, 1998

degasser in HTPB0.03 is taller than HTPB0.06 and air is supplied to them at different flow rates.

The top and walls of the degasser were sealed with rubber bungs through which liquid and gas inlet and outlet, temperature, dissolved oxygen and pH sensors were passed. The pH in the culture was measured with an Ingold pH glass probe and controlled by the automatic addition of CO_2 bubbles. The dissolved oxygen content above air saturation in seawater was measured by an Ingold polarographic probe. Temperature of the culture was measured using a PT-100 sensor. All the probes were connected to an ML-4100 control unit (New Brunswick Scientific, USA), attached to a computer for on-line data acquisition and pH control. The air and CO_2 mass flow supply to the culture were measured by two mass flow meters (Model 824 SIERRA Instruments Inc., Monterey, USA). The composition of gas inlet and outlet of the system were measured by a gas analyzer (Servomex Analyzer Series 1400). Both mass flow meter and gas analyzer were on line connected with a data log board (DaqBoard/112, IOTECH Inc., Ohio, USA) and 486 computer for data acquisition.

The solar receiver was submerged in a thermostatic water pool, the temperature of which was kept at 20°C using a Calorex 4000 heat pump (additional details can be seen in Figure 4.2). The walls of the pond were painted white to increase reflection of the light and thereby increase light availability inside the culture. For this reason, the photon-flux density of photosynthetically active radiation (PAR) on the pond surface was determined with an LI-190SA Quantum Sensor (LI-COR Radiation Sensors, Lincoln, EN, USA) whereas the irradiance inside the water was determined with a quantum scalar irradiance meter (QSL-100 Biospherical Instruments Inc., San Diego, CA). Scalar irradiance meters measure the photon flux from all directions and should be preferred for systems that operate independently of the geometry of the light field (Lassen and Jorgensen, 1994). Since photosynthesis does not depend on the direction the photons are travelling, scalar irradiance is the best measurement of the photon flux available for microalgae. The results showed that the irradiance inside the water was higher than incident radiation on the surface due to the reflection of incident radiation on the walls and bottom of the pond; this phenomenon is named albedo.

During the operation of the photobioreactors the influence of dilution rate, an operating variable, the incident radiation, an environmental variable, and the tube diameter, a design variable, were studied (Figure 4.3). The on-line data acquisition of dissolved oxygen, pH and temperature inside the culture allows for rapid characterization of the physiological state of the cells. The measurement of CO_2 and O_2 molar flow at the gas inlet and outlet, enables the estimation of the mass transfer capability of the system and the efficiency of CO_2 utilization. The measurement of the biomass concentration and fatty acid content and profile allow the estimation of the photosynthetic efficiency and the biomass and EPA productivity, in addition to the elaboration of mathematical growth models. In this way, the overall assessment of the reactor allows a determination of the optimal values of dilution rate, incident irradiance and tube diameter in order to increase the yield and scale up the system.

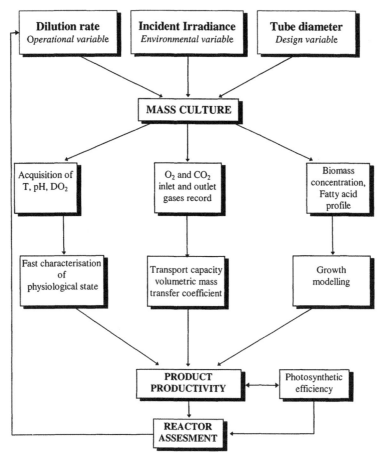

Figure 4.3 Assessment of photobioreactor. Operational variables, measurements and objectives relationship. Molina Grima *et al.*, 1998

Biomass productivity

Factors influencing the biomass productivity

Biomass productivity in any culture system depends on the degree to which the culture conditions match the requirements of the selected strain. Thus, microalgal mass culture systems have been developed for four decades, for which light availability inside the photobioreactor and temperature were found to be the main culture management factors in obtaining optimum system profitability, since mineral nutrient limitation is easily avoided (Little, 1953; Richmond, 1990). When culture temperature is kept within a narrow interval, light availability is the only factor determining growth rate of the microorganism (Lee and Low, 1991; Tredici and Materassi, 1992).

Figure 4.4 shows the relationship between the different factors influencing biomass productivity under outdoor conditions. The key factor is the growth rate which is a function of the light profile within the reactor and the light regime to which the cells are subjected. Once this function is obtained, it may be possible to get a correlation

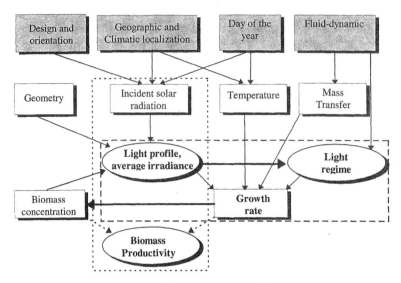

Figure 4.4 Relationship between different parameters that influence on biomass productivity in outdoor conditions. Molina Grima *et al.*, 1998

between biomass productivity and the average irradiance inside the culture (Iav). On the other hand, Iav is a function of the irradiance impinging of the reactor surface which is, at the same time, dependent on the geographic and environmental factors. The geographic location and day of the year determine the solar incident radiation and temperature to which the cells are exposed. While temperature can be kept within a narrow interval by using suitable thermostatic systems, solar radiation cannot be controlled. The incident solar radiation, which is a function of climatic and geographic parameters of the facility location (Incropera and Thomas, 1978), as well as the design and orientation of the photobioreactor (Lee and Low, 1991; Quiang and Richmond, 1996), determines the maximum energy available for growth. The incident solar radiation, along with the photobioreactor geometry and biomass concentration, determines a heterogeneous light profile due to mutual shading to which the cells are exposed inside the culture (Acién Fernández *et al.*, 1997; Molina Grima *et al.*, 1997). Their volumetric integration leads to estimation of the light availability or average irradiance inside the culture, to which the cell metabolism adapts and thus, their biochemical composition and growth rate (Acién Fernández *et al.*, 1998). However, the photobioreactor fluid-dynamics also determines the mass transfer and the light regime of the cells. The latter is affected by the time the cells spend in the zones of various irradiances and their frequency of movement in and out of these zones (Philipps and Myers, 1954; Terry, 1986; Grobbelaar, 1994). This light regime affects the behaviour of the cells, determining the extent of photolimitation and photoinhibition phenomena, and thereby, the efficiency of using solar radiation by the cultures. The biomass concentration is in itself influenced by growth rate, which is a function of the average irradiance, temperature, mass transfer and light regime to which the cells are exposed. Both growth rate and biomass concentration determine the final biomass productivity of the system (Figure 4.4). Thus, in an optimum system where no other limitation is imposed, a direct correlation between light availability, rate of photosynthesis and productivity may be expected.

In fact, it seems that other limitations not only limit growth through their direct effect, but also impose a limitation on the ability to utilize the absorbed solar energy. This may result in a photoinhibitory response further limiting the ability of the photosynthetic apparatus to operate at its maximal efficiency (Vonshak, 1996).

In order to estimate the biomass productivity along the year, three tasks must be undertaken: first, to estimate the incident solar radiation on the culture surface; second, to determine the light profile and average irradiance inside the culture as a function of incident radiation, geometry of the reactor and biomass concentration; and third, to determine the influence of incident radiation and average irradiance on the growth rate and finally in the biomass productivity. These aspects are discussed below.

Solar irradiance

The daily global solar radiation on the atmospheric surface, and the extraterrestrial radiation *Ho* (Figure 4.5), can be analytically estimated, depending on the time and the geographic location, using the equation (Liu and Jordan, 1960):

$$Ho = \left(\frac{24}{\pi} \cdot Isc\right) \cdot \left(1 + 0.033 \cdot \cos\left(\frac{360 \cdot N}{365}\right)\right)$$
$$\cdot \left(\cos\phi \cdot \cos\delta \cdot \sin\omega_s + \frac{2 \cdot \pi \cdot \omega_s}{360} \cdot \sin\phi \cdot \sin\delta\right)$$

(4.1)

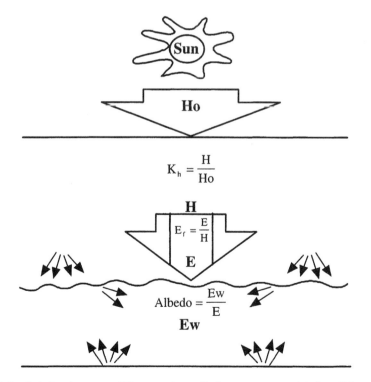

Figure 4.5 Relation between different solar radiation parameters. Molina Grima *et al.*, 1998

where Isc is the universal solar constant, $1353 \ \text{W m}^{-2}$ (Thekaekara and Drummond, 1971). Solar declination, δ, and the angle (ω_s) at sunrise hour can be calculated as functions of the day of the year, N, and the latitude, ϕ, as:

$$\delta = 23.45 \cdot \sin\left(360 \cdot \frac{284 + N}{365}\right) \tag{4.2}$$

$$\cos \omega_s = -\tan \delta \cdot \tan \phi \tag{4.3}$$

Extraterrestrial radiation, Ho, is attenuated by the higher atmospheric layers, and the clouds and dust in the lower layers. Thus, the daily clearness index, K_h, is defined as the ratio between the daily radiation (H) on a horizontal surface at ground level and the daily extraterrestrial radiation (H_o) (Figure 4.5). If K_h is known for a specific geographic location, the daily radiation on a horizontal surface can be determined from Ho. In addition, K_h enables the daily global direct (H_D) and diffuse (H_d) radiation impinging on a horizontal surface to be estimated (Liu and Jordan, 1960):

$$\frac{H_d}{H} = 1.390 - 4.027 \cdot K_h + 5.530 \cdot K_h^2 - 3.108 \cdot K_h^3 \tag{4.4}$$

In applications related to photoautotrophic growth, such as biomass production and natural illumination, the incident photosynthetically active radiation must be determined. Thus the daily incident photosynthetic radiation on a horizontal surface, E $(\mu\text{E m}^{-2}\,\text{d}^{-1})$, can be estimated as a constant fraction of the global incident radiation, H $(\text{kJ m}^{-2}\,\text{d}^{-1})$. The photosynthetic efficiency of solar radiation (E_f) which is defined as the ratio of photosynthetically active radiation (E) to the global incident radiation (H) on a horizontal surface (Incropera and Thomas, 1978), depends on the day and season. However, an additional effect to be considered is the increase in solar irradiance over the solar radiation incident on a horizontal surface due to the reflection off surrounding surfaces. This effect, or albedo, is specific to surface properties, such as colour and texture, and is defined as the ratio between incident radiation and the measured irradiance. The albedo is a function of the incident radiation parameters such as the angle of incidence and fraction of diffuse radiation, attaining values of up to 2 on clear days and approximately equal to 1 on cloudy days.

If the daily global radiation (H) is known, the hourly values (G) can be determined by the Collares–Rabl (1968) equation (Collares-Pereira and Rabl, 1968):

$$\frac{G}{H} = \frac{I}{E} = \left(\frac{\pi}{24}\right) \cdot (a + b \cdot \cos \omega) \cdot \left(\frac{\cos \omega \cdot \cos \omega_s}{\sin \omega_s - \omega_s \cdot \cos \omega_s}\right) \tag{4.5}$$

$$a = 0.409 + 0.5016 \cdot \sin(\omega_s - 60) \tag{4.6}$$

$$b = 0.6609 - 0.4767 \cdot \sin(\omega_s - 60) \tag{4.7}$$

where the hourly solar angle, ω, can be calculated as a function of the solar hour, h_s, by:

$$\omega = 15 \cdot (12 - hs) \tag{4.8}$$

Likewise, if the photosynthetic efficiency of solar radiation, E_f, is known, Equation 4.5 can be used to estimate hourly incident photosynthetic radiation, I, from the daily photosynthetic irradiance, E. Thus, by estimation of the albedo of the system, irradiance inside the pond, Iw, can be determined. If the coefficient of albedo for the surface used is known, the estimation of the daily photosynthetic irradiance inside the system, Ew, is possible, and together with Equation 4.5, the hourly irradiance on the photobioreactor surface can be estimated.

In this sense, for the specific location of the reactors and during the research period the parameters K_h and E_f were determined, in addition to the variation of albedo (Acién Fernández *et al.*, 1998). Results showed that constant values of $74 \pm 9\%$ and $1.74 \pm 0.07\,\mu\mathrm{E\,J^{-1}}$ for K_h and E_f, respectively, could be accepted for all the year. The albedo was experimentally determined, a mean value of 2 being obtained (Figure 4.6). The proposed solar irradiance model reproduced the measured values ($r^2 = 0.816$). Moreover, the equations used for solar irradiance also allowed the instantaneous values of incident global and photosynthetic radiation, and photosynthetic irradiance inside the thermostatic pond to be estimated (Figure 4.7). The results showed that the instantaneous values were symmetrically distributed during the day, reaching maximum values at solar noon, these being a function of solar hour and date (Figure 4.7). Notice that the method for estimating incident solar radiation yields averages, because it does not consider the weather conditions at every moment, but interpolates the irradiance as a function of a mean cloudiness or clearness index. In spite of this, the calculated irradiance adjusts to the measured values, and therefore, its application is feasible for system design.

Average irradiance inside tubular photobioreactors

The irradiance model proposed by Acién Fernández *et al.* (1997) was used to estimate the average irradiance inside tubular photobioreactors. The model determines the light path from the tube surface to any point inside the culture for each type of

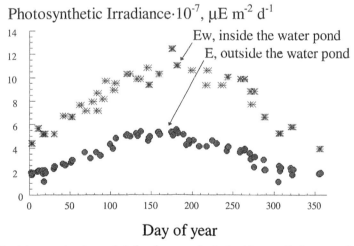

Photosynthetic Irradiance·10^{-7}, $\mu\mathrm{E\,m^{-2}\,d^{-1}}$

Day of year

Figure 4.6 Measured values of daily photosynthetic incident radiation on a horizontal surface, E, and daily photosynthetic irradiance inside the thermostatic pond, Ew, for various days of the year 1994, in Almería, Spain (36°48′N, 2°54′W). The value of the average albedo all through the year (albedo=2) was determined as a function of the diffuse to direct radiation ratio and the solar incidence angle (from Modelling of biomass productivity in tubular photobioreactors for microalgal cultures: effects of dilution rate, tube diameter, and solar irradiance, by Acién Fernández F.G., García Camacho F., Sánchez Pérez J.A., Fernández Sevilla J.M. and Molina Grima E., 1998, *Biotechnology and Bioengineering*, **58**(6), 605–16. Reprinted by permission of John Wiley & Sons, Inc.).

Photosynthetic irradiance·10^{-3}, µE m^{-2} s^{-1}

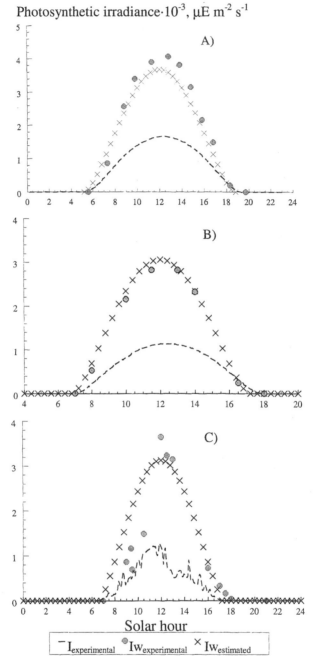

Figure 4.7 Variation of estimated and measured photosynthetic irradiance inside the pond, Iw, and the photosynthetic irradiance on the water surface of the thermostatic pond, I, with time of day for (A) 9 September 1994; (B) 10 January 1995; and (C) 2 November 1994 (in Modelling of biomass productivity in tubular photobioreactors for microalgal cultures: effects of dilution rate, tube diameter, and solar irradiance, by Acién Fernández F.G., García Camacho F., Sánchez Pérez J.A., Fernández Sevilla J.M. and Molina Grima E., 1998, *Biotechnology and Bioengineering*, **58**(6), 605–16. Reprinted by permission of John Wiley & Sons, Inc.).

incident radiation: direct, which has a defined direction determined by the sun position, and disperse, the sum of diffuse and that reflected by the pond walls, both of which inch in the culture surface from all directions. The analytical determination of the direct light path length, p_D, for any point inside the photobioreactor, defined in cylindrical co-ordinates by the radius r_i and the angle φ, and the determination of the direct external irradiance, $I_{D0}(\varepsilon)$, at each angle of the culture surface, ε, enables the estimation of the local direct irradiance inside the culture, $I_D(ri, \varphi)$, by using Lambert–Beer's law (Figure 4.8). The p_D and I_D values were estimated using the equations:

$$p_D = \frac{R \cdot \sin \varepsilon - ri \cdot \sin \varphi}{\cos \theta} \qquad (4.9)$$

$$I_D(ri, \varphi) = I_{D0}(\varepsilon) \cdot \exp(-Ka \cdot p_D \cdot C) \qquad (4.10)$$

The disperse radiation comes from all spatial directions, $I_{do}(\varepsilon)$, and irradiance is determined by integration over all angles of incidence, ε.

$$p_d = \sqrt{(ri \cdot \sin \varphi - R \cdot \sin \varepsilon)^2 + (ri \cdot \cos \varphi - R \cdot \cos \varepsilon)^2} \qquad (4.11)$$

$$I_d(ri, \varphi) = \frac{1}{2 \cdot \pi} \int_\varepsilon I_{d0}(\varepsilon) \cdot \exp(-Ka \cdot p_d(\varepsilon) \cdot C) \, d\varepsilon \qquad (4.12)$$

Direct and disperse contributions to solar irradiance on the culture surface, I_o, were determined as a function of angular position, ε, as $I_{D0}(\varepsilon)$ and $I_{d0}(\varepsilon)$, respectively (Acién Fernández *et al.*, 1997). Thus, by extension of these equations, it is possible to determine the light profiles and average irradiance inside the culture by:

$$I(ri, \varphi) = I_D(ri, \varphi) + \frac{1}{2 \cdot \pi} \int_\varepsilon I_d(ri, \varphi) \, d\varepsilon \qquad (4.13)$$

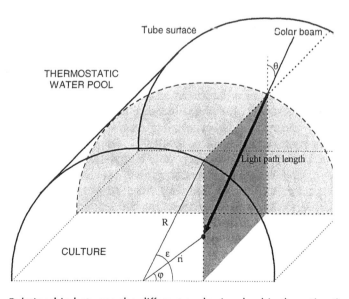

Figure 4.8 Relationship between the different angles involved in the estimation of punctual irradiance inside tubular external photobioreactors. Molina Grima *et al.*, 1998

$$Iav = \frac{1}{\pi \cdot R^2} \left\{ \left(\iint_{R\,\varphi} I_D r \, dr \, d\varphi \right) + \left(\frac{1}{2 \cdot \pi} \iiint_{R\,\varphi\,\varepsilon} I_d r \, dr \, d\varphi \, d\varepsilon \right) \right\} \qquad (4.14)$$

Estimation of mean radial velocity of the cells

In order to estimate the average radial velocity of the cells, the speed at which they move inside the culture, it can be approximately compared to the speed of the smallest turbulent eddies, v. This velocity can be calculated by means of the Kolmogorov theory of the isotropic turbulence (Kolmogorov, 1962) from the rate of turbulent energy dissipation per culture mass, ξ, and the kinematic viscosity of the culture, v, defined as the ratio between dynamic viscosity and density of the fluid, by:

$$v = (v \cdot \xi)^{1/4} \qquad (4.15)$$

At steady state, the energy dissipated by turbulence equals that supplied by the gas, so the average volume ξ may be calculated by:

$$\xi = \frac{P_G}{M} = \frac{gU_g}{1 + \dfrac{A_d}{A_r}} \qquad (4.16)$$

in which all the variables (the superficial gas velocity, U_g, the cross-sectional area of downcomer and riser, A_d and A_r, respectively) are known reactor parameters.

Influence of irradiance on culture surface and average irradiance on biomass productivity

The relationship between the irradiance on culture surface, the light availability inside the culture and the biomass productivity has been studied in *Phaeodactylum tricornutum* UTEX 640 cultures grown at different dilution rates and time of the year; experimental results are shown in Table 4.4. Figure 4.9 shows how *Iav* varies with dilution rate regardless of the tube diameter, hence indicating that the cultures adapt mainly to average irradiance. At $D = 0.025\,h^{-1}$, *Iav* increased only slightly with increasing *Iwm*; the mean value of *Iav* for the whole range was $84 \pm 9\,\mu E\,m^{-2}\,s^{-1}$. With smaller tube diameter and/or higher external irradiance, *Iwm*, the biomass concentration increased as expected for self-shaded light-limited growth (Table 4.4). However, the slight increase in *Iav* with *Iwm* points to some photoinhibition by external irradiance, although this is not very extensive due to the high biomass concentrations. At $D = 0.040\,h^{-1}$, *Iav* values are higher than at $D = 0.025\,h^{-1}$ because of the lower steady-state biomass concentrations. The greater light availability gives rise to higher specific growth rate as well as photoinhibition, because of the greater increase in *Iav* with *Iwm*. At $0.050\,h^{-1}$ dilution rate, the exponential increase in *Iav* with *Iwm* caused by the low steady state biomass concentration, underlines the existence of photoinhibition, and thus, the one-to-one relationship between μ and *Iav* no longer applies. In spite of photoinhibition, none of the experiments resulted in total culture washout. Instead, a new steady state was reached, indicating that not all culture was photoinhibited because any degree of photoinhibition of all cells would always lead to the total washout of the system. This effect might be explained by coexistence of photoinhibition and photo-limitation. Due to the light gradients inside the culture, both light-limited growth

Table 4.4 Quasi-steady state results for continuous cultures carried out in the two photobioreactors.

HTPB	D, h^{-1}	Month	C, g l^{-1}	P$_b$, g l^{-1} d^{-1}	Ew 10^{-7}, μE m^{-2} d^{-1}	Iwm, μE m^{-2} s^{-1}	Xp, %dw
0.06	0.025	Dec. 94	2.0 ± 0.1	0.50 ± 0.03	3.9 ± 0.4	914	2.23
0.06	0.025	Nov. 94	3.1 ± 0.1	0.78 ± 0.03	5.8 ± 0.6	1351	2.09
0.06	0.025	Jan. 95	3.3 ± 0.1	0.83 ± 0.03	6.5 ± 0.3	1503	1.81
0.06	0.025	Feb. 94	4.0 ± 0.2	1.01 ± 0.05	6.7 ± 0.4	1563	1.53
0.06	0.025	Oct. 94	5.3 ± 0.2	1.34 ± 0.05	7.1 ± 0.5	1643	1.55
0.06	0.025	Sep. 94	6.4 ± 0.2	1.61 ± 0.05	9.9 ± 0.6	2291	1.36
0.06	0.025	May 94	5.5 ± 0.3	1.39 ± 0.08	10.3 ± 0.5	2362	1.37
0.06	0.025	April 94	6.6 ± 0.3	1.66 ± 0.08	10.3 ± 0.4	2367	1.36
0.06	0.040	Dec. 94	2.5 ± 0.1	1.00 ± 0.08	4.4 ± 0.5	1032	0.92
0.06	0.040	Jan. 95	2.7 ± 0.2	1.08 ± 0.04	5.2 ± 0.3	1211	0.97
0.06	0.040	Feb. 94	3.3 ± 0.3	1.32 ± 0.08	6.6 ± 0.6	1543	0.88
0.06	0.040	May 95	4.2 ± 0.2	1.68 ± 0.08	7.9 ± 0.5	1807	0.71
0.06	0.040	Aug. 94	4.4 ± 0.1	1.76 ± 0.08	10.0 ± 0.4	2319	0.87
0.06	0.040	June 94	5.1 ± 0.2	2.04 ± 0.04	12.5 ± 0.4	2860	0.86
0.06	0.050	Oct. 94	2.5 ± 0.1	1.25 ± 0.08	5.2 ± 0.2	1225	0.83
0.06	0.050	Sep. 94	3.0 ± 0.1	1.50 ± 0.05	8.8 ± 0.6	2052	0.92
0.06	0.050	May 95	3.3 ± 0.1	1.65 ± 0.05	9.4 ± 0.7	2161	0.59
0.06	0.050	June 94	2.8 ± 0.2	1.40 ± 0.09	11.0 ± 0.04	2538	0.59
0.03	0.025	Dec. 94	4.3 ± 0.2	1.08 ± 0.05	3.9 ± 0.2	914	1.55
0.03	0.025	Nov. 94	4.9 ± 0.2	1.23 ± 0.05	5.8 ± 0.3	1348	1.47
0.03	0.025	Jan. 95	5.9 ± 0.2	1.49 ± 0.05	6.5 ± 0.4	1502	1.42
0.03	0.025	Oct. 94	6.8 ± 0.2	1.71 ± 0.05	6.6 ± 0.3	1545	1.21
0.03	0.025	Aug. 93	9.1 ± 0.3	2.29 ± 0.08	8.9 ± 0.3	2046	1.31
0.03	0.025	Sep. 94	8.5 ± 0.2	2.14 ± 0.05	9.9 ± 0.2	2290	1.31
0.03	0.040	Dec. 94	3.3 ± 0.1	1.32 ± 0.04	4.4 ± 0.3	1028	0.87
0.03	0.040	Jan. 95	4.2 ± 0.2	1.68 ± 0.08	5.2 ± 0.4	1208	0.76
0.03	0.040	Sep. 93	4.9 ± 0.1	1.96 ± 0.04	7.4 ± 0.7	1704	0.82
0.03	0.040	May 95	6.0 ± 0.2	2.40 ± 0.08	7.9 ± 0.6	1813	0.72
0.03	0.040	Aug. 93	6.9 ± 0.1	2.76 ± 0.04	9.3 ± 0.5	2160	0.74
0.03	0.400	Aug. 94	6.1 ± 0.2	2.44 ± 0.08	10.0 ± 0.3	2319	0.78
0.03	0.050	Oct. 94	3.4 ± 0.1	1.70 ± 0.05	5.2 ± 0.4	1220	0.84
0.03	0.050	Sep. 94	3.6 ± 0.1	1.80 ± 0.05	8.8 ± 0.3	2048	0.92
0.03	0.050	May 95	4.3 ± 0.2	2.12 ± 0.09	9.4 ± 0.4	2170	0.67

Values are the means of at least five consecutive days (from Molina Grima *et al.*, 1997; reprinted with permission from Kluwer Academic Publishers).

and photoinhibiting conditions could occur simultaneously, each one to a different extent depending on the external irradiance (Molina Grima *et al.*, 1996b; Acién Fernández *et al.*, 1998).

Photoinhibition degrades the microalgae capacity to harvest light. This has been attributed to the destruction of key components of the PSII photosystem (Samuelsson, 1983). Jensen and Knutsen (1993) have shown that this destruction is a reversible process in which degradation and regeneration of key components of the photosynthetic apparatus coexist, and the available amount of functional photosynthetic pigments is the result of an equilibrium between these two processes.

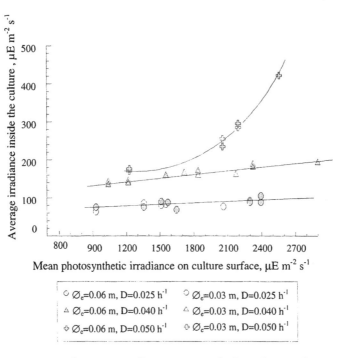

Figure 4.9 Variation of average irradiance, *Iav*, inside the culture with mean photosynthetic irradiance inside the pond, *Iwm*, for the various dilution rates and the photobioreactors used (from Modelling of biomass productivity in tubular photobioreactors for microalgal cultures: effects of dilution rate, tube diameter, and solar irradiance, by Acién Fernández F.G., García Camacho F., Sánchez Pérez J.A., Fernández Sevilla J.M. and Molina Grima E., 1998, *Biotechnology and Bioengineering*, **58**(6), 605–16. Reprinted by permission of John Wiley & Sons, Inc.).

Bearing this in mind, photoinhibition can be explained by assuming that damage to the photosystem only occurs at very high irradiances found only in a limited amount of the culture volume, while the regeneration process can take place throughout the culture. In this case, despite decreased efficiency for light harvesting, the growth rate may continue to show a hyperbolic relationship with *Iav* (Molina Grima *et al.*, 1996b; Acién Fernández *et al.*, 1998). In outdoor reactors, the phenomena of photolimitation and photoinhibition take place in two dimensions. During the mid-day hours, photoinhibition takes place due to the high irradiance (Richmond *et al.*, 1986; Vonshak and Guy, 1988). On the other hand, for any external irradiance, the existence of gradients inside the culture due to self shading causes a photolimited culture volume deeper inside the reactor and a photoinhibited culture volume nearer to the surface (Molina Grima *et al.*, 1996b; Acién Fernández *et al.*, 1998).

Moreover, extension of each one of the photolimitation and photoinhibition phenomena is a function of the light regime to which the cells are exposed (Molina Grima *et al.*, 1997). The light regime is a function of the light profile inside the culture and the mean radial velocity of the cells. The first is a function of the incident solar radiation, the system geometry and the biomass concentration, while the second is a function of the fluid-dynamic. The results showed that the light profile was very similar in both photobioreactors at each imposed dilution rate in spite of different incident irradiances, giving rise to similar values of average irradiance that showed that the outdoor cultures were mainly adapted to average irradiance. In addition, in keeping with the previous subsection, the estimated speed of the eddies was 0.014 and 0.016 m s^{-1} in HTPB0.06 and HTPB0.03, respectively, indicating that turbulence is similar in the two photobioreactors since the fluidodynamic conditions were constant throughout the study (Molina Grima *et al.*, 1997). Thus, due to the greater tube diameter, the cycle must be longer in HTPB0.06 than in HTP0.03, and therefore the period of exposure to light would be longer and photoinhibition very pronounced. The period of darkness would also be very long and photosynthesis therefore reduced, which might increase the contribution of photorespiration and increase the pigment content (Table 4.4).

Estimation of annual biomass productivity

Although many studies on the influence of environmental variables on biomass productivity have been carried out (Tamiya *et al.*, 1953; Banninster, 1979; Molina Grima *et al.*, 1994b, 1996b), very few have been performed under outdoor conditions. Some models have been proposed specifically for outdoor algal ponds (Incropera and Thomas, 1978; Sukenik *et al.*, 1986, 1991; Guterman *et al.*, 1990), but none of these is appropriate for tubular photobioreactors. Thus, Incropera and Thomas (1978) present a model for determining the mean irradiance in open ponds as a function of climatic parameters and geographic location. Starting from estimated irradiance values, a growth model having six parameters was proposed. The model did not consider photoinhibition, and only four of the six parameters it required were experimentally obtained. The predicted maximum productivity of about 34 g m^{-2} d^{-1} was not experimentally verified. Similarly, Sukenik *et al.* (1986, 1991) proposed a five-parameter model (determined under laboratory conditions) for estimation of biomass productivity in outdoor systems. The cells were assumed to adapt to mean irradiance inside the culture; however, photoinhibition was not taken into account and the results were not valid at low biomass concentrations. Again, the estimated maximum productivities of 10.2–27.2 g m^{-2} d^{-1}, were not experimentally verified. All these models were developed under the assumption that the results obtained from indoor laboratory experiments may be directly extrapolated to outdoor conditions. This assumption does not concur with experience. Using mean temperature and irradiance data, Guterman *et al.* (1990) proposed a growth model for open systems. Starting from the initial conditions, the model estimated the variation of culture parameters such as dissolved oxygen, pH and biomass concentration, using empirical relationships. Photoinhibition was not considered. The estimated maximum productivity was 38 g m^{-2} d^{-1}, at irradiances of 2500 µE m^{-2} s^{-1}. Thus, existing models do not account for photoinhibition. Accordingly, they estimate that the maximum productivity is reached at very low biomass concentrations, when the mutual-shading effect is negligible and the culture may be considered light-saturated.

In a previous work (Molina Grima *et al.*, 1994b), the relationship between specific growth rate, μ, and average irradiance, *Iav*, for light-limited growth was adjusted to the following hyperbolic expression:

$$\mu = \frac{\mu_{max} I_{av}^n}{I_k^n + I_{av}^n} \tag{4.17}$$

where n is a factor taking into account the abruptness of the μ vs. *Iav* curve in the transition from low to high irradiance conditions, analogous to Bannister's shape parameter (Bannister, 1979), and I_k is a constant representing the affinity microalgae have for light. However, due to the simultaneous existence of photolimitation and photoinhibition the parameters n and I_k, are not constant (Molina Grima *et al.*, 1996b), varying with incident irradiance. When the external irradiance increases, n decreases while I_k increases, due to an increase in the extension of the photoinhibition, that leads to a decrease in the biomass generation capacity of the system, because a higher average irradiance is necessary to maintain the cellular growth, since a fraction of the photosynthetic apparatus is inactive by photoinhibition (Molina Grima *et al.*, 1996b). By applying this model to experimental results obtained during three years, a generalized semi-empirical kinetic model was obtained (Acién Fernández *et al.*, 1998).

$$\mu = \frac{\mu_{max}\ Iav^{(n2+\frac{n3}{Ivm})}}{\left[Ik\left(1+\left(\frac{Io}{K_I}\right)^{n1}\right)\right]^{(n2+\frac{n3}{Ivm})}+Iav^{(n2+\frac{n3}{Ivm})}} \tag{4.18}$$

$\mu_{max} = 0.063\,\mathrm{h}^{-1}$, $I_k' = 94.3\,\mathrm{\mu E\,m^{-2}\cdot s^{-1}}$, $K_I = 3426\,\mathrm{\mu E\,m^{-2}\,s^{-1}}$, $n_1 = 3.04$, $n_2 = 1.209$, $n_3 = 514.6$.

In addition to Equation 4.18, the solar incident irradiance in any location can be analytically estimated in agreement with the subsection on solar irradiance (these topics are also extensively treated in Duffie and Beckman, 1980). In this form, by the conjunction of solar irradiance estimation, the model for estimating the average irradiance inside the culture showed in the subsection on factors influencing biomass productivity (Acién Fernández *et al.*, 1997), and the kinetic model (Equation 4.18), the biomass productivity along the year can be estimated. Figure 4.10 depicts a flowchart for estimating biomass productivity from the location of the reactor and the biokinetic parameters of the strain. This methodology could be extended to others types of tubular reactor and strains. Thus, the macromodel reproduces the experimental biomass productivity with an error lower than 20 per cent (Acién Fernández *et al.*, 1998), the simulated productivities all through the year being showed in Figure 4.11.

The biomass productivity varies differently in the two reactors over the annual cycle (Figure 4.11). Thus, in HTP0.06 the biomass productivity is lower during the winter than in summer at all imposed dilution rates, indicating that the cultures grown in this reactor are photolimited all year long. At $0.025\,\mathrm{h}^{-1}$, the minimum biomass productivity was obtained all year long due to the photolimitation effect. However, biomass productivity obtained at $0.050\,\mathrm{h}^{-1}$ is lower than that obtained at $0.040\,\mathrm{h}^{-1}$, in spite of the greater light availability since the effect of photoinhibition in this reactor is very strong. On the other hand, in the HTP0.03 maximum productivity was obtained during the winter at $0.050\,\mathrm{h}^{-1}$, $2.7\,\mathrm{g\,l^{-1}\,d^{-1}}$, but in summer the biomass productivities obtained at 0.040 and $0.050\,\mathrm{h}^{-1}$ were the same, $2.4\,\mathrm{g\,l^{-1}\,d^{-1}}$, due to photoinhibition.

The lower biomass productivity observed in HTP0.06 when the dilution rate was changed from $0.040\,\mathrm{h}^{-1}$ to $0.050\,\mathrm{h}^{-1}$ shows that the photoinhibition in HTP0.06 is stronger than in HTP0.03. Furthermore, photolimitation was also stronger in HTP0.06 than HTP0.03, because biomass productivity was lower. Thus, the differ-

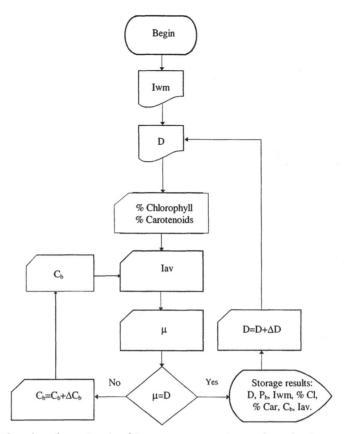

Figure 4.10 Flowchart for estimating biomass concentration and productivity starting from values of environmental variable, *Iwm*, and the operational variable *D* (from Modelling of biomass productivity in tubular photobioreactors for microalgal cultures: effects of dilution rate, tube diameter, and solar irradiance, by Acién Fernández F.G., García Camacho F., Sánchez Pérez J.A., Fernández Sevilla J.M. and Molina Grima E., 1998, *Biotechnology and Bioengineering*, **58**(6), 605–16. Reprinted by permission of John Wiley & Sons, Inc.).

ence in behaviour is due to the different light regimes to which the cells are exposed inside the culture. Therefore, although the flow rate of the culture was the same for all the assays, the double tube diameter of HTP0.06 caused the period of daylight during which the cells are exposed to high irradiance and darkness during which the cells remain in the dark to be longer in HTP0.06 than in HTP0.03.

The biomass productivity obtained in both photobioreactors is very high in comparison with similar culture systems, 2.7 g l^{-1} d^{-1} in HTP0.03 and 2.05 g l^{-1} d^{-1} in HTP0.06, due to the high incident irradiance. Thus, Gudin and Chaumont (1983), pioneers in the development of a tubular system for culturing *Porphyridium* spp, obtained a maximum productivity of 0.71 g l^{-1} d^{-1} with 0.03 m diameter tubes. Recently, Richmond *et al.* (1993) have developed a tubular reactor for *Anabaena siamensis* cultures. Using tube diameters of 2.8, 3.2 and 5.0 cm, they observed that the productivity increases when tube diameter diminishes, since light availability increases, reporting maximum productivities of 0.55 g l^{-1} d^{-1}. A new system of inclined panels developed by Hu *et al.* (1996) with *Spirulina platensis*, attained productivities of 4.43 g l^{-1} d^{-1} with 1.3 cm thick panels, while if thickness were increased

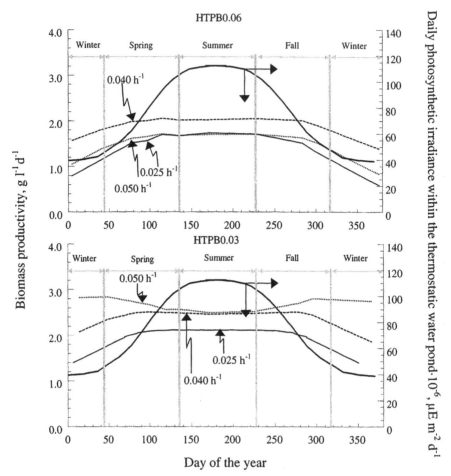

Figure 4.11 Variation of biomass productivity with the day of the year for chemostat cultures of *Phaeodactylum tricornutum* UTEX 640 in tubular photobioreactors. Data estimated from solar radiation equations and the proposed growth model. (A) Øe=0.060 m; (B) Øe=0.030 m. Solid lines denote the daily photosynthetic irradiance inside the pond (from Modelling of biomass productivity in tubular photobioreactors for microalgal cultures: effects of dilution rate, tube diameter, and solar irradiance, by Acién Fernández F.G., García Camacho F., Sánchez Pérez J.A., Fernández Sevilla J.M. and Molina Grima E., 1998, *Biotechnology and Bioengineering*, **58**(6), 605–16. Reprinted by permission of John Wiley & Sons, Inc.).

to 2.6 cm, productivity decreased to $2.1\,\mathrm{g\,l^{-1}\,d^{-1}}$, and to $0.34\,\mathrm{g\,l^{-1}\,d^{-1}}$ with a 10.4 cm panel. With this same strain, Torzillo *et al.* (1993), obtained $1.5\,\mathrm{g\,l^{-1}\,d^{-1}}$ biomass productivity using a two plane tubular reactor. Lee and Low (1992) have carried out cultures of *Chlorella pyrenoidosa* in a tubular photobioreactor with a 1.2 cm diameter tube, reaching productivities of $3.64\,\mathrm{g\,l^{-1}\,d^{-1}}$, the specific growth rate of this microalga being over $0.15\,\mathrm{h^{-1}}$ in contrast to the 0.06–$0.09\,\mathrm{h^{-1}}$ determined for *Phaeodactylum tricornutum* (Mann and Myers, 1968; Kaixian and Borowitzka, 1992). This high biomass productivity is attributed to two effects: first, the high incident irradiance measured on the culture surface; and second, the controlled temperature, which increased the daytime growth rate.

EPA production from *Phaeodactylum tricornutum* UTEX 640

Factors influencing the EPA content of the biomass

The fatty acid content of the biomass is influenced by the environmental conditions of the culture. Thus, parameters such as temperature, mineral nutrients, incident irradiance, light availability and light–dark cycles, determine the biochemical composition and fatty acid profile of the biomass (Cohen *et al.*, 1988; Bajpai and Bajpai, 1993; Molina Grima *et al.*, 1994a, 1994c, 1995).

Culture conditions

Siron *et al.* (1989) studied the influence of phosphate concentration in the culture medium on the fatty acid content and composition of *Phaeodactylum tricornutum*. Their results showed that fatty acid content varied from 11.2 to 4.9 per cent both with and without addition of phosphate, although cell-growth rate increased in the first case. Furthermore, the EPA proportion decreased from 26.8 to 6.9 per cent. On the other hand, Yongmanitchai and Ward (1991) studied the influence of mineral nutrients in the fatty acid composition of the biomass extensively, finding that EPA content increased from 1.4 to 3.5 per cent while palmitoleic acid decreased from 6.6 to 2.4 per cent when nitrate concentration was increased from 0.25 to 1.5 gl^{-1}. Similar variations were obtained by increasing phosphate, urea, vitamins B_1 and B_{12}, and CO_2 concentrations in the medium, while an increase in salinity and silica concentrations was observed to cause an adverse effect.

The EPA content of the biomass has been reported to be inversely proportional to temperature and light intensity (Iwamoto and Sato, 1986). The increase in the degree of unsaturation of membrane lipid fatty acids was related to the decrease in temperature and the increase cytoplasm and membrane fluidity. On the other hand, strong light intensity was observed to elevate the PUFA levels in many species of microalgae, possibly resulting from an increase in oxygen availability and hence an increase in oxygen-requiring lipid desaturation rates. However, for other species, such as *Phaeodactylum tricornutum*, *Cyclotella*, *Chaetoceros calcitrans*, *Thalassiosira pseudonana* and *Isochrysis galbana*, the opposite effect was observed (Kyle, 1991).

Nutrient, light and CO_2 availability are known to have an inverse effect on lipid content and composition. Thus, under limiting conditions, the growth rate is reduced and the cellular metabolism is redirected at the accumulation of triglycerides for energy storage, rather than the synthesis of phospholipids or proteins as is the case under optimum culture conditions and higher growth rates (Parrish and Wangersky, 1987), the composition of triglycerides being higher in palmitic (16:0) and palmitoleic (16:1) acids than the phospholipids, which are very rich in polyunsaturated fatty acids.

These results lead one to believe that growth rate determines the fatty acid content and their proportion in the biomass, with the fatty acid profile being richer in EPA at higher growth rates. Thus, the EPA proportion in batch cultures decreased from 21.3 to 4.0 per cent when the culture passed from the exponential to the linear growth phase, but palmitic (16:0) and palmitoleic (16:1) acids increased from 15.5 to 31.5 per cent, and from 24.1 to 43.0 per cent, respectively (Siron *et al.*, 1989). In the same way, Veloso *et al.* (1991) showed how after ten days of starvation, the fatty acid profile changed, the EPA proportion decreasing from 35 to 10 per cent, and palmitic and palmitoleic acids increasing from 10 to 20 per cent and 17 to 25 per cent respectively. Moreover, Reis *et al.* (1996) show how the proportion of EPA rose from 10.2 per cent to 25.3 per cent when the dilution rate went from 0.14 to 1.22 day^{-1}, palmitoleic acid decreasing from 43.3 to 21.0 per cent. In outdoor continuous cultures, the same behaviour was observed, the total fatty acid content decreasing with increased dilution rate due to the change in the cellular metabolism (Molina Grima *et al.*, 1994a). Thus, when dilution rate increases, the cellular metabolism varies, synthesizing proteins and structural lipids, and decreasing the synthesis of storage lipids. This decrease is different for each fatty acid due to the different cellular functions and location of each one in the cell. Henderson and Mackinlay (1989) reported that the short-chain fatty acids are mainly located in the triglycerides that make up the neutral lipids. Similarly, Fernández Sevilla (1995) studied the location of different fatty acids in the microalga *Isochrysis galbana*, observing that the polyunsaturated fatty acids are mainly located in the structural lipids, glycolipids and phospholipids, the fatty acid profile varying when growth rate is modified. Results obtained with *Phaeodactylum tricornutum* (Molina Grima *et al.*, 1994a) showed the same behaviour, the proportion of EPA increasing with dilution rate, while saturated and monounsaturated fatty acids decreased (Molina Grima *et al.*, 1994a).

The growth rate also determines the protein content in *Phaeodactylum tricornutum*, which was found by Kaixian and Borowitzka (1992) to be between 30 and 50 per cent depending on culture conditions. These results were corroborated by Molina Grima *et al.* (1994a) (Table 4.5), who reported that the higher percentage corresponded to the highest dilution rate, as observed in a comparison of fast- and slow-growing cell contents. In this context, lipid content also varies, but inversely to protein variation, a phenomenon which was related to the structural function of proteins, as well as to the role played by some kinds of, mainly neutral, storage lipids (Whyte, 1987).

Growth status (photolimitation–photoinhibition)

When microalgal cultures are operated under nutrient saturation conditions and controlled temperature, the growth rate is determined by the light availability or average irradiance inside the culture (Molina Grima *et al.*, 1994b). However, due to the existence of light gradients, the cells are exposed to very different irradiance levels, from maximum values on culture surface to near zero in the centre. Thus,

Table 4.5 Variation of biochemical composition (% dry weight) and dilution rate in outdoor chemostats of *Phaeodactylum tricornutum* UTEX 640.

	Dilution rate, h^{-1}		
	0.012	0.024	0.036
Proteins	28.24	33.02	41.41
Neutral lipids	4.84	1.65	0.60
Glycolipids	2.46	3.29	3.57
Phospholipids	1.10	1.33	2.31

From Molina Grima *et al.*, 1994a, with permission.

those cells on the culture surface exposed to the highest irradiance may be photo-inhibited (see subsection on the influence of irradiance), whereas cells in the centre are photolimited and the extent of each of these phenomena is responsible for the growth status of the culture. Just as $Iav-Iwm$ pairs (which determine the extent of the photolimitation–photoinhibition) determine the growth status of the cultures, the cultures themselves determine the biochemical composition of the biomass.

For this reason, the fatty acid content and profile of the biomass were found for all steady states reached in both photobioreactors. The effects of tube diameter (design variable), dilution rate (operating variable) and irradiance on the culture surface (environmental variable) were studied (Acién Fernández *et al.*, 1998). A statistical analysis of the results showed that the tube diameter has no influence on the amounts of the main fatty acids, this being a function of dilution rate and irradiance on the culture surface (Table 4.6). However, F-ratios were higher for irradiance on the culture surface than for dilution rate, because the effect of this variable is stronger. Thus, EPA in the biomass decreased in both photobioreactors from approximately 3.6 to 1.1 per cent, from the lowest ($900\,\mu\mathrm{E\,m^{-2}\,s^{-1}}$, $0.025\,\mathrm{h^{-1}}$) to the highest ($2800\,\mu\mathrm{E\,m^{-2}\,s^{-1}}$, $0.050\,\mathrm{h^{-1}}$) incident irradiances and imposed dilution rates (Figure 4.12). Due to the relationship between dilution rate, average irradiance and irradiance on the culture surface (Equation 4.18), the particular way in which each of the two types of irradiance influences the fatty acids was studied (Table 4.7). Although both types of irradiance affected the amounts of all the individual fatty acids, the influence of irradiance on the culture surface, Iwm, was greater than that of average irradiance, Iav, especially on EPA (Acién Fernández *et al.*, 1998b).

Table 4.6 Statistical variance analysis of the fatty acid content of the biomass with the photobioreactor diameter, dilution rate and irradiance on culture surface.

Fatty acid	Photobioreactor		Dilution rate		Irradiance on culture surface	
	F-ratio	P	F-ratio	P	F-ratio	P
14:0	0.932	ns	30.542	> 99.9	22.943	> 99.9
16:0	8.139	ns	22.626	> 99.9	27.171	> 99.9
16:1	7.323	ns	18.668	> 99.9	20.980	> 99.9
20:5n3	4.609	ns	3.811	> 99.5	19.443	> 99.9
Total	6.899	ns	13.203	> 99.9	28.194	> 99.9

Table 4.7 Statistical variance analysis of fatty acid content of the biomass with the average irradiance within the reactor and irradiance on culture surface.

	Irradiance on culture surface		Average irradiance	
	F-ratio	P	F-ratio	P
14:0	6.144	> 99.9	9.443	> 99.0
16:0	8.119	> 99.9	5.365	> 99.0
16:1	7.469	> 99.9	5.450	> 99.0
20:5n3	13.071	> 99.9	1.829	> 99.0
Total	12.386	> 99.9	4.080	> 99.0

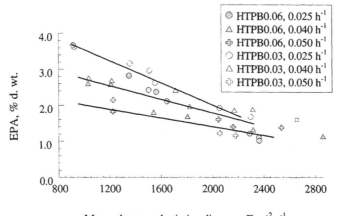

Figure 4.12 Variation of EPA content of the biomass with the mean incident irradiance and dilution rate in both photobioreactors. Molina Grima *et al.*, 1998

Likewise, in continuously illuminated cultures of *Isochrysis galbana* under laboratory conditions (Fernández Sevilla, 1995), the higher the irradiance on the culture surface was, the higher the photoinhibition effect and the lower the fatty acid content of the biomass, especially EPA. With other species of microalgae, similar results have been observed (Piorreck and Pohl, 1984; Sukenik *et al.*, 1989; Thompson *et al.*, 1990; Renaud *et al.*, 1991), but with *Porphyridium cruentum* (Cohen *et al.*, 1988) the relationship was inverse, perhaps due to a decrease in the growth rate with decreased irradiance on the culture surface. The lower fatty acid content obtained under higher incident irradiance can be related to the lower pigment content of the biomass, especially chlorophyll, due to a reduction in the size of the chloroplast (Fawley, 1984) and decrease in cell size with greater light availability. In the membranes of the photosynthetic organelles, which represent an important fraction of the cell, up to 75 per cent of the total membranes (Forde and Steer, 1976), consist of up to 90 per cent PUFAs.

A statistical analysis similar to that done for the fatty acid content was done for the biomass fatty acid profile. The fatty acid profile is not influenced by the irradiance on the culture surface or the photobioreactor tube diameter, whereas there is a close relationship between the fatty acid composition and dilution rate. In the same way, due to the photoadaptation of the cultures to average irradiance, the influence of average irradiance in fatty acid profile of the biomass was greater than that of irradiance on the culture surface (Acién Fernández *et al.*, 1998).

The mean values for each of the imposed dilution rates show how the fatty acid profile of the biomass changes with dilution rate, saturated and monounsaturated fatty acids decreasing and polyunsaturated fatty acids increasing, especially EPA (Table 4.8). These variations are in agreement with the increase in protein and decrease in lipids found for increased growth rates in *Phaeodactylum tricornutum* by Molina Grima *et al.* (1994a) (Table 4.5). Thus, the lower the growth rate, the longer the duplication time, and the microalgae store more neutral lipids rich in saturated and monounsaturated fatty acids (14:0, 16:0, 16:1). Since the need for structural biomolecules is lower, the lipids store the carbon fixed during photosynthesis. This is observed in the lower protein content. However, when growth rate

Table 4.8 Variation of the fatty acid profile of the biomass with dilution rate.

D, h^{-1}	14:0	16:0	16:1	20:5n3
0.0252	7.9 ± 0.3	17.2 ± 0.3	19.8 ± 0.6	41.9 ± 0.9
0.0400	6.7 ± 0.4	16.5 ± 0.3	18.8 ± 0.6	43.4 ± 0.9
0.0500	4.0 ± 0.5	13.7 ± 0.4	15.8 ± 0.8	50.0 ± 0.9

Values are the mean of results obtained in both photobioreactors at any irradiance on culture surface.

increases, duplication time decreases and the need for structural biomolecules increases, reducing storage of neutral lipids. This effect produces a reduction in the saturated and monounsaturated fatty acid content, and an increase in polyunsaturated fatty acids, which are the main constituents of cellular membranes (Kates and Volcani, 1966).

To summarize, growth status determines not only the biomass productivity, but the biochemical composition of the biomass as well. Thus, if the growth rate is high, the lipid content is low and mainly made up of structural lipids (rich in polyunsaturated fatty acids) (Figure 4.13). Therefore, an optimum growth rate (i.e., dilution rate in the continuous culture) must be balanced so that the lipid content is not too low, consisting mainly of structural lipids (i.e., PUFAs), thus facilitating the downstream process. On the other hand, regardless of the growth rate, the growth status of the culture (average irradiance, *Iav*-irradiance on culture surface, *Iwm*) influences the fatty acid content of the biomass: the more photoinhibiting the conditions (i.e., high *Iwm*) the lower the fatty acid content. Thus the cellular metabolism can be modified and directed through the synthesis of storage or structural lipids by varying the growth rate, fatty acid content being a function of irradiance conditions inside the culture and the extent of photolimitation–photoinhibition (Figure 4.13).

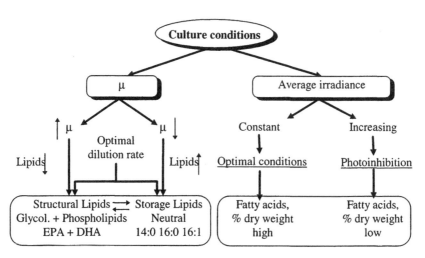

Figure 4.13 Relationship between different parameters that influence the fatty acids content and profile of the biomass under outdoor conditions. Molina Grima *et al.*, 1998

Estimation of annual EPA productivity

In both photobioreactors, maximum EPA productivity of 33–$50\,mg\,l^{-1}\,d^{-1}$ was reached at the same dilution rate of $0.040\,h^{-1}$ (Acién Fernández *et al.*, 1998). With low irradiance on the culture surface, EPA productivity is low due to the reduction in biomass productivity; however with strong irradiance, there is a decrease in EPA productivity due to the reduction in the EPA content of the biomass from photo-inhibition. Thus, the decrease in the EPA content in the biomass with irradiance on the culture surface and dilution rate might lead one to believe that the highest EPA productivity would be reached during the winter at the lowest dilution rate. However, since biomass productivity behaves in the opposite manner, increasing with dilution rate and irradiance on the culture surface, an optimum growth rate must be defined. By combining the above mentioned macromodel of biomass productivity (see section on estimation of annual biomass productivity) and the experimental variation in EPA content with irradiance on the culture surface (Figure 4.12), the variation in EPA productivity throughout the year can be simulated (Figure 4.14). Maximum productivity is reached in both photobioreactors at $0.040\,h^{-1}$, being higher in spring and autumn, and slightly decreasing during summer and winter. These results show that in both photobioreactors the photoinhibition observed during the summer reduces EPA content, because EPA productivity decreases, this being more noticeable at the $0.05\,h^{-1}$ dilution rate and in HTP0.06.

To date, little information has been published about EPA productivity in outdoor cultures. Cohen and Heimer (1992) reported EPA productivity of $2.3\,mg\,l^{-1}\,d^{-1}$ in outdoor cultures of *Porphyridium cruentum*. Molina Grima *et al.* (1994c) reported EPA productivity of about $8.2\,mg\,l^{-1}\,d^{-1}$ with *Isochrysis galbana* AL-II4 grown in tubular photobioreactors. Yongmanitchai and Ward (1991), using *Phaeodactylum tricornutum* as their EPA source under laboratory conditions, obtained EPA productivity of $19\,mg\,l^{-1}\,d^{-1}$, higher than the $3\,mg\,l^{-1}\,d^{-1}$ reported by the Nisshing Oil Mills with *Chlorella minutissima* at industrial scale. However, all values are lower than the $50\,mg\,l^{-1}\,d^{-1}$ obtained for *Phaeodactylum tricornutum* in the different tube diameters of two outdoor horizontal tubular photobioreactors (Figure 4.14). This productivity is similar to the $58.9\,mg\,l^{-1}\,d^{-1}$ recently reported by Hu *et al.* (1997) using *Monodus subterraneus* in a 2.8 cm deep outdoor flat inclined modular bioreactor.

Concluding remarks

For the industrial production of EPA from microalgae, all of the factors influencing the biomass productivity and EPA content of the biomass must be known. In this sense, solar radiation and temperature are the most important. Solar radiation, a function of the factory location, date and reactor design, determines the irradiance on the culture surface, which, in addition to attenuation by the biomass determines not only the light availability inside the culture (or average irradiance), but the extent of photolimitation–photoinhibition taking place in the culture. This relationship between solar radiation, irradiance on culture surface, average irradiance and growth rate has been established and a kinetic growth model obtained. This

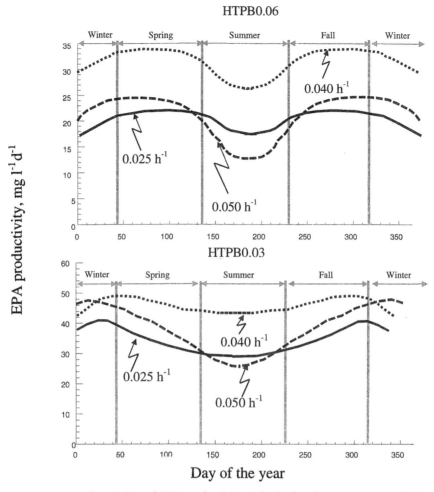

Figure 4.14 Annual variation of EPA productivity in both photobioreactors at each imposed dilution rate. Molina Grima *et al.,* 1998

model allows the influence of operating, design and environmental variables to be studied, and moreover, the system may be scaled up for the improvement of biomass productivity.

The product generation rate (in this case EPA) is the result of two factors, biomass productivity and EPA content of the biomass. The first has been previously modelled, whereas the second is a function of the growth status, that is, the extent of photolimitation and photoinhibition, and growth rate. The higher the growth rate, the lower the fatty acid content of the biomass, but the higher the polyunsaturated fatty acid proportion in the single cell oils. The stronger photoinhibition is (i.e., from the strong irradiance on the culture surface and high dilution rate), the lower the fatty acid content of the biomass. These results, in addition to the previous biomass productivity model, allow the annual variation in EPA productivity with dilution rate and reactor tube diameter to be determined, thus obtaining a new EPA production model. By applying this model, the system EPA productivity could be optimized throughout the year, at more than $40\,\mathrm{mg\,l^{-1}\,d^{-1}}$, and reactor scale-up and operating strategies can be established, thus serving as a powerful tool for the industrial exploitation of EPA production from microalgae.

Acknowledgements

The authors wish to express their gratitude to Dr Claude Gudin for his advice, the financial support of the Comisión Interministerial de Ciencia y Tecnología (CICYT) (BIO 95-0652) (Spain) for some of the work reported in this chapter and the contribution of the members of the marine Microalgae Biotechnology group of the University of Almería, who have worked with the authors on this subject.

Nomenclature

Albedo	measured irradiance to incident radiation in a system
A_r	cross section area of the riser, $\mathrm{m^{-2}}$
A_d	cross section area of the downcomer, $\mathrm{m^{-2}}$
C	biomass concentration
D	dilution rate, $\mathrm{h^{-1}}$
E	daily photosynthetic solar irradiance on a horizontal surface, $\mathrm{\mu E\,m^{-2}\,d^{-1}}$
E_f	photosynthetic efficiency of solar radiation, $\mathrm{\mu E\,J^{-1}}$
E_w	daily photosynthetic solar irradiance inside the water pond, $\mathrm{\mu E\,m^{-2}\,d^{-1}}$
G	hourly global radiation on a horizontal surface, $\mathrm{J\,m^{-2}\,h^{-1}}$
H	daily global radiation on a horizontal surface, $\mathrm{J\,m^{-2}\,d^{-1}}$
h	height of the riser, m
H_d	diffuse daily global radiation on a horizontal surface, $\mathrm{J\,m^{-2}\,d^{-1}}$
H_D	direct daily global radiation on a horizontal surface, $\mathrm{J\,m^{-2}\,d^{-1}}$
H_o	extraterrestrial global radiation, $\mathrm{J\,m^{-2}\,h^{-1}}$
h_s	hour of sunrise, h
I	hourly photosynthetic solar irradiance on a horizontal surface, $\mathrm{\mu E\,m^{-2}\,s^{-1}}$
Iav	average photosynthetic irradiance inside the culture, $\mathrm{\mu E\,m^{-2}\,s^{-1}}$
I_d	disperse irradiance inside the culture, $\mathrm{\mu E\,m^{-2}\,s^{-1}}$
I_D	direct irradiance inside the culture, $\mathrm{\mu E\,m^{-2}\,s^{-1}}$
I_{do}	disperse irradiance on the culture surface, $\mathrm{\mu E\,m^{-2}\,s^{-1}}$
I_{Do}	direct irradiance on the culture surface, $\mathrm{\mu E\,m^{-2}\,s^{-1}}$
I_k	constant of affinity for the light, $\mathrm{\mu E\,m^{-2}\,s^{-1}}$
I_{sc}	universal solar constant, $\mathrm{W\,m^{-2}}$
I_w	hourly irradiance inside the water pond, $\mathrm{\mu E\,m^{-2}\,s^{-1}}$
Iwm	mean hourly irradiance inside the water pond, $\mathrm{\mu E\,m^{-2}\,s^{-1}}$
Ka	extinction coefficient of the biomass, $\mathrm{m^2\,g^{-1}}$
K_h	daily clearness index
K_I	irradiance saturation constant, $\mathrm{\mu E\,m^{-2}\,s^{-1}}$
M	mass of the culture, kg
N	day of the year
Pb	biomass productivity, $\mathrm{g\,l^{-1}\,d^{-1}}$
p_D	direct light path length, m
p_d	disperse light path length, m

P_G energy supply by the expansion of the gas, w
R radius of the tube, m
r_i distance from the point to the centre of the tube, m
U_g superficial gas velocity, $m\,s^{-1}$
v velocity of the smallest eddies, $m\,s^{-1}$
x_p pigments content, per cent d. wt.

Greek letters

ϕ latitude of the location, degrees
δ solar declination, degrees
ω hourly solar angle, degrees
μ growth rate, h^{-1}
ξ energy supply by mass liquid, $J\,kg^{-1}$
v kinematic viscosity, $m^2\,s^{-1}$
θ solar incidence angle, degrees
ε angle between the inlet point of radiation in the tube and the horizontal, degrees
φ angle of position of the point inside the tube, degrees
ω_s hourly solar angle at sunrise, degrees

References

ACIÉN FERNÁNDEZ, F.G., GARCÍA CAMACHO, F., SÁNCHEZ PÉREZ, J.A., FERNÁNDEZ SEVILLA, J. M., and MOLINA GRIMA, E., 1997, A model for light distribution and average solar irradiance inside outdoor tubular photobioreactors for the microalgal mass culture, *Biotechnology and Bioengineering*, **55**(5), 701–714.

ACIÉN FERNÁNDEZ, F.G., GARCÍA CAMACHO, F., SÁNCHEZ PÉREZ, J.A., FERNÁNDEZ SEVILLA, J. M., and MOLINA GRIMA, E., 1998, Modelling of biomass productivity in tubular photo-bioreactors for microalgal cultures. Effects of dilution rate, tube diameter and solar irradiance, *Biotechnology and Bioengineering*, **58**(6), 605–616.

BAJPAI, P. and BAJPAI, P.K., 1993, Eicosapentaenoic acid (EPA) production from micro-organisms: a review, *J. Biotechnol.*, **30**, 161–183.

BANNISTER, T.T., 1979, Quantitative description of steady state, nutrient-saturated algal growth, including adaptation, *Limnol. Oceanograf.*, **24**(1), 79–96.

BARCLAY, W.R., MEAGER, K.M., and ABRIL, J.R., 1994, Heterotrophic production of long chain omega-3 fatty acids utilizing algae and algae-like micro-organisms, *J. Appl. Phycol.*, **6**, 123–129.

BOROWITZKA, M.A., 1988a, Vitamins and fine chemicals from microalgae, in *Microalgal Biotechnology*, Borowitzka and Borowitzka (Eds), Cambridge, Cambridge University Press, pp. 153–196.

BOROWITZKA, M.A., 1988b, Fats, oils and hydrocarbons, in *Microalgal Biotechnology*, Borowitzka and Borowitzka (Eds.), Cambridge, Cambridge University Press, pp. 257–287.

CHRISMADHA, T. and BOROWITZKA, M.A., 1994, Effect of cell density and irradiance on growth, proximate composition and eicosapentaenoic acid production of *Phaeodactylum tricornutum* grown in a tubular photobioreactor, *J. Appl. Phycol.*, **6**, 67–74.

COBELAS, M.A. and LECHAZO, J.Z., 1989, Lipids in microalgae. A review. I. Biochemistry, *Grasas Aceites*, **40**, 118–145.

COHEN, Z., 1986, Products from microalgae, in *Handbook of Microalgae*, Boca Raton, CRC Press, pp. 421–454.

COHEN, Z., 1990, The production potential of eicosapentaenoic and arachidonic acids by the red alga *Porphyridium cruentum*, *J. Am. Oil Chem. Soc.*, **67**, 916–920.

COHEN, Z., 1994, Production potential of eicosapentaenoic acid by *Monodus subterraneus*. *J. Am. Oil. Chem. Soc.*, **71**, 941–945.

COHEN, Z. and HEIMER, Y.M., 1992, Production of polyunsaturated fatty acids (EPA, ARA and GLA) by the microalgae *Porphyridium* and *Spirulina*, in *Industrial Applications of Single Cell Oils*, Kyle, D.J. and Ratledge, C. (Eds), American Oils Chemists' Society, pp. 243–273.

COHEN, Z., VONSHAK, A., BOUSSIBA, S. and RICHMOND, A., 1988, The effect of temperature and cell concentration on the fatty acid composition of outdoor cultures of *Porphyridium cruentum*, in *Algal Biotechnology*, Stadler, T., Mollion, J., Verdus, M.C., Karamanos, Y., Morvan, H. and Christiaen, D. (Eds), London, Elsevier Applied Science, pp. 421–429.

COHEN, Z., NORMAN, H.A. and HEIMER, Y.M., 1995, Microalgae as a source of omega-3 fatty acids, in Simopoulos, A.P. (Ed.), *Plants in Human Nutrition*, *World Rev. Nutr. Diet*, Basel, Karger, **77**, 1–31.

COLLARES-PEREIRA, M. and RABL, A., 1968, The average distribution of solar radiation – correlations between diffuse and hemispherical and between daily and hourly insolation values, *Solar Energy*, **11**, 112.

COOK, P.M., 1950, Large-scale culture of *Chlorella*, in Brunnel, J., Prescott, G.W. and Tiffany, L.H. (Eds), *The Culturing of Algae*, Yellow Springs, OH: Charles F. Kettering Foundations, pp. 53–74.

DeMORT, C.L., LOWRY, R., TINSLEY, I. and PHINNEY, H.K., 1972, The biochemical analysis of some estuarine phytoplankton species I. Fatty acid composition. *J. Phycol.*, **8**, 211–216.

DUFFIE, J.A. and BECKMAN, W.A., 1980, *Solar Engineering of Thermal Processes*, Wiley Interscience (Ed.), USA.

DYERBERG, J., 1986, Linoleate-derived polyunsaturated fatty acids and prevention of arthero-sclerosis, *Nutrition Reviews*, **44**, 125–134.

FAWLEY, M.W., 1984, Effects of light intensity and temperature interactions on growth characteristics of *Phaeodactylum tricornutum* (Bacillariophyceae), *J. Phycol.*, **20**, 67–72.

FERNÁNDEZ SEVILLA, J.M., 1995, Estudio del crecimiento simultáneamente fotolimitado y fotoinhibido de la microalga marina *Isochrysis galbana*. Productividad en ácidos grasos poliinsaturados *n*-3, Universidad de Almería, PhD Thesis, pp. 200–225.

FORDE, J. and STEER, M. W., 1976, The use of quantitative electron microscopy in the study of lipid composition of membranes, *J. Exp. Bot.*, **27**, 1137–1141.

GILLAN, F.T., McFADDEN, R., WETHERBEE, R. and JOHNS, R. B., 1981, Sterols and fatty acids of an Antarctic Sea ice diatom, *Stauroneis amphioxys*, *Phytochemistry*, **20**, 1935–1937.

GROBBELAAR, J.U., 1994, Turbulence in algal mass cultures and the role of light/dark fluctuations, *Journal of Applied Phycology*, **6**, 331–335.

GUDIN, C. and CHAUMONT, D., 1983, Solar biotechnology study and development of tubular solar receptors for controlled production of photosynthetic cellular biomass, in Palz W. and Pirrwitz, D. (Eds), *Proceedings of the Workshop and EC Contractors' Meeting in Capri*, Dordrecht: Reidel Publ. Co., pp. 184–193.

GUDIN, C. and THERPENIER, C., 1986, Bioconversion of solar energy into organic chemicals by microalgae, in A. Mizrahi (Ed.), *Advances in Biotechnological Processes*, **6**, Liss, New York 73–110.

GUDIN, C. and CHAUMONT, D., 1991, Cell fragility, the key problem of microalgae mass production in closed photobioreactors, *Bioresource Technol.*, **38**, 141–151.

GUTERMAN, H., VONSHAK, A. and BEN-YAAKOV, S., 1990, A macromodel for outdoor algal mass production, *Biotech. Bioeng.*, 35, 809–819.

HENDERSON, R.J. and MACKINLAY, E.E., 1989, Effect of temperature on lipid composition of the marine cryptomona *Chroomonas salina*, *Phytochemistry*, **28**, 2943–2948.

HU, Q., GUTTERMAN, H. and RICHMOND, A., 1996, Flat inclined modular photobioreactor (FIMP) for outdoor mass cultivation of photoautotrophes, *Biotechnology and Bioengineering*, **51**, 51–60.

HU, Q., HU, Z., COHEN, Z. and RICHMOND, A., 1997, Enhancement of eicosapentaenoic acid (EPA) and γ-linolenic acid (GLA) production by manipulating algal density of outdoor cultures of *Monodus subterraneus* (Eurotrymatophyta) and spirulina platiasis (cyanobacteria), *Eur. J. Phycol.*, **32**, 81–86.

INCROPERA, F.P. and THOMAS, J.F., 1978, A model for solar radiation conversion to algae in shallow pond, *Solar Energy*, **20**, 157–165.

IWAMOTO, H. and SATO, S., 1986, Production of EPA by freshwater unicellular algae, *J Am. Oil Chem. Soc.* **63**, 434.

JAREONKITMONGKOL, S., SHIMIZU, S. and YAMADA, H., 1993, Production of an eicosapentaenoic acid-containing oil by a Δ12 desaturase-defective mutant of *Morteriella alpina* 1S-4, *J. Am. Oil Chem. Soc.*, **70**, 119–123.

JENSEN, S. and KNUTSEN, G., 1993, Influence of light and temperature on photoinhibition of photosynthesis in *Spirulina platensis*, *J. Appl. Phycol.*, **5**, 495–504.

JOSEPH, J.D., 1975, Identification of 3,6,9,12,15-octadecapentaenoic acid in laboratory-cultured photosynthetic dinoflagellates, *Lipids*, 10395–10403.

KAIXIAN, Q. and BOROWITZKA, M.A., 1992, Light and deficiency effects on the growth and composition of *Phaeodactylum tricornutum*, *Appl. Biochem. Biotechnol.*, **38**, 93–103.

KATES, M. and VOLCANI, B.E., 1966, Lipid content of diatoms, *Biochemistry Biophysical Acta*, **116**, 264–278.

KOLMOGOROV, A.N., 1962, Refinement of previous hypothesis concerning the local structure of a viscous incompressible fluid at high Reynolds numbers, *J. Fluid Mech.*, **13**, 82–85.

KOON, T.C. and JOHNS, M.R., 1996, Screening of diatoms for heterotrophic eicosapentaenoic acid production, *Journal of Appl. Phytocol.*, **8**, 59–64.

KYLE, D., 1991, Specialty oils from microalgae: new perspectives, in *Biotechnology of Plant Fats and Oils*, Rattray, J. (Ed.), Champaign, IL: American Oil Chemists Society, pp. 130–143.

LASSEN, C. and JORGENSEN, B.B., 1994, A fiber optic irradiance microsensor (cosine collector): application for in situ measurements of absorption coefficients in sediments and microbial mats, *FEMS Microbiol. Ecol.*, **15**, 321–336.

LEE, Y.K., 1986, Enclosed bioreactors for the mass cultivation of photosynthetic microorganisms: the future trend, *TIBTECH*, **4**, 186–189.

LEE, Y.K. and LOW, C.S., 1991, Effect of photobioreactor inclination on the biomass productivity of an outdoor algal culture, *Biotech. Bioeng.*, **38**, 995–1000.

LEE, Y.K. and LOW, C. S., 1992, Productivity of outdoor algal cultures in enclosed tubular photobioreactors, *Biotech. Bioeng.*, **40**, 1119–1122.

LEWIN, J.C., LEWIN, R.A. and PHILPOTT, D.E., 1958, Observations on *Phaeodactylum tricornutum*. *J. Gen. Microbiol.*, **18**, 418–426.

LITTLE, A.D., 1953, Pilot plant studies in the production of *Chlorella*, in Burlew, J.S. (Ed.), *Algal Culture from Laboratory to Pilot Plant*, Carnegie Inst. of Washington, pp. 235–273.

LIU, B.Y.H. and JORDAN, R.C., 1960, The interrelationship and characteristic distribution of direct, diffuse and total solar radiation, *Solar Energy*, **7**, 53–65.

LÓPEZ ALONSO, D., MOLINA GRIMA, E., SÁNCHEZ PÉREZ, J.A., GARCÍA SÁNCHEZ, J.L. and GARCÍA CAMACHO, F., 1992, Isolation of clones of *Isochrysis galbana* rich in eicosapentaenoic acid, *Aquaculture*, **102**, 363–371.

LÓPEZ ALONSO, D., SEGURA DEL CASTILLO, C.I., MOLINA GRIMA, E. and COHEN, Z., 1996, First insights into improvement of eicosapentaenoic acid content in *Phaeodactylum tricornutum* (Bacillariophyceae) by induced mutagenesis, *J. Phycol.*, **32**, 339–345.

MANN, J.E. and MYERS, J., 1968, On pigments, growth and photosynthesis of *Phaeodactylum tricornutum*, *J. Phycol.*, **4**, 349–355.

MOLINA GRIMA, E., GARCÍA CAMACHO, F., SÁNCHEZ PÉREZ, J.A., ACIÉN FERNÁNDEZ, F.G., FERNÁNDEZ SEVILLA, J.M. and VALDÉS SANZ, F., 1994a, Effect of dilution rate on eicosapentaenoic acid productivity of *Phaeodactylum tricornutum* UTEX 640 in outdoor chemostat culture, *Biotechnology Letters*, **16**, 1035–1040.

MOLINA GRIMA, E., GARCÍA CAMACHO, F., SÁNCHEZ PÉREZ, J.A., FERNÁNDEZ SEVILLA, J.M., ACIÉN FERNÁNDEZ, F.G. and CONTRERAS GÓMEZ, A., 1994b, A Mathematical model of microalgal growth in light limited chemostat culture, *J. Chem. Tech. Biotechnol.*, **61**, 167–173.

MOLINA GRIMA, E., SÁNCHEZ PÉREZ, J.A., GARCÍA CAMACHO, F., FERNÁNDEZ SEVILLA, J.M. and ACIÉN FERNÁNDEZ, F.G., 1994c Effect of growth rate on EPA and DHA content of *Isochrysis galbana* in chemostat culture, *Appl. Microbiol. and Biotechnol.*, **41**, 23–27.

MOLINA GRIMA, E., SÁNCHEZ PÉREZ, J.A., GARCÍA CAMACHO, F., GARCÍA SÁNCHEZ, J.L. and FERNÁNDEZ SEVILLA, J.M., 1995, Variation of fatty acid profile with solar cycle in outdoor chemostat culture of *Isochrysis galbana* ALII-4, *J. Appl. Phycol.*, **7**, 129-134.

MOLINA GRIMA, E., ROBLES MEDINA, A., GIMÉNEZ GIMÉNEZ, A. and IBAÑEZ GONZÁLEZ, M.J., 1996a, Gram-scale purification of eicosapentaenoic acid (EPA 20:5n3) from wet *Phaeodactylum tricornutum* UTEX 640 biomass, *Journal of Applied Phycology*, **8**, 359–367.

MOLINA GRIMA, E., FERNÁNDEZ SEVILLA, J.M., SÁNCHEZ PÉREZ, J.A. and GARCÍA CAMACHO, F., 1996b, A study on simultaneous photolimitation and photoinhibition in dense microalgal cultures taking into account incident and averaged irradiances, *J. Biotechnol.*, **45**, 59–69.

MOLINA GRIMA, E., SÁNCHEZ PÉREZ, J.A., GARCÍA CAMACHO, F., FERNÁNDEZ SEVILLA, J.M. and ACIÉN FERNÁNDEZ, F.G., 1997, Productivity analysis of outdoor chemostat culture in tubular air-lift photobioreactors, *Journal of Applied Phycology*, **8**, 369–380.

NETTLETON, J.A., 1993, Are n-3 fatty acids essential nutrients for fetal and infant development? *J. Am. Diet. Assoc.*, **93**, 58–64.

O'BRIEN, D.J., KURANTZ, M.J. and KWOCZAK, R, 1993, Production of eicosapentaenoic acid by the filamentous fungus *Pythium irregulare*, *Appl. Microbiol. Biotechnol.*, **40**, 211–214.

OHTA, S., CHANG, T., AOZASA, O., KONDO, M. and MIYATA, H., 1992, Sustained production of arachidonic and eicosapentaenoic acids by the red alga *Porphyridium purpureum* cultured in a light/dark cycle, *J. Ferment. Bioeng.*, **74**, 398–402.

PARRISH, C.G. and WANGERSKY, P.J., 1987, Particulate and dissolved classes in cultures of *Phaeodactylum tricornutum* grown in cage culture turbidostats with a range of nitrogen supply rates, *Mar. Ecol. Prog. Ser.*, **35**, 119–128.

PHILIPPS, J.N. and MYERS, J., 1954, Growth rate of *Chlorella* in flashing light, *Plant Physiol.*, **29**, 152–161.

PIORRECK, M. and POHL, P., 1984, Formation of biomass, total protein, chlorophylls, lipids and fatty acids in green and blue-green algae during one growth phase, *Phytochemistry*, **23**, 217–223.

PIRT, S.J., LEE, Y.K., WALACH, M.R., PIRT, M.W., BALUZI, H.H.M. and BAZIN, M.J., 1983, A tubular bioreactor for photosynthetic production of biomass from carbon dioxide. Design and performance, *J. Chem. Tech. Biotechnol.*, **33B**, 35–58.

PLAKAS, S.M. and CUARINO, A.M., 1986, Omega-3 fatty acids and fish oils: is the news all good? *Proceedings of the Eleventh Annual Tropical and Subtropical Fisheries Conference of the Americas*, pp. 275–283.

POHL, P., 1982, Lipids and fatty acids of microalgae, in Zaborsky, O. (Ed.), CRC *Handbook of Biosolar Resources*, Boca Raton: CRC Press, **1**, 383–404.

QUIANG, H. and RICHMOND, A., 1996, Productivity and photosynthetic efficiency of *Spirulina platensis* as affected by light intensity, algal density and rate of mixing in a flat plate photobioreactor, *Journal of Applied Phycology*, **8**, 139–145.

RADMER, R.J., 1990, Omega-3 fatty acids from algae, in *Omega-3 Fatty Acids in Health and Disease*, Lees R.S. and Karel, M. (Eds), Marcel Dekker: New York, Food Sci. Technolo. **37**, 211–214.

RATCHFORD, I.A.J. and FALLOWFIELD, H.J., 1992, Performance of a flat plate, air-lift reactor for the growth of high biomass algal cultures, *J. Appl. Phycol.*, **4**, 1–9.

RATLEDGE, C., 1989, Microbial routes to lipids, *Biochem. Soc. Trans.*, **17**, 1139–1141.

REIS, A., GOUVEIA, L., VELOSO, V., FERNANDES, H.L., EMPIS, J.A. and NOVAIS, J.M., 1996, Eicosapentaenoic acid-rich biomass production by the microalga *Phaeodactylum tricornutum* in a continuous-flow reactor, *Bioresource Technology*, **55**, 83–88.

RENAUD, S.M., PARRY, D.L., LUONG-VAN, T., KUO, C., PADOVAN, A. and SAMMY, N., 1991, Effect of light intensity on the proximate biochemical and fatty acid composition of *Isochrysis* sp. and *Nannochloropsis oculata* for use in tropical aquaculture. *J. Appl. Phycol.*, **3**, 43–53.

RICHMOND, A., 1986, Microalgae of economic potential, in *CRC Handbook of Microalgal Mass Culture*, Richmond (ed.), Boca Ratón, Florida: CRC Press Inc., pp. 199–243.

RICHMOND, A., 1990, Large scale microalgal culture and applications, in *Progress in Phycological Research*, Vol. 7, Round and Chapman (Eds), Biopress Ltd, 269–329.

RICHMOND, A., BOUSSIBA, S., VONSHAK, A. and KOPEL, R., 1993, A new tubular reactor for mass production of microalgae outdoors, *J. Appl. Phycol.*, **5**, 327–332.

SAMUELSSON, B., 1983, *Science*, **220**, 568, in Yongmanitchai, W. and Ward, O.P., 1989, ω-3 fatty acids. Alternative sources of production, *Process Biochemistry*, 117–125.

SETO, A., WANG, H.L. and HESSELTINE, C.W., 1984, Culture conditions affect eicosapentaenoic acid content of *Chlorella minutissima*, *J. Am. Oil Chem. Soc.*, **61**, 892–894.

SIMONOPOULOS, A.P., 1991, Omega-3 fatty acids in health and disease and in growth and development, *Am. J. Clin. Nutr.*, **54**, 438–463.

SIRON, R., GIUSTI, G. and BERLAND, B., 1989, Changes in the fatty acid composition of *Phaeodactylum tricornutum* and *Dunaliella tertiolecta* during growth and under phosphorus deficiency, *Marine Ecology Progress Series*, **55**, 95–100.

SUKENIK, A., FALKOWSKI, P.G. and BENNETT, J., 1986, Potential enhancement of photosynthetic energy conversion in algal mass culture, *Biotech. Bioeng.*, **30**, 970–977.

SUKENIK, A., CARMELI, Y. and BERNER, T., 1989, Regulation of fatty acid composition by irradiance level in the eustigmatophyte *Nannochloropsis* sp., *J. Phycol.*, **25**, 686–692.

SUKENIK, A., LEVY, R.S., LEVY, Y., FALKOWSKI, P.G. and DUBINSKY, Z., 1991, Optimizing algal biomass production in an outdoor pond: a simulation model. *Journal of Applied Phycology*, **3**, 191–201.

SUZUKI, N., INOUE, A., SHIKANO, M., YAZAWA, K. and KONDO, K., 1991, Effect of arginine on the production of eicosapentaenoic acid (EPA) in EPA-elaborating bacterium SCRC-2738, *Bull. Jap. Soc.*, **57**, 1407.

TAMIYA, H., HASE, E., SHIBATA, K., MITUYA, A., IWAMURA, T., NIHEI, T. and SASA, T., 1953, Kinetics of growth of *Chlorella* with special reference to its dependence on quantity of available light and temperature, in Burlew, J.S. (Ed.), *Algal Culture from Laboratory to Pilot Plant*, Washington: Carnegie Inst. of Washington, pp. 204–232.

TERRY, K.L., 1986, Photosynthesis in modulated light: quantitative dependence of photosynthetic enhancement on flashing rate, *Biotechnology and Bioengineering*, **28**, 988–995.

TERRY, K.L. and RAYMOND, L.P., 1985, System design for the autotrophic production of microalgae, *Enzyme Microb. Technol.*, **7**, 474–481.

THEKAEKARA, M.P. and DRUMMOND, A.J., 1971, Standard values for the solar constant and its spectral components, *Nat. Phys. Sci.*, **6**, 229.

THOMPSON, P.A., HARRISON, P.J. and WHYTE, J.N., 1990, Influence of irradiance on the fatty acid composition of phytoplankton, *J. Phycol.*, **26**, 278–288.

TORZILLO, G., CARLOZZI, P., PUSHPARAJ, B., MONTAINI, E. and MATERASSI, R., 1993, A two plane tubular photobioreactor for outdoor culture of *Spirulina*, *Biotechnology Bioengineering*, **42**, 891–898.

TREDICI, M.R. and MATERASSI, R., 1992, From open ponds to vertical alveolar panels: the Italian experience of the development of reactors for the mass cultivation of phototrophic microorganisms. *J. Appl. Phycol.*, **4**, 221–231.

VELOSO, V., REIS, A., GOUVEIA, L., FERNANDES, H.L., EMPIS, J.A. and NOVAIS, J.M., 1991, Lipid production by *Phaeodactylum tricornutum*, *Bioresource Technology* **38**, 115–119.

VOLKMAN, J.K., JEFFREY, S.W., NICHOLS, P.D., 1989, Fatty acids and lipid composition of 10 species of microalgae used in mariculture, *J. Exp. Marine Biol. Ecol.*, **128**, 219–240.

VONSHAK, A., 1996, Environmental factors and photosynthesis: Their role in productivity of outdoor algal cultures, *7th International Conference. Opportunities from Micro- and Macroalgae*. Knysna, South Africa, p. 78.

VONSHAK, A. and GUY, R., 1988, Photoinhibition as a limiting factor in outdoor cultivation of *Spirulina platensis*, in *Algal Biotechnology* (Stadler, T., Mollion, J., Verdus, M.C., Karamanos, Y., Morvan, H. and Christiaen, D., eds), Elsevier Science, 365–370.

WATANABE, Y. and HALL, D.O., 1996, Photosynthetic production of the filamentous cyano-bacterium *Spirulina platensis* in a cone-shaped helical tubular photobioreactor, *Appl. Microbiol. Biotechnol.*, **44**, 693–698.

WHYTE, J.C., 1987, Biochemical composition and energy content of six species of phytoplank-ton used in mariculture of bivalves, *Aquaculture*, **60**, 231–241.

WIM VERMAAS, 1996, The time is ripe for applied molecular genetics in cyanobacteria, *7th International Conference. Opportunities from Micro- and Macroalgae*, Knysna, South Africa, p. 76.

WOOD, J.B., 1974, Fatty acids and saponifiable lipids, in Stewart W.D.P. (Ed.), *Algal Physiology and Biochemistry*, Oxford: Blackwell, pp. 236–265.

YONGMANITCHAI, W. and WARD, O.P., 1989, ω-3 fatty acids. Alternative sources of production, *Process Biochemistry*, **24**, 117–125.

YONGMANITCHAI, W. and WARD, O., 1991, Growth of and omega-3 fatty acid production by *Phaeodactylum tricornutum* under different culture conditions, *Appl. Env. Microbiol.*, **57**, 419–425.

ZIBOH, V.A., COHEN, K.A. and ELLIS, C.N., 1986, Effects of dietary supplementation of fish oil on neutrophil and epidermal fatty acids. Modulation of clinical source of psoriatic sub-jects, *Arch. Dermatol.*, **122**, 1277–1282.

5

Genetic improvement of EPA content in microalgae

DIEGO LÓPEZ ALONSO AND CLARA I. SEGURA DEL CASTILLO

Introduction

Polyunsaturated fatty acids (PUFAs) have been receiving increasing attention due to their involvement in human health. Several of them, specifically linoleic acid (LA, 18:2(n-6)), α-linolenic acid (ALA, 18:3(n-3)), γ-linolenic acid (GLA, 18:3(n-6)) and arachidonic acid (AA, 20:4(n-6)) are considered essential for human nutrition (WHO/FAO, 1977). On the other hand, eicosapentaenoic acid (EPA, 20:5(n-3)) has recently been found to be beneficial in coronary disease (Nelson, 1972; van Gent *et al.*, 1979; Bronsgeest-Schoute *et al.*, 1981; Dyerberg, 1986; Singh and Chandra, 1988; Simopoulos, 1991; Iacono and Dougherty, 1993; Nettleton, 1993), and its positive effect in certain types of cancer, and other aspects of human health has also been suggested (Cohen *et al.*, 1995). Finally, docosahexaenoic acid (DHA, 22:6(n-3)) is involved mainly in the development of the nervous system and retina (Innis, 1991; Brown, 1994).

EPA is currently obtained from fish oils, but this source may be problematic due to the accumulation of heavy metals present in seawater (Plakas and Guarino, 1986), fluctuation in fish captures, uncontrolled fatty acid variation among captures (Yongmanitchai and Ward, 1989), significant taste, odour, and stability problems, and the complicated process of extracting this fatty acid in pure form from fish oil (Barclay *et al.*, 1994).

Microalgae have been suggested as an alternative source of EPA (Kyle, 1989, 1991; Yongmanitchai and Ward 1989, 1991; Radmer, 1990; López Alonso *et al.*, 1992, 1993, 1994, 1996; Barclay *et al.*, 1994; Cohen *et al.*, 1995) because of their great potential for improving content and also because of the good oxidative stability of EPA from algal oil sources (Radmer, 1990).

Microalgal EPA content is modified by environmental factors like light, temperature, nutrients, etc. (Shifrin and Chisholm, 1981; Thompson *et al.*, 1990, 1992; Molina Grima *et al.*, 1992, 1993, 1994, 1995; García Sánchez *et al.*, 1994), but EPA content is also subject to genetic control as has been shown by López Alonso *et al.* (1994), supporting a genetic approach to improvement of EPA content in microalgae (López Alonso *et al.*, 1996) as suggested by de la Noue and de Paw (1988), Brown *et al.* (1990) and López Alonso *et al.* (1992).

Although there have been several industrial attempts to exploit microalgal EPA production commercially (e.g. Martek Corporation in USA, and a joint venture by

Scotia Pharmaceutica of the UK and Helyosynthése of France (Anonymous, 1994; for a short history see also Klausner, 1986), strain improvement is required to increase EPA productivity enough to make it economically feasible for industrial production (Kyle, 1989, 1991; Radmer, 1990). In this regard genetic improvement may be appropriate (Craig and Reichelt, 1986; Craig *et al.*, 1988; de la Noue and de Pauw, 1988; Brown *et al.*, 1990).

The genetic approach to strain improvement

There are three basic genetic approaches to improving product titres in industrial microorganisms: genetic engineering, hybridization, and mutation (Esser and Stahl, 1981; Baltz, 1986).

Genetic engineering

Recombinant DNA technology has proven to be a powerful tool in genetic research and some applications of genetic engineering to industrial microbiology have been successful, notably for the production of some human hormones (i.e. insulin and somatotropin). There have also been some suggestions for the application of genetic engineering to microalgal strain improvement (Craig and Reichelt, 1986; Craig *et al.*, 1988; Brown *et al.*, 1990; Dunahay *et al.*, 1992).

For the application of genetic engineering, the following minimum requirements must be met: first, the gene (or genes) involved in the product titre (e.g. in EPA biosynthesis) must be known; second, the gene (or genes) must be isolated and cloned; and third, a suitable vector to express the cloned gene (or genes) in a microbial or higher plant host must be found. In order to appreciate better the potential of genetic engineering with regard to EPA – or long-chain PUFA (LCPUFAs) – improvement, the current state of the art should be reviewed.

Browse, Somerville and colleagues have worked out the pathway of plant lipid biosynthesis, especially fatty acid desaturation (reviewed in Browse and Somerville, 1991; Ohlrogge *et al.*, 1991; Somerville and Browse, 1991). From their excellent work the plant lipid pathway and its tremendous complexity are now better understood. There are desaturases, acylases, trans-acylases, acyl-carrier proteins, 'elongases', and several other enzymes, obviously including the fatty acid synthetase complex located in the chloroplast. Moreover many of these enzymes exist in two forms, each one located in a different cell compartment – one in the chloroplast and the other in the endoplasmic reticulum. However, not much is known about LCPUFAs, since higher plants do not synthesize over 18-carbon PUFAs which are only synthesized by some eukaryotic microalgae and fungi (Radwan, 1991; Bajpai and Bajpai, 1992, 1993).

For this reason, information on the steps leading to the formation of LCPUFAs, which has been little studied, is scarce (Harwood, 1988; Harwood *et al.*, 1988; Schweizer, 1989). From a rather heterogeneous literature on fungi and microalgae (Nichols and Appleby, 1969; Moreno *et al.*, 1979; Kendrick and Ratledge, 1992; Higashi and Murata, 1993; Iareonkitmongkol *et al.*, 1993a, Ratledge, 1993; Arao and Yamada, 1994; Schneider and Roessler, 1994; Cohen *et al.*, 1995; Khozin *et al.*, 1997) a consensus emerges that attributes LCPUFA synthesis to the composition of an appropriate set of desaturases and 'elongases' (Bajpai and Bajpai, 1992, 1993; Ratledge, 1993; Cohen *et al.*, 1995). However, no one knows exactly how many desaturases are involved or how many enzymes are included under the single name of 'elongase' (Gurr and Harwood, 1991).

Some genes involved in fatty acid desaturation have been cloned, but all are from higher plants (Arondel *et al.*, 1992; Yadav *et al.*, 1993; Falcone *et al.*, 1994; Gibson *et al.*, 1994; Hitz *et al.*, 1994; Okuley *et al.*, 1994), and cyanobacteria (Reddy *et al.*, 1993; Sakamoto *et al.*, 1994); that is, they are not directly involved in LCPUFA synthesis.

Genetic transformation has been achieved in some cyanobacteria-transferring fatty-acid-related genes (Wada *et al.*, 1990; Reddy *et al.*, 1993; Hitz *et al.*, 1994; Sakamoto *et al.*, 1994). There have also been some attempts at looking for vectors for eukaryotic microalgae (Simon and Latorella, 1986; Dunahay *et al.*, 1992), but this is currently under development.

It may be concluded that, although there have been significant advances in micro-algal genetic engineering, the minimum requirements necessary for the application of genetic engineering to LCPUFA production have not yet been met.

On the other hand, a trait like EPA production is very unlikely to be under the control of a single gene. Instead, from the very partial knowledge of the complex pathway of EPA biosynthesis referred to above, it is highly probable that EPA production is a character with complex polygenic control. For both these reasons, the genetic engineering approach to EPA production is highly speculative.

Hybridization

In a broad sense, hybridization 'is any crossing of genetically different individuals leading to an offspring with a genotype different from that of either parent' (Esser and Stahl, 1981). The basis of hybridization is the process of genetic recombination that generates different gene combinations. Therefore, the aim of the hybridization method is to generate new gene combinations, selecting those useful strains with appropriate traits.

Hybridization is basically precluded to sexual microorganisms (eukaryotes), but several prokaryotes also have some kind of sexual reproduction (e.g. conjugation). The development of protoplast fusion techniques also currently permits a form of parasexual hybridization for prokaryotes and asexual eukaryotes (Matsushima and Baltz, 1986). This procedure has even been applied with success to some very common species of microalgae (Sivan *et al.*, 1992, 1995).

It is difficult to estimate to what extent this procedure has been applied in strain improvement by microbiological industries since the industrial policy of protecting proprietary rights prevents publication of positive results (Esser and Stahl, 1981). Conjugation has been used for strain improvement of *Streptomyces* and there is detailed information on the genetic manipulation of this industrial microorganism (Hopwood *et al.*, 1985). At any rate, genetic recombination becomes an important adjunct to mutagenesis once several mutant strains have been successfully established to combine different mutations (Baltz, 1986); therefore, hybridization is not currently applicable for EPA improvement.

Mutation and selection

The mutation and selection process is an iterative procedure that involves three basic processes: first, induction of genetic variability in a population of an organism by mutagenesis; second, small-scale cultivation of many mutagenized individuals; and third, screening of the individuals to identify improved strains (Rowlands, 1984; Queener and Lively, 1986).

Probably the most outstanding results of a mutational programme have been in the production of antibiotics. From the pioneering work (Backus and Stauffer, 1955; Alikhanian, 1962) penicillin production has dramatically increased over 400-fold (Elander and Demain, 1981; Queener and Lively, 1986). The mutational approach has also produced similar results in the production of other metabolites (Crueger and Crueger, 1990).

Mutation has the advantage of simplicity, often being the most direct and least expensive means of improving an industrial microorganism (Rowlands, 1984; Baltz, 1986; Queener and Lively, 1986). Therefore, despite the appearance of the powerful technology of recombinant DNA, there is considerable agreement about the importance of mutation in strain development for the microbiological industry (Aharonowitz and Cohen, 1981; Esser and Stahl, 1981; Jacobson, 1981; Queener and Lively, 1986) and also for microalgal strain improvement (Craig and Reichelt, 1986; Craig *et al.*, 1988; de la Noue and de Pauw, 1988; Brown *et al.*, 1990).

Mutation has been widely applied to microalgae with many different goals: for example, for the production of extracellular ammonia by cyanobacteria (Singh *et al.*, 1983; Kerby *et al.*, 1986; Spiller *et al.*, 1986), for the photoproduction of amino acids (Kerby *et al.*, 1987), and for obtaining resistant mutants with *Dunaliella tertiolecta* (Simon and Latorella, 1986). Shaish *et al.* (1991) have isolated mutants of *Dunaliella bardawil* with several-fold higher content of β-carotene.

With regard to PUFA production, several defective mutants of PUFA biosynthesis in the fungus *Morteriella alpina* have been reported (Jareonkitmongkol *et al.*, 1992, 1993a, 1993b), which are very promising for the elucidation of PUFA pathways and for potential commercial applications of PUFA production. Recently, the isolation of a mutant of the microalga *Nannochloropsis* rich in AA and deficient in EPA was reported (Schneider *et al.*, 1995).

Genetic improvement of EPA content in microalgae

Phenotypic selection

Phenotypic selection of the Prymnesiophyte *Isochrysis galbana* Parke was conducted in our laboratory for EPA and DHA content. The 'parent' culture had an EPA content of 2.4 per cent of dry weight, a DHA content of 0.9 per cent and a total lipid content of 16 per cent. Single cell cultures were obtained by isolation of the cells with a micropipette under a microscope (López Alonso *et al.*, 1992, 1993). This procedure produced 59 generation I isolates. The fatty acid pattern of this generation was roughly similar to the parent culture but there were significant among-isolate differences in fatty acid profile (Table 5.1). For example, EPA content ranged

Table 5.1 Statistical comparison for fatty acid content ($mg\,g^{-1}$) between two generations of isolates of *Isochrysis galbana*.

Fatty acid	Range		Mean			CV		
	I	II	I	II	K–W	I	II	*t*-test
14:0	9–33	11–33	21	21	0.01	28	23	−1.97*
16:0	12–48	19–57	27	30	1.70	27	30	1.06
16:1(n-7)	16–49	18–51	30	30	0.08	27	25	−0.75
18:4(n-3)	6–27	9–21	14	14	0.31	32	22	−3.76*
20:5(n-3)	18–66	36–78	42	53	17.61*	27	18	−5.14*
22:6(n-3)	6–28	10–38	15	20	30.46*	27	26	−0.41

*$0.05 > P > 0.01$.
(K–W = Kruskal–Wallis test; CV = coefficient of variation).

between 2.8 and 6.6 per cent d.w. (dry weight). The isolate richest in EPA was I41, which was selected to produce a new generation of isolates following the same single-cell isolation method. This time, 42 isolates were obtained (López Alonso *et al.*, 1993). The generation II isolates again showed inter-isolate variation for fatty acid content (Table 5.1), e.g., the EPA content ranged from 3.6 to 7.8 per cent d.w. After phenotypic selection, the change between generations was remarkable: average EPA content increased from 4.2 per cent in generation I up to 5.3 per cent in generation II, ($P < 0.001$ by Kruskal–Wallis test). DHA content also increased from 1.5 to 2.0 per cent ($P < 0.001$) (López Alonso *et al.*, 1993). It may be noted that EPA content was increased from 2.4 per cent in the 'parent' culture to over 5 per cent. The strain obtained, called ALII4, was genetically stable as demonstrated by several years of continuous cultivation. Therefore, phenotypic selection is seen to be a likely way to improve EPA content in microalgae. Indirectly, this also supports the genetic control of EPA content in microalgae (López Alonso *et al.*, 1992, 1993).

As the above isolation method was very monotonous, time-consuming, and inefficient, isolation was attempted in a solid medium (agar). *I. galbana* growth in agar was very slow and liquid-recovery cultures very poor. However, following isolation in agar, the so-called generation III was produced from the ALII42 isolate. This time, no increase in EPA content was obtained. This was probably due to the increasing reduction in genetic variation due to stringent phenotypic selection (one single 'progenitor' each time). Although sexual reproduction was probably present in *I. galbana* (Parke, 1949) and, therefore, genetic recombination, the increasingly narrow genetic base probably led to a virtual clone (i.e., genetically homogeneous strain). This suggestion was supported by the decrease in fatty acid variation as estimated by the coefficient of variation (CV). For example, The CV for EPA dropped from 27.4 per cent in generation I to 17.6 per cent in generation II, and 15.2 per cent in generation III. Therefore, continuous application of phenotypic selection probably requires cross-breeding between unrelated strains to maintain a suitable level of genetic variation. Alternatively, genetic variation could be induced by mutagenesis, then following a standard programme of mutation and selection such as this one.

The genetic component of fatty acid variation

Although there may be enough indirect evidence of the genetic control of fatty acid variation (López Alonso *et al.*, 1992), before initiating a selection programme, it must be granted that the genetic component for character variation is enough to sustain such a programme, because the probability of success strictly depends on the heritability of the character to be improved (Falconer, 1989). If the heritability of the character to be improved is high, that is, the proportion of the observed character variation is mainly due to the genome, the selection programme may be successful. Otherwise, if the character has very low heritability, such a programme is unlikely to succeed.

A study was conducted with five strains of *Isochrysis galbana*: PCC1, PCC507 and PCC506 from the Plymouth Culture Collection; IGC from the Carolina Biological Supply; and CCMP463 from the Provasoli-Guillard Center for Culture of Marine Phytoplankton (West Boothbay Harbor, Maine, USA) and two isolates (II25 and II26) of Generation II were included for comparison.

A similar study was carried out with five UTEX strains of *Phaeodactylum tricornutum* (UTEX640, UTEX642, UTEX646, UTEX2089, and UTEX2090), and, recently, with several strains of *Porphyridium cruentum* UTEX161 and several other *Porphyridium* strains kindly supplied by Dr Claude Gudin.

The analysis of variance of the inter-strain variation in the three species cited above showed that genetic variation may explain much of the variation in composition. Estimated heritability was 0.68–0.99, 0.25–0.96, and 0.41–0.64 for *I. galbana*, *P. tricornutum*, and *P. cruentum*, respectively. Heritability was always over 0.41 for EPA content (Table 5.2) indicating that there was a substantial proportion of inter-strain EPA variation of genetic origin, supporting our suggestion that genetic improvement of EPA content in microalgae by mutation and selection is feasible.

Mutation and selection

Two rounds of mutation and selection have been carried out in a *P. tricornutum* UTEX640 'parent' strain using ultraviolet irradiation as the mutational agent (López

Table 5.2 Heritability, the proportion of the total phenotypic variance of each fatty acid that can be interpreted as genetic, for three studied microalgae.

| Fatty acid | Microalgae | | |
	I. galbana	*P. tricornutum*	*P. cruentum*
16:0	0.97***	0.80***	0.41ns
16:1n7	0.98***	0.25ns	n.d.
18:2n6	n.d.	n.d.	0.42ns
18:4n3	0.92***	0.00ns	n.d.
20:4n6	n.d.	n.d.	0.42ns
20:5n3	0.99***	0.41ns	0.64**
22:6n3	0.92***	n.d.	n.d.

ns = not significant, *0.05 > P > 0.01, **0.01 > P > 0.001, ***0.001 > P. n.d. = not determined values in minor fatty acids.

Alonso *et al.*, 1996). The mutation programme followed a two-stage protocol adapted from Queener and Lively (1986) as illustrated in Figure 5.1. In the first stage of the first round, 65 colonies were successfully cultured after UV irradiation. The cultured colonies showed a highly varied fatty acid content. As expected mutational treatment significantly increased the amount of variation present in the wild type (Table 5.3). For instance, the coefficient of variation for EPA increased from 12.3 per cent in the wild type to as much as 26.2 per cent in the mutated population, also proving that UV light was able to modify fatty acid-related genes.

For the second stage of the protocol (Figure 5.1), four putative mutants were selected for their high EPA content (Table 5.4). The selected mutants, as well as the wild type, were recultured in quintuplicate and re-checked for their EPA content. Three of the four putative mutants had a higher EPA content than the wild type. Mutant I14 was the best EPA producer, its EPA content being 37 per cent higher

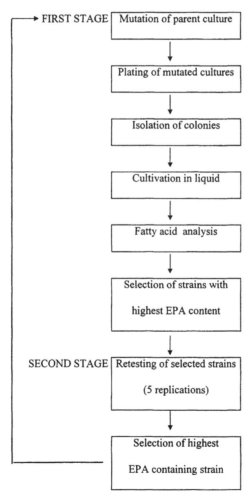

Figure 5.1 Protocol used for a two-stage mutation-selection programme for improving the EPA content in *Phaeodactylum tricornutum*.

Table 5.3 Summary statistics of the fatty acid content (mg g^{-1} dry matter) in *Phaeodactylum tricornutum* following the first mutagenesis with UV (n = 65 cultures).

				Fatty acid			
		14:0	16:0	16:1(n-7)	18:0	20:5(n-3)	TFA[a]
Average		2.9	11.6	13.4	2.3	12.2	55.7
	WT	(2.9)	(10.8)	(13.0)	(2.3)	(11.5)	(53.1)
Minimum		1.0	5.4	6.2	1.1	6.5	30.9
	WT	(2.0)	(6.6)	(8.4)	(1.8)	(8.9)	(35.3)
Maximum		5.0	24.9	22.2	7.8	21.3	117.7
	WT	(3.8)	(15.4)	(17.6)	(2.8)	(14.1)	(93.9)
CV[b]		30.6	34.3	28.9	40.1	26.2	28.7
	WT	(20.0)	(27.1)	(22.1)	(12.8)	(12.3)	(18.3)
I14		4.5	14.3	19.6	3.4	17.4	77.0
I16a		2.6	12.2	14.8	2.5	18.4	67.9
I16b		4.4	19.6	22.2	3.3	19.2	87.3
I20		4.6	18.5	20.6	3.2	18.2	84.1

Data for the 12 single-colony cultures of the wild type (WT) are shown in parenthesis. Single-culture fatty acid content of selected mutants is shown in the bottom half of the table.
[a]TFA = total fatty acid (mg g^{-1}).
[b]CV = coefficient of variation.

Table 5.4 Average (n = 5 cultures) fatty acid content (mg g^{-1} dry matter) of mutants selected from the first round of mutation expressed as a percentage of wild type content and ANOVA F-ratio of the inter-mutant comparison.

			Fatty acid			
Strain	14:0	16:0	16:1(n-7)	18:0	20:5(n-3)	TFA[a]
WT[b]	2.8	11.3	12.7	2.1	10.7	47.4
I14[c]	133	124	141	123	137	134
I16a[c]	94	80	101	109	111	98
I16b[c]	108	100	115	101	100	109
I20[c]	111	112	123	100	107	115
F-ratio	2.78ns	3.11*	4.39*	4.60**	6.21**	3.66*

Significant F-ratio is marked with an asterisk (ns = not significant, *0.05 > P > 0.01, **0.01 > P > 0.001).
[a]TFA = total fatty acid.
[b]WT = wild type; mg g^{-1} dry matter.
[c]Percentage of wild type.

than that of the wild type. It also had more of every other major fatty acid. An analysis of variance (ANOVA) of the fatty acid variation indicated that the among-strain differences were statistically significant for every major fatty acid except for 14:0. Therefore the putative mutants were truly different from the wild type with respect to their fatty acid content.

A new round of mutation–selection was started with mutant I14. A culture of this putative mutant was reexposed to UV irradiance, and 103 colonies were then successfully isolated. The level of variation in this second UV-treated population, as estimated by their CV was very similar to the first one (Tables 5.3 and 5.5). This is noticeable because the second set of 103 cultures originated from a single-colony culture and a great reduction in variation would have been expected. Thus, it seems that genetic variation, presumably erased by stringent selection, was restored by UV irradiation.

The ten cultures with the highest EPA content (Table 5.5) were selected and recultured in quintuplicate (second stage of Figure 5.1). Several putative mutants had EPA content higher than the wild type (Table 5.6) especially II242 with a 44 per cent increase over the wild type. Clustering the mutant strains by EPA content following an LSD range test (Sokal and Rohlf, 1981), strains II242 and II181 were identified as a well-differentiated group.

In conclusion, 37 per cent improvement was achieved after the first round of mutation–selection, and an additional 7 per cent of improvement after the second round. This means that a substantial increase in EPA content was achieved in *P. tricornutum* following a standard protocol of mutation and selection (Queener and Lively, 1986). Under suboptimal working conditions, the full process involved less than a year and a half, and the major expense was gas-chromatography analysis. Neither the biochemistry nor the genetics of the microorganism had to be known for these promising results, which should encourage the use of a similar approach with other microalgae or fungi.

Considerable confusion exists between this mutational approach and the use of mutation in basic science. Workers in basic biochemistry and genetics are looking for 'major' mutations, i.e. mutations which drastically alter the functioning of a gene (see for example the excellent works of Browse, Somerville and colleagues). In this

Table 5.5 Summary statistics of the fatty acid content (mg g^{-1} dry matter) in *Phaeodactylum tricornutum* following the second mutagenesis with UV (n – 103 cultures).

	Fatty acid					
	14:0	16:0	16:1(n-7)	18:0	20:5(n-3)	TFA[a]
Average	2.9	11.6	13.5	2.4	12.1	52.3
Minimum	1.1	4.3	7.6	1.1	5.6	24.1
Maximum	5.6	30.0	41.7	3.8	20.8	111.6
CV[b]	31.7	33.6	34.7	22.1	28.7	28.2
II181	5.0	17.3	20.7	2.6	20.8	79.8
II185b	5.2	19.9	18.8	2.2	18.2	82.3
II1824	5.3	20.0	23.0	3.2	19.2	85.6
II208	4.5	30.0	41.8	3.8	18.8	111.6
II2217	4.6	17.5	18.1	3.0	18.2	77.9
II242	4.9	19.8	19.6	2.8	17.1	79.6
II2418	4.0	22.3	26.9	2.4	18.4	90.7
II261	4.8	21.4	26.4	2.8	18.2	87.6
II2619	4.4	14.6	15.4	2.9	17.3	68.0
II266	4.7	18.6	26.0	2.7	17.4	83.4

Single-culture fatty acid content of selected mutants is shown in the bottom half of the table.
[a]TFA = total fatty acid (mg g^{-1}).
[b]CV = coefficient of variation.

Table 5.6 Average (n = 5 cultures) fatty acid content (mg g^{-1} dry matter) in mutants selected from the second round of mutation expressed as a percentage of wild type content and ANOVA F-ratio of the inter-mutant comparison.

	Fatty acid					
Strain	14:0	16:0	16:1(n-7)	18:0	20:5(n-3)	TFA[a]
WT[b]	3.5	9.4	22.5	4.7	15.9	67.1
II181[c]	133	118	110	137	125	115
II185b[c]	127	111	102	75	115	105
II1824[c]	84	94	84	108	93	92
II208[c]	82	105	80	56	119	88
II2217[c]	85	86	73	75	77	84
II242[c]	121	132	102	118	144	113
II2418[c]	91	84	87	26	99	86
II261[c]	94	94	98	167	115	103
II2619[c]	76	79	63	61	70	69
II266[c]	118	104	106	91	126	106
F-ratio	4.02***	3.21**	2.95**	1.05ns	2.71**	3.73***

Significant F-ratio is marked with an asterisk (ns = not significant, *0.05 > P > 0.01, **0.01 > P > 0.001, ***0.001 > P).
[a]TFA = total fatty acid.
[b]WT = wild type; mg g^{-1} dry matter.
[c]Percentage of wild type.

respect the appropriate protocol is one with high mortality (over 95 per cent or even 99 per cent) and a large number of colonies screened (in the order of thousands) per round of mutation (Calam, 1964, 1970; Hopwood, 1970; Fantini, 1975; Carlton and Brown, 1981). When faced with strain improvement, the desired mutations are 'leaky' or 'minor', with small cumulative effects on metabolite production. The appropriate protocol in this case is one independent of the mutational mortality (Calam, 1970; Fantini, 1975; Crueger and Crueger, 1990) and screening is of a relatively small number of colonies per round, but as many rounds as possible (Davies, 1964; Elander, 1966; Calam, 1970; Rowlands, 1984; Queener and Lively, 1986)

Conclusions

Both the poor scientific knowledge of the biochemistry and genetics of LCPUFAs, and the probable polygenic control of LCPUFA synthesis, currently make genetic engineering a highly speculative approach for strain-improving LCPUFA production by microalgae. On the other hand, the success of many mutational programmes for the production of many different compounds, the successful application of mutation and selection to some microalgae with many different targets, and the first insights on EPA improvement in microalgae by mutation lend support to standard mutational programmes as a way to improve LCPUFA content in microalgae and in other microorganisms, with the additional advantage of simplicity.

Acknowledgements

Several tables and Figure 5.1 are published with the kind permission of the Editor of the *Journal of Phycology* which is gratefully acknowledged. This work has been supported by the Comision Interministerial de Ciencia y Tecnología (CICYT) of the Spanish government (project BIO-91 0652), and by the Spanish dairy company, PULEVA.

References

AHARANOWITZ, Y. and COHEN, G., 1981, Producción microbiológica de fármacos, *Investigación y Ciencia*, **62**, 79–92.

ALIKHANIAN, S.I., 1962, Induced mutagenesis in the selection of microorganisms, *Adv. Appl. Microbiol.*, **4**, 1–50.

ANONYMOUS, 1994, Scotia links with French company to make algal lipids, *Lipid Technology*, **6**, 130–131.

ARAO, T. and YAMADA, M., 1994, Biosynthesis of polyunsaturated fatty acids in the marine diatom, *Phaeodactylum tricornutum*, *Phytochemistry*, **35**, 1177–1181.

ARONDEL, V., LEMIEUX, B, HWANG, I., GIBSON, S., GOODMAN, H.M. and SOMERVILLE, C.R., 1992, Map-based cloning of a gene controlling omega-3 fatty acid desaturation in *Arabidopsis*, *Science*, **258**, 1353–1354.

BACKUS, M.P. and STAUFFER, J.F., 1955, The production and selection of a family of strains in *Penicillium chrysogenum*, *Mycologia*, **47**, 429–463.

BAJPAI, P. and BAJPAI, P.K., 1992, Arachidonic acid production by microorganisms, *Biotech. App. Biochem.*, **15**, 1–10.

BAJPAI, P. and BAJPAI, P.K., 1993, Eicosapentaenoic acid (EPA) production from microorganisms: a review, *J. Biotechnol.*, **30**, 161–183.

BALTZ, R.H., 1986, Strain improvements: Introduction, in Demain, A.L. and Solomon, N.A. (Eds), *Manual of Industrial Microbiology and Biotechnology*, p. 154, Washington, DC: American Society for Microbiology.

BARCLAY, W.R., MEAGER, K.M. and ABRIL, J.R., 1994, Heterotrophic production of long chain omega-3 fatty acids utilizing algae and algae-like microorganisms, *J. Appl. Phycol.*, **6**, 123–129.

BRONSGEEST-SCHOUTE, H.C., VAN GENT, C.M., LUTEN, J.B. and RUITER, A., 1981, The effects of various intakes of ω3 fatty acids on the blood lipid composition in healthy human subjects, *Am. J. Clin. Nutr.*, **34**, 1752–1757.

BROWN, P., 1994, Baby formula needs essential extras, *New Scientist*, November, 1952.

BROWN, L., DUNAHAY, T. and JARVIS, E., 1990, Applications of genetics to microalgae production, *Dev. Ind. Microbiol.*, **31**, 271–274.

BROWSE, J. and SOMERVILLE, C., 1991, Glycerolipid synthesis: biochemistry and regulation, *Annu. Rev. Plant. Physiol. Plant Mol. Biol.*, **42**, 467–506.

CALAM, C.T., 1964, The selection, improvement, and preservation of micro-organisms, *Prog. Ind. Microbiol.*, **5**, 1–53.

CALAM, C.T., 1970, Improvement of micro-organisms by mutation, hybridization and selection, in Norris, J.R. and Ribbons, D.W. (Eds), *Methods in Microbiology*, Vol. 3A, pp. 435–459, New York: Academic Press, Inc.

CARLTON, B.C. and BROWN, B.J., 1981, Gene mutation, in *Manual of Methods for General Bacteriology*, Chapter 13, pp. 22–42, Washington, DC: American Society for Microbiology.

COHEN, Z., NORMAN, H.A. and HEIMER, Y.M., 1995, Microalgae as a source of omega3 fatty acids, in: Simopoulos, A.P. (Ed.), *Plants in Human Nutrition, World Rev. Nutr. Diet*, Vol. 77, pp. 1–31, Basel: Karger.

CRAIG, R. and REICHELT, B.Y., 1986, Genetic engineering in algal biotechnology, *TIBTECH*, November, 280–285.

CRAIG, R., REICHELT, B.Y. and REICHELT, J.L., 1988, Genetic engineering of microalgae, in Borowitzka, M.A. and Borowitzka, L.J. (Eds), *Microalgae Biotechnology*, pp. 416–455, Cambridge: Cambridge University Press.

CRUEGER, W. and CRUEGER, A., 1990, *A Textbook of Industrial Microbiology*, 2nd edn, Sunderland: Sinauer Associates, Inc.

DAVIES, O.L., 1964, Screening for improved mutants in antibiotic research, *Biometrics*, **20**, 576–591.

DE LA NOUE, J. and DE PAUW, N., 1988, The potential of microalgal biotechnology: a review of production and uses of microalgae, *Biotech. Adv.*, **6**, 725–770.

DUNAHAY, T.G., JARVIS, E.J., ZEILER, K.G., ROESSLER, P.G. and BROWN, L.M., 1992, Genetic engineering of microalgae for fuel production, *App. Biochem. Biotech.*, **34/35**, 331–339.

DYERBERG, J., 1986, Linolenate-derived polyunsaturated fatty acids and prevention of atherosclerosis, *Nutrit. Rev.*, **44**, 125–134.

ELANDER, R.P., 1966, Two decades of strain development in antibiotic-producing microorganisms, *Dev. Ind. Microbiol.*, **7**, 61–73.

ELANDER, R.P. and DEMAIN, A.L, 1981, Genetics of microorganisms in relation to industrial requirements, in Rehm, H-J. and Reed, G. (Eds), *Biotechnology*, Vol. 1, Microbial Fundamentals, pp. 235–277, Weinheim: Verlag Chemie GmbH.

ESSER, K. and STAHL, U., 1981, Hybridization, in Rehm, H.-J. and Reed, G. (Eds), *Biotechnology*, Vol. 1, Microbial Fundamentals, pp. 305–329, Weinheim: Verlag Chemie GmbH.

FALCONE, D.L., GIBSON, S., LEMIEUX, B. and SOMERVILLE, C., 1994, Identification of a gene that complements an Arabidopsis mutant deficient in chloroplast ω6 desaturase activity, *Plant Physiol.*, **106**, 1453–1459.

FALCONER, D.S., 1989, *Introduction to Quantitative Genetics*, 3rd edn, Harlow: Longman Scientific & Technical.

FANTINI, A.A., 1975, Strain development, *Methods Enzimol.*, **43**, 24–41.

GARCÍA SÁNCHEZ, J.L., MOLINA GRIMA, E., GARCÍA CAMACHO, F., SÁNCHEZ PÉREZ, J.A. and LÓPEZ ALONSO, D., 1994, Estudio de macronutrientes para la producción de PUFAs a partir de la microalga marina *Isochrysis galbana*, *Grasas y Aceites*, **45**, 323–331.

GIBSON, S, ARONDEL, V., IBA, K. and SOMERVILLE, C., 1994, Cloning of a temperature-regulated gene encoding a chloroplast ω-3 desaturase from *Arabidopsis thaliana*, *Plant Physiol.*, **106**, 1615–1621.

GURR, M.I. and HARWOOD, J.L., 1991, *Lipid Biochemistry: An Introduction*, 4th edn, London: Chapman & Hall.

HARWOOD, J.L., 1988, Fatty acid metabolism, *Ann. Rev. Plant Physiol. Plant Mol. Biol.*, **39**, 101–188.

HARWOOD, J.L., PETTITT, T.P. and JONES, A.L., 1988, Lipid metabolism, in Rogers, L.J. and Gallon, J.R. (Eds), *Biochemistry of the Algae and Cyanobacteria*, pp. 49–67, Oxford: Clarendon Press.

HIGASHI, S. and MURATA, N., 1993, An in vivo study of substrate specificities of acyl-lipid desaturases and acyltransferases in lipid synthesis in *Synechocystis* PCC6803, *Plant Physiol.*, **102**, 1275–1278.

HITZ, W.D., CARLSON, T.J., BOOTH, J.R., KINNEY, A.J., STECCA, K.L. and YADAV, N.S., 1994, Cloning of a higher-plant plastid ω-6 fatty acid desaturase cDNA and its expression in a cyanobacterium, *Plant Physiol.*, **105**, 635–641.

HOPWOOD, D.A., 1970, The isolation of mutants, in Norris, J.R. and Ribbons, D.W. (Eds), *Methods in Microbiology*, Vol. 3A, pp. 363–433, New York: Academic Press, Inc.

HOPWOOD, D.A., BIBB, M.J., CHATER, K.F., KIESER, T., BURTON, C.J., KIESER, H.M., *et al.*, 1985, *Genetic Manipulation in* Streptomyces: *A Laboratory Manual*, Norwich: John Innes Foundation.

IACONO, J.M. and DOUGHERTY, R.M., 1993, Effects of polyunsaturated fats on blood pressure, *Annu. Rev. Nutr.*, **13**, 243–260.

INNIS, S.M., 1991, Essential fatty acids in growth and development, *Prog. Lipid Res.*, **30**, 39–103.

JACOBSON, G.K., 1981, Mutations, in Rehm, H.-J. and Reed, G. (Eds), *Biotechnology*, Vol. 1, Microbial Fundamentals, pp. 279–304, Weinheim: Verlag Chemie GmbH.

JAREONKITMONGKOL, S., SHIMIZU, S. and YAMADA, H., 1992, Fatty acid desaturation-defective mutants of an arachidonic-acid-producing fungus, *Mortierella alpina* 1S-4, *J. Gen. Microbiol.*, **138**, 997–1002.

JAREONKITMONGKOL, S., SAKURADANI, E. and SHIMIZU, S., 1993a, A novel delta5-desaturase-defective mutant of *Mortierella alpina* 1S-4 and its dihomo-gamma-linolenic acid productivity, *Appl. Environ. Microbiol.*, **59**, 4300–4304.

JAREONKITMONGKOL, S., SHIMIZU, S. and YAMADA, H., 1993b, Production of an eicosapentaenoic acid-containing oil by a D12 desaturase-defective mutant of *Morteriella alpina* 1S-4, *J. Am. Oil Chem. Soc.*, **70**, 119–123.

KENDRICK, A. and RATLEDGE, C., 1992, Lipids of selected molds grown for production of n-3 and n-6 polyunsaturated fatty acids, *Lipids*, **27**, 15–20.

KERBY, N.W., MUSGRAVE, S.C., ROWELL, P., SHESTAKOV, S.V. and STEWART, W.D.P., 1986, Photoproduction of ammonium by immobilized mutant strains of *Anabaena variabilis*, *Appl. Microbiol. Biotechnol.*, **24**, 42–46.

KERBY, N.W., NIVEN, G.W., ROWELL, P. and STEWART, W.D.P., 1987, Photoproduction of aminoacids by mutant strains of N$_2$-fixing cyanobacteria, *Appl. Microbiol. Biotechnol.*, **25**, 547–552.

KHOZIN, I., ADLERSTEIN, D., BIGOGNO, C., HEIMER, Y.M. and COHEN, Z., 1997, Elucidation of the biosynthesis of eicosapentaenoic acid in the microalga *Porphyridium cruentum*, *Plant Physiol.*, **114**, 1–8.

KLAUSNER, A., 1986, Algaculture: food for thought, *Bio/Technology*, **4**, 947–953.

KYLE, D., 1989, Market applications for microalgae, *J. Am. Oil Chem. Soc.*, **66**, 648–653.

KYLE, D., 1991, Specialty oils from microalgae: new perspectives, in Rattray, J. (Ed.), *Biotechnology of Plant Fats and Oils*, pp. 130–143, Champaign: American Oil Chemists' Society.

LÓPEZ ALONSO, D., MOLINA GRIMA, E., SÁNCHEZ PÉREZ, J.A., GARCÍA SÁNCHEZ, J.L. and GARCÍA CAMACHO, F., 1992, Isolation of clones of *Isochrysis galbana* rich in eicosapentaenoic acid, *Aquaculture*, **102**, 363–371.

LÓPEZ ALONSO, D., SÁNCHEZ PÉREZ, J.A., GARCÍA SÁNCHEZ, J.L., GARCÍA CAMACHO, F. and MOLINA GRIMA, E., 1993, Improvement of the EPA content in isolates of *Isochrysis galbana*, *J. Mar. Biotech.*, **1**, 147–149.

LÓPEZ ALONSO, D., SEGURA DEL CASTILLO, C.I., GARCÍA SÁNCHEZ, J.L., SÁNCHEZ PÉREZ, J.A. and GARCÍA CAMACHO, F., 1994, Quantitative genetics of fatty acid variation in *Isochrysis galbana* (Prymnesiophyceae) and *Phaeodactylum tricornutum* (Bacillariophyceae), *J. Phycol.*, **30**, 553–558.

LÓPEZ ALONSO, D., SEGURA DEL CASTILLO, C.I, MOLINA GRIMA, E. and COHEN, Z., 1996, First insights into improvement of eicosapentaenoic acid content in *Phaeodactylum tricornutum* (Bacillariophyceae) by induced mutagenesis, *J. Phycol.*, **32**, 339–345.

MATSUSHIMA, P. and BALTZ, R.H., 1986, Protoplast fusion, in Demain, A.L. and Solomon, N.A. (Eds), *Manual of Industrial Microbiology and Biotechnology*, pp. 170–183, Washington, DC: American Society for Microbiology.

MOLINA GRIMA, E., SÁNCHEZ PÉREZ, J.A., GARCÍA SÁNCHEZ, J.L., GARCÍA CAMACHO, F. and LÓPEZ ALONSO, D., 1992, EPA from *Isochrysis galbana*. Growth conditions and productivity, *Process Biochem.*, **27**, 299–305.

MOLINA GRIMA, E., SÁNCHEZ PÉREZ, J.A, GARCÍA CAMACHO, F., GARCÍA SÁNCHEZ, J.L. and LÓPEZ ALONSO, D., 1993, n-3 PUFA productivity in chemostat cultures of microalgae, *Appl. Microbiol. Biotechnol.*, **38**, 599–605.

MOLINA GRIMA, E., SÁNCHEZ PÉREZ, J.A., GARCÍA CAMACHO, F., GARCÍA SÁNCHEZ, J.L., ACIÉN FERNÁNDEZ, G. and LÓPEZ ALONSO, D., 1994, Outdoor culture of *Isochrysis galbana* ALII4 in a closed tubular photobioreactor, *J. Biotechnol.*, **37**, 159–166.

MOLINA GRIMA, E., SÁNCHEZ PÉREZ, J.A., GARCÍA CAMACHO, F., ROBLES MEDINA, A., GIMÉNEZ GIMÉNEZ, A. and LÓPEZ ALONSO, D., 1995, The production of polyunsaturated fatty acids by microalgae: from strain selection to product purification, *Process Biochem.*, **30**, 711–719.

MORENO, V.J., MORENO, J.E.A. and BRENNER, R.R., 1979, Biosynthesis of unsaturated fatty acids in the diatom *Phaeodactylum tricornutum*, *Lipids*, **14**, 15–19.

NELSON, A.M., 1972, Diet therapy in coronary disease, *Geriatrics*, **27**, 103–116.

NETTLETON, J.A., 1993, Are n-3 fatty acids essential nutrients for fetal and infant development? *J. Am. Diet. Assoc.*, **93**, 58–64.

NICHOLS, B.W. and APPLEBY, R.S., 1969, The distribution and biosynthesis of arachidonic acid in algae, *Phytochemistry*, **8**, 1907–1915.

OHLROGGE, J.B., BROWSE, J. and SOMERVILLE, C.R. 1991, The genetics of plant lipids, *Biochim. Biophys. Acta*, **1082**, 1–25.

OKULEY, J., LIGHTNER, J., FELDMANN, K., YADAV, N., LARK, E. and BROWSE, J., 1994, Arabidopsis FAD2 gene encodes the enzyme that is essential for polyunsaturated lipid synthesis, *Plant Cell*, **6**, 147–158.

PARKE, M., 1949, Studies of marine flagellates, *J. Mar. Biol. Ass. UK*, **28**, 255–286.

PLAKAS, S.M. and GUARINO, A.M., 1986, Omega-3 fatty acids and fish oils: is the news all good? *Proceedings of the Eleventh Annual Tropical and Subtropical Fisheries Conference of the Americas*, 275–283.

QUEENER, S.W. and LIVELY, D.H., 1986, Screening and selection for strain improvement, in Demain, A.L. and Solomon, N.A. (Eds), *Manual of Industrial Microbiology and Biotechnology*, pp. 155-169, Washington, DC: American Society for Microbiology.

RADMER, R.J., 1990, Omega-3 fatty acids from algae, in Lee, R.S. and Karel, M. (Eds), *Omega-3 Fatty Acids in Health and Disease*, pp. 211–214, New York: Marcel Dekker, Inc.

RADWAN, S.S., 1991, Sources of C20-polyunsaturated fatty acids for biotechnological use, *Appl. Microbiol. Biotechnol.*, **35**, 421–430.

RATLEDGE, C., 1993, Single cell oils – have they a biotechnological future? *TIBTECH*, **11**, 278–284.

REDDY, A.S., NUCCIO, M.L., GROSS, L.M. and THOMAS, T.L., 1993, Isolation of a delta-6-desaturase gene from the cyanobacterium *Synechocystis* sp. strain PCC 6803 by gain-of-function expression in *Anabaena* sp. strain PCC 7120, *Plant Mol. Biol.*, **27**, 293–300.

ROWLANDS, R.T., 1984, Industrial strain improvement: mutagenesis and random screening procedures, *Enzyme Microb. Technol.*, **6**, 3–10.

SAKAMOTO, T., WADA, H., NISHIDA, I., OHMORI, M. and MURATA, N., 1994, Delta-9 acyl-lipid desaturases of cyanobacteria, *J. Biol. Chem.*, **269**, 25576–25580.

SCHNEIDER, J.C. and ROESSLER. P., 1994, Radiolabeling studies of lipids and fatty acids in *Nannochloropsis* (Eustigmatophyceae), an oleaginous marine alga, *J. Phycol.*, **30**, 594–598

SCHNEIDER, J.C., LIVNE, A., SUKENIK, A. and ROESSLER, P.G., 1995, A mutant of *Nannochloropsis* deficient in eicosapentaenoic acid production, *Phytochemistry*, **40**, 807–814.

SCHWEIZER, E., 1989, Biosynthesis of fatty acids and related compounds, in Ratledge, C. and Wilkinson, S.G. (Eds), *Microbial Lipids*, Vol. 2., pp. 3–50, London: Academic Press.

SHAISH, A., BEN-AMOTZ, A. and AVRON, M., 1991, Production and selection of high β-carotene mutants of *Dunaliella bardawil* (Chlorophyta), *J. Phycol.*, **27**, 652–656.

SHIFRIN, N.S. and CHISHOLM, S.W., 1981, Phytoplankton lipids: interspecific differences and effects of nitrate, silicate, and light-dark cycles, *J. Phycol.*, **17**, 374–384.

SIMON, R.D. and LATORELLA, A.H., 1986, Approaches to the genetics of *Dunaliella*, in Cramer, J. (Ed.), *Algal Biomass Technologies*, Vol. 83, pp. 15–20, Berlin: Nova Hedwigia Beih.

SIMOPOULOS, A.P., 1991, Omega-3 fatty acids in health and disease and in growth and development, *Am. J. Clin. Nutr.*, **54**, 438–463.

SINGH, G. and CHANDRA, R.K., 1988, Biochemical and cellular effects of fish and fish oils, *Progress in Food and Nutrition*, **12**, 371–419.

SINGH, H.N., SINGH, R.K. and SHARMA, R., 1983, An L-methionine-D, L-sulfoximine mutant of the cyanobacterium *Nostoc muscorum* showing inhibitor-resistant gamma-glutamyl-transferase, defective glutamine synthetase and producing extracellular ammonia during N_2 fixation, *FEBS Lett.*, **154**, 10–14.

SIVAN, A., VAN MOPPES, D. and ARAD (MALIS), S., 1992, Protoplast production from the unicellular red alga *Porphyridium* sp. (UTEX 637) by an extracellular bacterial lytic preparation, *Phycologia*, **31**, 253–261.

SIVAN, A., THOMAS, J.-C., DUBACQ, J.-P., VAN MOPPES, D. and ARAD, S., 1995, Protoplast fusion and genetic complementation of pigment mutations in the red microalga *Porphyridium* sp., *J. Phycol.*, **31**, 167–172.

SOKAL, R.R. and ROHLF, F.J., 1981, *Biometry*, San Francisco: W.H. Freeman.

SOMERVILLE, C. and BROWSE, J. 1991, Plant lipids: metabolism, mutants, and membranes, *Science*, **252**, 80–87.

SPILLER, H., LATORRE, C., HASSAN, M.E. and SHANMUGAM, K.T., 1986, Isolation and characterization of nitrogenase-derepressed mutant strains of cyanobacterium *Anabaena variabilis*, *J. Bact.*, **165**, 412–419.

THOMPSON, P.A., HARRISON, P.J. and WHYTE, J.N.C., 1990, Influence of irradiance on the fatty acid composition of phytoplankton, *J. Phycol.*, **26**, 278–288.

THOMPSON, P.A., GUO, M.-X., HARRISON, P.J. and WHYTE, J.N.C., 1992, Effects of variation in temperature. II. On the fatty acid composition of eight species of marine phytoplankton, *J. Phycol.*, **28**, 488–497.

VAN GENT, C.M., LUTEN, J.B, BRONGEEST-SCHOUTE, H.C. and RUITER, 1979, Effect on serum lipid levels of omega-3 fatty acids, of ingesting fish-oil concentrate. *Lancet*, December, 1249–1250.

WADA, H., GOMBOS, Z. and MURATA, N., 1990, Enhancement of chilling tolerance of a cyanobacterium by genetic manipulation of fatty acid desaturation, *Nature*, **347**, 200–203.

WHO/FAO, 1977, Dietary fats and oils in human nutrition, Report of an expert consultation, Rome: FAO.

YADAV, N.S., WIERZBICKI, A., AEGERTER, M., CASTER, C.S., PÉREZ-GRAU, L., KINNEY, A.J., *et al*, 1993, Cloning of higher plant omega-3 fatty acid desaturases, *Plant Physiol.*, **103**, 467–476.

YONGMANITCHAI, W. and WARD, O.P., 1989, Omega-3 fatty acids: alternative source of production, *Process Biochem.*, **24**, 117–125.

YONGMANITCHAI, W. and WARD, O.P., 1991, Screening of algae for potential alternative sources of eicosapentaenoic acid, *Phytochemistry*, **30**, 2963–2967.

6

Recovery of algal PUFAs

E. MOLINA GRIMA, A. ROBLES MEDINA AND A. GIMÉNEZ GIMÉNEZ

Introduction

Fish oils are the main source of PUFAs. Fish such as cod, menhaden, herring, anchovy and sardine contain a high proportion of fat and are used to produce PUFAs. Fish oils derived PUFAs have several problems: an unpleasant odour, contamination with heavy metals (Plakas and Guarino, 1986), presence of cholesterol (Cohen and Cohen, 1991; Bajpai and Bajpai, 1993), variable production and a complex fatty acid profile (up to 40 different fatty acids may be present). In contrast PUFAs from microalgae lack these disadvantages, having no unpleasant odour, heavy metals, or cholesterol; stable production (cultured under controlled conditions); and a simpler fatty acid composition facilitates purification. To illustrate these differences we have compared the fatty acid compositions of cod liver oil and the microalgae *Isochrysis galbana* (Robles Medina *et al.*, 1995a), *Phaeodactylum tricornutum* (Cartens *et al.*, 1996) and *Porphyridium cruentum* (Giménez Giménez *et al.*, 1998) (Table 6.1).

The interrupted double bonds of methylene of PUFAs are very easily attacked by oxygen to form hydroperoxides with conjugated double bonds. Formation of these hydroperoxides can be detected by UV absorption at 233–234 nm and 268–275 nm, due, respectively, to diene and triene conjugated double bonds (Mehlenbacher, 1970). Once PUFA oxidation has started, it can be propagated by a series of autocatalytic reactions. The oxidation rate is accelerated by light, heat, oxygen, moisture and the presence of catalysts, such as metal ions (Lundberg and Järvi, 1966). The exposure of PUFAs to these agents must be minimized during the process of their isolation and therefore must be done under an inert atmosphere (nitrogen or argon).

Once PUFAs or lipids are obtained, they must be protected against oxidation during storage. For temporary storage, diluting the sample with a solvent such as petroleum ether (low oxygen solubility), may greatly reduce the rate of oxidation. Ideally, the samples should be sealed under vacuum in ampoules with minimum headspace. Extended storage of several years requires temperatures of −40 to −80°C. Antioxidant additives can effectively slow oxidation of PUFAs. Effective antioxidants are tertiary butylhydroquinone (TBHQ), butylated hydroxytoluene (BHT) and octyl gallate, at concentrations of 0.01–0.02 per cent. In view of potential toxicity of these synthetic antioxidants, use of natural ones obtained from the resi-

Table 6.1 Fatty acid composition (percentage of total fatty acids) of cod liver oil, *Isochrysis galbana* (Robles Medina *et al.*, 1995a), *Phaeodactylum tricornutum* (Cartens *et al.*, 1996) and *Porphyridium cruentum* (Giménez Giménez *et al.*, 1998) microalgae and total fatty acid (FA) percentage of biomass dry weight.[a]

Fatty acid	Cod liver oil	*I. galbana*	*P. tricornutum*	*P. cruentum*
14:0	4.04	10.10	5.44	1.90
16:0	12.54	20.30	15.47	26.60
16:1n-7	6.50	21.40	17.29	1.80
16:2n-4	1.01	–	1.41	–
16:3n-4	0.13	–	1.18	–
16:4n-1	0.61	–	6.85	–
18:0	2.92	0.65	0.32	0.80
18:1n-9	18.49	1.40	0.51	1.00
18:1n-7	0.30	3.60	0.76	1.00
18:2n-6	1.51	0.90	2.21	6.20
18:3n-6	0.19	0.20	0.56	–
18:3n-3	0.82	1.20	0.33	1.00
18:4n-1	0.44	–	–	–
18:4n-3 (SA)	2.51	6.40	0.35	0.50
20:0	0.22	–	–	–
20:1n-9	12.45	0.15	–	–
20:2n-6	2.66	–	–	–
20:4n-6 (AA)	2.69	0.70	3.37	23.00
20:4n-3	0.79	–	0.37	–
20:5n-3 (EPA)	12.45	22.60	29.83	23.90
22:1n-11	6.66	0.10	–	–
22:3n-3	0.43	–	–	–
22:4n-6	0.38	1.30	0.25	–
22:5n-3	1.53	0.20	–	–
22:6n-3 (DHA)	9.91	8.40	0.81	0.20
FA (% d.w.)[a]	–	9.50	10.50	4.13

nous residues of the first phase of pollen digestion by bees, has been suggested (Méndez *et al.*, 1993).

Typically, a bioseparation involves: removal of insolubles, isolation of products, purification and polishing (Belter *et al.*, 1988). The first step is usually by filtration or centrifugation. Microalgae cultivated in photobioreactors are recovered from the culture broth by one of these two techniques, but centrifugation is more commonly used. For example, Molina Grima *et al.* (1996) harvested *Phaeodactylum tricornutum* by centrifugation to obtain a wet paste containing 21 per cent dry biomass. The latter was processed immediately to extract the lipid fraction.

The isolation of products is generally a relatively non-specific step to remove materials of widely divergent properties as compared to the desired product. Appreciable concentration usually follows. Adsorption and solvent extraction are typical operations in this step. To obtain PUFAs from microalgal biomass the first step is the extraction of the lipids and/or fatty acids by solvents.

The crude extract is purified by highly selective methods such as chromatography and precipitation. Because of the complexity of the mixtures, recovery of highly pure PUFAs is a multistep process. Initially a gross fractionation may be used to concentrate the unsaturated fatty acids or the one with the longest chain. A second step is then used to recover the PUFA of interest. The most developed separation techniques include formation of urea inclusion compounds, low temperature fractional crystallization, salt solubility methods, liquid–liquid fractionation and liquid chromatography. However, not all these methods are suited to preparative or production scale. These techniques usually fractionate the fatty acid based on the number of double bonds or chain length; few of these methods are capable of providing an individual fatty acid from a natural source. The urea inclusion compound method is the procedure generally of choice for concentration of PUFA. However the most efficient technique for purifying fatty acids is liquid chromatography. Some other potential by useful method are also noted below.

'Polishing' is a final purification step that in many cases, depending on the end use of the product, may not be required. It can be performed in such operations as crystallization or drying. PUFAs can also be purified by some of the techniques mentioned above (e.g., low temperature fractional crystallization and liquid chromatography techniques).

Low temperature fractional crystallization may be used to concentrate or to purify PUFAs. Usually the fatty acids containing three or more double bonds are concentrated in the filtrate fraction and are re-concentrated by further crystallizations at lower temperatures. When PUFAs become major components, they crystallize together with a small amount of more saturated fatty acids, remaining in the PUFA solution. Therefore, there is a practical limit to the extent to which the PUFAs can be concentrated in the filtrate. For example, crystallization of the esters of arachidonic acid (AA) from the mixed acids of beef suprarenal phosphatides in acetone at $-20°C$ or in methanol at $-70°C$ produced a filtrate containing 50–60 per cent methyl arachidonate (Shinowara and Brown, 1940). Concentration and chilling to $-70°C$ resulted in a final filtrate containing 70–75 per cent AA esters.

PUFAs may also be crystallized and purified by recrystallization from their solutions in organic solvents after they have been obtained in a highly concentrated form. Final purification by crystallization is therefore generally applied to products that are at least 98 per cent pure. Minute impurities may have a profound effect on physical properties such as melting point; therefore purification by crystallization is mandatory for using the final products as physical standards (Privett, 1968).

Fractionation of metal salts of fatty acids in solvents is one of the oldest methods reported. The solubility of a number of fatty acid salts in organic solvents depends on the nature of the metal ion, the chain length, double bond number, temperature, as well as another factors. Several solvent–salt combinations have been described (Markley, 1964). The most used method is based on the solubility of lithium soaps in acetone and alcohol. The lithium salts of polyenoic fatty acids are soluble in 95 per cent acetone whereas less unsaturated acids are relatively insoluble in this medium. Thus polyenoic acids may be separated from monoenoic and saturated fatty acids. Toyama and co-workers used the lithium salt–alcohol method for isolating octadecatrienoic acid from sardine oil (Markley, 1964).

A simple isolation procedure to purify docosahexaenoic acid (DHA) and eicosapentaenoic acid (EPA) is the iodolactonization method. This process is based on the principle that the rate of formation of iodo γ-lactone derived from DHA, is significantly higher than that of iodo δ-lactone derived from EPA. This offers an opportunity for obtaining pure DHA from fatty acid concentrates. Wright *et al.* (1987) and Corey and Wright (1988) reported a simple extractive isolation of DHA and EPA, respectively, from cod liver oil. Gaiday *et al.* (1991) prepared pure AA from a mixture of AA, DHA and other fatty acids from sardine oil and rat liver using this reaction.

Among the liquid–liquid fractionation techniques, the extraction of PUFAs with aqueous silver nitrate solutions and with acetone at low temperature seems the most recently used. The former is a comprehensive method for selective extraction of PUFAs (Teramoto *et al.*, 1994). The ethyl and methyl esters of PUFAs (from fish oils) could be extracted from various organic solvents (hexane, heptane, dodecane, benzene, toluene, etc.) into aqueous silver nitrate solutions. The PUFA ethyl ester distribution ratio depended on the number of double bonds and increased drastically with a decrease in temperature or by addition of a water-soluble alcohol such as methanol, ethanol, and *n*-propanol to aqueous solution. Solvent extraction of ethyl esters of PUFAs, such as EPA and DHA, was also performed with an aqueous silver nitrate solution using a microporous hydrophobic hollow-fibre membrane extractor (Kubota *et al.*, 1997). It was demonstrated that EPA and DHA ethyl esters could be satisfactory separated from ethyl esters of lower fatty acids such as oleic and palmitic acids. The permeation rate of fatty acid ethyl esters was controlled by the diffusion across the membrane and the laminar boundary layer in the inner and outer sides of the hollow fibre.

Moffat *et al.* (1993) have concentrated polyunsaturated triglycerides by extraction with acetone at low temperature. The method involves rapid solidification of fish oil droplets in liquid nitrogen followed by extraction with acetone at $-60°C$. Triglycerides containing only the saturated fatty acids or a combination of saturated and monoenoic acids were totally removed. Significant increases were observed in EPA and DHA concentration. The minimum combined percentage of these two fatty acids was 41.7 per cent (25.8 per cent in original oil) for menhaden oil, while a maximum of 57.4 per cent (34.4 per cent in original oil) was recorded for enriched anchovy oil. Free fatty acids have also been concentrated by this process, but this requires further investigation.

Besides high pressure liquid chromatography, some authors fractionated fatty acids using the open column chromatography. Silicic acid column chromatography is a simple, rapid method for separating highly polyunsaturated triglycerides and is applicable to large-scale preparation (Hayashi and Kishimura, 1993). This technique has been used to separate triglycerides from a raw lipid extract according to the differences in their degree of unsaturation. EPA-enriched triglycerides were separated from raw scallops, sardine and squid lipids by column chromatography silicic acid using 5 per cent diethyl ether in an *n*-hexane eluent. Five to six fractions were obtained, the last containing considerable amounts of EPA (47–58 per cent for the scallop and 32–40 per cent for the sardine and squid). On the other hand, reverse-phase open column chromatography is cheaper than the high-pressure reverse-phase chromatography. Cohen and Cohen (1991) obtained up to 93 per cent pure EPA fractions from a mixture of fatty acid methyl esters of the glycolipid fraction of *Porphyridium cruentum*.

Supercritical fluid extraction (SFE) has a promising future in the selective extraction of PUFAs and in their purification. For food application CO_2 is the solvent of choice because it is non-toxic, non-flammable, readily available and relatively inexpensive. Extraction with CO_2 is carried out at moderate temperatures (critical temperature and pressure are $31°C$ and 7.3 MPa), which limits autoxidation, decomposition and/or polymerization of the PUFAs present in fish oil.

Furthermore the inert atmosphere of CO_2 excludes oxygen, the cause of autoxidation. Supercritical fluid technology has been used for the extraction and production of PUFA concentrates from fish oil and for the enrichment of PUFAs by supercritical fluid chromatography (SFC). A very comprehensive review was published by Mishra *et al.* (1993).

Extraction with supercritical (SC) CO_2 at pressures of 13.8–24.1 MPa and temperatures of 24–80°C causes the oil esters to separate, mostly according to chain length, but also somewhat according to the degree and kind of unsaturation present. At higher temperatures, SC CO_2 becomes less dense and solubility of the esters is decreased; the shorter chain components become more soluble than the longer chain ones, thereby enhancing the selectivity of the process. Passing SC CO_2 through an extraction vessel charged with a mixture of fish oil esters, may be used to obtain fractions containing as much as 57–64 per cent DHA and/or 19–23 per cent EPA, starting with fish oil esters containing only 8–10 per cent of these fatty acids (Stout and Spinelli, 1987). Nilsson *et al.* (1988) attempted the fractionation of a mixture of urea-pretreated fatty acid esters from menhaden oil by isobaric fractionation with a temperature gradient in a column. A product containing EPA and DHA in purities exceeding 90 per cent was obtained.

The enrichment of PUFA by SFC is more efficient than by SFE because of the introduction of another auxiliary phase. The methyl esters of sardine fatty acids were extracted with SC CO_2 at 8 MPa and 40°C (Higashidate *et al.*, 1990). SFC was carried out in a silica gel column (12.1 × 1 cm) coated with silver nitrate. An increasing pressure programming sequence was used in the presence of ethanol as a modifier of CO_2. Five fractions were collected sequentially. The fraction III contained 93 per cent EPA and no DHA, whereas fraction V had 82 per cent DHA and with no EPA (EPA and DHA content in the starting material were 12 and 13 per cent, respectively). The order of elution was in accordance with interactions of the C = C double bond with silver ions in the stationary phase.

Despite the advantages, supercritical fluid technology has been little used with algae. Choi *et al.* (1987) extracted lipids from the freeze-dried green algae *Scenedesmus obliquus*. SC CO_2 was first used to extract some lipids and then the residue was re-extracted with 15 per cent ethanol as a modifier in SC CO_2. The first extraction yielded neutral lipids with some glycolipids; ethanol increased the solubility of polar lipids. The extraction of lipids from *Skeletonema costatum* (5–10 per cent oil, dry basis) and *Ochromonas danica* (30 per cent) with SC CO_2 was investigated by Polak *et al.* (1989). The extraction was carried out at 17–31 MPa and 40°C. The solubility of lipids from freeze-dried algae reached a maximum at 24 MPa. Chlorophyll did not dissolve under the conditions of extraction.

For medical or dietetic purposes, PUFAs may be administered as free fatty acids (FFA) or triacylglycerols (TAG), both forms being similarly absorbed (El Boustani *et al.*, 1987; Lawson and Hughes, 1988). However TAG would probably be the chemical form of choice for pharmaceutical preparations. Perhaps the best way to obtain highly PUFA-enriched TAG is by lipase-catalyzed reactions. TAG enriched PUFAs may be synthesized by enzymatic esterification of PUFAs with glycerol, interesterification of fats with PUFAs (highly pure or concentrates), and hydrolysis of marine oils. To date no work has been done using PUFAs or lipids from algae to obtain TAG enriched in PUFAs.

A greater TAG yield from PUFA, using immobilized lipase from *Candida antarctica* or *Rhizomucor miehei*, has been obtained by esterification of glycerol. More than

95 per cent of PUFA (EPA, DHA and AA) was converted to TAG by esterification at 50–60°C with shaking and dehydration for 24 hours. Pure tridocosahexaenoyl, trieicosapentaenoyl and triarachidonyl glycerol were isolated after passing the product through a basic aluminium oxide column (Kosugi and Azuma, 1994). A PUFA enriched TAG yield of 84.7 per cent containing 27.4 per cent EPA and 45.1 per cent DHA was obtained from cod liver oil PUFA concentrate using lipase from *Candida antarctica* (Esteban Cerdán *et al.*, 1998).

Interesterification is based on exchanging the fatty acid components of a TAG or a TAG mixture with either free fatty acids (FFA) (acidolysis) or the fatty acids from other TAG or monoesters. Lipase catalyzed interesterification of cod liver oil with PUFA concentrate (39 per cent EPA and 24 per cent DHA), using the immobilized lipase lipozyme from *Mucor miehei* achieved a TAG product of 31 per cent EPA and 20 per cent DHA, when equilibrium was reached (Haraldsson *et al.*, 1989). Using immobilized lipase IM60 from *Mucor miehei* in an organic solvent, 10–18 per cent of EPA or DHA were incorporated into various edible oils (Zu-Yi and Ward, 1993).

Enzymatic hydrolysis of oils produced a mixture of monoacylglycerols, diacylglycerols, TAG and FFA. Most of the work in this area aims to increase the PUFA content of the remaining glycerides. Tuna oil (30.3 per cent DHA and 8.2 per cent EPA) was treated with *Geotrichum candidum* and with *Candida cylindracea* lipases. By repeated hydrolysis, the *G. candidum* lipase produced glycerides containing 46.6 per cent DHA and 10.9 per cent EPA, while the *C. cylindracea* lipase produced glycerides containing 58.5 per cent DHA and 7.1 per cent EPA (did not concentrate EPA). Moreover the TAG content in glycerides produced with *G. candidum* lipase (85.5 per cent) was higher than with *C. cylindracea* lipase (74.2 per cent) (Shimada *et al.*, 1994).

Lipase catalyzed reactions also can be used to purify PUFAs. DHA was purified by a two-step enzymatic method that consisted of hydrolysis of tuna oil and selective esterification of the resulting FFA. *Pseudomonas* sp. lipase showed stronger activity on the hydrolysis of DHA ester than on the EPA ester and was suitable for preparation of FFA rich in DHA; 83 per cent of DHA in tuna oil was recovered in the FFA fraction at 79 per cent hydrolysis. The selective esterification of FFA was carried out with *Rhizopus delemar* lipase and lauryl alcohol; thus the DHA content in the FFA fraction was raised to 91 per cent in 88 per cent yield by two esterifications (Shimada *et al.*, 1997a). An identical procedure was used to obtain a 93.7 per cent pure γ-linoleic acid (GLA) from borage oil (Shimada *et al.*, 1997b).

A general overview of lipid and fatty acid extraction

Lipid classes in microalgae: fractionation of lipids

The algae produce a great variety of fatty acids and lipids. The main fatty acids are saturates and the *cis* isomers of unsaturates, with 12 to 22 carbons and up to six double bonds. Extraction of lipids from algae requires attention to their polarity. Polarity is related to distribution of lipids within the algal cell and association of lipids to non-lipid components. Thus lipids present in algae can be classified into neutral and polar lipids that are subdivided as indicated in Table 6.2. PUFAs may be present in TAG, phospholipids and glycolipids. Triglycerides are usually regarded as

Table 6.2 Classification of lipids (Pohl and Zurheide, 1982).

| Neutral lipids | Polar lipids | |
	Phospholipids	Glycolipids
Triglycerides	Phosphatidylcholine	Sulphoquinosyldiglyceride
Wax esters	Phosphatidylethanolamine	Monogalactosyldiglyceride
Hydrocarbons	Phosphatidylserine	Digalactosyldiglyceride
Free fatty acids	Phosphatidylglycerol	
Sterols	Phosphatidylinositol	

energy storage products, whereas phospholipids and glycolipids are structural lipids contained in the cell walls.

The lipid fractionation procedure depends largely on the particular classes of lipids present. Usually, most animal and microbial cellular lipids consist of about 60–85 per cent polar lipids, whereas seed lipids have a considerably higher proportions of neutral lipids (Kates, 1988). Lipid extracts rich in polar lipids are commonly fractionated by countercurrent distribution, ion exchange chromatography and adsorption chromatography on silicic acid and aluminium oxide. Adsorption chromatography on silicic acid is specially suitable. This technique ensures the separation of non-polar from polar lipids and separation of the latter into fractions. The crude extract is applied to a silica gel column (250–400 mesh) and solvents with increasing polarity (e.g. chloroform, acetone and methanol) are used to elute increasingly polar lipids: neutral lipids, low-polarity lipids (e.g. galactolipids) and polar lipids (such as phospholipids) (Kates, 1988). For microalgae, the expected composition of the various solvent fraction is:

- Chloroform: hydrocarbons, carotenoids, chlorophylls, sterols, triglycerides, wax esters, long-chain alcohols, aldehydes and free fatty acids.
- Acetone: mono and digalactosyl diacylglycerols, cerebrosides and sulpholipids.
- Methanol: phospholipids and traces of glycolipids.

As an example, the fractionation of the lipid extract from *I. galbana* and *P. tricornutum* into neutral lipids and polar lipids is shown in Table 6.3. In both microalgae, a greater proportion of PUFAs are found in the polar lipid fraction. Consequently,

Table 6.3 Neutral and polar lipid contents and EPA distribution of *I. galbana* (Molina Grima et al., 1994) and *P. tricornutum* (Cartens et al., 1996).

| Lipids | *I. galbana* | | *P. tricornutum* | |
	% of total lipids	% of total EPA	% of total lipids	% of total EPA
Neutral	26.5	28.6	23.2	15.1
Glycolipid	–	–	49.1	56.7
Phospholipid	–	–	27.7	28.2
Polar	73.5	71.4	76.8	84.9

some authors use fractionation into neutral and polar fractions as a first step in recovery of highly pure PUFAs (Cohen and Cohen, 1991).

Pretreatment of the samples

Lipids from microorganisms may generally be extracted in the wet state directly after harvesting. The cells do not need to be homogenized since they are readily broken by suspending in the extracting solvent. In some cases, cell breakage may be necessary to allow better penetration of the solvent into the cell and increase the lipid yield. This may be accomplished using one of the established cell-disruption techniques such as sonication, homogenization, freezing and grinding.

Cell disruption has been reviewed by Chisti and Moo-Young (1986) and a recent comprehensive study of disruption in bead mills has also been reported (Garrido *et al.*, 1994). For large-scale cell disruption of algae a mechanical shearing in a bead mill appears most suitable. Bead mills have been used to disintegrate the algae *Scenedesmmus obliquus* and *Spirulina platensis* (Hedenskog and Ebbinghaus, 1972).

Ultrasonification is another potentially useful disruption method. Ultrasounds at high acoustic intensities disrupt microbial cells in suspension (Chisti and Moo-Young, 1986). Dunstan *et al.* (1992) extracted lipids from green algae (Chlorophyceae and Prasinophyceae) with chloroform:methanol:water (1:2:0.8 v/v/v) mixture and between each extraction the samples were sonicated at 20°C.

Extraction from freeze-dried samples has also been used. Freeze-drying breaks up the cells and turns the algal material into a loose, fine powder, making homogenization unnecessary (Ahlgren and Merino, 1991). However, direct extraction of fatty acid from wet *P. tricornutum* biomass (after harvesting by centrifugation) with 96 per cent ethanol produced only slightly lower yields (90 per cent) than those obtained from lyophilized biomass (96 per cent); therefore cost of extraction may be reduced by omitting lyophilization (Molina Grima *et al.*, 1996).

Extraction of lipids and fatty acids

Extraction methods should be fast, efficient and gentle in order to reduce degradation of the lipids or fatty acids. The extraction solvents should be inexpensive, volatile (for ready removal later), free from toxic or reactive impurities (to avoid reaction with the lipids), able to form a two-phase system with water (to remove non-lipids), and be poor extractors of undesirable components (e.g., proteolipid protein and small molecules). Effectiveness of solvents for the different classes of lipids should also be considered.

For complete extraction, all the linkages between the lipids and other non-lipid cell components must be broken and, at the same time, the disruption agents used must not degrade the lipids. There are three main types of associations in which lipids participate: (a) hydrophobic or van der Waals interactions, in which neutral or non-polar lipids, such as glycerides, are bound by relatively weak forces through their hydrocarbon chains to other lipids and to hydrophobic regions of proteins; (b) hydrogen bonding and electrostatic association by which polar lipids are bound to proteins; and (c) covalent association, although this type of interaction is less frequent. The energy of the weak hydrophobic interactions which link the stored lipids, never exceeded 2 kcal/mole, so they may be disrupted by non-polar organic solvents, such as chloroform, hexane or ether (Chuecas and Riley, 1969). Hydrogen bonds of membrane-associated polar lipids have an energy of 0.2–12 kcal/mole and may be disrupted only by polar organic solvents, such as methanol, ethanol and other alcohols, and also by water (a solvent having a high dielectric constant). To extract lipids linked by stronger electrostatic forces such as ionic bonds, it is necessary to shift the pH value somewhat towards the acidic or alkaline region (Zhukov and Vereshchagin, 1981; Kates, 1988).

Lipids are retained in living matter also by mechanical confinement. For example, the poor permeability of the cell walls to solvents hinders extraction. In such cases the lipid yield may be increased by adding a small amount of water to the extractant. Water causes swelling of the cellular structures rich in polysaccharides, thereby facilitating access of extracting solvents to the lipids. Thus the presence of water in the extractant is absolutely necessary for the quantitative extraction of polar lipids (Zhukov and Vereshchagin, 1981).

Some biological materials contain enzymes that degrade lipids during extraction. Typically, alcohol-containing solvent mixtures inactivate many of the lipid-degrading phosphatidases and lipases (Zhukov and Vereshchagin, 1981; Kates, 1988). Alcohol also aids disruption of lipid–protein complexes and dissolution of the lipids. However, the alcoholic solvents also extract some cellular contaminants such as sugars, amino acids, salts, hydrophobic proteins and pigment. Therefore crude alcoholic extract must be treated to remove these water-soluble contaminants (Kates, 1988).

Methods of purifying crude lipid extract are usually based on differences in affinity of the polar lipids and the contaminants for a certain solvent. The crude extract may be treated with apolar solvents such as chloroform (Folch *et al.*, 1957, Bligh and Dyer, 1959; Kates, 1988), hexane (Molina Grima *et al.*, 1994) or diethyl ether, in which the non-lipid contaminants are less soluble. These procedures do not completely extract most polar lipids (e.g., proteolipids) because of their low solubility in these solvents. Other less used purification methods are molecular adsorption on silicic acid or alumina and the use of ion-exchange and gel filtration (Zhukov and Vereshchagin, 1981). None of these methods guarantee recovery of completely pure lipids.

An efficient extraction system for microorganisms is a 2:1 v/v mixture of chloroform:methanol (Folch *et al.*, 1957; Kochert, 1978). In a simplified and improved procedure, Bligh and Dyer (1959) used a single-phase extractant, chloroform:methanol:water (1:2:0.8 by v/v/v) for extraction of lipids from fresh tissues. The single phase extraction mixture was diluted with chloroform and water to yield a biphasic system (2:2:1.8 ratio of chloroform:methanol:water) chosen in such a manner that the lower layer was almost pure chloroform and contained the purified lipids; the upper layer was nearly all methanol–water and contained the non-lipids. The extraction was fast and gentle (room temperature). Lipid extraction was complete and the separation of lipids and non-lipids was nearly quantitative. That method is applicable to a variety of materials (animal or plant tissue, as microorganisms), but with microorganisms (including microalgae), homogenization is not necessary and the extraction takes almost two hours at room temperature (Kates, 1988). This procedure is routinely used for lipids extraction from microalgae. Among the disadvantages of the procedure are the toxicity and harmability of the solvents (chloroform and methanol) (Hara and Radin, 1978).

Hexane:isopropanol (3:2 v/v) is a low-toxicity solvent. Lipid extraction with this system followed by a wash of the extract with aqueous sodium sulphate to remove the non-lipid contaminants has some advantages over the Bligh and Dyer system. The extract contains less non-lipid material, interference in processing by proteolipid protein contamination is avoided, the two phases separate rapidly during the washing step, the solvent density is low enough to permit centrifugation of the homogenate as an alternative to filtration, the solvents are cheaper, the solvents can be eliminated by vacuum evaporation to dryness, and the washed extract can be directly applied to a chromatographic column with UV detection. Hara and Radin (1978) used this system to extract lipids from rat or mouse brain; the yield was very similar to that with chloroform:methanol:water (1:2:0.8 v/v/v). However, the hexane:isopropanol system gave substantially lower lipid yields when used with microalgae (Ahlgren and Merino, 1991; Molina Grima *et al.*, 1994).

Butanol and ethanol are also low-toxicity solvents that are relatively cheap, sufficiently volatile and only slightly carcinogenic, but the lipid extract needs to be further purified, for example, by a chloroform:methanol:water phase separation. Nagle and Lemke (1990) compared the extraction efficiency of these solvents and hexane:isopropanol (2:3 v/v) with the Bligh and Dyers (1959) system (control method). Tests were performed on the diatom *Chaetoceros muelleri*. The most effective system was butanol (90 per cent of yield obtained with chloroform:methanol: water 1:2:0.8 v/v/v); the yield with hexane:isopropanol (2:3 v/v) was 78 per cent and with ethanol the yield was 73 per cent.

Recovery of lipids from the microalga *I. galbana* using various solvent systems is detailed in Table 6.4. This single-phase extractant was used, first, to extract lipids from *I. galbana* lyophilized biomass and then to purify them by adding hexane and water to form a biphasic system in a proportion chosen in such a manner that the lower layer is practically 100 per cent hexane and contains the purified lipids and the upper layer, nearly all ethanol–water, contains the non-lipids. Although a high lipid yield (79.6 per cent) was obtained, a high proportion of lipids (27.4 per cent) remained in the ethanol–water phase together with the non-lipid contaminants (Table 6.4), implying that a greater consumption of hexane is required to extract them. As shown in Table 6.4 the lipid yield increased with increasing alcohol content or alcohol polarity in the extraction solvent mixture. These results agree with the fractionation behaviour *I. galbana* lipids shown in Table 6.3 (Molina Grima *et al.*, 1994).

To obtain free fatty acids from lipidic extracts, the extracts must be saponified. An alkali added to the extraction mixture can extract the fatty acids directly (Molina Grima *et al.*, 1994) (Table 6.5). Such direct saponification during extraction from the biomass is faster and less expensive than separate extraction followed by saponification, but more severe operating conditions are necessary (1 h at 60°C or 8 h at room temperature). The resulting fatty acid extraction yields (Table 6.5) are somewhat lower than for lipid extraction with the same system (Table 6.4).

Biocompatible systems such as ethanol (96 per cent) and hexane:ethanol (96 per cent) (1:2.5 v/v) have also been used to extract fatty acids by direct saponification from lyophilized *P. tricornutum* (Cartens *et al.*, 1996). Ethanol (96 per cent) (96 per cent yield) was more efficient than hexane:ethanol (96 per cent) (1:2.5 v/v) (86 per cent) because of the higher polarity of the former.

Table 6.4 Yields (%) of extracts and raffinates (in brackets) obtained by lipid extraction[a] from *I. galbana* (Molina Grima et al., 1994).

				Solvent system			
Fatty acid	Chloroform: methanol: water (1:2:0.8 v/v/v)	Hexane: ethanol (96%) (1:2.5 v/v)	Hexane: ethanol (96%) (1:0.9 v/v)	Butanol	Ethanol (96%)	Ethanol (96%): water (1:1 v/v)	Hexane: isopropanol (1:1.5 v.v)
18:4n3	108.9 (1.2)	46.5 (30.8)	49.0 (11.6)	70.7	91.3	82.0	74.1
20:5n3	101.4 (0.7)	49.4 (24.8)	53.2 (11.0)	67.3	82.2	65.7	67.0
22:6n3	92.9 (0.5)	51.0 (17.4)	49.6 (9.9)	60.8	83.0	32.3	55.1
Total n3 PUFAs	100.0 (0.7)	49.5 (24.2)	49.2 (10.8)	66.2	84.0	60.5	66.0
Total fatty acids	92.9 (0.9)	52.2 (27.4)	49.5 (8.3)	70.4	84.4	63.3	66.0

[a]76 ml of solvent system per gram of lyophilized biomass, 1h, room temperature.

Table 6.5 Yields (%) of extracts obtained by direct saponification[a] from *I. galbana* (Molina Grima et al., 1994).

	Saponification systems										
	Hexane:ethanol (96%) (1:2.5 v/v)		Hexane:ethanol (96%) (1:0.9 v/v)		Butanol	Ethanol (96%)		Ethanol (96%):water (1:1 v/v)		Hexane:isopropanol (1:1.5 v/v)	
Fatty acid	1h, 60°C	8h, room temperature	1h, 60°C	8h, room temperature	1h, 60°C	1h, 60°C	8h, room temperature	1h, 60°C	8h, room temperature	1h, 60°C	8h, room temperature
18:4n3	82.3	91.4	60.0	65.6	17.4	81.0	77.3	55.9	57.6	94.9	90.7
20:5n3	79.9	79.0	48.2	46.1	7.5	80.2	72.9	45.1	44.5	63.9	66.5
22:6n3	72.3	68.8	40.0	33.1	6.1	74.1	68.8	37.6	37.4	46.5	48.0
Total n3 PUFAs	79.3	78.7	47.4	47.1	8.5	79.0	72.7	46.4	47.6	60.9	66.1
Total fatty acids	81.0	75.0	48.0	47.1	9.0	79.8	79.2	46.5	45.5	62.0	61.1

[a]76 ml of solvent–saponification system per gram of lyophilized biomass.

PUFA concentration and purification techniques

PUFA concentration by urea fractionation

The theory and practice of fractionation of organic compounds with urea has been extensively reviewed (Smith, 1952; Schlenk, 1954; Swern, 1964). When urea crystallizes from a solution containing straight-chain organic molecules, it precipitates as hexagonal crystals that enclose the molecules. The included compound is attached to urea by van der Waals forces, London dispersion forces or induced electrostatic attractions (Swern, 1964) and they are held together merely by their shape, size and geometry. Therefore, the main application of the urea–inclusion-compound method is the separation of straight-chain compounds from branched or cyclic compounds (Smith, 1952; Schlenk, 1954).

When urea crystallizes from a solution of fatty acids of varying degrees of unsaturation, the saturated and monounsaturated fatty acids are included in crystals while the PUFAs remain in the solution. Therefore, the urea technique fractionates fatty acids mainly according to the degree of their unsaturation. The presence of double bonds introduces steric irregularities in the fatty acid chain with consequent imbalance of the optimum intermolecular distances, diminishing the net attractive force stabilizing the inclusion compound. The tendency of fatty acids and esters to combine with urea decreases with increasing unsaturation and decreasing chain length (Abu-Nasr and Holman, 1954).

This technique is the most employed to obtain concentrates of PUFAs with four or more double bonds for several reasons: it allows handling of large quantities of material in simple equipment; biocompatible solvents such as ethanol can be used; milder conditions are required (e.g. room temperature); the separation is more efficient than with other methods such as fractional crystallization or selective solvent extraction; and its cost is lower. Urea fractionation is a versatile process, because the fractionation characteristics can be altered simply by changing the amounts of either solvent or urea (Abu-Nasr and Holman, 1954; Gruger, 1960; Ratnayake *et al.*, 1988; Robles Medina *et al.*, 1995a, 1995b).

The stability constant is the reciprocal of the equilibrium constant and is specially useful in predicting separation behaviour. Values of stability constant have been noted by Swern (1964) for several fatty acids. Based on values of the stability constants it is relatively easy to obtain almost quantitative yields of urea complexes when the chain length of the fatty acid or ester is about eight or more carbon atoms. Progressive introduction of double bonds reduces the stability constant by a factor of about 10–20 per double bond. Stearic, oleic, and linoleic acid complexes have stability constant of 5000, 480 and 53, respectively. As a rough empirical rule, a stability factor of about 20 or higher is needed to obtain high yields.

Urea inclusion complexing is carried out mostly in methanol and ethanol, especially the former, because it affords greater concentration ratios (Robles Medina *et al.*, 1995b; Cartens *et al.*, 1996). During complexing, the urea concentration should be near saturation; concentration of recovered PUFAs decreases with decreasing urea concentration. When the fatty acid solvent itself is an adduct former, its equilibrium with urea competes with the desired reaction; therefore, use of higher hydrocarbons or acetone should be avoided (Schlenk, 1954).

The urea:fatty acid ratio (U:FA) and temperature, both of which are strongly related, are the most influential variables affecting the degree of PUFA concentration. The variation in PUFA concentration and yield with the U:FA ratio is different when the temperature is low than when it is high (Robles Medina *et al.*, 1995a, 1995b). At higher temperatures (over −12°C) the PUFA concentration increases with U:FA. A U:FA of 3:1 or 4:1 is the most appropriate (Table 6.6). Similar results have been reported by others (Wille *et al.*, 1987; Ratnayake *et al.*, 1988; Traitler *et al.*, 1988). This is reinforced by the fact that urea binds fatty acids in a ratio, different for each fatty acid, of approximately 3:1 by weight. An increase in this ratio favours the formation of urea compounds as their stability increases (Schlenk, 1954; Robles Medina *et al.*, 1995a, 1995b). Below −12°C the concentration factor and the yield

Table 6.6 Influence of urea:fatty acid ratio (U:FA) on fatty acid composition (% of total fatty acids) of urea concentrates obtained at 4°C from fatty acid extract from cod liver oil.

Fatty acid	Cod liver oil	Urea:fatty acid ratio			
		1:1	2:1	3:1	4:1
14:0	4.2	2.7	0.7	0.5	0.7
16:0	10.6	2.0	0.2	0.5	0.0
16:1n-7	7.8	9.6	6.9	2.5	3.2
18:0	2.6	0.9	0.1	0.0	0.0
18:1n-9	17.0	17.6	3.2	2.9	0.7
18:1n-7	4.6	5.9	1.4	1.0	0.0
18:2n-6	1.5	2.0	1.6	0.7	0.7
18:3n-6	0.2	0.2	0.4	0.5	0.5
20:0	0.2	0.2	0.2	0.0	0.1
18:3n-3	0.8	1.1	1.0	0.6	0.6
20:1n-9	10.8	9.0	1.3	0.6	0.8
18:4n-3 (SA)	2.4	3.3	6.3	8.0	8.5
20:3n-6	0.1	0.1	0.1	0.2	0.2
22:0	0.1	0.0	0.0	0.0	0.0
20:4n-6	0.5	0.8	1.0	0.9	1.0
22:1n-11	8.3	3.8	0.4	0.0	0.0
22:1n-9	0.1	0.0	0.0	0.0	0.0
20:5n-3 (EPA)	9.4	13.0	22.6	24.8	25.6
24:0	0.0	0.0	0.0	0.0	0.0
22:4n-6	0.5	0.7	1.5	1.7	1.8
22:5n-3	1.2	1.6	2.4	1.4	1.6
22:6n-3 (DHA)	11.0	15.8	45.4	58.2	59.9
Total saturated	17.6	5.8	1.2	0.9	0.8
Total monounsaturated	48.5	45.9	13.2	7.0	4.6
Total saturated, monounsaturated	66.1	51.7	14.5	8.0	5.4
Total SA, EPA, DHA	22.7	32.1	74.3	91.0	94.0
Total fatty acid yield	100	33.3	26.4	20.8	19.8
SA, EPA, DHA yield	100	47.1	86.2	83.2	82.1

Solvent methanol (Robles Medina *et al.*, 1995a).

diminish when U:FA increases, suggesting that PUFAs also form urea compounds at these low temperatures (Robles Medina *et al.*, 1995a, 1995b).

The tendency to form urea compounds increases with decreasing temperature; the optimum temperature depends on the particular PUFA. The PUFA concentration factor is maximum at about 4°C (Table 6.7). This temperature is particularly suitable for concentrating SA and DHA. The higher the temperature, the lower the tendency of fatty acids to form urea compounds, and therefore the PUFA concentration factors are lower. PUFAs form inclusion compounds also at lower temperatures; however, when EPA is the desired fatty acid, crystallization at higher temperatures (20–28°C) is preferable because concentrations and yields are higher (Ratnayake *et*

Table 6.7 Influence of temperature on fatty acid composition (% of total fatty acids) of urea concentrates obtained with a 4:1 urea:fatty acid ratio from fatty acid extract from cod liver oil.

Fatty acid	Cod liver oil	Temperature (°C)									
		−36	−28	−20	−12	−4	4	12	20	28	36
14:0	4.2	3.0	3.7	0.7	1.6	1.5	0.7	1.0	0.3	0.9	0.4
16:0	10.6	1.5	2.7	0.3	0.0	0.0	0.0	0.0	0.2	0.3	0.0
16:1n-7	7.8	10.8	9.3	5.4	0.0	2.9	3.2	1.4	0.8	2.5	8.6
18:0	2.6	0.6	0.3	0.9	1.6	1.3	0.0	0.8	2.1	0.9	0.6
18:1n-9	17.0	21.2	17.7	3.9	0.7	0.6	0.7	0.7	0.5	0.7	3.5
18:1n-7	4.6	6.9	6.6	2.2	0.0	0.0	0.0	0.6	0.0	0.0	1.6
18:2n-6	1.5	2.1	1.8	1.6	0.0	0.0	0.7	0.7	0.7	1.7	2.9
18:3n-6	0.2	0.4	0.2	0.8	0.0	0.0	0.5	0.5	0.6	0.4	0.4
18:3n-3	0.8	0.2	0.9	1.9	0.6	0.6	0.6	0.6	0.8	1.3	1.7
20:1n-9	10.8	1.1	1.5	1.2	0.0	0.0	0.8	0.2	0.6	0.4	0.4
18:4n-3 (SA)	2.4	8.2	8.4	9.2	13.1	11.5	8.5	8.8	8.5	8.1	7.2
20:4n-6	0.5	0.8	0.6	2.7	0.0	0.5	1.0	0.8	1.2	1.1	1.2
22:1n-11	8.3	3.3	4.5	0.3	0.6	0.6	0.0	0.0	0.0	0.3	0.3
22:1n-9	0.1	0.0	0.8	0.0	0.0	0.0	0.0	0.0	0.0	0.0	0.0
20:5n-3 (EPA)	9.4	12.4	11.2	16.4	17.7	14.9	25.6	26.1	29.5	28.7	25.4
22:4n-6	0.5	0.8	0.7	1.2	1.7	1.3	1.8	2.3	1.8	1.7	1.4
22:5n-3	1.2	1.6	1.4	1.6	0.0	0.5	1.6	1.2	1.8	2.5	2.8
22:6n-3 (DHA)	11.0	14.5	18.0	31.8	58.5	54.7	59.9	41.5	45.1	38.2	31.6
Total saturated	17.6	5.1	6.8	1.9	4.5	2.7	0.8	2.0	2.7	2.1	1.0
Total monounsaturated	48.5	43.1	40.4	12.9	1.3	4.1	4.6	2.9	1.9	3.9	14.3
Total SA, EPA DHA	22.7	35.0	37.7	57.4	89.3	81.2	94.0	76.4	83.0	75.0	64.2
Total fatty acid yield	100	9.2	10.3	11.3	13.6	19.1	19.8	22.3	22.5	24.7	26.4
SA, EPA, DHA yield	100	14.1	17.1	28.6	53.7	68.2	82.1	75.0	82.3	81.6	74.6

Solvent methanol (Robles Medina *et al.*, 1995a).

al., 1988; Robles Medina *et al.*, 1995a, 1995b). Attention to such differences is essential when the urea method is applied to extracts having different fatty acid profiles. *I. galbana* is rich in SA, EPA and DHA (Table 6.1) and therefore its extract should be crystallized at 4°C (Robles Medina *et al.*, 1995a), however *P. tricornutum* is rich only in EPA (Table 6.1), and therefore crystallization of its extract at 28°C is better (Cartens *et al.*, 1996).

The urea complexation conditions and recoveries of several authors are summarized in Table 6.8, where it is observed that the concentration factors (f) are similar in all cases, although they tend to get smaller as the PUFA concentration in the initial solution increases (Robles Medina *et al.*, 1995b; Cartens *et al.*, 1996). The major drawback of the urea method as a PUFA purification process is the low PUFA recovery, because some PUFAs, such as EPA, form a very high percentage of urea compounds (Robles Medina *et al.*, 1995a, 1995b; Cartens *et al.*, 1996).

PUFA purification by high pressure liquid chromatography

Relatively little work exists on analysis and separation of lipids and fatty acids by HPLC; fatty acid analysis by gas chromatography is generally preferred. The most suitable HPLC detector for fatty acid methyl esters is the differential refractometer, although it has several disadvantages, in that it is relatively insensitive to fatty acid concentrations at the nanogram level (Lie Ken Jie, 1980), gradient elution is not feasible and the detector is very sensitive to temperature and pressure fluctuations.

Because most fatty acids do not absorb UV radiation at the wavelength used in many HPLC instruments (254 nm), their conversion to UV-absorbing derivatives is necessary. However, conjugated unsaturated fatty acids or acids containing aromatic moieties are UV-absorbing and, therefore, can be readily detected by this method. Some fatty acids generally absorb in the 190 to 210 nm range, which interferes with mobile phases such as chloroform and acetone. The 192 nm wavelength is quite PUFA-selective, with sensitivity increasing with the number of double bonds (Aveldano *et al.*, 1983). Robles Medina *et al.* (1995a) obtained highly pure EPA and DHA spectra (Figure 6.1), using 217 nm for the detection of these PUFAs. Shukla (1988) demonstrated that the triglycerides can be analyzed quantitatively at 220 nm, at which wavelength the absorption is fundamentally due to the $C = O$ group.

Silver nitrate HPLC

The long-chain fatty methyl esters can be separated on the basis of number, position, and geometric configuration of double bonds with a silver nitrate-impregnated silica column (Scholfield, 1979; Özcimder and Hammers, 1980). Silver nitrate HPLC is a rapid semi-preparative prefractionation method for highly unsaturated fatty esters with 3–6 double bonds. It is based on the formation of a reversible charge–transfer complex involving the silver ion and double bonds. The greater the number of double bonds, the more strongly the acid is held up and the greater is the elution time. As an example, Özcimder and Hammers (1980) separated fatty acid methyl esters of cod liver oil using a 5 per cent Ag (I) stationary phase; hexane containing 0.4 per cent acetonitrile was the mobile phase. In the resulting chromatogram, all of the esters with an equal number of carbon atoms overlapped. The authors concluded

Table 6.8 PUFA concentrations in fatty acid extract (x_E) and urea concentrate (x_U) (% of total fatty acids), concentration factors (f) and PUFA recovery yields (R_A), obtained for SA, EPA and DHA by the urea method.

Reference and conditions[b]	SA x_E^c	SA x_U^d	SA f^e	SA R_A^f	EPA x_E^c	EPA x_U^d	EPA f^e	EPA R_A^f	DHA, GLA[a] x_E^c	DHA, GLA[a] x_U^d	DHA, GLA[a] f^e	DHA, GLA[a] R_A^f
Haagsma et al. (1982), 3:1, 4°C, methanol	2.4	10	4.2	100	12.1	22.6	2.3	60	11.7	44.6	3.8	100
Wille et al. (1987), 3:1, 15°C, methanol	1.7	7.2	4.2	84.7	13.9	36.8	2.6	53	13.6	40.9	3	60.2
Ackman et al. (1988), 2.6:1, 5°C, ethanol	0.8	5.1	6.4	–	5.4	32.5	6.0	–	3.9	29.2	7.5	64
Rattayake et al. (1988), 3:1, 1°C, ethanol	3.6	11.2	3.1	100	12.6	32.9	2.6	91	4.4	11.1	2.5	88
Cohen et al. (1993), 4:1, −15°C, methanol	–	–	–	–	–	–	–	–	21.5	79.7	3.7[a]	–
Cohen and Cohen (1991), 4:1, −15°C, methanol	–	–	–	–	47.5	81.9	1.7	–	–	–	–	–
Trailer et al. (1988), 3:1, 4°C, methanol	3.7	15.5	4.2	–	–	–	–	–	18.2	79.6	4.4[a]	–
Fujita and Makuta (1983), 1.5:1, 35°C, ethanol	1.7	61	3.6	56	14.5	57.3	4.0	51	4.3	17.5	4.1	49
Robes Medina et al. (1995b)												
4:1, 4°C, methanol	2.4	8.5	3.6	71.6	9.4	25.6	2.7	54.2	10.9	59.8	5.5	100
4:1, 28°C, methanol	2.4	8.1	3.4	84.9	9.4	28.7	3.1	75.6	10.9	38.2	3.5	85.9
4:1, 4°C, ethanol	2.4	7.0	3.0	74.1	9.4	21.5	2.3	57.9	10.9	51.4	4.7	100
4:1, 28°C, ethanol	2.4	6.6	2.8	92.4	9.4	24.3	2.6	86.1	10.9	30.2	2.8	91.3

[a] GLA: γ-linolenic acid.
[b] Fatty acid:urea ratio, crystallization temperature and urea solvent, respectively.
[c] PUFA concentrations in fatty acid extract (% of total fatty acids).
[d] PUFA concentration in urea concentrate (% of total fatty acids).
[e] concentration factors ($f = x_U/x_E$).
[f] PUFA recovery yields.

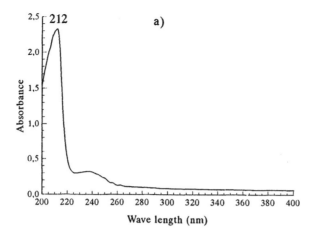

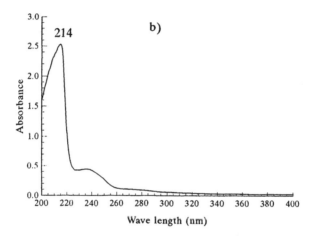

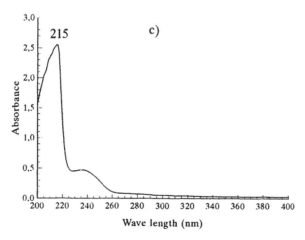

Figure 6.1 Ultraviolet spectra of (*a*) stearidonic acid (SA, 18:4n-3); (*b*) eicosapentaenoic acid (EPA, 20:5n-3); and (*c*) docosahexaenoic acid (DHA, 22:6n-3) Sigma standard (Sigma Chemical Co., St Louis, MO) (reprinted from Robles Medina *et al.* (1995a) with kind permission from the American Oil Chemists' Society).

that the method was suitable for separating PUFAs, but not for separating saturated, mono or di-unsaturated fatty esters.

An important application of silver nitrate chromatography is the separation of geometrical *cis-* and *trans-*isomers. The *trans-*isomers are eluted before the corresponding *cis-*isomers, owing to the lesser influence of steric effects and to the release of the intramolecular strain on the formation of the *cis* double bond–silver complex (Hartley, 1973).

Fatty acids that are poorly resoluted in reverse-phase chromatography may also be separated on a silver nitrate column because the number of double bonds is the only basis for the separation. In reverse-phase separation the effect of double bonds is sometimes compensated by that of the chain length.

Moffat *et al.* (1993) separated PUFA triglycerides from sardine oil by Ag-HPLC. The first region of the chromatogram was comprised of triglycerides containing only saturated fatty acids and those containing one, two or three monoenoic acids. The second region contained triglycerides with increasing proportions of PUFAs with the progressive introduction of more highly unsaturated fatty acids, including EPA and DHA. The third region was comprised of highly unsaturated triglycerides with retention times equivalent to those of trilinolenin and triarachidonin.

Reverse-phase HPLC

Reverse-phase (RP) chromatographic separations are largely based on the sorption of hydrophobic moieties of the soluble molecules from a polar solvent to an apolar sorbent. The stationary phase is typically the octadecylsilyl (C_{18}) group, linked to a silanol surface by covalent bond, although octylsilyl (C_8) is also used (Henke and Schubert, 1980; Aveldano *et al.*, 1983). Bonded phase C_8 allows for faster separations of very long-chain fatty acids (Aveldano *et al.*, 1983).

The separation of fatty acids and esters depends on the chain length, and on the degree and type of unsaturation. The retention time increases with increasing chain length; a linear relationship has been observed between the carbon number and the logarithm of the retention volume. Owing to the interactions between the mobile phase (polar) and double bond, the fatty acids or esters containing the largest number of double bonds have the shortest retention times.

Cis- and *trans-*isomers have been successfully separated by RP-HPLC. Owing to the influence of steric effects on the eluent–double bond interaction, the *cis* unsaturated fatty acid eluted earlier from the column than did the corresponding *trans-*isomers (Jordi, 1978). In RP separation, free fatty acids generally separate more rapidly because they are more polar than the corresponding esters, which are more strongly retained. Therefore, when separating complex mixtures, separation as free fatty acids is preferred.

To evaluate the difficulty of an HPLC separation, the capacity factor (k') of each component needs to be calculated from the chromatogram:

$$k' = (t_R - t_0)/t_0 \tag{6.1}$$

where t_R is the retention time of a solute and t_0 is the elution time of the mobile phase. k' represents the ratio between the quantity of solute retained in the stationary phase and that retained in the mobile phase at thermodynamic equilibrium. The separation factor α is the ratio of the capacity factors k' of two compounds. From a practical point of view, α can be used to assess the difficulty of the separation (the lower the α value, the greater the difficulty of separation; $\alpha = 1$ indicates total over-lapping). Note, however, that α values are based on thermodynamic considerations, whereas kinetics determines the peak width and the extent of the overlapping. Resolution, however, does take into account both thermodynamics and kinetics and therefore it is a more useful parameter for characterizing the efficiency of a chromatographic system. Resolution is calculated as:

$$R = 2(t_{R2} - t_{R1})(\Delta t_1 - \Delta t_2) \tag{6.2}$$

where t_{R1} and t_{R2} are the retention times of two compounds and Δt_1 and Δt_2 are the corresponding peak widths, measured at the base line. Resolution may be used as the determining parameter for optimal separation and scale-up.

The commonly used mobile phases are the methanol–water and acetonitrile–water mixtures in which the proportions of the organic phase may range from over 60 per cent to 100 per cent. Acetonitrile separations take longer and some fatty acids may be difficult to dissolve in that solvent (Aveldano *et al.*, 1983). However, acetonitrile seems to be more selective than methanol and yields sharper peaks, probably because of its lower viscosity. Methanol produces similar separations but with slightly higher retention times, smaller separation factors (α) and broader peaks. Fatty acids for use in clinical trials should be purified using biocompatible solvents. Ethanol–water mixture may be used, but flow rate should be lower than when using methanol–water or acetonitrile–water. This is because the greater viscosity of ethanol leads to greater pressure drops than with the other solvents. Under similar operating conditions, resolution is lower than with methanol–water, because the equilibrium of PUFAs between the stationary and mobile phases is nearer the mobile phase with ethanol–water (Robles Medina *et al.*, 1995a).

The k' values of fatty acids are affected by changes in the composition of the mobile phase, hence the order of elution may be affected. Addition of water to the acetonitrile or methanol increases the retention time; with acetonitrile this increase is greater for saturated fatty acids than for unsaturated ones. A logarithmic relationship exits between the k' of a fatty acid and the percentage of acetonitrile or methanol in the mobile phase. Furthermore, k' increases with chain length and diminishes with the double bond number. As a rule, strong hydrophobicity (longer chain length or less unsaturation) increases retention time while the solvent strength becomes weaker. That is, a decrease in the strength of the eluent improves resolution (Aveldano *et al.*, 1983; Robles Medina *et al.*, 1995a).

Preparative scale chromatography in needed to isolate material for further testing, synthesis, commercialization and physiological or nutritional research. PUFA purity in preparative fractions should be as great as possible, usually 90–95 per cent. The initial developmental chromatography is performed in analytical columns to evaluate the parameters k', α and R. The process is then scaled up in columns that permit larger load while retaining the yield and purity of the analytical scale. Maintenance of purity in the scaled up process requires that the concentration profiles of the solutes should not be dramatically altered. This can be done by scaling-up those chromatograms in which the resolution (R) between the peaks of interest is about one and maintaining this resolution constant at preparative level. For $R = 1$, separation is around 98 per cent and for $R = 1.5$ the components are fully separated (Valcarcel Cases and Gómez Hens, 1990). Chromatography is best scaled up by increasing the column diameter (i.e., increase capacity) and while keeping the surface velocity, the column length and the particle size the same as at the small scale (to preserve the retention times, peak width and pressure drop) (Belter *et al.*, 1988). In this approach, the flow rate and the sample loading at preparative scale are calculated by multiplying the values corresponding to analytical scale by the factors F and m, respectively. These geometric factors are calculated thus:

$$m = L_2 D_2^2 / (L_1 D_1^2) \qquad\qquad (6.3)$$

$$F = D_2^2 / D_1^2 \qquad\qquad (6.4)$$

where L and D are depth backing and column diameters, respectively. The subscripts 1 and 2 denote analytical and preparative scales, respectively.

This procedure was used by Robles Medina *et al.* (1995a) for scaling up an analytical separation for obtaining highly pure PUFA from cod liver oil and the marine microalga *I. galbana*. The analytical column was a Beckman, RP C_{18} (5 µm particle size, 4.6 mm column diameter, 25 cm length). The analytical chromatographic separation with a resolution of around 1 was scaled up to semi-preparative level (RP C_{18}, 5 µm particle size, 10 mm column diameter, 25 cm length) to produce SA, EPA and DHA fractions of 90–95 per cent purity. The highest yields, purest fractions and largest loads were obtained with methanol:water (1 per cent acetic acid) 80:20 w/w (Table 6.9). The best fraction had an EPA yield of 99.3 per cent and purity of 94.3 per cent. A 100 per cent yield of 96 per cent pure EPA was obtained from *I. galbana*. Similar scale-up procedures were used by Cartens *et al.* (1996) and Molina Grima *et al.* (1996) for producing highly pure EPA from the diatom *P. tricornutum*. Giménez Giménez *et al.* (1998) carried out a scale-up to preparative level using a compression radial cartridge Waters, RP C_{18} (37–55 µm particle size, 4.7 cm column diameter, 30 cm length). Arachidonic acid (AA) and EPA were obtained from the microalga *Porphyridium cruentum*; the EPA and AA yields and purities are shown in Table 6.10 and Figure 6.2. The main impurities in AA were palmitoleic and linoleic acids (16:2n-1 and 18:2n-6, respectively), but these two fatty acids are well separated at the semi-preparative scale (Table 6.10). This could be due to different particle size, and different porosities and different concentration of active C18 sites of the adsorbent at the two scales; it should be noted that the scale-up criterion noted above (Equations 6.3 and 6.4) implied maintaining the same stationary phase at the two scales. Thus, for example, an increase in the particle size increased the broadness of the peak, which implies larger overlap and reduced recovery yield and/or purity.

HPLC-based methods are specially capable of yielding high purities as reported by several authors (Table 6.11). The yields obtained by the most of authors cited in Table 6.11 are relatively lower because generally the column loads were excessive. A low yield is of concern especially when the PUFA concentrate is from an expensive source such as marine microalgae.

Integrated process for obtaining highly pure PUFA from algae

Figure 6.3 shows a flowchart of the process developed at the University of Almería (Molina Grima *et al.*, 1996) for purifying PUFAs from various sources including cod liver oil and the marine microalgae *I. galbana*, *P. tricornutum* and *P. cruentum*. The biomass obtained in an outdoor photobioreactor (1) is concentrated by centrifugation (2) and the wet paste is conserved at −20°C (3) until needed. Fatty acids are

Table 6.9 Scale-up of PUFA isolation from a concentrate of cod liver oil by HPLC.

| | Analytical scale[a] | | | | Semi-preparative scale[b] | | | | | | | | |
| | | | Resolution | | | | SA | | EPA | | DHA | | |
Mobile phase	Flow rate (ml/min)	Load (mg)	SA-EPA	EPA-DHA	Flow rate (ml/min)	Load (mg)	Purity (%)	Yield (%)	Purity (%)	Yield (%)	Purity (%)	Yield (%)
Methanol:water (90:10)	0.75	0.416	1.19	1.40	3.5	1.940	93.2	93.3	88.1	89.8	80.5	84.4
Methanol:water (85:15)	0.75	1.386	0.80	1.01	3.5	6.650	85.5	81.5	93.6	97.6	84.9	95.4
Methanol:water (80:20)	0.60	2.218	0.94	1.32	3.0	10.53	91.9	98.4	94.4	99.3	84.6	96.6
Ethanol:water (80:20)	0.30	0.138	1.11	1.06	1.5	0.693	87.3	87.7	87.4	88.6	72.2	81.7
Ethanol:water (75:25)	0.30	0.416	1.74	0.96	1.5	1.386	84.5	78.2	90.3	79.7	79.6	70.5
Ethanol:water (70:30)	0.40	0.564	2.05	0.71	2.0	2.771	69.7	76.4	91.3	85.1	79.5	100

Experimental conditions as noted (Robles Medina et al., 1995a).
[a] Column Beckman, RP C_{18}, 5 μm particle size, 8 nm pore, 4.6 mm i.d. × 25 cm.
[b] Column Beckman, RP C_{18}, 5 μm particle size, 8 nm pore, 10 mm i.d. × 25 cm.

Table 6.10　Fatty acid composition (% of total fatty acids) of PUFA concentrate and HPLC fractions obtained from *Porphyridium cruentum* microalga.

| | | HPLC fractions | | | |
| | | Semi-preparative | | Preparative | |
Fatty acid	Concentrate	EPA	AA	EPA	AA
16:0	0.2	2.1	2.2	–	–
16:1n7	1.1	–	–	2.6	4.8
16:2n4	0.7	–	–	2.1	2.0
16:3n4	0.5	–	–	–	–
16:4n1	0.7	–	–	–	–
18:1n9	0.4	–	–	–	–
18:2n6	4.7	–	–	–	6.1
18:3n3	1.1	1.6	–	0.9	–
18:4n3	1.1	–	–	–	–
20:4n6 (AA)	34.1	2.0	92.0	–	81.4
20:5n3 (EPA)	42.0	94.3	–	94.3	3.6
22:6n3	1.1	–	–	–	–
R(%)[a]		64.6	55.4	53.4	49.5

Semi-preparative column (1 cm i.d. × 25 cm; load 5 mg) and in a preparative cartridge (4.7 × 30 cm, load 133 mg) (Giménez Giménez *et al.*, 1998).
[a]EPA and AA yields.

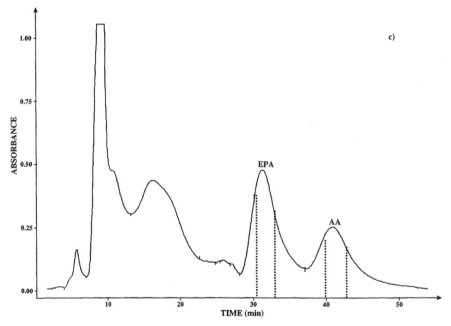

Figure 6.2　HPLC fractionation of *Porphyridium cruentum* PUFA concentrate to obtain EPA and AA. Mobile phase methanol:water (1% AcH) (80:20 w/w) Preparative cartridge (4.7 cm diameter and 30 cm length), RP C_{18}. Flow rate 66.0 ml/min. Mass load 133 mg of PUFA concentrate shown in Table 6.10 Purity of EPA fraction is also shown in Table 6.10 (reprinted from Giménez Giménez *et al.* (1998) with kind permission from Kluwer Academic Publishers).

Table 6.11 Comparison of experimental conditions and purities of PUFA-enriched fractions obtained by HPLC as reported by several authors.

	Tokiwa et al. (1981)	Wille et al. (1987)	Perrut (1988)	Grant (1988)	Traitler et al. (1988)	Traitler et al. (1988)	Cohen and Cohen (1991)	Robles Medina et al. (1995a)	Cartens et al. (1996)	Molina Grima et al. (1996)
Source	Fish oil	Fish oil	Fish oil	Fish oil	Seed oil	Seed oil	*P. tricornutum*	*I. galbana*	*P. tricornutum*	*P. tricornutum*
Concentrate	Urea method	Urea method	Urea method	Urea method	Urea method	Urea method	Urea method	Urea method	Urea method	Urea method
Sample	Methyl esters	Fatty acids	Ethyl esters	Deriv. F.A.	Fatty acids	Fatty acids	Methyl esters	Fatty acids	Fatty acids	Fatty acids
Load (g)	10	90	136	30	2	100	0.001	0.01	0.037	0.635
SA (%)	–	7.2	–	–	16.6	16.6	–	8.5	–	–
EPA (%)	30	36.8	46.7	–	–	–	81.9	25.6	55.2	50.5
DHA (%)	30	40.9	30.5	–	–	–	–	59.9	–	–
GLA[a]					80.2	80.2				
Column	Reverse C18	Reverse C18	Reverse C18	Reverse C18	Reverse C18	Reverse C18	Reverse C18[b]	Reverse C18	Reverse C18	Reverse C18
D × L (cm)	5.7 × 30	20 × 60	30 × 30	10 × 60	5.7 × 60 (two column)	20 × 60		1 × 25	1 × 25	4.7 × 30
Particle size (µm)	50–100	55–105	12–45	15–30	55–105	55–105		5	5	37–55
Mobile phase Composition	THF[c]: methanol: water	Methanol: water	Methanol: water	Ethanol:water	Methanol: water	Methanol: water	Acetonitrile: water	Methanol: water	Methanol: water	Methanol: water
Composition	25:55:20 v/v/v	90:10 w/w	90:10 w/w	80:20 w/w	90:10 w/w	90:10 w/w	Gradient	80:20 w/w	80:20 w/w	80:20 w/w
Flow (ml/min)	150	1200	3333	225	150	2000	5	3	3	66
Separation (min)	50	50	19	160	20	20	5	70	90	60
Detector	RI[d]	RI[d]	RI[d]	UV at 245 nm	RI[d]	RI[d]	GC[e]	UV at 217 nm	UV at 217 nm	UV at 217 nm
Fractions SA (%)	–	93.1	–	–	46.3	78.3	–	94.8 (100%)[f]	–	–
EPA (%)	91.0 (66.7%)[f]	85.6	91–96	97	–	–	97.3	96.0 (99.6%)[f]	93.4 (85%)[f]	95.8 (97.3%)[f]
DHA (%)	85.5 (53.3%)[f]	83.1	75–85	92	–	–	–	94.9 (94.0%)[f]	–	–
GLA (%)	–	–	–	–	95.4	95.7	–	–	–	–

[a]γ-linolenic acid; [b]elution at atmospheric pressure; [c]tetrahydrofuran; [d]refractive index; [e]gas chromatography; [f]values in brackets correspond to fatty acid yields for HPLC separation.

131

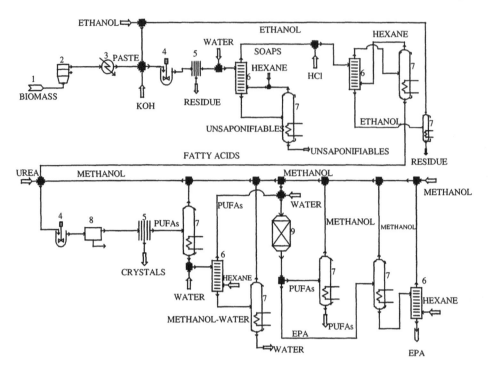

Figure 6.3 Schematic flowsheet for EPA purification from wet microalgal biomass. (1) culture from photobioreactor; (2) centrifuge; (3) freezer (this step can be omitted if the paste is used immediately after centrifuging); (4) stirred tank under N_2 or Ar atmosphere; (5) filter; (6) solvent extraction; (7) vacuum distillation; (8) crystallizer; (9) preparative HPLC (reprinted from Molina Grima *et al.* (1996) with kind permission from Kluwer Academic Publishers).

extracted by direct saponification of wet biomass with KOH–ethanol (96 per cent) (Molina Grima *et al.*, 1996) (4) and the mixture is filtered (5) to remove the biomass residue. Before extracting the unsaponifiable lipids with hexane (6), water is added to shift the equilibrium distribution of unsaponifiables to the hexane phase. Hexane is recovered by evaporation (7). The ethanolic solution of soaps is acidified with HCl and free fatty acids are extracted with hexane (6). Ethanol is recovered by rectification (7) and the solution of fatty acids in hexane is concentrated by vaporization of hexane under vacuum (7). Free fatty acids are added to an urea–methanol solution (4) and the urea–fatty acids adducts are crystallized (8) (Robles Medina *et al.*, 1995a, 1995b); the crystals are separated by filtration and the filtrate, which contains the PUFAs, is concentrated by vaporization of methanol under vacuum (7). Methanol is recycled and PUFAs are extracted from the hydromethanolic phase with hexane (6). The PUFA mixture is then loaded in the HPLC column (9) and the EPA fraction is separated from the other PUFAs (Molina Grima *et al.*, 1996). The mobile phase (methanol:water (1 per cent acetic acid) 80:20 w/w) is recovered from the PUFAs and EPA fractions (7). Finally EPA is extracted from the hydromethanolic solution with hexane and conserved in this solvent under argon atmosphere at −20°C.

The results obtained using the depicted process for PUFA recovery from *I. gal-bana* biomass are summarized in Table 6.12. The low overall EPA yield (43.3 per cent) is due to the low EPA yield in the urea concentration step (54.3 per cent).

Table 6.12 Fatty acid composition (% of total fatty acids) (A) and yield (B) of *I. galbana* biomass, fatty acid extract, urea concentrate and HPLC fractions (Robles Medina *et al.*, 1995a).

A

Fatty acid	Biomass	Extract[a]	Urea concentrate[b]	HPLC fractions[c] SA	EPA	DHA
14:0	10.1	10.7	0.3	–	–	–
16:0	20.3	18.9	0.2	1.1	–	–
16:1n-7	21.4	23.3	4.3	–	–	–
18:0	0.7	0.5	1.5	1.1	–	–
18:1n-9	1.4	1.7	0.2	–	–	–
18:1n-7	3.6	3.2	0.8	–	–	–
18:2n-6	0.9	0.9	0.2	–	–	1.4
18:3n-6	0.2	0.2	0.4	–	2.1	–
20:0	0.0	0.1	0.2	–	–	–
18:3n-3	1.2	1.3	0.8	–	2.0	–
20:1n-9	0.2	0.3	0.3	–	–	–
18:4n-3 (SA)	6.4	7.3	22.6	94.8	–	–
20:3n-6	0.4	0.2	0.2	–	–	–
22:0	0.0	0.1	0.1	–	–	–
20:4n-6	0.7	0.7	1.1	–	–	3.7
22:1n-11	0.1	0.1	0.1	–	–	–
22:1n-9	0.0	0.1	0.1	–	–	–
20:5n-3 (EPA)	22.6	22.4	39.4	–	96.0	–
24:0	0.0	0.1	0.1	–	–	–
22:4n-6	1.3	1.3	3.6	–	–	–
22:5n-3	0.2	0.1	0.1	–	–	–
22:6n-3 (DHA)	8.4	6.8	23.4	–	–	94.9

B

Yield (%)	SA	EPA	DHA
Extract[a]	92.5	80.1	65.2
Urea concentrate[b]	71.6	54.3	100
SA fraction[c]	100	–	–
EPA fraction[c]	–	99.6	–
DHA fraction[c]	–	–	94.0
Overall yield (%)	66.2	43.3	61.3

[a]Direct saponification of biomass: hexane:ethanol (96%) (1:2.5 v/v), 8 h, room temperature.
[b]Urea:fatty acid ratio 4:1, crystallization temperature 4°C, solvent methanol.
[c]Column Beckman, RP C_{18}, 5 µm particle size, 8 nm pore, 10 mm i.d. × 25 cm. Mobile phase methanol:water 80:20 w/w, flow rate 3 ml/min. Mass load 9.5 mg.

Therefore, in more recent work, we have attempted to recover EPA without the urea step. Instead, the fatty acid extract was fractionated by HPLC (Table 6.13). Most of the EPA (98.3 per cent) in the initial *P. tricornutum* biomass was recovered at high purity (93.6 per cent) (Cartens *et al.*, 1996). However, the greater complexity of the initial fatty acid extract meant that only a smaller amount could be loaded to ensure high yields and purities. When the simpler urea PUFA concentrate (55 per cent EPA)

Table 6.13 Fatty acid composition (% of total fatty acids) (A) and yield (B) of *P. tricornutum* extract and HPLC fraction (Cartens *et al.*, 1996).

A

Fatty acid	Extract[a]	HPLC fraction[b]
14:0	5.9	3.9
16:0	16.7	0.4
16:1n-7	19.0	0.4
16:2n-4	7.4	0.5
16:3n-4	6.4	0.1
16:4n-1	1.9	–
18:2n-6	2.4	0.2
18:3n-3	0.2	0.2
20:4n-6 (AA)	3.6	–
20:4n-3	0.3	0.1
20:5n-3 (EPA)	31.9	93.6
24:0	1.5	–
22:6n-3 (DHA)	0.8	–
Others	2.0	0.6

B

Yield (%)	EPA
Extract	98.3
HPLC fraction	100.0
Overall yield (%)	98.3

[a]Direct saponification of biomass: ethanol (96%), 1 h, 60°C.
[b]Column Beckman, RP C_{18}, 5 μm particle size, 8 nm pore, 10 mm i.d. × 25 cm. Mobile phase methanol:water 80:20 w/w, flow rate 3 ml/min. Mass load 12.3 mg.

was loaded (Table 6.14) a loading of 37.4 mg was feasible at the semi-preparative column; however when the more complex fatty acid extract (32 per cent EPA) was loaded (Table 6.13), only 12.3 mg could be loaded. 14:0 and 16:1n-7 were the most important fatty acid impurities in the EPA fraction (Table 6.13) and both were separated well from PUFA by the urea method. These results demonstrate the importance of the fatty acid profile in HPLC purification. Despite the high EPA content in the fatty acid mixture loaded, the omission of the urea step is not recommended.

This extraction and purification process has also been applied to obtain highly pure EPA from *P. tricornutum* at preparative scale (4.7 × 30 cm cartridge). Table 6.15 shows the results of the HPLC purification step. The gas chromatograms of the *P. tricornutum* fatty acid extract, urea concentrate and HPLC EPA fraction are shown in Figure 6.4. On the other hand, Figure 6.5 shows that the UV spectrum of the isolated EPA fraction is very similar to that of the standard (Figure 6.1a), demonstrating that there was no degradation of EPA during down-stream processing (Kates, 1988; Molina Grima *et al.*, 1996).

Table 6.16 shows a preliminary estimation of the cost of obtaining EPA from wet *P. tricornutum* biomass. For comparison, we have introduced the cost of producing EPA from cod liver oil. Table 6.16A shows that this process allows production of

Table 6.14 Fatty acid composition (% of total fatty acids) (A) and yield (B) of *P. tricornutum* extract, urea concentrate and HPLC fraction (Cartens *et al.*, 1996).

A

Fatty acid	Extract[a]	Urea concentrate[b]	HPLC fraction[c]
14:0	5.9	0.3	–
16:0	16.7	0.0	–
16:1n-7	19.0	6.9	2.9
16:2n-4	7.4	10.5	–
16:3n-4	6.4	12.3	2.1
16:4n-1	1.9	2.1	–
18:2n-6	2.4	2.1	1.6
18:3n-3	0.2	1.0	–
20:4n-6 (AA)	3.6	5.9	–
20:4n-3	0.3	1.4	–
20:5n-3 (EPA)	31.9	55.2	93.4
24:0	1.5	0.0	–
22:6n-3 (DHA)	0.8	1.5	–
Others	2.0	0.8	0.0

B

Yield (%)	EPA
Extract	98.3
Urea concentrate	78.6
HPLC fraction	85.0
Overall yield (%)	65.7

[a] Direct saponification of biomass: ethanol (96%), 1 h, 60°C.
[b] Urea:fatty acid ratio 4:1, crystallization temperature 28°C, solvent methanol.
[c] Column Beckman, RP C_{18} 5 μm particle size, 8 nm pore, 10 mm i.d. × 25 cm. Mobile phase methanol:water 80:20 w/w, flow rate 3 ml/min. Mass load 37.4 mg.

2.46 g per day of 95.8 per cent pure EPA from 100 g *P. tricornutum* microalga. At the relatively small scale used, Molina Grima *et al.* (1996) estimated a production cost of $188 per gram EPA. This cost is about two times lower than that from cod liver oil. This is because cod liver oil initially has only 9 per cent EPA, while *P. tricornutum* extract has 36.3 per cent, which means EPA concentrate may be up to 50.5 per cent versus 36 per cent for cod liver oil. The PUFA concentrate loading into the column is lower as is its EPA content. Thus with the same mobile phase cost, it is possible to obtain more EPA from each load of *P. tricornutum* concentrate, thus lowering the final cost of the product (Molina Grima *et al.*, 1996).

Table 6.17 shows a comparison of purity and yields of EPA obtained from the microalgae *I. galbana*, *P. tricornutum* and *P. cruentum* (Giménez Giménez *et al.*, 1998). The results obtained with *P. cruentum* were the least satisfactory because this microalga also contains AA, which overlaps with EPA in the HPLC fractionation and, therefore, the purity and/or yield are reduced. *P. tricornutum* is clearly the best source when only EPA is required but *P. cruentum* also contains AA and *I. galbana* is also appropriate for producing DHA.

Table 6.15 Fatty acid composition (% of total fatty acids) of *P. tricornutum* urea concentrate and HPLC fractions obtained at preparative level (Molina Grima *et al.*, 1996).

Fatty acid	Urea concentrate[a]	EPA fraction[b]	
		1	2 (EPA)
14:0	0.1		
16:0	0.1		
16:1n-7	0.7	0.8	0.5
16:2n-4	6.9	39.5	2.3
16:3n-4	20.7		
16:4n-1	6.0		
18:2n-6	0.5		
18:3n-3	–		
20:4n-6 (AA)	3.0		
20:4n-3	0.5		
20:5n-3 (EPA)	50.5	35.5	95.8
24:0	0.0		
22:6n-3 (DHA)	4.7		0.3
Fraction amount (g)			0.326
EPA yield			97.3

[a] Urea:fatty acid ratio 4:1, crystallization temperature 28°C, solvent methanol.
[b] Cartridge (4.7 cm i.d. × 30 cm), RP C_{18}, flow rate 66 ml/min, mass load 0.635 g.

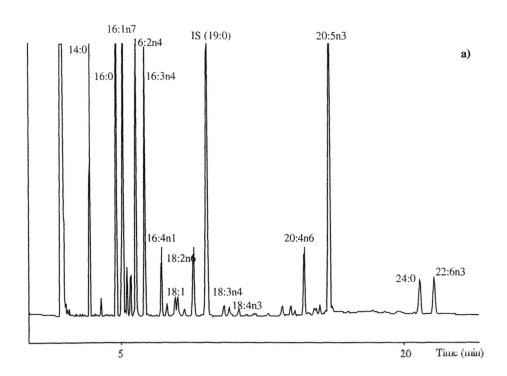

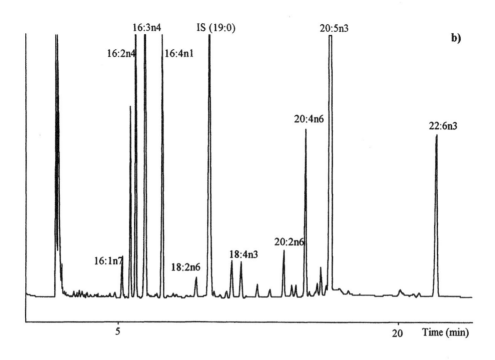

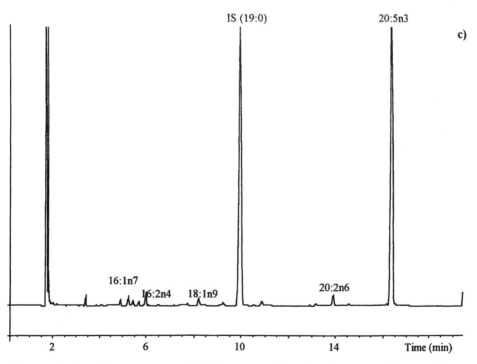

Figure 6.4 Gas chromatograms of (a) fatty extract from *P. tricornutum* biomass; (b) urea concentrate; and (c) HPLC EPA fraction.

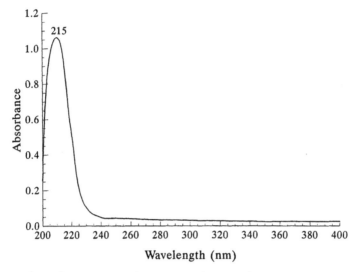

Figure 6.5 Ultraviolet spectrum of HPLC EPA fraction shown in Table 6.14 (from Molina Grima *et al.* (1996) with kind permission from Kluwer Academic Publishers).

Concluding remarks

Microalgae are an excellent source of PUFAs because the fatty acid profile is simpler than fish oil, production conditions can be controlled and the algal species can be selected according to the required PUFA. These factors also facilitate purification. A downstream process for obtaining highly pure PUFAs from microalgae must provide high yields and not degrade the PUFAs. An easy downstream process has been developed by Molina Grima *et al.* (1996) which consists of: concentration of biomass from culture medium by centrifugation; extraction of fatty acids by direct saponification of wet biomass with KOH–ethanol; extraction of unsaponifiables with hexane; separation of non-lipid contaminants by extraction of fatty acid with hexane; concentration of PUFAs by the urea method; and isolation of the PUFA of interest by preparative reverse phase HPLC. By this procedure PUFAs as SA, EPA and DHA of purities higher than 94 per cent have been obtained from the microalgae *P. tricornutum* and *I. galbana*.

Acknowledgements

The assistance of Professor Yusuf Chisti with the preparation of this chapter is greatly acknowledged. The investigation included in this chapter was supported in part by research grants from the Comisión Interministerial de Ciencia y Tecnología (CICYT), BIO 95-0652, Plan Andaluz de Investigación II (3339) (Spain) and the European Union project BRPR CT97 0537.

Table 6.16 Preliminary estimation for small scale production of the costs of obtaining EPA from wet *P. tricornutum* biomass (A) and cod liver oil (B) (Molina Grima *et al.*, 1996).

A

Process	Yields and purities	Costs (US$)/g EPA			
		Consumables[a]	Amortization[b]	Labour[c]	Relative cost (%)
Culture and centrifugation	Basis of calculation: Biomass produced in day: 100g 9.4% of fatty acid in dry weight	8.5	20.3		23
Direct extraction of Fatty acids	Product: 8.51g of fatty acid (36.3% EPA) Fatty acid yield: 90.6%	48.0	4.5		35
PUFA concentration by the urea method	Product: 4.81 g of PUFAs (50.5% EPA) EPA yield: 78.6%	0.8	3.7	55.7	10
EPA purification by HPLC	Product: 2.46 g EPA (95.8% EPA) EPA yield: 97.3%	34.6	11.8		32
Total costs/g EPA			**188**		

B

Cost of oil	Basis of calculation: 10 g of cod liver oil 9.4% EPA	0.115			0.03
Saponification	Product: 9.5 g of fatty acid (9.4% EPA) Fatty acid yield: 95%	2.7	12.2		21
PUFA concentration by the urea method	Product: 1.95 g of PUFAs (36.0% EPA) EPA yield: 78.6%	2.7	12.2	185	21
EPA purification by HPLC	Product: 0.74 g EPA[d] (94% EPA) EPA yield: 99%	114.9	39.2		58
Total costs/g EPA		**369**			

[a] 20% of the price of the solvents used + power costs (5% of the price of the recovery solvents) divided by 2.46 g (*P. tricornutum*) or by 0.74 g (cod liver oil) of EPA.
[b] US$256 500 (total cost of facilities used in the process to obtain EPA from *P. tricornutum*) or US$121 500 (total cost of facilities to obtain EPA from cod liver oil) divided by 10 years (260 days of operation per year were considered) and by 2.46 g (*P. tricornutum*) or by 0.74 g (cod liver oil) of EPA.
[c] Two full technicians (overall salary $68.50/person/day) divided by 2.46 g (*P. tricornutum*) or by 0.74 g (cod liver oil) of EPA.
[d] The controlling step of the process is the seven loads a day it is able to make in the preparative column.

Table 6.17 Comparison of purity and yields of EPA obtained from the microalgae *Isochrysis galbana* (Robles Medina *et al.*, 1995a), *Phaeodactylum tricornutum* (Molina Grima *et al.*, 1996) and *Porphyridium cruentum* (Giménez Giménez *et al.*, 1998).

	Extraction	Urea concentration	HPLC purification	Overall yield and purity
I. galbana				
Purity	22.4[a]	39.4[d]	96.0[f]	96.0
Yield	80.1[a]	54.3[d]	99.6[f]	43.3
P. tricornutum				
Purity	36.3[b]	50.5[e]	95.8[g]	95.8
Yield	90.6[b]	78.6[e]	97.3[g]	69.3
P. cruentum				
Purity	21.7[c]	42.0[e]	94.3[h]	94.3
Yield	69.4[c]	67.7[e]	53.4[h]	25.1

[a] 76 ml of hexane:ethanol (96%) 1:2.5 v/v with 1.6 g KOH per gram of lyophilized biomass; room temperature and 8 h.

[b] 11.9 ml of ethanol (96%) with 1.6 g KOH per gram of wet biomass (57 ml/g of biomass); 60°C and 1 h.

[c] As in [a] but at 60°C and 1 h.

[d] Urea:fatty acid ratio 4:1 w/w, crystallization temperature 4°C and solvent methanol.

[e] As in [d] but using 28°C as crystallization temperature.

[f] Semi-preparative column, Beckman, RP, C_{18}, 10 mm i.d. × 25 cm. Mobile phase methanol:water (1%AcH) 80:20 w/w, flow rate 3 ml/min and mass load 9.5 mg.

[g] Preparative cartridge, Waters, RP, C_{18}, 4.7 cm i.d. × 30 cm. Mobile phase methanol:water (1%AcH) 80:20 w/w, flow rate 66 ml/min and mass load 635 mg.

[h] As in [g] but with a mass load of 133 mg.

References

ABU-NASR, A.M. and HOLMAN, R.T., 1954, Highly unsaturated fatty acids. II. Fractionation by urea inclusion compounds, *J. Am. Oil Chem. Soc.*, **31**, 16–31.

ACKMAN, R.G., RATNAYAKE, W.M.N. and OLSSON, B., 1988, The 'basic' fatty acid composition of Atlantic fish oil: potential similarities useful for enrichment of polyunsaturated fatty acids by urea complexes, *J. Am. Oil Chem. Soc.*, **65**, 136–138.

AHLGREN, G. and MERINO, L., 1991, Lipid analysis of freshwater microalgae: a method study, *Arch. Hydrobiol.*, **121**, 295–306.

AVELDANO, M.I., VAN ROOLLINS, M. and HORROCKS, L.A., 1983, Separation and quantitation of free fatty acids and fatty acids methyl esters by reverse phase high pressure liquid chromatography, *J. Lipid Res.*, **24**, 83–93.

BAJPAI, P. and BAJPAI, P.K., 1993, Eicosapentaenoic acid (EPA) production from microorganisms: a review, *J. Biotechnol.*, **30**, 161–183.

BELTER, P.A., CUSSLER, E.L. and HU, W.-S., 1988, Elution chromatography, in *Bioseparation. Downstream Processing for Biotechnology*, pp. 209–213, Wiley-Interscience Publication, John Wiley & Sons, Inc.

BLIGH, E.G. and DYER, W.J., 1959, A rapid method of total lipid extraction and purification, *Can. J. Biochem. Physiol.*, **37**, 911–917.

CARTENS, M., MOLINA GRIMA, E., ROBLES MEDINA, A., GIMÉNEZ GIMÉNEZ, A. and IBÁÑEZ GONZÁLEZ, M.J., 1996, Eicosapentaenoic acid (20:5n-3) from the marine microalga *Phaeodactylum tricornutum*, *J. Am. Oil Chem. Soc.*, **73**, 1025–1031.

CHISTI, Y. and MOO-YOUNG, M., 1986, Disruption of microbial cells for intracellular products. *Enzyme Microb. Technol.*, **8**, 194–204.

CHOI, K.J., NAKHOST, Z., KRUKONIS, V.J. and KAREL, M., 1987, Supercritical fluid extraction and characterization of lipids from algae *Scenedesmus obliquus*, *Food Biotechnology*, **1**, 263–271.

CHUECAS, L. and RILEY, J.P., 1969, Component fatty acids of the total lipids of some marine phytoplankton, *J. Mar. Biol. Ass. UK*, **49**, 97–116.

COHEN, Z. and COHEN, S., 1991, Preparation of eicosapentaenoic acid (EPA) concentrate from *Porphyridium cruentum*, *J. Am. Oil Chem. Soc.*, **68**, 16–19.

COHEN, Z., REUNGJITCHACHAWALI, M., SIANGDUNG, W. and TANTICHAROEN, M., 1993, Production and partial purification of γ-linolenic acid and some pigments from *Spirulina platensis*, *J. Appl. Phycol.*, **5**, 109–115.

COREY, E.J. and WRIGHT, S.W., 1988, Convenient method for the recovery of eicosapentaenoic acid from cod liver oil, *J. Org. Chem.*, **53**, 5980–5981.

DUNSTAN, G.A., VOLKMAN, J.K., JEFFREY, S.W. and BARRETT, S.M., 1992, Biochemical composition of microalgae from the green algal classes Chlorophyceae and Prasinophyceae. 2. Lipid classes and fatty acids, *J. Exp. Mar. Biol. Ecol.*, **161**, 115–134.

EL BOUSTANI, S., COLETTE, C., MONNIER, L., DESCOMPS, B., CRASTES DE PAULET, A. and MENDY, F., 1987, Enteral absorption in man of eicosapentaenoic acid in different chemical forms, *Lipids*, **22**, 711–714.

ESTEBAN CERDÁN, L., ROBLES MEDINA, A., GIMÉNEZ GIMÉNEZ, A., IBÁÑEZ GONZÁLEZ, M.J. and MOLINA GRIMA, E., 1998, Synthesis of PUFA enriched triglycerides by lipase catalyzed esterification, *J. Am. Oil Chem. Soc.*, in press.

FOLCH, J., LEES, M. and STANLEY, G.M., 1957, A simple method for the isolation and purification of total lipids from animal tissues, *J. Biol. Chem.*, **226**, 497–509.

FUJITA, T. and MAKUTA, M., 1983, Method of purifying eicosapentaenoic acid and its esters, USA patent No. 4,377,526.

GAIDAY, N.V., IMBS, A.B., KUKLEV, D.V. and LATYSHEV, N.A., 1991, Separation of natural polyunsaturated fatty acids by means of iodolactonization, *J. Am. Oil Chem. Soc.*, **68**, 230–233.

GARRIDO, F., BANERJEE, Y., CHISTI, Y. and MOO-YOUNG, M., 1994, Disruption of a recombinant yeast for the release of β-galactosidase, *Bioseparation*, **4**, 319–328.

GIMÉNEZ GIMÉNEZ, A., IBÁÑEZ GONZÁLEZ, M.J., ROBLES MEDINA, A., MOLINA GRIMA, E., GARCÍA SALAS, S. and ESTEBAN CERDÁN, L., 1998, Downstream processing and purification of eicosapentaenoic (20:5n-3) and arachidonic acid (20:4n-6) from the microalga *Porphyridium cruentum*, *Bioseparation*, in press.

GRANT, G., 1988, Multigram isolation of omega-3-fatty acids from fish oil using preparative HPLC, *American Laboratory*, May, 80–85.

GRUGER, E.H. Jr, 1960, Methods of separation of fatty acids from fish oils with emphasis on industrial applications, 11th Annual Conference, Pacific Fisheries Technologists, Gerhart, Oregon, pp. 31–40.

HAAGSMA, N., VAN GENT, C.M., LUTEN, J.B., DE JONG, R.W. and VAN DOORN, E., 1982, Preparation of an ω-3 fatty acid concentrate from cod liver oil, *J. Am. Oil Chem. Soc.*, **59**, 117–118.

HARA, A. and RADIN, N. S., 1978, Lipid extraction of tissues with a low-toxicity solvent, *Analyt. Biochem.*, **90**, 420–426.

HARALDSSON, G.G., HÖSKULDSSON, P.A., SIGURDSSON, S.T., THORSTEINSSON, F. and GUDBJARNASON, S., 1989, The preparation of triglycerides highly enriched with ω-3 polyunsaturated fatty acids via lipase catalyzed interesterification, *Tet. Lett.*, **30**, 1671–1674.

HARTLEY, F.R., 1973, Thermodynamic data for olefin and acetylene complexes of transition metals, *Chem. Rev.*, **73**, 163–190.

HAYASHI, K. and KISHIMURA, H., 1993, Separation of eicosapentaenoic acid-enriched trigly-cerides by column chromatography on silicic acid, *Bull. Fac. Fish. Hokkaido Univ.*, **44**, 24–31.

HEDENSKOG, G. and EBBINGHAUS, L., 1972, Reduction of the nucleic acid content of single-cell protein concentrates, *Biotechnol. Bioeng.*, **14**, 447–457.

HENKE, H. and SCHUBERT, J., 1980, HPLC of fatty acid esters of mono- and polyhydric alcohols, Part 1: Analytical separation, *J. High Res. Chrom. & Chrom. Comm.*, **3**, 69–78.

HIGASHIDATE, S., YAMAUCHI, Y. and SAITO, M., 1990, Enrichment of eicosapentaenoic acid and docosahexaenoic acid esters from esterified fish oil by programmed extraction-elution with supercritical carbon dioxide, *J. Chrom.*, **515**, 295–303.

JORDI, H.C., 1978, Separation of long and short chain fatty acids as naphthacyl and substituted phenacyl esters by high performance liquid chromatography, *J. Liquid. Chrom.*, **1**, 215–230.

KATES, M., 1988, Separation of lipid mixtures, in Burdon, R.H. and Van Knippenberg, P.H. (Eds), *Techniques of Lipidology: Isolation, Analysis and Identification of Lipids*, 2nd printing, 2nd edition, pp. 186–278, Amsterdam, Netherlands: Elsevier Science Publishers BV.

KOCHERT, G., 1978, Quantitation of the macromolecular components of microalgae, in Hellebust, J. and Crage, S. (Eds), *Handbook of Phycological Methods. Physiological and Biochemical Methods*, pp. 189–195, London: Cambridge University Press.

KOSUGI, Y. and AZUMA, N., 1994, Synthesis of triacylglycerol from polyunsaturated fatty acid by immobilized lipase, *J. Am. Oil Chem. Soc.*, **71**, 1397–1403.

KUBOTA, F., GOTO, M., NAKASHIO, F. and HANO, T., 1997, Separation of polyunsaturated fatty acids with silver nitrate using a hollow-fiber membrane extractor, *Separation Sci. Technol.*, **32**, 1529–1541.

LAWSON, L.D. and HUGHES, B.G., 1988, Human absorption of fish oil fatty acids as triacyl-glycerols, free acids, or ethyl esters, *Biochem. Biophysic. Res. Commun.*, **152**, 328–335.

LIE KEN JIE, M.S.F., 1980, The characterization of long-chain fatty acids and their derivatives by chromatography, *Adv. Chromatogr.*, **18**, 1–57.

LUNDBERG, W.O. and JÄRVI, P., 1966, Peroxidation of polyunsaturated fatty compounds, in Holman, R.T., Lundberg, W.O. and Malkin, T. (Eds), *Progress in the Fats and Other Lipids*, Vol 9, pp. 265–275, London: Pergamon Press Ltd.

MARKLEY, K.S., 1964, Technique of separation. A. Distillation, salt solubility, low-temperature crystallization, in Markley, K.S. (Ed.), *Fatty Acids, their Chemistry, Properties, Production, and Uses*, 2nd edition, Part 3, pp. 1983–2123, Interscience Publishers.

MEHLENBACHER, V.C., 1970, Espectroscopía de absorción ultravioleta, in Análisis de grasas y aceites, Ed. Urmo, pp. 538–543, Bilbao, Spain.

MÉNDEZ, E., LUCAS, M.E., JACHMANIAN, I., and GROMPONE, M.A., 1993, Estabilización de aceite de pescado con propóleo, *5° Congreso Latinoamericano sobre Procesamiento de Grasas y Aceites*. Venezuela.

MISHRA, V.K., TEMELLI, F. and OORAIKUL, B., 1993, Extraction and purification of ω-3 fatty acids with emphasis on supercritical fluid extraction – a review, *Food Res. International*, **26**, 217–226.

MOFFAT, C.F., McGILL, A.S., HARDY, R. and ANDERSON, R.S., 1993, The production of fish oils enriched in polyunsaturated fatty acid-containing triglycerides, *J. Am. Oil Chem. Soc.*, **70**, 133–138.

MOLINA GRIMA, E., ROBLES MEDINA, A, GIMÉNEZ GIMÉNEZ, A., SÁNCHEZ PÉREZ, J.A., GARCÍA CAMACHO, F. and GARCÍA SÁNCHEZ, J.L., 1994, Comparison between extraction of lipids and fatty acids from microalgal biomass, *J. Am. Oil Chem. Soc.*, **71**, 955–959.

MOLINA GRIMA, E., ROBLES MEDINA, A., GIMÉNEZ GIMÉNEZ, A. and IBÁÑEZ GONZÁLEZ, M.J., 1996, Gram-scale purification of eicosapentaenoic acid (EPA, 20:5n-3) from wet *Phaeodactylum tricornutum* UTEX 640 biomass, *J. Appl. Phycol.*, **8**, 359–367.

NAGLE, N. and LEMKE, P., 1990, Production of methyl ester fuel from microalgae, *Appl. Biochem. Biotechnol.*, **24/25**, 355–361.

NILSSON, W.B., GAUGLITZ, E.J. Jr, HUDSON, J.K., STOUT, V.F. and SPINELLI, J., 1988, Fractionation of menhaden oil ethyl esters using supercritical fluid CO_2, *J. Am. Oil Chem. Soc.*, **65**, 109–117.

ÖZCIMDER, M. and HAMMERS, W.E., 1980, Fractionation of fish oil fatty acid methyl esters by means of argentation and reverse-phase high-performance liquid chromatography, and its utility in total fatty acid analysis, *J. Chrom.*, **187**, 307–317.

PERRUT, M., 1988, Purification of polyunsaturated fatty acid (EPA and DHA) ethyl esters by preparative high performance liquid chromatography, *LC-GC International*, **1**, 58–62.

PLAKAS, S.P. and GUARINO, A.M., 1986, Omega-3 fatty acids and fish oils: is the news all good? *Proceedings of the Eleventh Annual Tropical and Subtropical Fisheries Conference of the Americas*, compiled by Ward, D.R. and Smith, B.A., individual papers edited by respective authors, Texas Agricultural Extension Service, Marine Service Program.

POHL, P. and ZURHEIDE, F., 1982, Fat production in freshwater and marine algae, in Hoope, H.A., Levring, T. and Tanaka, Y. (Eds), *Marine Algae in Pharmaceutical Science*, Vol. 2, pp. 65–80, New York: Walter de Gruyter & Co.

POLAK, J.T., BALABAN, M., PEPLOW, A. and PHILIPS, A.J., 1989, Supercritical carbon dioxide extraction of lipids from algae, in Johnston, K.P. and Penninger, J.M.L. (Eds), *Supercritical Fluid Science and Technology*, ACS Symposium Series, No. 406, pp. 449–487.

PRIVETT, O. S., 1968, Preparation of polyunsaturated fatty acids from natural sources, in Holmnan R.T. (Ed.), *Progress in the Chemistry of Fats and Other Lipids*, Vol. 9, Chapter 11, Part 3, pp. 409–452, Pergamon Press.

RATNAYAKE, W.M.N., OLSSON, B., MATTHEWS, D. and ACKMAN, R.G., 1988, Preparation of omega-3 PUFA concentrates from fish oils via urea complexation, *Fat Sci. Technol.*, **10**, 381–386.

ROBLES MEDINA, A., GIMÉNEZ GIMÉNEZ, A., GARCÍA CAMACHO, F., SÁNCHEZ PÉREZ, J.A., MOLINA GRIMA, E. and CONTRERAS GÓMEZ, A., 1995a, Concentration and purification of stearidonic, eicosapentaenoic, and docosahexaenoic acids from cod liver oil and the marine microalga *Isochrysis galbana*, *J. Am. Oil Chem. Soc.*, **72**, 575–583.

ROBLES MEDINA, A., GIMÉNEZ GIMÉNEZ, A., MOLINA GRIMA, E. and GARCÍA SÁNCHEZ, J.L., 1995b, Obtención de concentrados de ácidos grasos poliinsaturados por el método de los compuestos de inclusión de urea, *Grasas y Aceites*, **42**, 174–182.

SCHLENK, H., 1954, Urea inclusion compounds of fatty acids, in *Progress in the Fats and other Lipids*, Vol. 2, pp. 243–267, London: Pergamon Press.

SCHOLFIELD, C.R., 1979, Silver nitrate-high performance liquid chromatography of fatty methyl esters, *J. Am. Oil Chem. Soc.*, April, 510–511.

SHIMADA, Y., MARUYAMA, K., OKAZAKI, S., NAKAMURA, M., SUGIHARA, A. and TOMINAGA, Y., 1994, Enrichment of polyunsaturated fatty acids with *Geotrichum candidum* lipase, *J. Am. Oil Chem. Soc.*, **71**, 951–954.

SHIMADA, Y., MARUYAMA, K., SUGIHARA, A., MORIYAMA, S. and TOMINAGA, Y., 1997a, Purification of docosohexaenoic acid from tuna oil by a two-step enzymatic method: hydrolysis and selective esterification, *J. Am. Oil Chem. Soc.*, **74**, 1441–1446.

SHIMADA, Y., SUGIHARA, A., SHIBAHIRAKI, M., FUJITA, H., NAKANO, H., NAGAO, T., TERAI, T. and TOMINAGA, Y., 1997b, Purification of γ-linoleic acid from borage oil by a two-step enzymatic method, *J. Am. Oil Chem. Soc.*, **74**, 1465–1470.

SHINOWARA, G.Y. and BROWN, J.B., 1940, Studies on the chemistry of the fatty acids. VI. The application of crystallization methods to the isolation of arachidonic acid, with a comparison of the properties of this acid prepared by crystallization and by debromination. Observations on the structure of arachidonic acid, *J. Biol. Chem.*, **134**, 331–340.

SHUKLA, V., 1988, Recent advances in the high performance liquid chromatography of lipids, *Prog. Lipid Res.*, **27**, 5–38.

SMITH, A.E., 1952, The crystal structure of the urea-hydrocarbon complexes, *Acta Cryst.*, **5**, 224–235.

STOUT, V.F. and SPINELLI, J., 1987, Polyunsaturated fatty acids from fish oils, USA patent No. 4,675,132.

SWERN, D., 1964, Techniques of separation. Urea complexes, in Markley, K.S. (Ed.), *Fatty Acids. Their Chemistry, Properties, Production, and Uses*, Part 3, second edition, pp. 2309–2358, New York: Interscience Publishers.

TERAMOTO, M., MATSUYAMA, H., OHNISHI, N., UWAGAWA, S. and NAKAI, K., 1994, Extraction of ethyl and methyl esters of polyunsaturated fatty acids with aqueous silver nitrate solutions, *Ind. Eng. Chem. Res.*, **33**, 341–345.

TOKIWA, S., KANAZAWA, A. and TESHINA, S., 1981, Preparation of eicosapentaenoic and docosahexaenoic acid by reversed phase high performance liquid chromatography, *Bull. Jap. Soc. Sci. Fish.*, **47**, 675.

TRAITLER, H., WILLE, H.J. and STUDER, A., 1988, Fractionation of blackcurrant seed oil, *J. Am. Oil Chem. Soc.*, **65**, 755–760.

VALCARCEL CASES, M. and GÓMEZ HENS, A., 1990, Técnicas analíticas de separación, pp. 373–376, Barcelona: Ed. Reverté S.A.

WILLE, H.J., TRAITLER, H. and KELLY, M., 1987, Production of polyenoic fish oil fatty acids by combined urea fractionation and industrial scale preparative 'HPLC', *Revue Française des corps gras.*, **34**, 69–73.

WRIGHT, S.W., KUO, E.Y. and COREY, E.J., 1987, An effective process for the isolation of docosahexaenoic acid in quantity from cod liver oil, *J. Org. Chem.*, **52**, 4399–4401.

ZHUKOV, A.V. and VERESHCHAGIN, A.G., 1981, Current techniques of extraction, purification and preliminary fractionation of polar lipids of natural origin, *Adv. Lipids Res.*, **18**, 247–282.

ZU-YI LI and WARD, O.P., 1993, Enzyme catalyzed production of vegetable oils containing omega-3 polyunsaturated fatty acid, *Biotechnol. Lett.*, **15**, 185–188.

7

Microalgal carotenoids

SYNNØVE LIAAEN-JENSEN AND EINAR SKARSTAD EGELAND

Introduction

Carotenoids are yellow to red isoprenoid polyene pigments, widely distributed in nature, where they serve particular functions (Isler, 1971; Krinsky, 1994; Britton *et al.*, 1995a; Frank *et al.*, 1997). They are produced *de novo* by all photosynthetic organisms, including microalgae (Goodwin, 1980). Of the more than six hundred naturally occurring carotenoids structurally identified today (Straub, 1987; Britton *et al.*, 1995a), more than one hundred different carotenoids are encountered in microalgae.

Whereas large scale cultivation of *Dunaliella* sp. and *Haematococcus* sp. for carotenoid production on a technical scale is dealt with in other chapters, the purpose of the present chapter is to survey the different types of carotenoids encountered in microalgae, and to focus on microalgae as potential sources for the isolation of particular carotenoids.

The annual production of phytoplankton in the oceans has been estimated to be 4×10^{10} metric tons of organic matter in terms of dry weight, of which around 0.1 per cent or several million tons, constitutes carotenoids (Fox, 1974). Whereas microalgae from natural phytoplankton blooms or from cultivation are only likely to serve as a natural source for their major carotenoids, the structures of minor microalgal carotenoids could also serve as models for chemical synthesis and challenge the cloning of carotenogenic genes for future biotechnological application.

In this chapter on microalgal constituents, seaweeds are considered as macroalgae and are excluded. Since this review is selective rather than exhaustive examples are frequently chosen from the authors' own research.

Structures of microalgal carotenoids

Major carotenoids and general structural features

Microalgal carotenoids exhibit great structural diversity. *De novo* biosynthesis of some of these carotenoids is restricted to algae. Complete chemical structures of common carotenoids from microalgae are presented in Figure 7.1. These structures serve to characterize the general structural features of microalgal carotenoids.

New, rational IUPAC names are used here for cyclic carotenes including β,β-carotene (**1**, formerly β-carotene) and β,ε-carotene (**2**, formerly α-carotene), whereas trivial names are maintained when available for the oxygen containing xanthophylls. Complete rational names including stereochemical notations are compiled elsewhere (Straub, 1987; Britton *et al.*, 1995a). The numbering of the carotenoid skeleton is indicated on a general carotenoid structure in Figure 7.2.

In Figure 7.1, β,β-carotene (**1**) and β,ε-carotene (**2**) represent common dicyclic C_{40}-carotenes. Zeaxanthin (**3**) and lutein (**4**) are simple carotene 3,3′-diols formally derived from β,β-carotene (**1**) and β,ε-carotene (**2**). Epoxidic carotenoids are usually 5,6(5′,6′)-epoxides as demonstrated for violaxanthin (**5**) and diadinoxanthin (**6**). This structural feature is also seen in more complex structures (**11–14**). Acetylenic carotenoids, produced *de novo* only by microalgae, generally contain a 7,8(7′,8′)-dehydro linkage. Diatoxanthin (**7**) and alloxanthin (**8**) are simple mono- and diacetylenic diols respectively. Again this structural feature is also encountered in more complex structures (**9,10**). The unique C_{37}-skeletal nor-carotenoids, formally lacking three in-chain carbon atoms deserve particular attention, with peridinin (**11**) as prominent example besides pyrrhoxanthin (**10**). Allenic carotenoids represent another special structural feature with fucoxanthin (**14**) and peridinin (**11**) as the most important representatives. These carotenoids are the two produced in largest quantity in the oceans. Other allenic carotenoids are neoxanthin (**12**) and vaucheriaxanthin (**13**). Allenic carotenoids frequently carry an acetate function in the 3-position, exemplified by fucoxanthin (**14**) and peridinin (**11**). Various carotenoid ketones are common, including conjugated, simple 4(4′) ketones such as echinenone (**15**), canthaxanthin (**16**) and the important feed ingredient for salmon fishes, astaxanthin (**17**). Finally glycosidic carotenoids, with a single exception (Aakermann *et al.*, 1993) restricted to blue–green algae (Cyanobacteria), are illustrated with the monocyclic myxoxanthophyll (**21**) and the aliphatic carotenoid oscillaxanthin (**22**).

The structural modifications described above may in general be referred to as of two types: end group modifications (most examples); and in-chain modifications involving substitution or other structural changes of the polyene chain (**6–11, 13–14, 18–20**). It should be pointed out that besides the various structural features focused on above, the common algal carotenoids contain from zero [β,β-carotene (**1**)] up to five [fucoxanthin (**14**), peridinin (**11**)] chiral centres and one chiral axis (allenic carotenoids) in their carotenoid skeleton, complicating a chemical synthesis. However, the chemical synthesis of both optically active peridinin (**11**, Yamano and Ito, 1993), fucoxanthin (**14**, Yamano *et al.*, 1995) and neoxanthin (**12**, Baumeler and Eugster, 1992) have recently been achieved in non-technical scale.

Stereochemical aspects of carotenoids, comprising *E/Z* (*trans/cis*) isomerization of the polyene chain, and chirality including allenic isomerism have recently been discussed (Liaaen-Jensen, 1997).

Modern studies on chemical conversion to a number of interesting carotenoid products have been effected with particularly fucoxanthin (**14**, Bonnett *et al.*, 1969; Haugan *et al.*, 1992, Haugan and Liaaen-Jensen, 1994b), peridinin (**11**, Kjøsen *et al.*, 1976), neoxanthin (**12**) and violaxanthin (**5**) (Haugan and Liaaen-Jensen, 1994c), and prasinoxanthin (**20**, Foss *et al.*, 1984).

CAROTENES

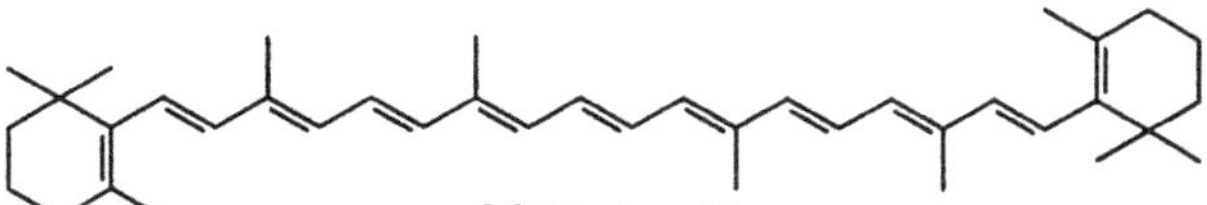

β,β-Carotene (1)

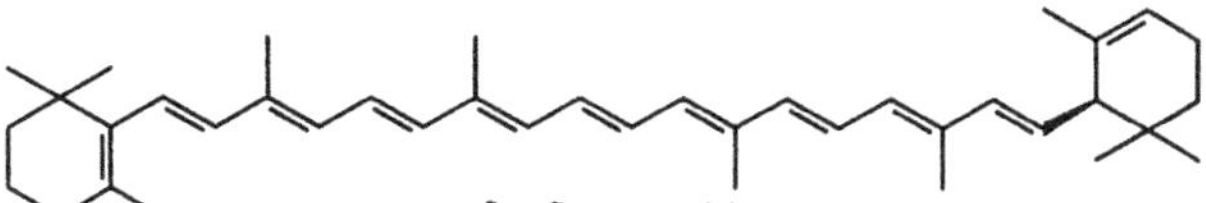

β,ε-Carotene (2)

SIMPLE CAROTENOLS

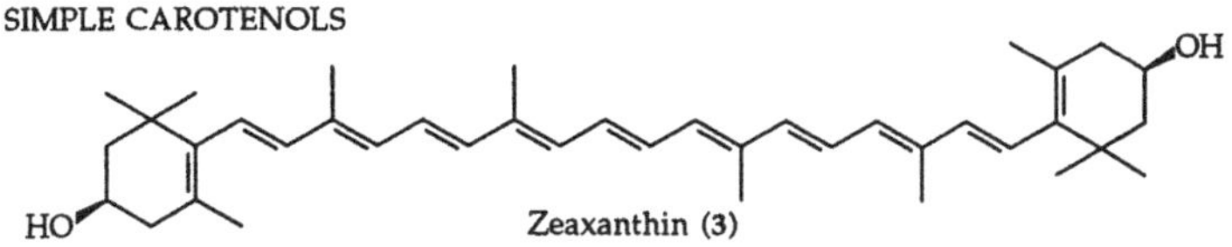

Zeaxanthin (3)

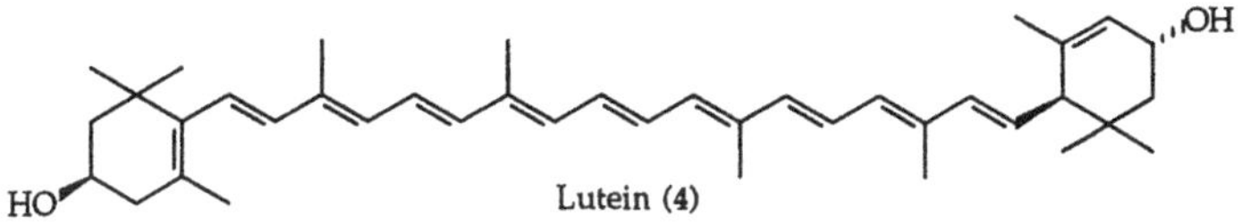

Lutein (4)

EPOXIDES

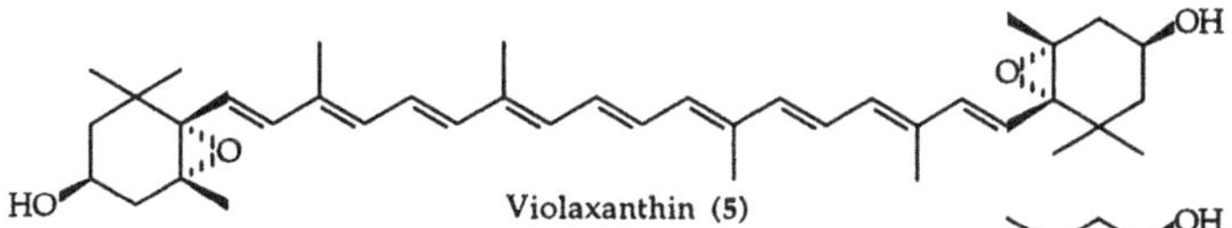

Violaxanthin (5)

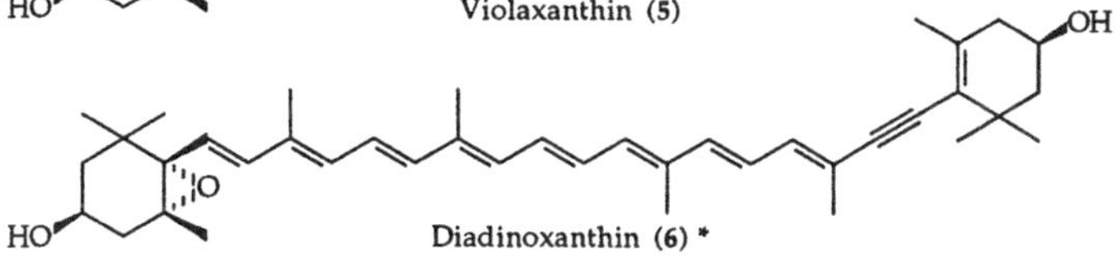

Diadinoxanthin (6) *

ACETYLENIC

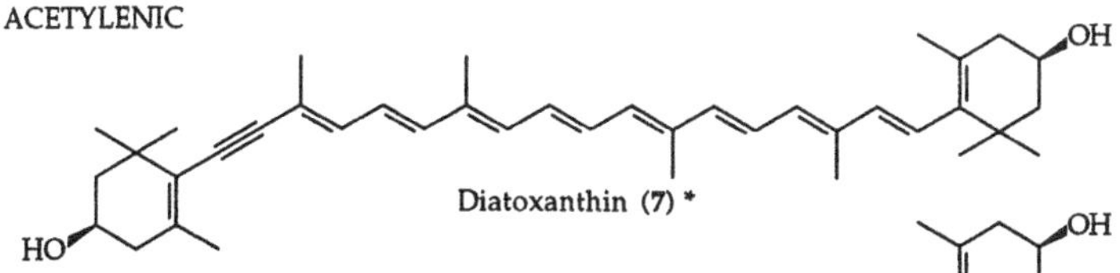

Diatoxanthin (7) *

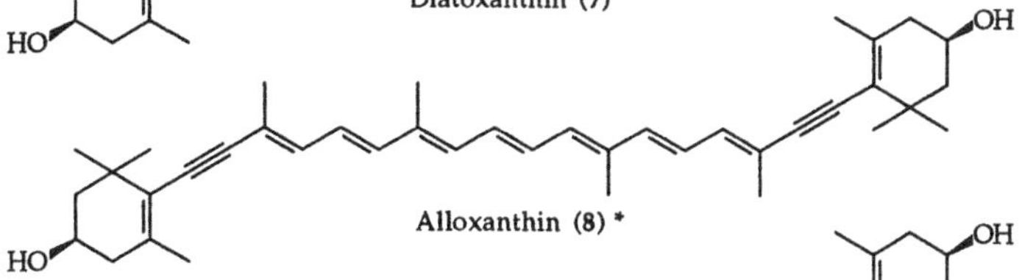

Alloxanthin (8) *

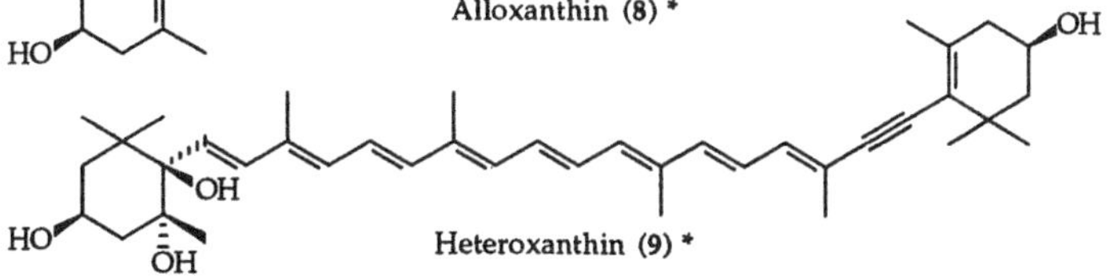

Heteroxanthin (9) *

*) Produced *de novo* only by algae.

Figure 7.1a Common microalgal carotenoids.

C$_{37}$-SKELETAL

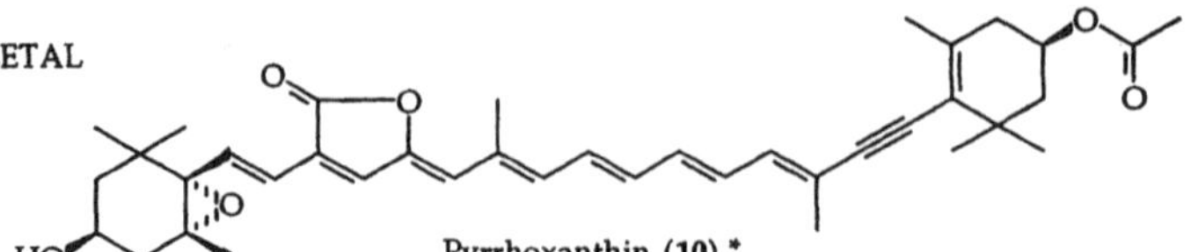

Pyrrhoxanthin (10) *

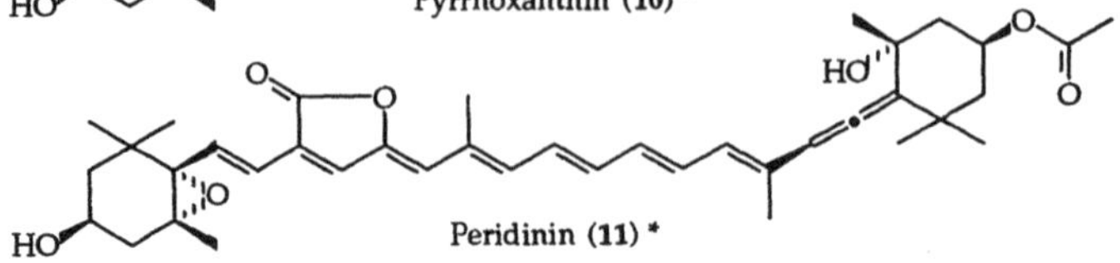

Peridinin (11) *

ALLENIC

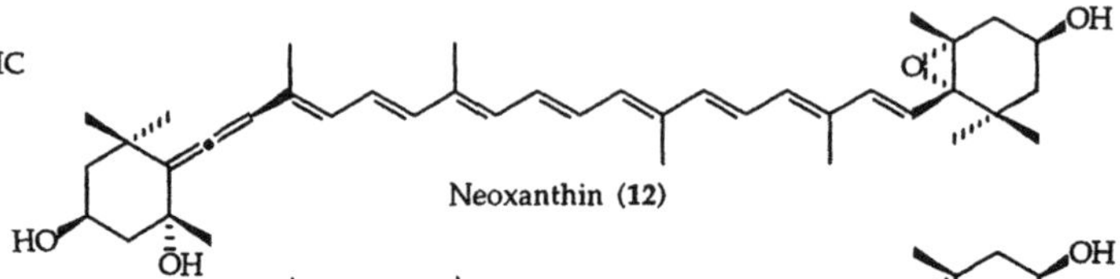

Neoxanthin (12)

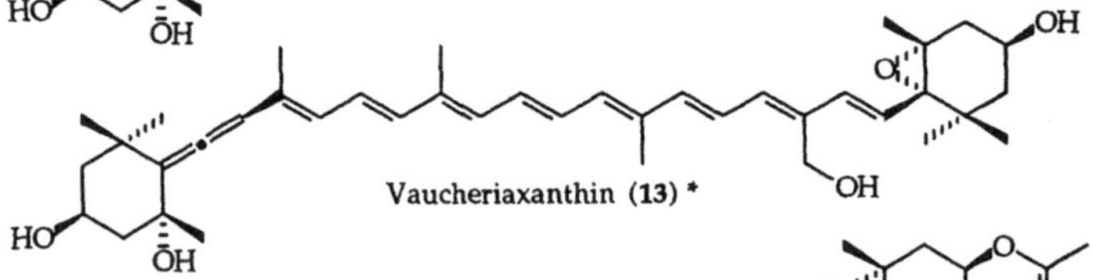

Vaucheriaxanthin (13) *

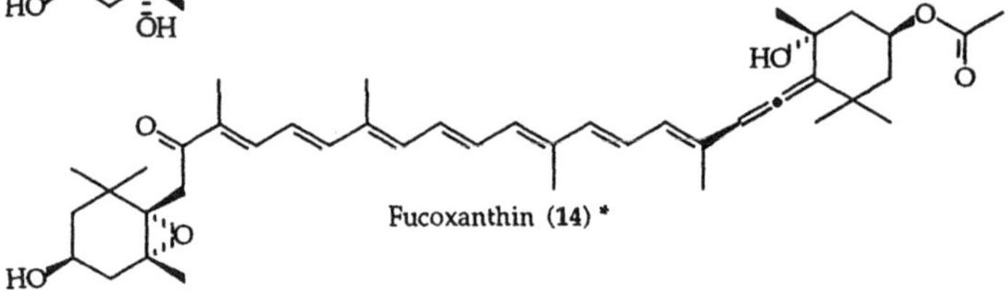

Fucoxanthin (14) *

KETONES

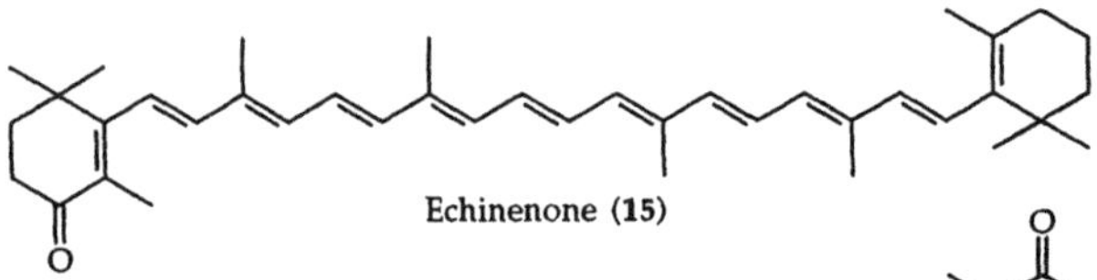

Echinenone (15)

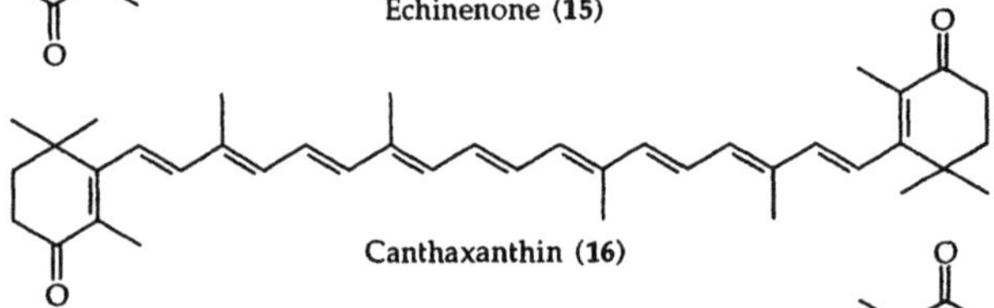

Canthaxanthin (16)

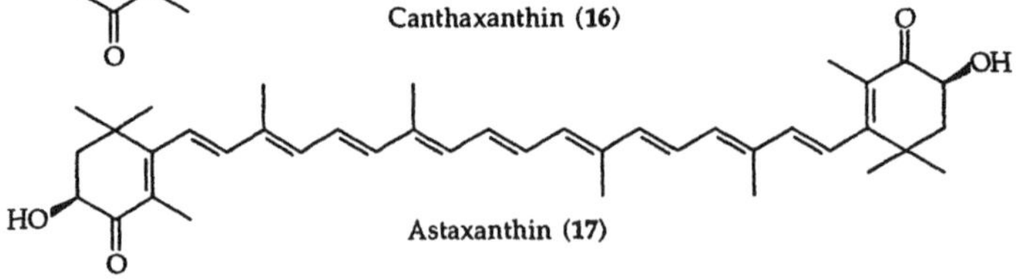

Astaxanthin (17)

*) Produced *de novo* only by algae.

Figure 7.1b Common microalgal carotenoids.

KETONES cont.

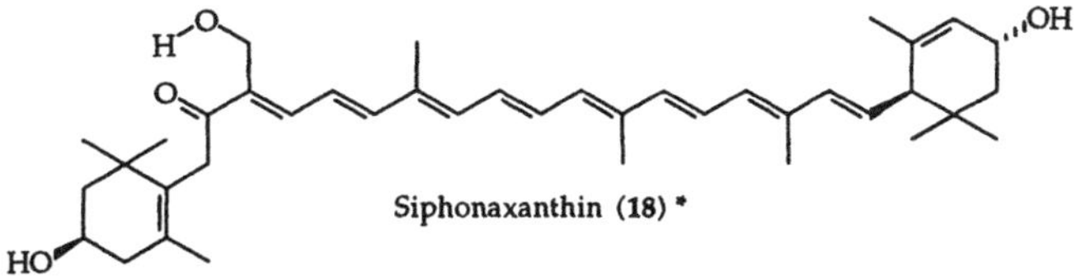

Siphonaxanthin (18) *

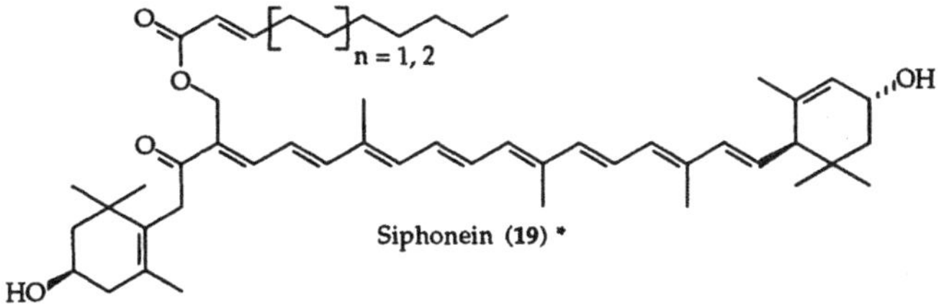

Siphonein (19) *

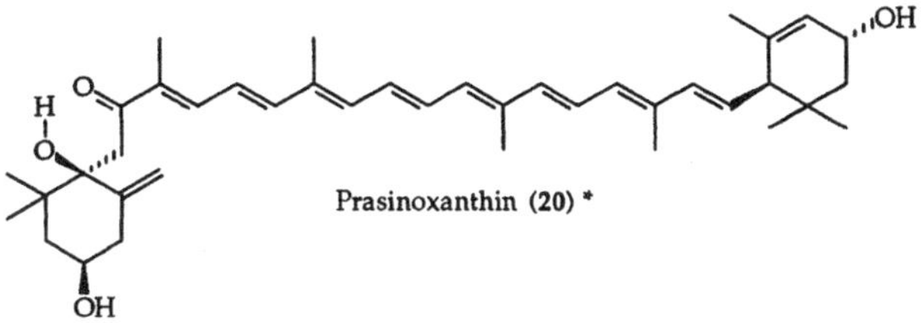

Prasinoxanthin (20) *

GLYCOSIDIC

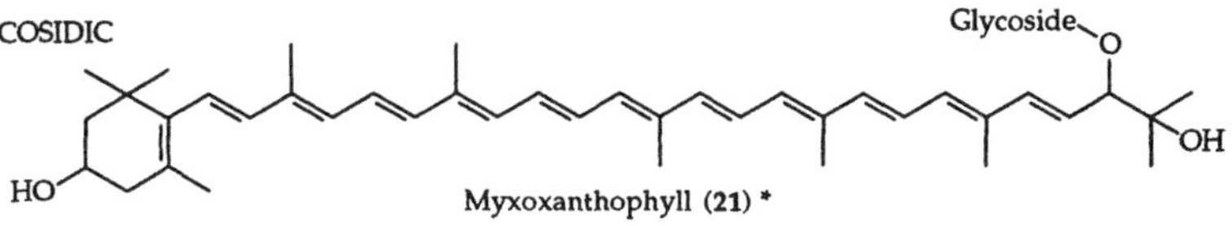

Myxoxanthophyll (21) *

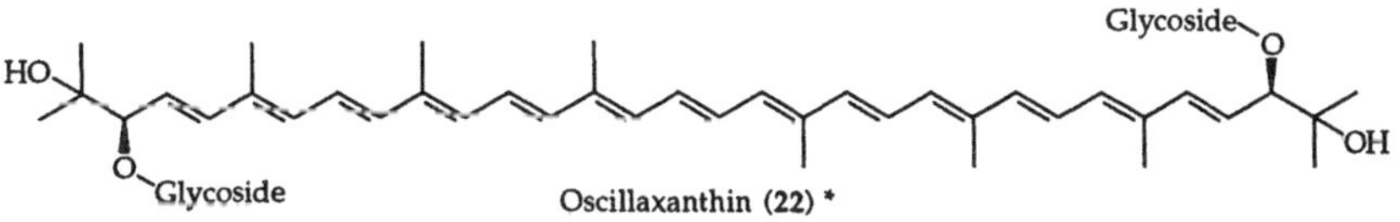

Oscillaxanthin (22) *

*) Produced *de novo* only by algae.

Figure 7.1c Common microalgal carotenoids.

Distribution of microalgal carotenoids

The distribution of individual carotenoids in the various microalgal classes will now be commented on, with reference to the classification system of Christensen (1980–94, see Table 7.1). The carotenoid end groups listed in Figure 7.2 will be referred to in the text and in Table 7.2.

Nostocophyceae (blue–green algae/cyanobacteria; and prochlorophytes)

These algae can be divided into two types on the basis of their pigment composition. Type 1 does not produce chlorophyll *b*.

Table 7.1 Algal classification (Christensen, 1980–94).

Kingdom	Division	Class	Other class name[a]	English names	Approx. no. of species[b]
Procaryota	Cyanophyta	Nostocophyceae	Cyanophyceae + Prochlorophyceae; Cyanobacteria	Blue–green algae + prochlorophytes; cyanobacteria	2000
Eucaryota	Rhodophyta	Bangiophyceae	Rhodophyceae; Bangiophyceae + Florideophyceae	Red algae	5000–5500
	Chromophyta	Cryptophyceae		Recoiling algae, cryptophytes	200
		Dinophyceae		Dinoflagellates, peridinians	2000 (+ 2000 fossil species)
		Chrysophyceae		Golden–brown algae	1000
		Pelagophyceae	(also included in Chrysophyceae)	Pelagophytes	?
		Synurophyceae	(also included in Chrysophyceae)	Synurophytes	?
		Fucophyceae	Phaeophyceae	Brown algae	1500–2000
		Diatomophyceae	Bacillariophyceae	Diatoms	100 000
		Tribophyceae	Xanthophyceae + Raphidophyceae; Xanthophyceae + Chloromonadophyceae	Yellow–green algae	600 (Xanthophyceae)
		Eustigmatophyceae		Eustigmatophytes	12
		Prymnesiophyceae	Haptophyceae	Prymnesiophytes (haptophytes)	500
		Glaucocystophyceae	Glaucophyceae	Glaucocystophytes	3[c]
	Chlorophyta	Prasinophyceae	Micromonadophyceae	Prasinophytes	180
		Pedinophyceae	Loxophyceae (also included in Prasinophyceae)	Pedinophytes	?
		Charophyceae[d]	(also included in Chlorophyceae)	Charophytes	4000–6000
		Chlorophyceae[e]		Green algae, chlorophytes	2000

Table 7.1 Algal classification (Christensen, 1980–94) (contd.).

Kingdom	Division	Class	Other class name[a]	English names	Approx. no. of species[b]
	Euglenophyta	Euglenophyceae		Euglenoids, euglenophytes	800
	Chlorarachniophyta	Chlorarachniophyceae		Chlorarachniophytes	2

[a]The present algal class is not always equal to the classes presented here. ' + ' means that the given class names together cover the class name given by Christensen (1980–94).
[b]Hoek et al. (1995).
[c]Christensen (1980–94) divides this class into four families, hence the number of species should be more than three.
[d]Hoek et al. (1995) divide this class into the classes Klebsormidiophyceae, Zygnematophyceae and Charophyceae.
[e]Hoek et al. (1995) divide this class into the Chlorophyceae, Ulvophyceae, Cladophorophyceae, Bryopsidophyceae, Dasycladophyceae, Trentepohliaphyceae and Pleurastrophyceae, and suggest further that their Chlorophyceae should be divided into a new Chlorophyceae, Chlamydophyceae and maybe also Oedogoniophyceae class.

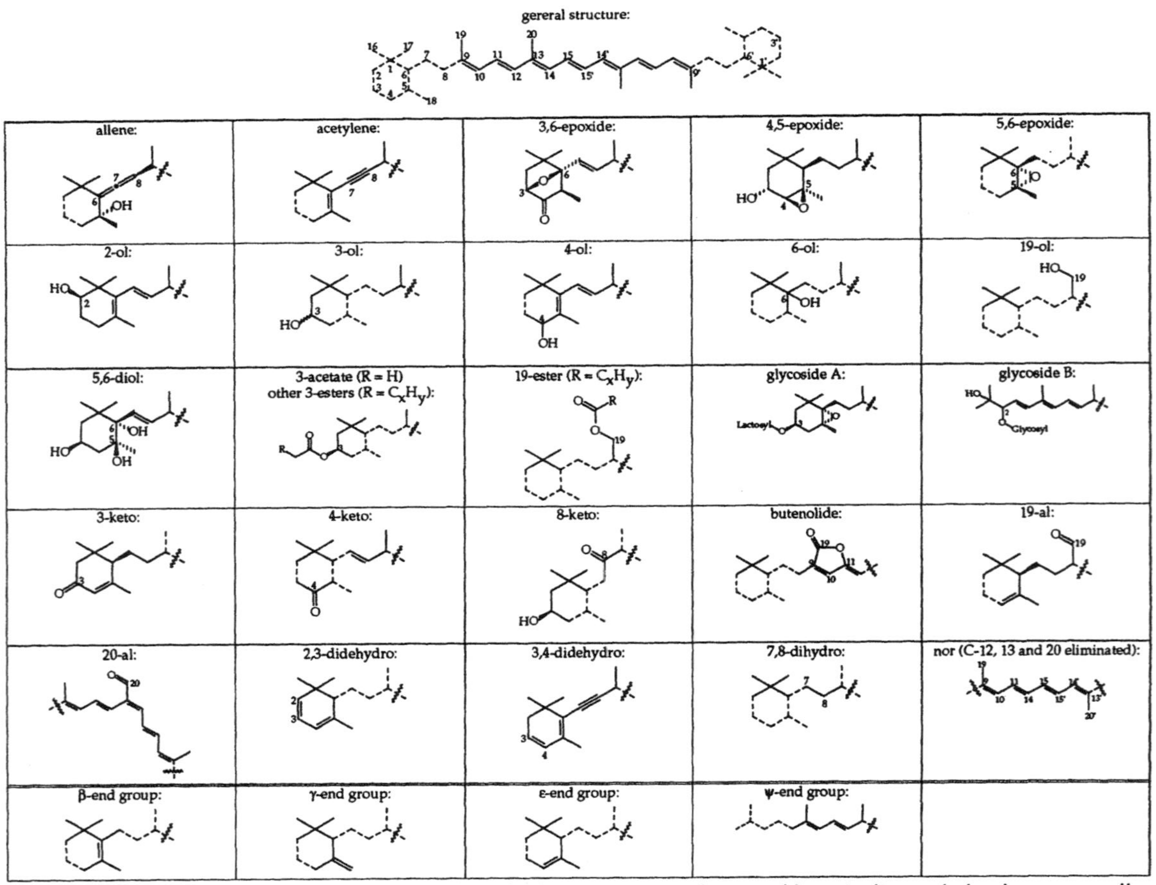

Figure 7.2 Structural elements in algal carotenoids. Chirality is included when unequivocal. Dotted lines indicate skeletal structure allowing for structural modifications.

152

Type 1 (blue–green algae)

Carotenoids with ψ-end group are common, whereas carotenoids with ε-end group are absent (Goodwin, 1980). For end group designations according to IUPAC nomenclature (Britton *et al.*, 1995a) see Figure 7.2. Carotenoid glycosides and 4-keto-type carotenoids are encountered. Epoxidic carotenoids are generally absent, except for a minor furanoid, potential rearrangement product (Hertzberg and Liaaen-Jensen, 1971; Rowan, 1989). β,β-Carotene (**1**) and echinenone (**15**) are always present.

Type 2 (prochlorophytes)

In this small group zeaxanthin (**3**) and carotenes (**1, 2**) are dominant (Foss *et al.*, 1987; Goericke and Repeta, 1992).

Bangiophyceae (red algae)

The red algae comprise a small number of microalgae, in the order Porphyridiales (Christensen, 1980–94). A few species are reported to produce β,β-carotene (**1**) and its hydroxylated derivatives cryptoxanthin (**23**) and zeaxanthin (**3**) (Goodwin, 1980). Red macroalgae are reported to produce β,ε-carotene (**2**) and lutein (**4**) in addition to **1**, **3** and **23** (Goodwin, 1980). The origin of fucoxanthin (**14**) isolated from red macroalgae is disputed (Liaaen-Jensen, 1978).

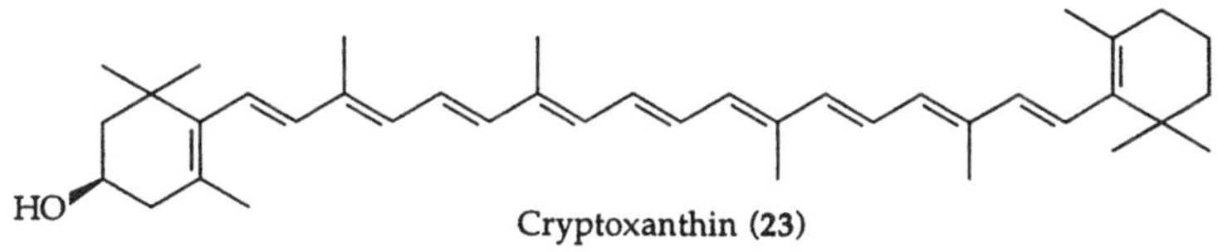

Cryptoxanthin (**23**)

Cryptophyceae (recoiling algae, cryptophytes)

Species of this class contain acetylenic carotenoids and carotenoids with ε-end groups, whereas epoxidic carotenoids are absent. The diacetylenic alloxanthin (**8**), besides the monoacetylenic monadoxanthin (**24**) and crocoxanthin (**25**) are common (Bjørnland and Liaaen-Jensen, 1989).

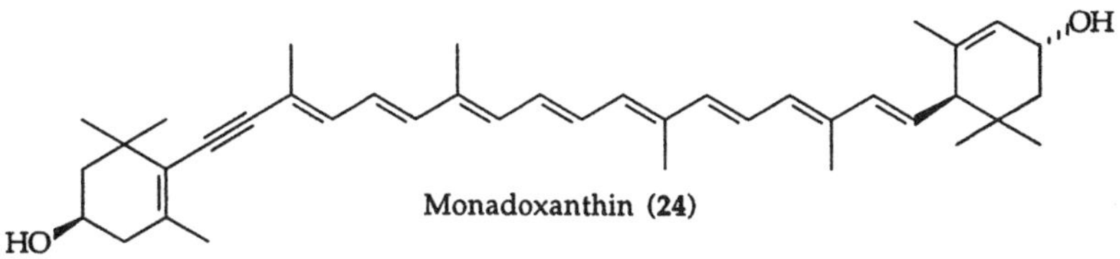

Monadoxanthin (**24**)

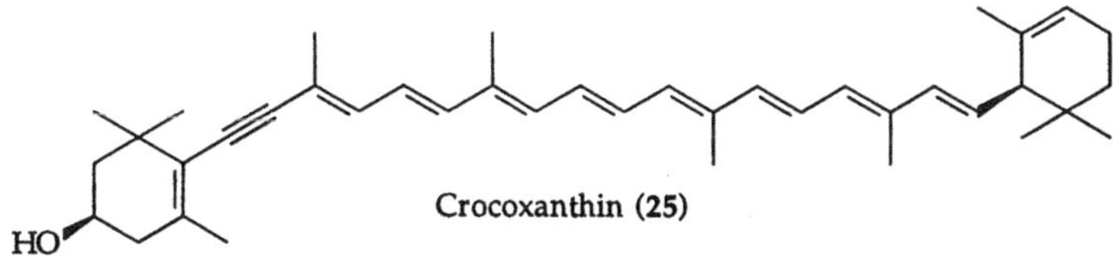

Crocoxanthin (**25**)

Dinophyceae (dinoflagellates, peridinians)

The dinoflagellates may be divided into three pigment types according to their carotenoid composition. Carotenoids with ε-end group are generally absent, whereas carotenoid acetates and epoxidic, allenic and acetylenic carotenoids are common (Bjørnland and Liaaen-Jensen, 1989; Young, 1993). Dinoflagellates generally contain β,β-carotene (**1**) and the monoacetylenic C_{40}-carotenoids diatoxanthin (**7**) and diadinoxanthin (**6**).

Type 1 (Rowan, 1989; Liaaen-Jensen, 1998)

Peculiar to this major carotenoid type are nor-carotenoids (C_{37} skeletal), and in-chain modifications to carotenoid lactones. Peridinin (**11**) is the major carotenoid. A minor, complex carotenoid disaccharide P457 (13'-*cis*-7',8'-dihydroneoxanthin-20'-al 3'-β-lactoside, **26**) occurring in several species, is the only carotenoid glycoside encountered in eucaryote algae (Aakermann *et al.*, 1993; Englert *et al.*, 1995).

P457 (**26**)

Type 2 (Rowan, 1989; Liaaen-Jensen, 1998)

Major carotenoids are fucoxanthin (**14**) or its 19'-butanoyloxy (**27**) or 19'-hexanoyloxy (**28**) derivatives.

19'-Butanoyloxyfucoxanthin (**27**), n = 1
19'-Hexanoyloxyfucoxanthin (**28**), n = 2

Type 3

This carotenoid type is represented by two species (Watanabe *et al.*, 1987, 1990; Elbrächter and Schnepf, 1996), containing both chlorophyll *a* and *b* together with β,β-carotene (**1**), but no fucoxanthin (**14**) or peridinin (**11**). Prasinoxanthin (**20**) is reported from one species (Elbrächter and Schnepf, 1996). A single species is reported to have incorporated intact cells of a green flagellate (class Pedinophyceae) (Sweeney, 1976).

Chrysophyceae (golden–brown algae)

Epoxidic and allenic carotenoids are generally occurring. The major carotenoid is fucoxanthin (**14**), occasionally co-occurring with the 19′-esters **27** and **28**, and β,β-carotene (**1**) (Bjørnland and Liaaen-Jensen, 1989). The ε-end group is only encountered in carotenes. This class may also be divided into two carotenoid types.

Type 1

Species belonging to this carotenoid type have zeaxanthin (**3**) and violaxanthin (**5**) as minor carotenoids.

Type 2

The acetylenic carotenoids diatoxanthin (**7**) and diadinoxanthin (**6**) are accompanying carotenoids.

Diatomophyceae (diatoms)

Diatoms generally produce monoacetylenic and allenic carotenoids with fucoxanthin (**14**) as the major carotenoid. Diatoxanthin (**7**) and diadinoxanthin (**6**) are generally occurring in addition to β,β-carotene (**1**). 4-Ketocarotenoids may occur as minor constituents (Bjørnland and Liaaen-Jensen, 1989; Rowan, 1989).

Tribophyceae (yellow–green algae)

To this algal class are transferred the previous class Xanthophyceae as well as most members of the Raphidophyceae. Monoacetylenic, allenic and epoxidic carotenoids are dominant, whereas carotenoids with ε-end groups are absent. Diadinoxanthin (**6**), diatoxanthin (**7**) and heteroxanthin (**9**) are always present (Goodwin, 1980). Vaucheriaxanthin (**13**), as diester, is reported from some species (Rowan, 1989).

Eustigmatophyceae (eustigmatophytes)

Acetylenic carotenoids and carotenoids with ε-end groups are absent. Allenic and epoxidic carotenoids are common with vaucheriaxanthin (**13**), as free tetrol or ester-

ified, frequently as the major carotenoid. Zeaxanthin (**3**) and the epoxides violaxanthin (**5**) and antheraxanthin (**29**) are also common. Again 4-ketocarotenoids
may occur as minor components (Rowan, 1989; Young, 1993).

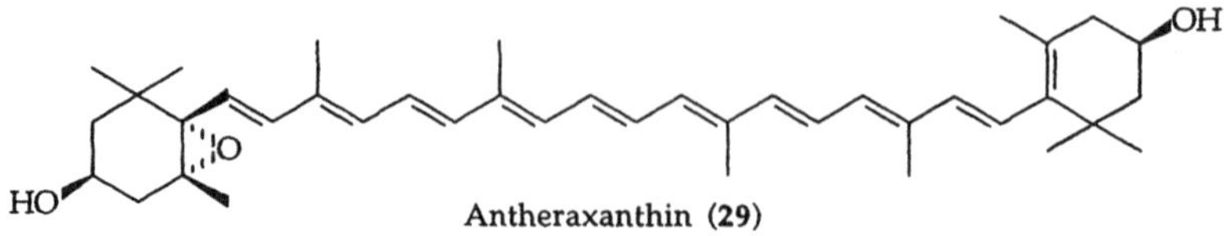

Antheraxanthin (**29**)

Prymnesiophyceae (prymnesiophytes) (haptophytes)

Acetylenic, allenic and epoxidic carotenoids are characteristic for this algal class.
Fucoxanthin (**14**), or its 19′-esters (**27, 28**), is the major carotenoid, whereas diatoxanthin (**7**) and diadinoxanthin (**6**) are generally present. Besides the common β,β-
carotene (**1**), the presence of β,ε-carotene (**2**) has been reported as the only carotenoid with ε-end group (Bjørnland and Liaaen-Jensen, 1989).

Prasinophyceae (prasinophytes)

No acetylenic carotenoids have been encountered, whereas allenic and epoxidic
carotenoids, as well as carotenoids with ε-end groups are common. On the basis
of their carotenoid signature the prasinophytes may be divided into three carotenoid
types (Egeland *et al.*, 1997).

Type 1

Higher plant type carotenoids are consistent with the carotenoid profile of this algal
group. Lutein (**4**) is the major carotenoid and zeaxanthin (**3**), violaxanthin (**5**) and
neoxanthin (**12**) are other general xanthophylls. Besides β,β-carotene (**1**) and β,ε-
carotene (**2**), the monocyclic β,ψ-carotene (**30**) and the aliphatic carotene lycopene
(**31**) are encountered.

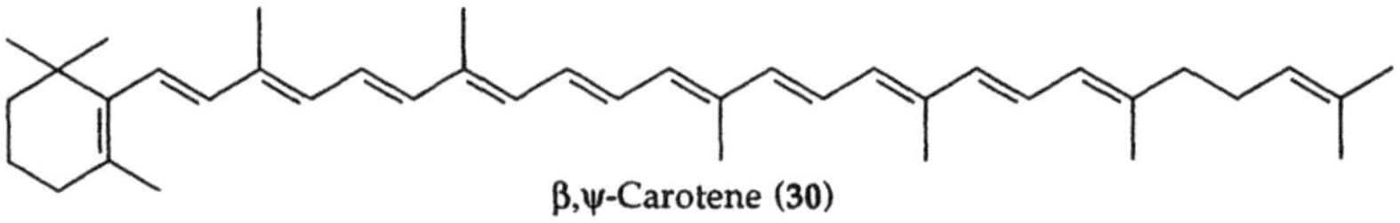

β,ψ-Carotene (**30**)

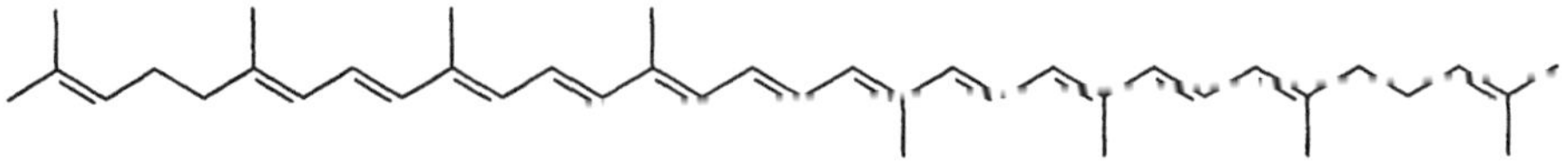

Lycopene (**31**)

Type 2

In addition to the carotenoids encountered in carotenoid type 1, siphonein (**19**), 6'-hydroxysiphonein (**32**) (= Xanthophyll K2; Ricketts, 1970; Egeland *et al.*, 1997), loroxanthin (**33**) or its 19-dodecyl ester (**34**), are present besides some minor unidentified carotenoids. Lutein (**4**) is frequently absent.

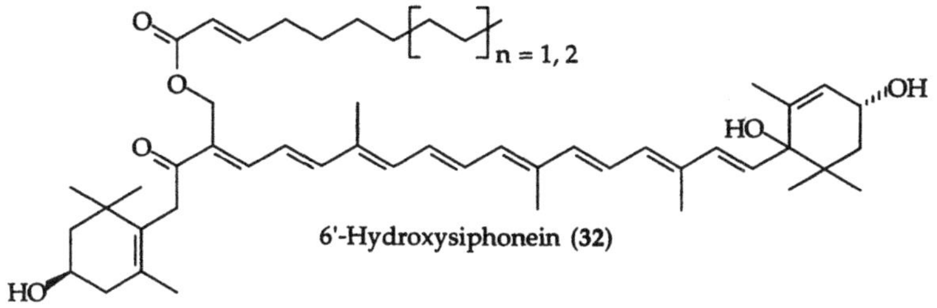

6'-Hydroxysiphonein (**32**)

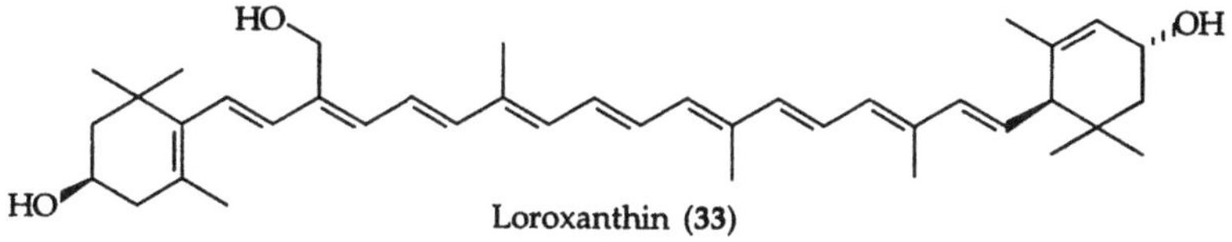

Loroxanthin (**33**)

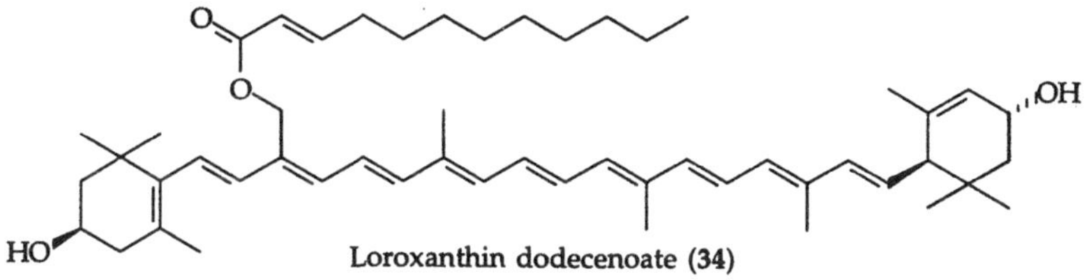

Loroxanthin dodecenoate (**34**)

Type 3

Prasinoxanthin (**20**) is the major carotenoid, and the butenolide uriolide (**35**) and the conjugated aldehyde micromonal (**36**) are usually present. Several other carotenoids with unusual structural elements have been encountered in smaller amounts from different species, e.g. dihydroprasinoxanthin epoxide (**37**), anhydrouriolide (**38**), 3'-dehydrouriolide (**39**) and micromonol (**40**) and structurally related carotenoids (see Egeland *et al.*, 1997).

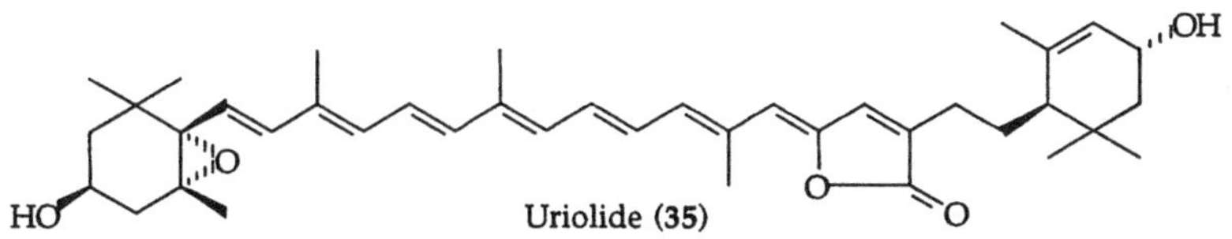

Uriolide (**35**)

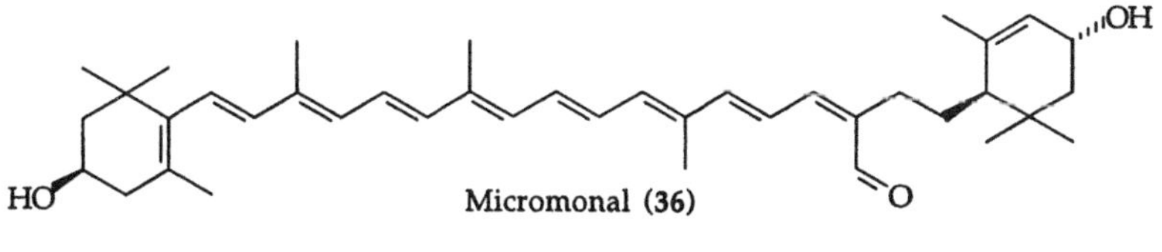

Micromonal (**36**)

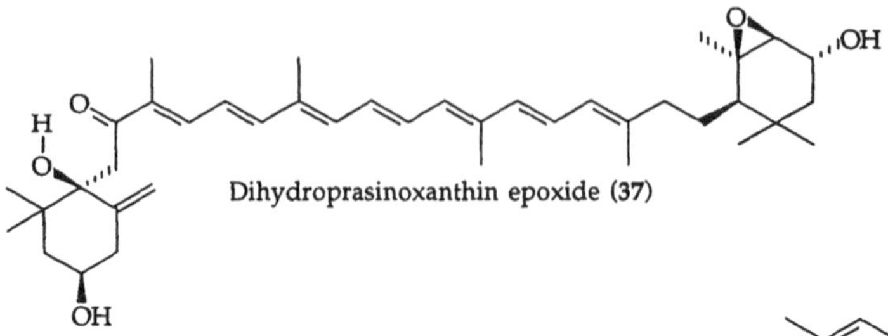

Dihydroprasinoxanthin epoxide (37)

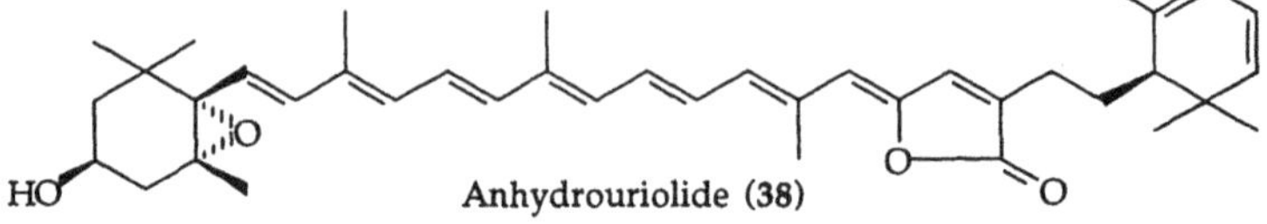

Anhydrouriolide (38)

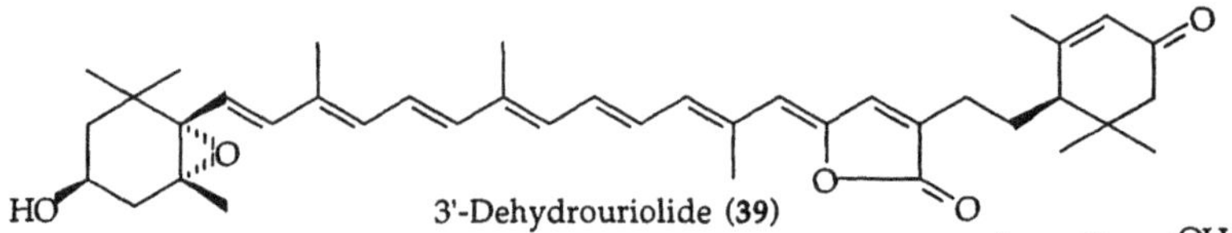

3'-Dehydrouriolide (39)

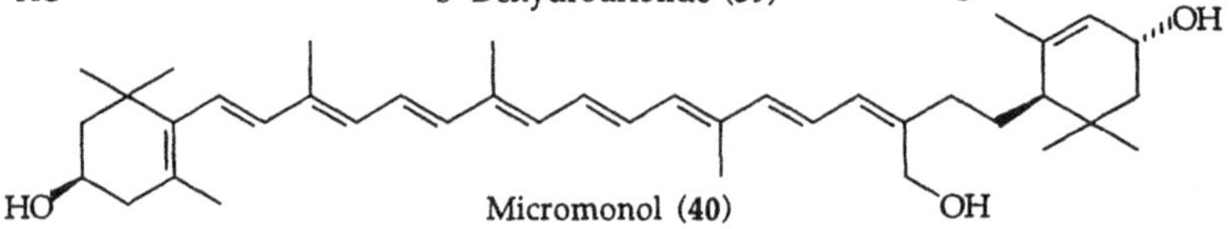

Micromonol (40)

Chlorophyceae (green algae, chlorophytes)

The carotenoids resemble those of green tissue in higher plants. Acetylenic carotenoids are absent in green algae. Epoxidic carotenoids and one allenic carotenoid (neoxanthin, **12**) are present. Carotenoids with ε-end groups are abundant and lutein (**4**) is usually the major carotenoid. In addition β,ε-carotene (**2**), β,β-carotene (**1**), zeaxanthin (**3**) and the epoxides antheraxanthin (**29**) and violaxanthin (**5**) are generally occurring (Goodwin, 1980; Young, 1993).

Particularly in species living in deep waters loroxanthin (**33**) and siphonaxanthin (**18**)/siphonein (**19**) may replace lutein (**4**) to a variable extent (Yokohama *et al.*, 1992). The unusual β,β-carotene-2-ol (**41**) and β,β-carotene-2,2'-diol (**42**) have been obtained from a *Trentepohlia* sp. (Kjøsen *et al.*, 1972).

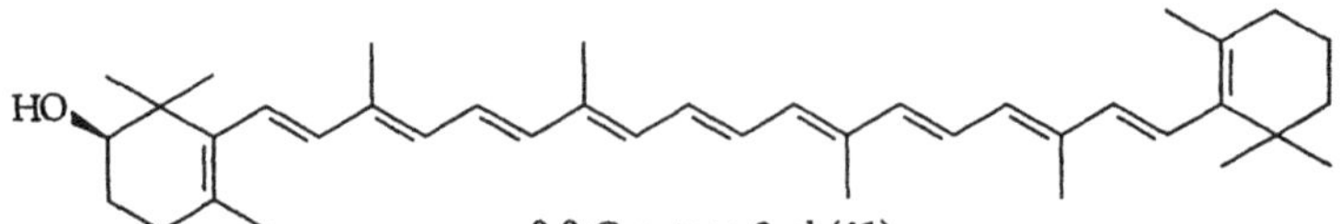

β,β-Carotene-2-ol (41)

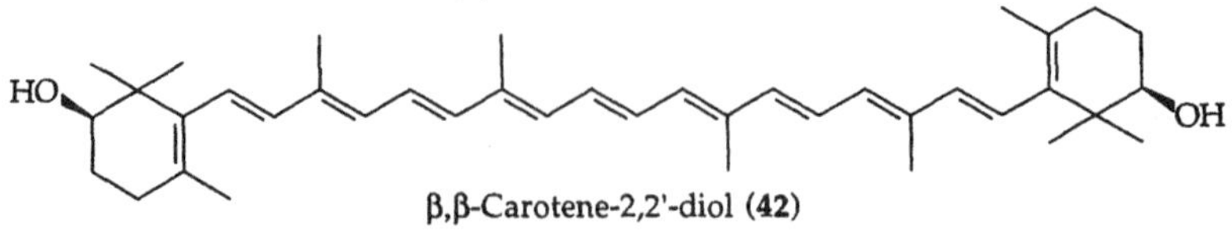

β,β-Carotene-2,2'-diol (42)

β,β-Carotene (**1**) or 4-ketocarotenoids including canthaxanthin (**16**) and astaxanthin (**17**) are produced as so-called secondary carotenoids under special conditions (Rowan, 1989; Grung *et al.*, 1994a); see below and also Chapter 8.

Euglenophyceae (euglenoids, euglenophytes)

Acetylenic, allenic and epoxidic carotenoids are common in euglenoids. Thus the acetylenic diatoxanthin (**7**) and diadinoxanthin (**6**) are general carotenoids besides the carotenes β,β-carotene (**1**) and β,ε-carotene (**2**) (Rowan, 1989). Siphonein (**19**) is reported from some species. Two pigment types are distinguished on the basis of the additional carotenoids present.

Type 1

Heteroxanthin (**9**) is frequently occurring. Highly unsaturated acetylenic carotenoids such as the carotene 3,4,7,8,3',4',7',8'-octadehydro-β,β-carotene (**43**) and 7,8,3', 4',7',8'-hexadehydro-β,β-carotene-3-ol (**44**) have been characterized (Fiksdahl and Liaaen-Jensen, 1988).

3,4,7,8,3',4',7',8'-Octadehydro-β,β-carotene (**43**)

7,8,3',4',7',8'-Hexadehydro-β,β-carotene-3-ol (**44**)

Type 2

This type is so far represented by a single species, *Eutreptiella gymnastica*, in which carotenoids with a particular bicyclic end group of oxabicyclo[2.2.1]heptane type, including eutreptiellanone (**45**), are unique (Fiksdahl *et al.*, 1984; Bjørnland *et al.*, 1986).

Eutreptiellanone (**45**)

Other classes

Pelagophyceae (pelagophytes)

In this newly created algal class a few species have been incorporated. These algae are reported to produce fucoxanthin (**14**), diatoxanthin (**7**), diadinoxanthin (**6**) and sometimes 19'-butanoyloxyfucoxanthin (**27**) (Saunders *et al.*, 1997)

Synurophyceae (synurophytes)

This class is now separated from the Chrysophyceae (Andersen, 1987), and some species may have been analyzed and reported as Chrysophyceae.

Glaucocystophyceae (glaucocystophytes)

The pigment composition is reported to be as for red algae (Hoek *et al.*, 1995).

Pedinophyceae (pedinophytes)

Two species examined for carotenoids had β,β-carotene (**1**), β,ψ-carotene (**30**), lutein (**4**), violaxanthin (**5**) and neoxanthin (**12**) as major carotenoids (Ricketts, 1967).

Chlorarachniophyceae (chlorarachniophytes)

One species is reported to produce β,β-carotene (**1**), lutein (**4**), zeaxanthin (**3**), violaxanthin (**5**), neoxanthin (**12**) and loroxanthin (**33**) dodecenoate (Sasa *et al.*, 1992).

Charophyceae (charophytes)

The carotenoid composition resembles that of the green algae (Goodwin, 1980).

Fucophyceae (brown algae)

The carotenoids in this algal class have been extensively studied. The brown algae display a rather constant carotenoid distribution pattern with fucoxanthin (**14**) as the major carotenoid (Haugan and Liaaen-Jensen, 1994a). However, this class does not contain microalgae.

Overview

The above considerations on the carotenoid complement in the individual algal classes are summarized in Tables 7.2 and 7.3. In Table 7.3 the occurrence of the most common microalgal carotenoids **1–22**, listed in Figure 7.1, is illustrated. This table may serve to localize good sources for the major microalgal carotenoids, keeping in mind that macroalgae, e.g. brown algae for fucoxanthin (**14**), or higher plants, e.g. for lutein (**4**) and neoxanthin (**12**), may represent alternative, more convenient sources for particular carotenoids.

In Table 7.2 a different, chemosystematic approach is used. Here the ability of the various algal classes to synthesize carotenoids with particular structural features, cf. Figure 7.2, is shown in an updated version. For future isolation of carotenogenic genes from microalgae, Table 7.2 may serve as a useful guide. The variation in biosynthetic routes leading from an aliphatic C_{40}-carotene precursor to the various sophisticated xanthophylls is indirectly reflected by the structural features of the biosynthetic end products.

Isolation, identification and structure elucidation

Isolation

Recommended procedures for the isolation of carotenoids from various natural sources, including dinoflagellates and brown seaweeds, are available in a recent, comprehensive monograph (Britton *et al.*, 1995a). Since carotenoids are unstable towards light, heat, oxygen, peroxides and acids, special precautions have to be paid attention to in order to avoid isolation artefacts. Some carotenoids, including prasinoxanthin (**20**), fucoxanthin (**14**) and peridinin (**11**), and esters in general are base labile, and a saponification in order to remove chlorophylls cannot be included. In spite of precautions the formation of *Z* (*cis*)-isomers during the isolation is the most common artefact. However, in some cases *Z*-isomers are naturally occurring.

The isolation generally consists of solvent extraction at room temperature, saponification, chromatography by column (CC), thin layer (TLC) or high performance (HPLC). Chromatographic topics have recently been covered in detail (Britton *et al.*, 1995a). On the analytical scale HPLC (normal or reversed phase) is superior. As an example, the HPLC chromatogram of a total pigment extract of *Pseudoscourfieldia marina* is presented in Figure 7.3.

In a new, comprehensive treatment on phytoplankton pigments in oceanography (Jeffrey *et al.*, 1997b), 11 microalgae from ten different algal classes with a previously well-defined carotenoid complement are recommended as sources for authentic, analytical (HPLC) samples of mixed microalgal carotenoids.

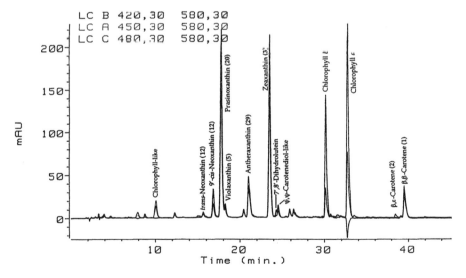

Figure 7.3 Chromatographic separation of carotenoids from the prasinophyte *Pseudoscourfieldia marina* by reversed-phase HPLC [reversed phase RP-18 5μm 220 × 46mm column (Brownlee Labs 0711-0017), eluent (gradient) 0min.; 1:4 1M ammonium acetate:methanol; 30min.: 7:3 methanol:acetone; 50min.: 3:5:2 methanol:acetone:hexane; flow 1.25 ml/min.] (Egeland *et al*, 1996).

Table 7.2 Structural elements of carotenoids present in algae from different algal classes (macroalgae included).

Structural element	Nst 1	Nst 2	Bng	Crp	Din 1	Din 2	Din 3	Crs 1	Crs 2	Syn	Pel	Fuc	Dia	Trb	Eus	Prm	Glu	Prs 1	Prs 2	Prs 3	Ped	Cha	Chl	Eug 1	Eug 2	Clc
Allere	?				+	+	?	+	+	?	+	+	+	+	+	+		+	+	+	+	+	+	+	+	+
Acetylene		+		+	+	+		+	+	?	+	?	+	+	+	+	+			+	+	+		+	+	
3,6-Epoxide																										
4,5-Epoxide																				±					+	
5,6-Epoxide	±	±	±		+	+	?	+	+	?	+	+	+	+	+	+	?	+	+	+	+	+	+	+	+	+
2-Ol									+														±			
3-Ol	+	+	+	+	+	+	+	+	+	?	+	+	+	+	+	+	+	+			+	+	+	+	+	
4-Ol	±	±																+			+	+	±			+
6-Ol							+							+	+											
19-Ol														+				+	+	+			±			
5,6-Diol																			±	±			±	±		
3-Acetate			?		+			+	+	?	+	+	+	±	+	+										
Other 3-Esters								±	±					±	±								±	±		
19-Ester						±		±	?	?	+			±	±	±			+				+	+	+	+
Glycosides	+			±																						
3-Keto													±			±				±			±			
4-Keto	+	+	+	±	±			+	+				±		±	±			±	±			±	±	+	
8-Keto			?			+	+			?	+	+	+			+			+	+			±	±	+	
Butenolide			?		+		?													+						
19-Al							?													+						
20-Al					±															±						
2,3-Didehydro																				±				±		
3,4-Didehydro																				+						
7,8-Dihydro				±			?																		+	
Nor	?		?	+																						

162

Table 7.2 Structural elements of carotenoids present in algae from different algal classes (macroalgae included) (contd.).

Structural element	Nst 1	Nst 2	Bng	Crp	Din 1	Din 2	Din 3	Crs 1	Crs 2	Syn	Pel	Fuc	Dia	Trb	Eus	Prm	Glu	Prs 1	Prs 2	Prs 3	Ped	Cha	Chl	Eug 1	Eug 2	Clc
β-End group	+	+	+	+	+	+	+	+	+	?	+	+	+	+	+	+	+	+	+	+	+	+	+	+	+	+
γ-End group						±														+						
ε-End group		±	+	+	±	±		±	±	?	±	?	±			±	?	+	+	+	+	±	±	+	+	+
ψ-End group	+				±			±								±	±	±			±	±	±			

Nst: Nostocophyceae, Bng: Bangiophyceae, Crp: Cryptophyceae, Din: Dinophyceae, Crs: Chrysophyceae, Syn: Synurophyceae, Pel: Pelagophyceae, Fuc: Fucophyceae, Dia: Diatomophyceae, Trb: Tribophyceae, Eus: Eustigmatophyceae, Prm: Prymnesiophyceae, Glu: Glaucocystophyceae, Prs: Prasinophyceae, Ped: Pedinophyceae, Cha: Charophyceae, Chl: Chlorophyceae, Eug: Euglenophyceae, Clc: Chlorarachniophyceae. The different carotenoid types are given by numbers.

± indicates the presence of this structural element in minor carotenoids and/or in carotenoids occurring only in a few species.

Table 7.3 The occurrence of major and/or characteristic carotenoids in various algal classes.

Carotenoid	Nst 1	Nst 2	Bng	Crp	Din 1	Din 2	Din 3	Crs 1	Crs 2	Syn	Pel	Fuc	Dia	Trb	Eus	Prm	Glu	Prs 1	Prs 2	Prs 3	Ped	Cha	Chl	Eug 1	Eug 2	Clc
β,β-Carotene (1)	+	±	+	+	+	+	+	+	+	?	+	+	+	+	+	+	+	+	+	+	+	+	+	+	+	+
β,ε-Carotene (2)	±	+	+[a]	+	+	±	?	±	±	?	?	+	+	+	±	±	+	+	+	+	+	+	+	+	+	+
Zeaxanthin (3)	+	+	±	±	+	±	?	±	±	?	+	±	+	±	+	+	±	±	±	±	+	±	+	+	+	+
Lutein (4)			+[a]													+		+	+	+	+	+	+	+	+	+
Violaxanthin (5)							?		?	?		±	+	+		+	+	+	+	+	+	+	+	+	+	+
Diadinoxanthin (6)					+	+		+	+	+	+	+	+	+		+								+	+	+
Diatoxanthin (7)					+	+		+	+	+	+	+	+	+		+								+	+	+
Alloxanthin (8)			+																				±	±		
Heteroxanthin (9)														+												
Pyrrhoxanthin (10)				±	+																					
Peridinin (11)				+	+																					
Neoxanthin (12)						?		±		?	+	±	±	?	±	+		+	+	+	+	+	+	±	±	+
Vaucheriaxanthin (13)												+	+	±	+											
Fucoxanthin (14)					+	+		+	+	?	+	+	+													
Echinenone (15)	+																					±				
Canthaxanthin (16)	+																					±	±			
Astaxanthin (17)					±															±		±	±			
Siphonaxanthin (18)																						±	±	±		
Siphonein (19)																				+		+	±		+	
Prasinoxanthin (20)							±												+	+						
Myxoxanthophyll (21)	+																									
Oscillaxanthin (22)	+																									

Nst: Nostocophyceae, Bng: Bangiophyceae, Crp: Cryptophyceae, Din: Dinophyceae, Crs: Chrysophyceae, Syn: Synurophyceae, Pel: Pelagophyceae, Fuc: Fucophyceae, Dia: Diatomophyceae, Trb: Tribophyceae, Eus: Eustigmatophyceae, Prm: Prymnesiophyceae, Glu: Glaucocystophyceae, Prs: Prasinophyceae, Ped: Pedinophyceae, Cha: Charophyceae, Chl: Chlorophyceae, Eug: Euglenophyceae, Clc: Chlorarachniophyceae. The different carotenoid types are given by numbers.
± indicates the presence of this structural element in minor carotenoids and/or in carotenoids occurring only in a few species.
[a] Only reported from macroalgae.

Identification

The identification of a carotenoid requires at least: a visible absorption spectrum (VIS); co-chromatography tests in two different systems with an authentic sample; and a mass spectrum, at least of a quality demonstrating the molecular ion. Identifications based on retention times by HPLC and VIS spectra only are tentative.

For full characterization the following are needed: VIS spectra for demonstration of the chromophore; complete mass spectrum; ^{1}H NMR spectrum including a COSY experiment in order to define spin–spin coupled neighbours; and circular dichroism (CD spectrum) for chirality, whereas an infrared spectrum (IR) and ^{13}C NMR spectrum will provide supporting evidence. Z (*cis*)-configured double bonds are identified from a combination of VIS and ^{1}H NMR data. A modern monograph discussing the application of spectroscopic methods in the carotenoid field is recommended (Britton *et al.*, 1995b).

As an example, the ^{1}H NMR spectrum of siphonein (**19**) with assignments (Egeland *et al.*, 1997) is presented in Figure 7.4.

Data for individual, naturally occurring carotenoids are available through references in *Key to Carotenoids* (Straub, 1987) and a recent supplement (Britton *et al.*, 1995a). VIS and HPLC profiles for major phytoplankton carotenoids have recently been compiled elsewhere (Jeffrey *et al.*, 1997a).

Structure elucidation

If a carotenoid has not previously been fully characterized and structurally identified, the structure elucidation is carried out by a combination of spectroscopic techniques mentioned above and chemical derivatization. Examples on structure elucidation have recently been discussed (Britton *et al.*, 1995b) and details available in the original literature. In the hands of experts the structure elucidation of a carotenoid, including stereochemistry, may be achieved on the microgram scale ($\leq 100\,\mu$g). The complete assignment of stereochemistry is a challenge.

Natural versus synthetic

As an alternative to isolation from microalgae some of these carotenoids are available by chemical synthesis (Isler, 1971; Britton *et al.*, 1996). In spite of a different belief amongst laymen there is no difference between a so-called natural and synthetic carotenoid provided that they have the same structure including the same stereochemistry (chirality and E/Z-configuration). The stereochemistry is of decisive importance and may affect the biological activity of the carotenoids including provitamin A activity, absorption by human and animals, colouring effects, etc.

Biological aspects

Biosynthetic pathways

A review on carotenoid biosynthesis has recently appeared (Porra *et al.*, 1997) and an extensive overview is now available (Britton, 1998). Important recent progress will briefly be commented on here.

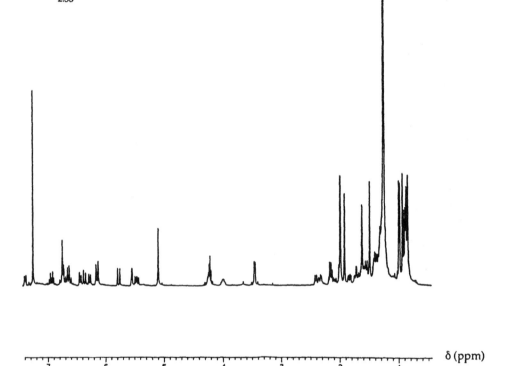

Figure 7.4 ¹H NMR spectrum (1D) of siphonein (19) from the prasinophyte *Pyramimonas amylifera* in CDCl₃. Signal assigments were confirmed by COSY experiments.

Whereas the general biosynthetic pathway to coloured carotenoids via phytoene (**46**) as the first C_{40}-precursor has so far been assumed to proceed via the classical acetate/mevalonate pathway, a novel pyruvate/glyceraldehyde 3-phosphate route has been established in eubacteria (Rohmer *et al.*, 1996), as well as a green alga (Schwender *et al.*, 1996).

Phytoene (**46**)

For a long time little progress has been made in the elucidation of biosynthetic pathways leading to many complex xanthophylls as terminal products in the synthesis of microalgal carotenoids. However, the xanthophyll cycle, linked to photoprotection, in which a reversible epoxidation of zeaxanthin (3) to antheraxanthin (29) and violaxanthin (3) takes place, has been further established. A similar relationship between the acetylenic analogues diatoxanthin (7) and diadinoxanthin (6) is claimed in microalgae (see Young *et al.*, 1997).

Plausible biosynthetic routes to microalgal carotenoids are mainly based on structural relationships, including chirality, and the assumption of simple chemical step reactions (Liaaen-Jensen, 1978; Goodwin, 1980; Haugan and Liaaen-Jensen, 1994a; Egeland *et al.*, 1997).

However, recombinant DNA technology has entered the scene. Great progress is currently being seen in studies on the molecular genetics of the carotenoid biosynthetic pathway in plants and algae; for a recent discussion see Hirschberg *et al.* (1997) and Hirschberg (1998). Transformation of genes from a green alga to a blue–green alga, resulting in the synthesis of astaxanthin (17) in a genetically engineered cyanobacterium, has been accomplished. Carotenogenic enzymes are currently being isolated (Bramley, 1997; Sandmann, 1997), although not yet from microalgae.

Due to the general progress in cloning of carotenogenic genes and in enzymology the elucidation of biosynthetic pathways to complex microalgal carotenoids is expected in the near future.

Regulation of carotenoid biosynthesis

Light intensity, nutrients, temperature, pH and inhibitors may effect the qualitative and quantitative composition of carotenoids in microalgae (Goodwin, 1980), cf. Chapter 8. A recent illustration is a study on a prasinophycean phytoplankton acclimated to three different irradiations (Egeland *et al.*, 1996).

Primary/secondary carotenoids

Primary carotenoids are those synthesized under normal, favourable growth conditions. So-called secondary carotenoids are produced under nitrogen starvation at stressed conditions. Secondary carotenoids are located outside the chloroplasts and are either β,β-carotene (1) or 4-keto type carotenoids including astaxanthin (17), cf. Chapter 8 and Grung *et al.* (1992).

Food chain and sediment aspects

Microalgal carotenoids are resorbed and frequently structurally modified by animals in the food web. The reader is referred to a recent discussion (Liaaen-Jensen, 1998). The chemical fate of microalgal carotenoids in sediments is another interesting aspect (see Damsté and Koopmans (1997) for an overview).

Function

Carotenoids serve well-established functions in microalgae. Major functions are: light harvesting with subsequent energy transfer to chlorophylls; and photoprotection.

The light harvesting complex (LHC) consists of carotenoid–chlorophyll–protein complexes, first crystallized and structure determined by X-ray analysis by Kühlbrandt *et al.* (1994). Light harvesting proteins in algae comprising diatoms (Owens and Wold, 1986; Apt *et al.*, 1994) and dinoflagellates (Hiller *et al.*, 1995) are under investigation. Studies on the LHC of a prasinophycean alga (Wilhelm *et al.*, 1997) demonstrated the presence of seven identified xanthophylls including prasinoxanthin (**20**), superseding the situation in higher plants whose LHC PII binds three different xanthophylls, lutein (**4**), violaxanthin (**5**) and neoxanthin (**12**) (Bassi *et al.*, 1993). The number of individual carotenoids known to be active in photosynthesis is thus increasing.

Other special functions are being demonstrated. As an example the enrichment of lycopene (**31**) and β,ψ-carotene (**30**) in the eye-spot of a flagellate green alga has recently been demonstrated and rationalized in terms of a screening effect of the retinal-based photoreceptor (Grung *et al.*, 1994b).

Summary

Microalgae are capable of producing more than one hundered different carotenoids, some of which are peculiar to algae. Whereas microalgae from natural phytoplankton blooms or from cultivation may serve as natural sources for their major carotenoids, the structures of sophisticated minor microalgal carotenoids might also serve as models for chemical synthesis, and challenge the cloning of carotenogenic genes for future biotechnological application.

The present status concerning the structures and distribution of individual microalgal carotenoids in the various algal classes, based on the systematics of Christensen (1980–94) has been evaluated. The occurrence of some 20 major carotenoids in the various microalgal classes has been tabulated, and the ability within these classes to synthesize carotenoids with particular structural elements was shown in an updated tabulation.

General procedures for the isolation, identification and structure elucidation of microalgal carotenoids have briefly been described with selected illustrations and key references. The term natural versus synthetic is debated. Biological aspects, including biosynthetic pathways, regulation of carotenoid biosynthesis, primary/secondary carotenoids, food chains and sediments, and function of carotenoids have briefly been described, emphasizing recent progress, particularly in recombinant DNA technology.

References

AAKERMANN, T., GUILLARD R.R.L. and LIAAEN-JENSEN, S., 1993, Algal carotenoids 55. Structure elucidation of (3S,5R,6R,3'S,5'R,6'S)-13'-*cis*-7',8'-dihydroneoxanthin-20'-al 3'-

β-D-lactoside (P457). Part 1. Reisolation, derivatization and synthesis of model compounds, *Acta Chemica Scandinavica*, **47**, 1207–1213.

ANDERSEN, R.A., 1987, Synurophyceae classis nov., a new class of algae, *American Journal of Botany*, **74**, 337–353.

APT, K.E., BHAYA, D. and GROSSMANN, A.R., 1994, Characterization of genes encoding the light-harvesting proteins in diatoms: biogenesis of the fucoxanthin chlorophyll *a/c* protein complex, *Journal of Applied Phycology*, **6**, 225–230.

BASSI, R., PINEAU, B., DAINESE, P. and MARQUANDT, J., 1993, Carotenoid-binding proteins of photosystem II, *European Journal of Biochemistry*, **212**, 297–303.

BAUMELER, A. and EUGSTER, C.H., 1992, Synthese von (6*R*, all-*E*)-Neoxanthin und verwendten Allen-Carotinoiden, *Helvetica Chimica Acta*, **75**, 773–790.

BJØRNLAND, T. and LIAAEN-JENSEN, S., 1989, Distribution patterns of carotenoids in relation to chromophyte phylogeny and systematics, in Green, J.C., Leadbeater, B.S.C. and Diver, W.L. (Eds), *The Chromophyte Algae: Problems and Perspectives*, pp. 37–60, Oxford: Clarendon Press.

BJØRNLAND, T., BORCH, G. and LIAAEN-JENSEN, S., 1986, Additional oxabicyclo[2.2.1]heptane carotenoids from *Eutreptiella gymnastica*, *Phytochemistry*, **25**, 201–205.

BONNETT, R., MALLAMS, A.K., SPARK, A.A., TEE, J.L. and WEEDON, B.C.L., 1969, Carotenoids and related compounds. Part 20. Structure and reactions of fucoxanthin, *Journal of Chemical Society C*, 429–454.

BRAMLEY, P.M., 1997, The regulation and genetic manipulation of carotenoid biosynthesis in tomato fruit, *Pure and Applied Chemistry*, **69**, 2159–2162.

BRITTON, G., 1998, Overview of carotenoid biosynthesis in Britton, G., Liaaen-Jensen, S. and Pfander, H. (Eds), *Carotenoids, Vol. 3, Biosynthesis and Metabolism*, pp. 13–147, Basel: Birkhäuser Verlag.

BRITTON, G., LIAAEN-JENSEN, S. and PFANDER, H. (Eds), 1995a, *Carotenoids, Vol. 1A, Isolation and Analysis*, Basel: Birkhäuser Verlag.

BRITTON, G., LIAAEN-JENSEN, S. and PFANDER, H. (Eds), 1995b, *Carotenoids, Vol. 1B, Spectroscopy*, Basel: Birkhäuser Verlag.

BRITTON, G., LIAAEN-JENSEN, S. and PFANDER, H. (Eds), 1996, *Carotenoids, Vol. 2, Synthesis*, Basel: Birkhäuser Verlag.

CHRISTENSEN, T., 1980–94, *Algae. A Taxonomic Survey*, Odense: AiO Print.

DAMSTÉ, J.S.S. and KOOPMANS, M.P., 1997, The fate of carotenoids in sediments: an overview, *Pure and Applied Chemistry*, **69**, 2076–2084.

EGELAND, E.S., JOHNSEN, G. and LIAAEN-JENSEN, S., 1996, Variable carotenoid composition in a prasinophycean phytoplankton acclimated to three different radiances, *Abstract of Poster Presentations, 11th International Symposium on Carotenoids*, Leiden, p. 92.

EGELAND, E.S., GUILLARD, R.R.L. and LIAAEN-JENSEN, S., 1997, Additional carotenoid prototype representatives and a general chemosystematic evaluation of carotenoids in Prasinophyceae (Chlorophyta), *Phytochemistry*, **44**, 1087–1097.

ELBRÄCHTER, M. and SCHNEPF, E., 1996, *Gymnodinium chlorophorum*, a new, green, bloom-forming dinoflagellate (Gymnodiniales, Dinophyceae) with a vestigal prasinophyte endosymbiont, *Phycologia*, **35**, 381–393.

ENGLERT, G., AAKERMANN, T., SCHIEDT, K. and LIAAEN-JENSEN, S., 1995, Structure elucidation of the algal carotenoid (3*S*,5*R*,6*R*,3′*S*,5′*R*,6′*S*)-13′-*cis*-7′,8′-dihydroneoxanthin-20′-al 3′-β-lactoside (P457). Part 2, NMR-studies, *Journal of Natural Products*, **58**, 1675–1682; erratum (1997) **60**, 536.

FIKSDAHL, A. and LIAAEN-JENSEN, S., 1988, Diacetylenic carotenoids from *Euglena viridis*, *Phytochemistry*, **27**, 1447–1450.

FIKSDAHL, A., BJØRNLAND, T. and LIAAEN-JENSEN, S., 1984, Algal carotenoids with novel end groups, *Phytochemistry*, **23**, 649–655.

FOSS, P., GUILLARD, R.R.L. and LIAAEN-JENSEN, S., 1984, Prasinoxanthin – a chemosystematic marker for algae, *Phytochemistry*, **23**, 1629–1633.

Foss P., Lewin, R.A. and Liaaen-Jensen, S., 1987, The carotenoids of *Prochloron* sp. (Prochlorophyta), *Phycologia*, **26**, 142–144.

Fox, D.L., 1974, Biochromes. Occurrence, distribution and comparative biochemistry of prominent natural pigments in the marine world, in *Biochemical and Biophysical Perspectives in Marine Biology*, pp. 169–211, London: Academic Press.

Frank, H.A., Chynwat, V., Desamers, R.Z.B., Farhoosh, R., Erickson, J. and Bautista, J., 1997, On the photophysics and photochemical properties of carotenoids and their role as light-harvesting pigments in photosynthesis, *Pure and Applied Chemistry*, **69**, 2117–2124.

Goericke, R. and Repeta, D.J., 1992, The pigments of *Prochlorococcus marinus*: the presence of divinyl chlorophyll *a* and *b* in a marine procaryote, *Limnology and Oceanography*, **37**, 425–433.

Goodwin, T.W., 1980, *The Biochemistry of Carotenoids, Vol. 1, Plants*, 2nd edn, London: Chapman and Hall.

Grung, M., D'Souza, F.M.L., Borowitzka, M. and Liaaen-Jensen, S., 1992, Algal carotenoids 51. Secondary carotenoids 2. *Haematococcus pluvialis* aplanospores as a source of (3*S*,3'*S*)-astaxanthin esters, *Journal of Applied Phycology*, **4**, 165–171.

Grung, M., Metzger, P., Berkaloff, C. and Liaaen-Jensen, S., 1994a, Studies on the formation and localization of primary and secondary carotenoids in the green alga *Botryococcus braunii*, including the regreening process, *Comparative Biochemistry and Physiology*, **107B**, 265–272.

Grung, M., Kreimer, G., Calenberg, M., Melkonian, M. and Liaaen-Jensen, S., 1994b, Carotenoids in the eyespot apparatus of the flagellated green alga *Spermatozopsis similis*: adaptation to the retinal based photoreceptor, *Planta*, **193**, 38–45.

Haugan, J.A. and Liaaen-Jensen, S., 1994a, Algal carotenoids 54. Carotenoids of brown algae, *Biochemical Systematics and Ecology*, **22**, 31–41.

Haugan, J.A. and Liaaen-Jensen, S., 1994b, Blue carotenoids. Part 1. Novel oxonium ions derived from fucoxanthin, *Acta Chemica Scandinavica*, **48**, 68–75.

Haugan, J.A. and Liaaen-Jensen, S., 1994c, Blue carotenoids. Part 2. The chemistry of the classical colour reaction for common carotenoid epoxides, *Acta Chemica Scandinavica*, **48**, 152–159.

Haugan, J.A., Englert, G. and Liaaen-Jensen, S., 1992, Algal carotenoids 50. Alkali lability of fucoxanthin – reactions and products, *Acta Chemica Scandinavica*, **46**, 614–624.

Hertzberg, S. and Liaaen-Jensen, S., 1971, The carotenoids of blue–green algae, *Phytochemistry*, **10**, 3121–3127.

Hiller, R.G., Wrench, P.M. and Sharples, F.P.S., 1995, The light-harvesting chlorophyll a-c-binding protein of dinoflagellates: a putative polyprotein, *FEBS Letters*, **363**, 175–178.

Hirschberg, J., 1998, Molecular biology of carotenoid biosynthesis, in Britton, G., Liaaen-Jensen, S. and Pfander, H. (Eds), *Carotenoids*, Vol. 3, Biosynthesis and metabolism, pp. 149–191, Basel; Birlehäuser.

Hirschberg, J., Cohen, M., Harker, M., Lotan, T., Mann, V. and Pecker, I., 1997, Molecular genetics of the carotenoid biosynthetic pathway in plants and algae, *Pure and Applied Chemistry*, **69**, 2151–2158.

Hoek, C.v.d., Mann, D.G. and Jahns, H.M., 1995, *Algae. An Introduction to Phycology*, Cambridge: Cambridge University Press.

Isler, O., 1971, *Carotenoids*, Basel: Birkhäuser Verlag.

Jeffrey, S.W., Mantoura, R.F.C. and Bjørnland, T., 1997a, Data for the identification of 47 key phytoplankton pigments, in Jeffrey, S.W., Mantoura, R.F.C. and Wright, S.W. (Eds), *Phytoplankton Pigments in Oceanography: Guidelines for Modern Methods*, pp. 447–559, Paris: UNESCO Publishing.

Jeffrey, S.W., Mantoura, R.F.C. and Wright, S.W. (Eds), 1997b, *Phytoplankton Pigments in Oceanography: Guidelines to Modern Methods*, Paris: UNESCO Publishing.

KJØSEN, H., ARPIN, N. and LIAAEN-JENSEN, S., 1972, Algal carotenoids 6. The carotenoids of *Trentepohlia iolithus*. Isolation of β,β-carotene-2,2′-diol, *Acta Chemica Scandinavica*, **26**, 3053–3067.

KJØSEN, H., NORGÅRD, S., LIAAEN-JENSEN, S., SVEC, W.A., STRAIN, H.H., WEGFAHRT, P., RAPOPORT, H. and HAXO, F. T., 1976, Algal carotenoids 15. Structural studies on peridinin. Part 2. Supporting evidence, *Acta Chemica Scandinavica*, **B30**, 157–164.

KRINSKY, N.I., 1994, The biological properties of carotenoids, *Pure and Applied Chemistry*, **66**, 1003–1010.

KÜHLBRANDT, W., WANG, D. and FUJIYOSHI, Y., 1994, Atomic model of plant light-harvesting complex by electron crystallography, *Nature*, **367**, 614–621.

LIAAEN-JENSEN, S., 1978, Marine carotenoids, in Scheuer, P. (Ed.), *Marine Natural Products. Chemical and Biological Perspectives*, Vol. 2, pp. 1–73, New York: Academic Press.

LIAAEN-JENSEN, S., 1997, Stereochemical aspects of carotenoids, *Pure and Applied Chemistry*, **69**, 2027–2038.

LIAAEN-JENSEN, S., 1998, Carotenoids in food chains, in Britton, G., Liaaen-Jensen, S. and Pfander, H. (Eds), *Carotenoids, Vol. 3, Biosynthesis and Metabolism*, pp. 217–247, Basel: Birkhäuser Verlag.

OWENS, T.G. and WOLD, E.R., 1986, Light harvesting function in the diatom *Phaeodactylum tricornutum* I. Isolation and characterization of pigment-protein complexes, *Plant Physiology*, **80**, 732–738.

PORRA, R.J., PFÜNDEL, E.E. and ENGEL, N., 1997, Metabolism and function of photosynthetic pigments, in Jeffrey, S.W., Mantoura, R.F.C. and Wright, S.W. (Eds), *Phytoplankton Pigments in Oceanography: Guidelines for Modern Methods*, pp. 85–126, Paris: UNESCO Publishing.

RICKETTS, T.R., 1967, The pigments of the phytoflagellates, *Pedinomonas minor* and *Pedinomonas tuberculata*, *Phytochemistry*, **6**, 19–24.

RICKETTS, T.R., 1970, The pigments of the Prasinophyceae and related organisms, *Phytochemistry*, **9**, 1835–1842.

ROHMER, M., SEEMANN, M., HORBACH, S., BRINGER-MEYER, S. and SAHM, H., 1996, Glyceraldehyde 3-phosphate and pyruvate as precursors of isoprenoid units in an alternative non-mevalonate pathway for terpenoid biosynthesis, *Journal of American Chemical Society*, **118**, 2564–2566.

ROWAN, K.S., 1989, *Photosynthetic Pigments of Algae*, Cambridge: Cambridge University Press.

SANDMANN, G., 1997, High level expression of carotenogenic genes for enzyme purification and biochemical characterization, *Pure and Applied Chemistry*, **69**, 2163–2168.

SASA, T., TAKAICHI, S., HATAKEYAMA, N. and WATANABE, M.M., 1992, A novel carotenoid ester, loroxanthin dodecenoate, from *Pyramimonas parkeae* (Prasinophyceae) and a chlorarachniophycean alga, *Plant Cell Physiology*, **33**, 921–925.

SAUNDERS, G.W., POTTER, D. and ANDERSEN, R.A., 1997, Phylogentic affinities of the Sarcinochrysidales and Chrysomeridales (Heterokonta) based on analyses of molecular and combined data, *Journal of Phycology*, **33**, 310–318.

SCHWENDER, J., SEEMANN, M., LICHTENTHALER, H.K. and ROHMER, M., 1996, Biosynthesis of isoprenoids (carotenoids, sterols, prenyl side-chains of chlorophylls and plastoquinone) via a novel pyruvate/glyceraldehyde 3-phosphate non-mevalonate pathway in the green alga *Scenedesmus obliquus*, *Biochemical Journal*, **316**, 73–80.

STRAUB, O., 1987, *Key to Carotenoids*, 2nd edn., Pfander, H, Gerspacher, M., Rychener, M. and Schwabe, R. (Eds), Basel: Birkhäuser Verlag.

SWEENEY, B.M., 1976, *Pedinomonas noctilucae* (Prasinophyceae), the flagellate symbiotic in *Noctiluca* (Dinophyceae) in Southeast Asia, *Journal of Phycology*, **12**, 460–464.

WATANABE, M.M., TAKEDA, Y., SASA, T., INOUYE, I., SUDA, S., SAWAGUCHI, T. and CHIHARA, M., 1987, A green dinoflagellate with chlorophylls *a* and *b*: morphology, fine structure of the chloroplast and chlorophyll composition, *Journal of Phycology*, **23**, 382–389.

WATANABE, M.M., SUDA, S., INOUYE, I., SAWAGUCHI, T. and CHIHARA, M., 1990, *Lepidodinium viride* gen. et sp. nov. (Gymnodiniales, Dinophyta), a green dinoflagellate with a chlorophyll *a* and *b*-containing endosymbiont, *Journal of Phycology*, **26**, 741–751.

WILHELM, C., KOLZ, S., MEYER, M., SCHMITT, A., ZUBER, H., EGELAND, E.S. and LIAAEN-JENSEN, S., 1997, Refined carotenoid analysis of the major light-harvesting complex of *Mantoniella squamata*, *Photosynthetica*, **33**, 161–171.

YAMANO, Y. and ITO, M., 1993, First total synthesis of (±)-peridinin, (±)-pyrrhoxanthin and the optically active peridinin, *Journal of Chemical Society, Perkin Transactions*, 1, 1599–1610.

YAMANO, Y., TODE, C. and ITO, M., 1995, Carotenoids and related polyenes. Part 3. First total synthesis of fucoxanthin and halocynthiaxanthin using oxo-metallic catalyst, *Journal of Chemical Society, Perkin Transactions*, 1, 1895–1904.

YOKOHAMA, Y., HIRATA, T., MISONOU, T., TANAKA, J. and YOKOCHI, H., 1992, Distribution of green light-harvesting pigments, siphonaxanthin and siphonein, and their precursors in marine green algae, *Japanese Journal of Phycology (Sôrui)*, **40**, 25–31.

YOUNG, A.J., 1993, Occurrence and distribution of carotenoids in photosynthetic systems, in Young, A. and Britton, G. (Eds), *Carotenoids in Photosynthesis*, pp. 16–71, London: Chapman & Hall.

YOUNG, A.J., PHILLIP, D., RUBAN, A.V., HORTON, P. and FRANK, H.A., 1997, The xanthophyll cycle and carotenoid-mediated dissipation of excess excitation energy in photosynthesis, *Pure and Applied Chemistry*, **69**, 2125–2130.

8

Production of astaxanthin by *Haematococcus*

YUAN-KUN LEE AND DAO-HAI ZHANG

Introduction

Astaxanthin (3,3′-dihydroxy-β,β-carotene-4,4′-dione) was first described in aquatic crustaceans as an oxidized form of β-carotene, which gives the marine invertebrates (lobsters, crabs and shrimps) and fish (salmon and trout) the distinctive orange–red colour (Johnson and An, 1991). It was later found that the red pigment is a widely distributed carotenoid in the animal kingdom, especially in marine animals. However, only few animals can synthesize it *de novo* from other carotenoids and most of them acquire it in their food.

Some microorganisms are able to synthesize astaxanthin (Andrewes *et al.*, 1976; Grung *et al.*, 1992; Yokoyama *et al.*, 1994). The unicellular green alga, *Haematococcus* has been reported to be a promising microorganism for commercial production of astaxanthin (Johnson and An, 1991; Nelis and De Leenheer, 1991). Under optimal growth conditions, the cells are green and ellipsoid, the two flagella provide motility, and cells divide rapidly. The cell wall is separated from the protoplast by a region filled with watery jelly and traversed by cytoplasmic threads (Santos and Mesquita, 1984). *Haematococcus* produces chlorophyll *a* and *b*, and primary carotenoids normally found in the chlorophyta and in the chloroplasts of higher plants, namely β-carotene, lutein, violaxanthin, neoxanthin and zeaxanthin (Ricketts, 1970). Under certain growth conditions, the vegetative cells begin to synthesize astaxanthin, at the same time undergoing changes in cell physiology and morphology, resulting in the formation of large red palmelloid cells and akinetes. Morphological changes include the loss of two anterior flagella and formation of encysted cells (Elliot, 1934). The reduced photosynthetic activity of photosynthetic system II (PSII) and loss of cytochrome f (Cyt-f) in chloroplasts are the two main physiological characteristics in the cyst cells of *Haematococcus* (Zlotnik *et al.*, 1993; Shi *et al.*, 1995). In the red cells, the level of Cyt-f is greatly decreased and the linear electron flow from PSII to photosynthetic system I (PSI) is impaired. The respiration rate of red cells is much higher and O_2 evolution declines.

Astaxanthin (Table 8.3, K) exists in *Haematococcus* cells in the forms of mono- and di-esters which account for up to 95 per cent of total secondary carotenoids.

These pigments are present in lipid globules outside the chloroplast. Their functions in cells are complex and protection from photodamage by reducing the amount of light available to the light-harvesting pigment protein complex has been proposed (Yong and Lee, 1991; Hagen *et al.*, 1994). The pigments may provide a physico-chemical protective barrier, preventing photodynamic free-radical-mediated damage of genetic materials as they are accumulated around the nucleus. Astaxanthin and its precursors are more effective than β-carotene as quenchers of singlet oxygen, and thus serve the function of stabilizing membrane systems and act as buffers for oxidative reactions (Hagen *et al.*, 1993). In this case, astaxanthin could be an end product of an enzymatic antioxidation process (Lu *et al.*, 1996).

Some *Haematococcus* species can accumulate up to 6–8 per cent (w/w) of asta-xanthin in biomass (Tsavalos *et al.*, 1992); however, large scale production of asta-xanthin by *Haematococcus* is hampered by problems associated with its peculiar cell cycle, production process design and scale-up. Its slow growth rate, intolerance to high temperature and irradiance have limited the prospect of *Haematococcus* as producer of astaxanthin in outdoor culture systems. These biological and technical difficulties in the mass cultivation of *Haematococcus* must be addressed before com-mercial production of astaxanthin using *Haematococcus* is possible.

A major improvement in the productivity of microbial fermentation processes is generally ascribed to the development of superior strains via genetic manipulation. The cDNA involved in astaxanthin biosynthesis from *H. pluvialis* cells was isolated and was successively expressed in *Escherichia coli* containing carotenoid biosynthesis genes from *Erwinia uredovora* (Kajiwara *et al.*, 1995). Cloning of biosynthetic genes of astaxanthin and expression in *E. coli* pave the way for the use of various biotech-nological techniques in the production of this ketocarotenoid.

In the following sections, the various aspects of cell growth and astaxanthin accumulation in *Haematococcus* will be reviewed. The function, location and syn-thetic pathway of astaxanthin in *Haematococcus* cells will be summarized. Special attention will be paid to the environmental factors that influence the cell growth and astaxanthin accumulation, and the progress in molecular biology concerning the cloning of carotenoid genes and their expression. Practical aspects of mass produc-tion of astaxanthin will be evaluated.

Astaxanthin-producing microalgae

Most microalgae are able to accumulate secondary carotenoids. Goodwin (1980), Borowitzka (1988) and Nelis and De Leenheer (1991) have shown that a wide range of microalgae have the ability to synthesize high levels of astaxanthin or other keto-carotenoids when exposed to stress conditions (Table 8.1). The astaxanthin accumu-lated in algae occurs as the 3S and 3'S isomers (Renstrom and Liaaen-Jensen, 1981), in the form of mono- and di-esters (Table 8.3, L, M). The ratio of mono- and di-esters appears to be dependent upon the age of the culture (Harker *et al.*, 1996a). The content and composition of ketocarotenoids are different among the algae (Table 8.2).

Cell cycle of *Haematococcus*

The cell cycle of *Haematococcus* is not well understood. Upon transfer to favourable culture conditions, biflagellated motile zoospores are released from thick-wall aki-

Table 8.1 Astaxanthin producing microalgae.

Algal species	Reference
Acetabularia mediterranea	Czeczuga (1986)
Ankistrodesmus spp.	Czygan (1970)
Brachiomonas simplex	Goodwin (1952b)
Chlamydomonas nivalis	Viala (1966)
Chlorella fusca var. *rubescens*	Czygan (1968)
Chlorella zofingiensis	Kessler and Czygan (1965)
Chlorococcum infusionium	Czygan (1968)
Chlorococcum multinucleatum	Czygan (1968)
Chlorococcum olofaciens	Czygan (1968)
Chlorococcum wimmeri	Brown *et al.* (1967)
Chlorococcum sp.	Zhang *et al.* (1997)
Eremosphaera viridis	Vechtel *et al.* (1992a)
Euglena rubida	Czeczuga (1974)
Euglena sanguinea	Grung *et al.* (1990)
Fritschiella tuberosa	Eugster (1979)
Haematococcus droebakensis	Czygan (1968)
Haematococcus lacustris	Donkin (1976)
Haematococcus pluvialis	Goodwin and Jamikorn (1954)
Hydrodictyon reticulatum	Dersch (1960)
Protosiphon botryoides	Berkaloff and Kader (1975)
Scenedesmus spp.	Czygan (1964)
Scotiella spp.	Czygan (1968)
Trachelomonas volvocina	Green (1963)
Oocystis minuta	Tsavalos *et al.* (1992)
Crucigennella rectangularis	Tsavalos *et al.* (1992)
Muriellopsis sphaericum	Tsavalos *et al.* (1992)
Neospongiococcum spp.	Deason *et al.* (1977)
Selenastrum gracile	Czygan (1970)
Spongiochloris typica	McLean (1967)

netes within 15 hours and begin to grow, divide and disperse. This could be considered the dispersion phase. Lee and Ding (1994) reported that the motile cells stopped dividing after five cell doublings irrespective of the growth rate, and this was not due to nutrient limitation, because the biomass continued to increase for many folds in the next growth phase. As a result, it is difficult to achieve high cell density in *Haematococcus* culture, even in rich culture medium. Soon after the zoospores had ceased dividing, the ellipsoidal cells are transformed into non-motile spherical cells (palmelloid). The non-motile cells continue to grow in biomass after their formation and divide by forming aplanospores for as long as 200 hours in nutrient rich medium (Lee and Ding, 1994). The increase in biomass of *Haematococcus* cultures is largely contributed by the growing non-motile cells, despite their slower rate of division (ca. 0.03 h^{-1}) as compared to the motile cells (ca. 0.085 h^{-1}), and this could be regarded as the production phase. During the production phase, biomass and astaxanthin content increase rapidly (Borowitzka *et al.*, 1991; Boussiba and Vonshak, 1991; Kobayashi *et al.*, 1991). Cells of senescent cultures develop thick walls and turn into truly resting akinetes. It has been suggested that sexual reproduction does occur in *Haematococcus* cultures based on the

Table 8.2 The secondary carotenoid composition of selected microalgae (expressed as a percentage of total carotenoids).

Carotenoids	Euglena sanguinea[a]	Oocystis minuta[d]	Muriellopsis sphaericum[d]	Crucigenella rectangularis[d]	Haematococcus pluvialis[bd]		H. pluvialis[c]	H. sp.[f]	H. sp.[a]	H. sp.[e]
β-Carotene	–	2	2	8	4	–	1	5	3	<1
Echinenone	–	–	–	–	–	–	–	4	–	<1
Carthaxanthin	–	–	–	–	–	–	–	4	–	1
Adonirubin										
Esters	–	3	1	3	6	9	3	–	3	<1
Astaxanthin monoester	–	51	62	61	74	69	76	46	71	31
Astaxanthin diesters	91	36	28	24	10	20	7	34	76	65
Astaxanthin free	–	5	2	2	1	–	1	1	3	<1
3'-Hydroxy echinenone	–	–	–	–	–	–	–	–	–	–
Lutein/zeaxanthin	1	3	5	2	9	1	7	6	4	1
Violaxanthin	–	–	–	–	–	–	2	–	1	–
Neoxanthin	–	–	–	–	–	–	1	1	–	–
Diatoxanthin	6	–	–	–	–	–	–	–	–	–
Unidentified	2	–	–	–	–	–	–	1	2	–
Carotenoids (% dry weight)	–	–	–	–	–	–	–	1	2	3

[a]Grung et al. (1990); [b]Kobayashi et al. (1991); [c]Renstrom and Liaaen-Jensen (1981); [d]Tsavalos et al. (1992); [e]Harker et al. (1996a); [f]Grung et al. (1992).

measurement of cellular DNA content (Lee and Ding, 1994); this needs to be further verified.

Pathway of astaxanthin synthesis

Like all the terpenoids, carotenoids are biosythesized from mevalonic acid (MVA) (Table 8.3, A) by a head-to-tail condensation of two isoprene isomers, isopentenyl pyrophosphate (IPP) and dimethylallyl pyrophosphate (DMPP), to form successively geranylpyrophosphate (GPP) (Table 8.3, B), farnesylpyrophosphate (FPP) (Table 8.3, C) and the C_{20} terpenoid geranylgeranylpyrophosphate (GGPP) (Table 8.3, D). The prephytoene pyrophosphate (PPPP) is formed from GGPP by tail-to-tail condensation and is transformed to phytoene (Table 8.3, E) (Ruddat and Garder, 1983). Four successive dehydrogenation steps yield lycopene, and further cyclization steps lead to the formation of β-carotene from γ-carotene. Further details can be obtained from several review articles (Goodwin, 1971; Spurgeon and Porter, 1983; Britton, 1990, 1993).

The biosynthetic pathway of ketocarotenoids has not yet been fully elucidated, but there is a general agreement that β-carotene serves as precursor of the secondary carotenoids, as reported in yeast and other algae (Johnson and Lewis, 1979; Grung *et al.*, 1992). However, the astaxanthin formation in yeast and algae differs in some steps. In yeast, especially in *Phaffia rhodozyma*, astaxanthin is formed as the following steps: β-carotene (Table 8.3, F) → echinenone (Table 8.3, G) → hydroxyechinenone (Table 8.3, H) → phoenicoxanthin (adonirubin) (Table 8.3, J) → (3R,S′R) astaxanthin (Table 8.3, K) (Girard *et al.*, 1994). In algae, the identification of small but significant amounts of echinenone and canthaxanthin (Table 8.3, I) in *Haematococcus lacustris* (Donkin, 1976) and *H. pluvialis* (Grung *et al.*, 1992) implies that (3S,3′S) astaxanthin is synthesized via canthaxanthin. The inhibitory effect of diphenylamine to the synthesis of echinenone and canthaxanthin from β-carotene further confirmed that the pathway of astaxanthin synthesis in *Haematococcus* is different from that in *Phaffia* (Lu *et al.*, 1995). The cloning of gene encoding β-C-4-oxygenase from *H. pluvialis* suggested that echinenone/ canthaxanthin synthesis was determined by the enzyme ketolase (Lotan and Hirschberg, 1995). It was predicted that an additional enzyme, which is similar to Crt Z gene found in *Agrobacterium* (Misawa *et al.*, 1995b), converted canthaxanthin to astaxanthin by hydroxylation reaction in the C-3 position. Figure 8.1 summarizes the pathway of astaxanthin synthesis in *Haematococcus*, including the enzymes concerned and similar genes identified in bacteria. Recently, Yokoyama *et al.* (1994) reported that astaxanthin could be produced by the marine bacterium *Agrobacterium aurautiacum* and that astaxanthin was formed from β-carotene through either canthaxanthin or zeaxanthin depending on the culture conditions (Yokoyama and Miki, 1995).

Astaxanthin typically occurs as fatty acid esters (Renstrom and Liaaen-Jensen, 1981; Grung *et al.*, 1992). Diesters are dominated in cells. The fatty acids in astaxanthin monoesters were reported to be C16:0, C18:0, C20:0 and C18:1 in *Haematococcus* and C16:0, C14:0, C16:3, C18:1, C18:0, C18:2, C18:4, C20:4 and C20:5 in *Eremosphaera viridis* with C18:1 as the major fatty acid (Grung *et al.*, 1990). Since the astaxanthin esters could not be easily absorbed by animals

Table 8.3 Structure of main carotenoids and precursors.

Number	Name	Structure
A	MVA	
B	GPP	
C	FPP	
D	GGPP	
E	Phytoene	
F	β-Carotene	
G	Echinenone	
H	3-Hydroxyechinenone	
I	Canthaxanthin	
J	Adonirubin	
K	Astaxanthin	
L	Astaxanthin monoester	
M	Astaxanthin diester	

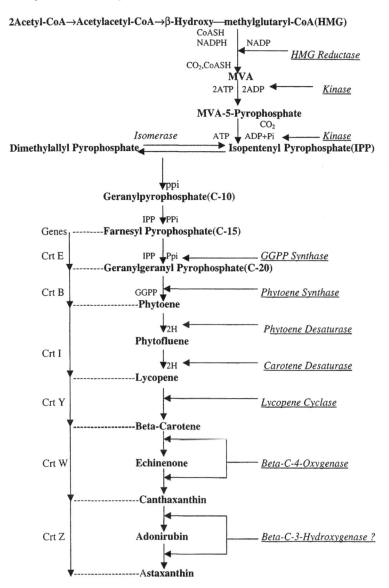

Figure 8.1 Pathway of astaxanthin synthesis in microalgae.

(Storebakken *et al.*, 1987), this resulted in reduced bioavailability of the pigment to the fish (Sommer *et al.*, 1991).

Regulation of biosynthesis of carotenoids leading to astaxanthin

Light stimulates the formation of carotenoids. This phenomenon was first described in fungal species (Rau, 1976) and was proposed as a derepression of carotenoid operon. However, opposite results were also observed in some cases (Sutter, 1970), which may be explained by photodestruction of carotenoids.

Other compounds, such as β-ionone, also enhance the light-induced carotenogenesis (Ciegle, 1965). β-Ionone is not incorporated into the product in spite of the structural similarity to β-carotene. A regulation in the enzymatic activity involved in the conversion of 5-phosphomevalonate to dimethylallylpyrophosphate has been suggested (Rao and Modi, 1977). A two- to threefold increase in β-carotene production in the presence of dimethylformamide, succinimide or isonicotynoilhydrazine together with β-ionone has been demonstrated with *Blakeslea trispora* (Ninet *et al.*, 1969). Several compounds have specific inhibitory effects on various sites in the dehydrogenation and cyclization steps of the carotene pathway. Diphenylamine (DPA) was shown to inhibit the dehydrogenation step (phytoene dehydrogenase), leading to the accumulation of more saturated phytoene intermediates (Goodwin, 1952a). N-heterocyclic compounds, e.g., nicotine, pyridine and imidazole inhibit the cyclase and cause accumulation of lycopene in cells (Davies, 1973). Phenylpyridazinone herbicides (norflurazon) strongly inhibited the phytoene synthetase and led to accumulation of phytoene precursors (Sandmann, 1980). Therefore, it is possible to employ specific inhibitors to induce accumulation of defined products.

Carotenogenesis is controlled primarily by certain genes that encode for the enzymes of the carotenoid pathway. Knowledge of the genetics of carotenogenesis was mainly gained from studies in fungi, such as *Phycomyces* (Cerda-Olmedo, 1985), bacteria, such as *Erwinia* sp. (Misawa *et al.*, 1990; Hundle *et al.*, 1993; To *et al.*, 1994) and the cyanobacterium *Synechococcus* (Chamovitz *et al.*, 1992; Cunningham *et al.*, 1993). By preparing a series of mutants with different colours as well as albino mutant in *Phycomyces*, the Car B gene, which determines the dehydrogenation from phytoene to lycopene was demonstrated (Ootak, 1973). Car R gene determines lycopene cyclase which converts lycopene to γ- and β-carotene (Torres-Martinez *et al.*, 1980). Car R mutants are red and accumulate large amounts of lycopene. Recessive mutation of Car S gene, which normally shuts off the pathway, leads to the abolition of the negative feedback control (Murillo, 1976). This end-product inhibition of pathway is mediated by a complex of β-carotene and the products of Car S gene (Bejarano, 1988). Gene Car RA is responsible for the cyclase. Mutants that had lost both Car S and Car RA genes are always white and lack all carotenoids, except a trace of lycopene. Another class of mutants, e.g. the white Car A mutants, has active cyclase and performs dehydrogenation, but makes only small amounts of β-carotene (Torres-Martinez *et al.*, 1980). The white Car A mutants are rendered yellow by the introduction of Car S gene (Lopez-Diaz and Cerda-Olmedo, 1980). Several lines of evidences suggest that the Car A mutants are defective in the transfer of substrates along the carotene pathway (Murillo, 1980). Many genes (Crt I) involved in the desaturation step, in converting phytoene to lycopene, have been obtained from bacteria (Misawa *et al.*, 1990; Lang *et al.*, 1994; Linden *et al.*, 1994) and fungi (Sandmann and Misawa, 1992). The genes (Crt Y) for lycopene cyclase have been isolated from *Erwinia* sp. (Hundle *et al.*, 1993) and *Synechococcus* sp. (Cunningham *et al.*, 1993). Recently, Misawa *et al.* (1995b) successfully isolated the carotenoid biosynthetic gene cluster which contains five carotenogenic genes, namely, Crt W, Crt Z, Crt Y, Crt I and Crt B, with the same orientation and elucidated the gene control of astaxanthin synthesis. The Crt W and Crt Z genes mediate the oxygenation reaction from β-carotene to astaxanthin. This is in agreement with the claim that in *Haematococcus*, formation of astaxanthin from β-carotene involves two genes, ketolase and hydroxylase (Breitenbach *et al.*, 1996).

Location of astaxanthin in *Haematococcus* cells

Secondary carotenoids are in general accumulated in lipid bodies outside the chloroplast in algal cells (Vechtel *et al.*, 1992a). However, in the alga *Eremosphaera viridis*, astaxanthin is also detected in the light-harvesting protein of PSI, suggesting that this carotenoid provides protection to the photosynthetic pigment protein complex against photosensitized damage (Vechtel *et al.*, 1992b), involving a mechanism similar to that observed in the primary carotenoids (Siefermann-Harms, 1987).

It has been verified that the astaxanthin accumulation occurs in the cytoplasm of palmella and resting cysts of *Haematococcus* (Boussiba and Vonshak, 1991; Yong and Lee, 1991). The initial site of astaxanthin deposition is around the nucleus (Pockock, 1961; Santos and Mesquita, 1984) forming a red centre and a green periphery. Upon exposure to higher light intensity, the central red patch thins out and spreads towards the periphery of the cell, eventually filling the whole cytoplasm, completely surrounding the reticulate chloroplasts (Yong and Lee, 1991). When the source of illumination is switched off, the dispersed carotenoids gradually migrate back to the middle of the cytoplasm. The time taken for complete dispersion of carotenoids ranges from 20 to 60 minutes. The relocation of pigments after illumination has been discontinued is generally longer, ranging from one to two hours. The cytoplasm of old cysts is completely filled with astaxanthin. The highest cellular content of astaxanthin in *Haematococcus* is strain dependent and independent of growth parameters (Lee and Soh, 1991).

Factors affecting astaxanthin accumulation and cell growth

Light intensity

Light intensity showed significant effect on cell growth and level of astaxanthin accumulation in the cells. It was shown that at low light intensity ($< 37 \, \mu mol \, m^{-2} \, s^{-1}$) the survival rate of the *Haematococcus* cells was high and higher biomass could be obtained (Harker *et al.*, 1996a). The optimum photon flux density (PFD) corresponding to the maximum level of algal biomass production was approximately 50 $\mu mol \, m^{-2} \, s^{-1}$ (Harker *et al.*, 1995). In contrast, Lu *et al.* (1994) reported that cell growth of *Haematococcus* was saturated at about 90 $\mu mol \, m^{-2} \, s^{-1}$; irradiance higher than 130 $\mu mol \, m^{-2} \, s^{-1}$ strongly inhibited cell growth. High light intensity caused relatively large quantities of astaxanthin to be accumulated in the cells of *Haematococcus*. Boussiba and Vonshak (1991) reported that light intensity of 170 $\mu mol \, m^{-2} \, s^{-1}$ was the inducing light intensity for astaxanthin production in *H. pluvialis*. High light intensity resulted in high rate of cell mortality, whereas the cells that did survive contained large quantities of astaxanthin. However, others reported that under the illumination of 90 $\mu mol \, m^{-2} \, s^{-1}$, cells accumulated 500 pg/cell of astaxanthin in 20 days (Harker *et al.*, 1996a). Further studies by the same authors showed that the optimum light intensity for the induction of astaxanthin synthesis was in the range of 1550–1650 $\mu mol \, m^{-2} \, s^{-1}$ (PAR) (Harker *et al.*, 1995). These studies suggested that it is technically difficult to achieve high astaxanthin productivity (specific rate of astaxanthin accumulation × cell density) in outdoor cultures. A PFD of 130 $\mu mol \, m^{-2} \, s^{-1}$ corresponds to the solar irradiance measured before 8.00 am and after 6.30 pm in the summer on the ground. The zoospores of *Haematococcus* cultures have a very narrow window of time to grow and divide. Soon, their growth is inhibited by high light intensity and they transform into palmelloids. Astaxanthin begins to accumulate at this stage, but in order that the cells are able to divide rapidly again, the cell cycle needs to be advanced to the stage of akinete before zoospores are released.

Studies by chemostat cultures of *Haematococcus* suggested that the rate of astaxanthin accumulation in the growing cultures is determined by photon flux density. However, the specific rate of astaxanthin accumulation in non-growing cells is no longer a function of photon flux density, but is determined by the growth rate of the cultures during the growth phase (Lee and Soh, 1991). The astaxanthin accumulation was also affected by light quantity and quality (Kobayashi *et al.*, 1992a). It was observed that the carotenoid content in *H. pluvialis* correlated proportionally to the light quantity (light intensity × net illumination time) and carotenoid formation was more efficiently enhanced under blue light than red light. The highest possible cellular astaxanthin content is however strain dependent and independent of culture parameters (Lee and Soh, 1991). Grown on an acetate-based medium, dark-grown cultures of *H. pluvialis* were shown to accumulate traces of astaxanthin and the cell growth rates were significantly lower than the illuminated cultures (Droop, 1955).

Nutrient (nitrogen and phosphorus) limitation

The synthesis of secondary carotenoids in microalgae appears to be dependent upon a large number of environmental factors; of particular importance is depletion of a nutrient source in medium. The early reports by Goodwin and Jamikorn (1954) and Droop (1955) showed that N-deficiency induced astaxanthin formation and they suggested that the C/N balance in the medium determined the degree of carotene formation. Studies by Kakizono *et al.* (1992) confirmed that a high ratio of C/N greatly stimulated carotenogenesis in *Haematococcus*. Under low nitrogen concentration, algal growth was limited and high levels of astaxanthin accumulated in the cells. Levels of total carotenoids (of which astaxanthin represents >95 per cent) in excess of 300 pg per cell were measured after 30 days of cultivation (Harker *et al.*, 1996a). Borowitzka *et al.* (1991) showed that low nitrogen concentration stimulated the formation of red palmelloid cells and factorial experiments further confirmed that the main factor leading to carotenoid accumulation was nitrogen starvation and this was enhanced by low phosphate concentration. A similar phenomenon was also found in *Chlorella zofingiensis* in which the maximal accumulation of astaxanthin occurred in cells starved of nitrogen (Rise *et al.*, 1994). In contrast, Boussiba *et al.* (1992) reported that nitrogen was an important requirement for astaxanthin synthesis in the alga, and suggested that nitrogen was required in the synthesis of proteins supporting massive accumulation of astaxanthin. The best nitrogen source for biomass production was identified as urea, followed by potassium nitrate (Harker *et al.*, 1995). The use of ammonium salt as a nitrogen source resulted in particularly poor levels of biomass and cell death because of the rapid acidification of the medium (Borowitzka, 1988). Previous investigation with *H. pluvialis* also showed that nitrate nitrogen was preferred to the ammonium nitrogen (Proctor, 1957), although Stross (1963) noted that exponentially growing cells at acid pH preferred ammonium nitrogen. The optimum concentration of nitrate for growth is between 0.5 and 1.0 g/l and the concentration of phosphate about 0.1 g/l (Borowitzka *et al.*, 1991), whereas the optimum concentration of urea required for cell growth is approx. 2.5–3.0 mM (Harker *et al.*, 1995).

Phosphate starvation had been reported previously to act as a trigger for the accumulation of astaxanthin (Boussiba and Vonshak, 1991; Harker *et al.*, 1996a). However, Borowitzka *et al.* (1991) suggested that high phosphate concentrations stimulated the production of astaxanthin in the cells.

The above observations may not be contradictory. It is likely that any factor (e.g. phosphate starvation) that caused a reduction in the specific rate of cell division of a *Haematococcus* culture would result in net accumulation of astaxanthin in the cells, while P and N are essential nutrients which determine the rate of synthesis and final concentration of astaxanthin.

Metal ions and oxidative stress

Studies of Kobayashi *et al.* (1991) showed that the basal medium enriched with Fe^{2+} and Mn^{2+} resulted in cessation of cell growth, morphological transformation and formation of astaxanthin in *Haematococcus* cells. The final carotenoid content was 20 mg g^{-1} dry cell. Harker *et al.* (1996a) reported that growth in the presence of low levels of ferrous ion produced little stimulative effect for astaxanthin synthesis; at high concentration of Fe^{2+}, which inhibited cell growth, the cells could accumulate astaxanthin and reached about 50 pg/cell in 30 days. Other metals, including Mn^{2+} and Cd^{2+} increased the carotenoid content on a cellular basis but algal growth was severely inhibited. Of all the metals tested, only the addition of Fe^{2+} to the growth medium resulted in an increase in carotenoid levels on a unit culture volume basis. However, the addition of EDTA-chelated $FeCl_3.6H_2O$ did not produce any significant difference in astaxanthin formation as reported by Borowitzka *et al.* (1991).

As the ferrous form of iron is known to give rise to free radical formation via the Fenton reaction, free radicals may play a role in astaxanthin formation. Several reports have determined that radicals could initiate the cellular process of carotenoid synthesis. Kobayashi *et al.* (1993) reported that increased astaxanthin formation caused by Fe^{2+} was due to the formation of hydroxyl radical (HO·) and the addition of potassium iodide, which could scavenge HO·, inhibited the astaxanthin biosynthesis. Other active oxygen species including $^1O_2$, H_2O_2, peroxyl radicals and superoxide anion radicals also stimulate the formation of astaxanthin in this alga, although the highest content of astaxanthin in cells is identical for the different active oxygen generators. The combination of several active oxygen generators further enhances the astaxanthin formation and the highest content of carotenoids can reach 600 pg/cell (Tjahjono *et al.*, 1994a). The enhanced astaxanthin formation by these oxidative stresses is also inhibited by pretreatment with actinomycin D or cycloheximide, but is unaffected by the same inhibitors after the encystment was induced, indicating that active oxygen species may be required for post-translational activation of the carotenoid biosynthesis in the cyst cells.

The enhanced formation of carotenoids by free radicals was also reported in *Dunaliella* (Ben-Amotz and Avron, 1983) and in *Phaffia rhodozyma* (Schroeder and Johnson, 1995). However, only when the yeast cells were exposed to exogenous $^1O_2$, not peroxyl radicals or H_2O_2, could the carotenoid synthesis be stimulated. High dissolved oxygen partial pressure in a photosynthetic *Haematococcus* culture is also an effective means to enhance astaxanthin synthesis (Lee and Ding, 1995).

The steady state astaxanthin contents in chemostat cultures of *Haematococcus* were found to be proportional to the dissolved O_2 partial pressure in the culture medium in the range of approx. 0–50 per cent O_2. The studies imply that synthesis of astaxanthin in *Haematococcus* could be regulated by light via photosynthetically generated O_2 and photo-oxidative reactions.

Salt stress

Borowitzka *et al.* (1991) reported that salinity of higher than 1 per cent (w/v) NaCl could be lethal to *H. pluvialis*, while the palmella cell formation was enhanced. Similarly, Boussiba and Vonshak (1991) reported that exposing *H. pluvialis* to stress by adding 0.8 per cent of NaCl, caused complete cessation of growth and induced a massive accumulation of astaxanthin. In contrast, Cordero *et al.* (1996) reported that the best condition for astaxanthin production was 0.2 per cent NaCl, at which higher than 3 per cent of astaxanthin in dry biomass and 18.0 mg/l culture were obtained. Higher than 0.2 per cent NaCl resulted in a decrease of astaxanthin formation. The results obtained by Harker *et al.* (1995) also showed that the NaCl concentration of approx. 25–30 mM was optimal for astaxanthin synthesis. However, the highest content of astaxanthin in cells (approx. 500 pg/cell) was found at 100 mM NaCl (0.58 per cent w/v), accompanied by a high rate of cell mortality (Harker *et al.*, 1996a). In their study, the optimum concentration of NaCl that gave maximum astaxanthin per unit volume of culture was 40 mM, which was in agreement with the results of Spencer (1989), who used 50 mM NaCl to induce astaxanthin formation in *H. pluvialis*.

The addition of KCl to the algal cultures, even at low concentration (i.e. 40 mM) resulted in a relatively high rate of cell mortality, although the astaxanthin formation was also enhanced and reached 350 pg per cell on the thirtieth day (Harker *et al.*, 1996a).

Temperature and pH

The optimum growth temperature for *Haematococcus* is relatively low. Borowitzka *et al.* (1991) reported that *H. pluvialis* MUR-1 grew best at the temperature of 15°C. At 25°C, growth was severely inhibited and at 35°C, the culture died. Similarly, Harker *et al.* (1995) showed that the optimum growth temperature was approx. 14–15°C. In contrast, Lu *et al.* (1994) reported that the optimal growth temperature of *H. pluvialis* was between 24°C and 28°C with a growth rate of about 0.03 h^{-1} calculated from biomass. The results from Ding and Lee (1994) also confirmed that the *H. lacustris* cells were relatively insensitive to temperature between 21 and 27°C with the optimum at 25°C. Immobilization of cells could improve the thermotolerance of *Haematococcus* cells up to 32°C with the growth rate of 0.011 h^{-1} (Ding and Lee, 1994). Increasing temperature induced the formation of palmelloid cells (Borowitzka *et al.*, 1991). Astaxanthin formation was significantly affected by culture temperature. When the *H. pluvialis* was cultivated at 30°C, astaxanthin formation reached more than 200 pg/cell and 20mg/l, which was threefold higher than at 20°C (Tjahjono *et al.*, 1994a).

The cell yield of *Haematococcus* is slightly higher in the cultures of pH 6.0 than that in pH 7.5, whereas more red palmelloid cells are formed at pH 7.5 than at pH 6.5 in acetate medium (Borowitzka *et al.*, 1991).

Function of secondary carotenoids including astaxanthin

Photoprotective function

Primary carotenoids are accessory pigments in light harvesting complexes, absorbing incident solar radiation at the 450–570 nm region of the visible spectrum; the light energy is transferred to the chlorophylls (Roman, 1989). In addition, they are also involved in at least two major physiological processes: protecting chlorophylls and the thylakoid membrane against photo-oxidative damage by absorbing excessive energy; and quenching triplet state chlorophyll molecules and scavenging singlet oxygen and other toxic oxygen species formed within the chloroplast (Figure 8.2) (Siefermann-Harms, 1987). The active mechanisms are compatible with their location in the thylakoid membrane as pigment–protein complexes. However, under conditions in which the primary carotenoids cannot deal with the excessive energy, some species of microalgae such as *Chlorella zofingiensis* (Rise *et al.*, 1994), *Haematococcus* sp. (Grung *et al.*, 1992), *Dunaliella salina* (Cifuentes, 1992), *Chlamydomonas nivalis* (Weiss, 1983), *Neospongiococcum* spp. (Deason *et al.*, 1977) and *Spongiochloris typica* (McLean, 1968) may synthesize secondary

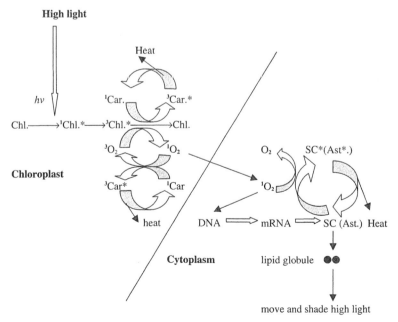

Figure 8.2 Schematic representation of the photoprotective role of astaxanthin in *Haematococcus cells*. *Excitation state; SC, secondary carotenoids; Ast., astaxanthin and its esters.

carotenoids that are not directly connected to the photosynthetic pigment apparatus of thylakoids, e.g., canthaxanthin and astaxanthin.

The function of the secondary carotenoids in the cells is not fully understood. It is assumed that secondary carotenoids function as passive photo-protectants by absorbing photons and reducing the amount of light available to the light-harvesting pigment protein complex of PSII and therefore minimizing the risk of photoinhibition (Yong and Lee, 1991; Brigate *et al.*, 1993; Hagen *et al.*, 1994). Esterification of astaxanthin with fatty acids represents a possible mechanism by which the astaxanthin is concentrated within the lipid body and maximizes its photo-protective efficiency due to the possibility of moving in the cytol in response to light intensity, as described in *Haematococcus* (Yong and Lee, 1991). Centrally located astaxanthin in *Haematococcus* disperses towards the periphery of cells under light induction and moves back towards the centre after illumination is discontinued. The protective activities of the extra-chloroplastic astaxanthin in *Haematococcus* may also include providing a physico-chemical protective barrier, thus preventing photodynamic free-radical-mediated damage in cells. In the cells, astaxanthin functions as stabilizer of membrane systems and acts as buffer for oxidative reactions to quench the radical reactions. The possible function of secondary carotenoids (astaxanthin) in cells is summarized in Figure 8.2. Exposure of cells to light intensity that causes injury to living cells resulted in the degradation of chlorophyll and primary carotenoids and also the deesterification of astaxanthin esters, which leads to increased free astaxanthin (Rise *et al.*, 1994).

Biological functions

The biological functions of secondary carotenoids have been widely studied in recent years. Most works concentrated on the antioxidant activity, inhibition of mutagenesis, enhancement of immune response and inhibition of tumour development (Santamaria, 1988). It is possible that these effects are somehow related, directly or indirectly, to the antioxidant effect of secondary carotenoids. Among the carotenoids, astaxanthin and its precusors are effective antioxidant molecules. Several research works have confirmed that the β-keto-carotenoids possess higher antioxidant activity than β-carotene (Terao, 1989) and α-tocopherol (Miki, 1991), as a lipophilic oxygen quencher. Terao (1989) examined the antioxidant activity of various carotenoids in a system consisting of free radical oxidation of methyl linoleate and the formation of methyl linoleate hydroperoxides. Of these carotenoids, canthaxanthin and astaxanthin which possess oxo groups at the 4 and 4' positions in the β-ionone ring retarded the hydroperoxide formation more efficiently than β-carotene and zeaxanthin. Di Mascio *et al.* (1989) also confirmed the antioxidant activity of canthaxanthin and astaxanthin in a system where singlet oxygens were generated. β-Keto-carotenoids acted as effective radical-trapping antioxidants that inhibited lipid peroxidation of rat liver microsomal membranes (Palozza and Krinsky, 1992). The protective effect in peroxyl radical-mediated peroxidation of phosphatidylcholine liposomes suggested that xanthophyll can function as chain-breaking antioxidant in the peroxidation of membrane phospholipids (Boey *et al.*, 1992).

The immunomodulated functions of ketocarotenoids have also been studied and the results showed that these carotenoids may help to maintain the membrane receptors that are essential to immunological function, and mediate the release of immunomodulatory lipids such as prostaglandins and leukotrienes (Bendich, 1988). Similar research in mice was conducted by Jyonouchi *et al.* (1993) and found that astaxanthin enhanced the *in vitro* and *in vivo* antibody production in response to T-dependent antigens. The chemopreventive effects of astaxanthin and canthaxanthin on carcinogenesis have been determined recently (Tanaka *et al.*, 1994). Their work revealed that the oral administration of astaxanthin significantly suppressed the development of transitional carcinomas in the urinary bladder of OH-BBN treated mice. The antitumour property of astaxanthin was partly due to its antiproliferative effect on carcinogen-exposed epithelium.

Improvement of cell growth and astaxanthin synthesis

Precursors

Haematococcus is able to grow mixotrophically on acetate (Droop, 1961). Acetate markedly enhances the growth rate and also induces the formation of red palmella cells and aplanospores (Borowitzka *et al.*, 1991). At the initial acetate concentration of 15 mM, the specific growth rate of *H. pluvialis* was found to be 0.24 d^{-1} for heterotrophic cultures and 0.58 d^{-1} for mixotrophic cultures (Kobayashi *et al.*, 1992b). A low concentration of acetate significantly stimulates the astaxanthin accumulation, whereas a high concentration of acetate ($> 0.1 \, g/l$) inhibits both the cell growth and astaxanthin synthesis (Cordero *et al.*, 1996).

Pyruvate is able to support faster cell growth and carotenoid synthesis than acetate (Kobayashi *et al.*, 1991). The combination of pyruvate and acetate further improves the cell growth and carotenoid synthesis. Other precursors, malonate and dimethylacrylate, also seem to promote carotenoid synthesis. However, the intermediates of the TCA cycle such as succinate, fumarate, malate and oxaloacetate all have an inhibitory effect at the level of 10 mM (Kobayashi *et al.*, 1991).

Immobilization

The immobilization of *Haematococcus* cells is one way to increase their thermotolerance. Cells entrapped in alginate beads were able to grow at 32°C with the growth rate of 0.011 h^{-1}, at which temperature the free *H. lacustris* cells were not able to survive (Ding and Lee, 1994). The entrapment of *Haematococcus* cells in Ca-alginate provides a protective and stable microenvironment which allows the cells to grow at a faster specific growth rate.

Mutagenesis and protoplast fusion

Mutagenesis, one of the important methods in strain improvement for specific product formation, is still the most effective way in microalgal biotechnology. Several bleaching herbicides such as norflurazon and fluridone are the specific inhibitors for phytoene desaturase which is the key enzyme for carotenoid synthesis in plant and algae. The herbicides were used to select resistant mutants of *H. pluvialis* after being treated with EMS (Tjahjono *et al.*, 1994b). The selected colour mutants could accumulate astaxanthin in the presence of the inhibitors. However, the selected mutants could not synthesize more astaxanthin than the wild type in the absence of the inhibitors.

Protoplast fusion technique was used to generate polyploidy hybrid cells of *Haematococcus*, as it was assumed that astaxanthin content would increase with the increment of gene copies for carotenogenesis. The herbicide resistant protoplasts generated by proteinase K were fused (Tjahjono *et al.*, 1993) and the fusants Fs6.1 and Fs6.5 were found to accumulate more than 120 pg astaxanthin per cell, threefold higher than the parental strain (Tjahjono *et al.*, 1994b).

Molecular biology of *Haematococcus*

Carotenoid ketolase gene (bkt) from *Haematococcus* has been cloned (Kajiwara *et al.*, 1995; Lotan and Hirschberg, 1995). The cloned gene for β-C-4 oxygenase (β-carotene ketolase) catalyzed the conversion of β-carotene to canthaxanthin via echinenone. This enzyme did not convert zeaxanthin to astaxanthin, indicating that in *Haematococcus*, astaxanthin is synthesized via canthaxanthin and therefore an additional enzyme is predicted (Lotan and Hirschberg, 1995). Northern blot analysis showed that the mRNA for astaxanthin formation was present only in the cyst cells of *H. pluvialis*, suggesting that genes for astaxanthin synthesis in the green cell were not activated and only transcribed in the cyst cells (Kajiwara *et al.*, 1995). It could also be interpreted as synthesis of astaxanthin is induced via transcription of genes activated by growth parameters such as nutrient starvation which retarded cell growth leading to formation of cyst cells. The successful expression of genes for astaxanthin synthesis in *E. coli* (Breitenbach *et al.*, 1996) is the first step towards the biotechnological production of astaxanthin with transgenic organisms.

Commercial production and application of astaxanthin from *Haematococcus*

Haematococcus cultures may accumulate as much as 6–8 per cent dry weight of astaxanthin (Tsavalos *et al.*, 1992). However, mass cultivation of *Haematococcus* for commercial production of astaxanthin is hindered by its slow growth rate, low growth temperature and light tolerance, and different requirements for cell growth and astaxanthin production. A two-stage production process, initially for biomass production which is followed by an astaxanthin accumulation stage was proposed (Harker *et al.*, 1996b). For the biomass production of *Haematococcus*, the photoreactor should be operated under low light intensity or high cell density to avoid the undesirable effect of photoinhibition. Low liquid velocity is desirable because of the fragility of the vegetative cells. At the stage for astaxanthin production, high light intensity or other stress factors such as high concentration of NaCl and oxygen are necessary. High liquid velocity is necessary to prevent sinking of large palmelloid and cyst cells.

To improve on the astaxanthin productivity, high cell density is imperative. However, it is very difficult to maintain high growth rate and high biomass in the outdoor system, as the *Haematococcus* has a complex life cycle and the culture is easily contaminated and overgrown by fast growing algae or eliminated by protozoal predators (Spencer, 1989).

For product formulation, there are several problems to be overcome. The red cyst cells have a thick and rigid cell wall made up of sporopollenin-like polymer, which is resistant to animal digestion (Good and Chapman, 1979). Therefore, bioavailability of astaxanthin in the red cells to aquacultured animals is low. In addition, the free astaxanthin and esters are easily oxidized during extraction to form astacene, which is a biologically unimportant pigment. A suitable antioxidant should be added to prevent oxidation of astaxanthin. The commercial product Algaxan Red has been successfully used for the pigmentation of salmon (Spencer, 1989).

The commercial application of astaxanthin products from *Haematococcus* is mainly focused on crustaceans and fish, for inducing desirable coloration of their eggs, flesh and skin (Sommer *et al.*, 1992; Choubert and Heinrich, 1993). The astaxanthin can produce native-identical pigmentation and it is efficiently deposited (Torrissen, 1986; Storebakken *et al.*, 1987) and stable during processing of fish flesh (Skrede and Storebakken, 1987). When supplying diets containing astaxanthin-rich *Haematococcus* cells to rainbow trout, deposition of carotenoids and visual enhancement of the colour of flesh in trout were observed (Sommer *et al.*, 1991). However, the highest level of coloration in trout flesh was found in groups fed with synthetic astaxanthin. The low coloration efficiency of *Haematococcus* astaxanthin was probably due at least in part to the high percentage of esterified carotenoids which reduced bioavailability (Renstrom and Liaaen-Jensen, 1981), and were less efficiently deposited in salmonids (Storebakken *et al.*, 1987).

Conclusion and future work

Haematococcus has been considered the best candidate natural source of astaxanthin, as it can accumulate up to 6–8 per cent astaxanthin in biomass. Physiological and environmental factors, which stimulate astaxanthin synthesis, have been investigated. The enzymes involved in its synthetic pathway have also been identified. The knowledge of carotenogenic enzymes is of considerable practical importance, as they are target sites for improvement by genetic manipulation. Although the cellular astaxanthin content could be increased, the relatively low optimal growth temperature, low growth rate and sensitivity to photoinhibition limit the mass production of *Haematococcus* and astaxanthin in outdoor culture systems.

The recent rapid advances in molecular genetics of carotenoid biosynthesis in bacteria lead to progress in gene cloning in microalgae. The gene clusters for β-carotene synthesis from bacteria have been constructed in plasmids (Misawa *et al.*, 1995a) and the plasmids have been used for cloning of the ketolase gene from *Haematococcus* (Kajiwara *et al.*, 1995; Breitenbach *et al.*, 1996). The successful expression of the astaxanthin-synthetic gene in bacteria (*E. coli*) demonstrates the prospect of producing astaxanthin in suitable genetically engineered micro-organisms.

References

ANDREWES, A.G., PHAFF, H.J. and STARR, M.P., 1976, Carotenoids of *Phaffia rhodozyma*, a red pigmented fermenting yeast, *Phytochemistry*, **15**, 1003–1007.

BEJARANO, E.R., 1988, End-product regulation of carotenogenesis in *Phycomyces*, *Arch. Microbiol.*, **1560**, 209–214.

BEN-AMOTZ, A. and AVRON, M., 1983, Accumulation of metabolites by halotolerant algae and its industrial potential, *Ann. Rev. Microbiol.*, **37**, 95–119.

BENDICH, A., 1988, Symposium conclusions: biological action of carotenoids, *J. Nutr.*, **119**, 112-115.

BERKALOFF, C. and KADER, J.C., 1975, Variation of the lipid composition during formation of cysts in the green alga *Protosiphon botryoides*, *Phytochemistry*, **14**, 2353–2355.

BOEY, P.L., NAGAO, A., TERAO, J., TAKAMA, K., SUZUKI, T. and TAKAMA, K., 1992, Antioxidant activity of xanthophylls on peroxyl radicals mediated phospholipid peroxidation, *Biochim. Biophys. Acta*, **1126**, 178–184.

BOROWITZKA, M.A., 1988, Vitamins and fine chemicals from microalgae, in Borowitzka, M.A. and Borowitzka, L.J. (Eds), *Micro-algal Biotechnology*, pp. 153–196. Cambridge: Cambridge University Press.

BOROWITZKA, M.A., HUISMAN, J.M. and OSBORN, A., 1991, Cultures of the astaxanthin-producing green algae *Haematococcus pluvialis*. I. Effect of nutrient on growth and cell type, *J. Appl. Phycol.*, **3**, 295–304.

BOUSSIBA, S. and VONSHAK, A., 1991, Astaxanthin accumulation in the green alga *H. pluvialis*, *Plant Cell Physiol.*, **32**, 1077–1082.

BOUSSIBA S., LU, F. and VONSHAK, A., 1992, Enhancement and determination of astaxanthin accumulation in green alga *Haematococcus pluvialis*, *Methods in Enzymol.*, **213**, 386–391.

BREITENBACH, J., MISAWA, N., KAJIWARA, S. and SANLMANN, G., 1996, Expression in *E. coli* and properties of the carotene ketolase from *H. pluvialis*, *FEMS Microbiology Letters*, **140**, 241–246.

BRIGATE, R.R., ONDRUSEK, M.E. and KENNICUTT, M.C., 1993, Evidence for a photoprotective function for secondary carotenoids of snow algae, *J. Phycol.*, **29**, 427–434.

BRITTON, G., 1990, Carotenoid biosynthesis – an overview, in Krinsky, N.I., Mathews-Roth, M.M. and Taylor, R.F. (Eds), *Carotenoids: Chemistry and Biology*, pp. 167–184. New York: Plenum Press.

BRITTON, G., 1993, The biosynthesis of carotenoids, in Young, A. and Britton, G. (Eds), *Carotenoids in Photosynthesis*, pp. 96–123, Suffolk: St Edmundsbury Press.

BROWN, T.T., RICHARDSON, F.L. and VAUGHN, M.L., 1967, Development of red pigmentation of *Chlorococcum wimmeri* (Chlorophyta: Chlorococcales), *Phycologia*, **6**, 167–184.

BUBRICK, P., 1991, Production of astaxanthin from *Haematococcus*, *Bioresource Technol.*, **38**, 237–239.

CERDA-OLMEDO, E., 1985, Carotene mutants of phycomyces, *Methods in Enzymology*, **110**, 220–243.

CHAMOVITZ, D., MISAWA, N., SANDMANN, G. and HIRSCHBERG, J., 1992, Molecular cloning and expression in *E. coli* of a cyanobacterial gene coding for phytoene synthase, a carotenoid biosynthesis enzyme, *FEBS Lett.*, **296**, 305–310.

CHOUBERT, G. and HEINRICH, O., 1993, Carotenoid pigments of the green alga, *Haematococcus*: assay on rainbow trout, *Oncorhynchus mykiss*, pigmentation in comparison with synthetic astaxanthin and canthaxanthin, *Aquaculture*, **112**, 217–226.

CIEGLE, A., 1965, Microbial carotenogenesis, *Adv. Appl. Microbiol.*, **7**, 1–34.

CIFUENTES, A.S., 1992, Growth and carotenogenesis in eight strains of *Dunaliella salina* Teodoresco from Chile, *J. Appl. Phycol.*, **4**, 111–118.

CORDERO, D., OTERO, A., PATINO, M., ARREDONDO, B.O. and FABREGAS, J., 1996, Astaxanthin production from the green alga *Haematococcus pluvialis* with different stress conditions, *Biotechnol. Lett.*, **18**, 213–218.

CUNNINGHAM, F.X., Jr, CHAMOVITZ, D., MISAWA, N., GANTT, E., and HIRSCHBERG, J., 1993, Cloning and functional expression in *E. coli* of a cyanobacterial gene for lycopen cyclase, the enzyme that catalyses the biosynthesis of β-carotene, *FEBS Lett.*, **328**, 130–138.

CZECZUGA, B., 1974, Carotenoids in *Euglena rubida* Mainx, *Comp. Biochem. Physiol.*, **48B**, 349–354.

CZECZUGA, B., 1986, Characteristic carotenoids in some phytobenthos species in the coastal area of the Adriatic Sea, *Acta Soc. Bot. Pol.*, **55**, 601–609.

CZYGAN, F.C., 1964, Canthaxanthin als sekundar-carotenoid in einiger grünalgen, *Experientia*, **20**, 573–575.

CZYGAN, F.C., 1968, Sekundar-carotenoide in grünalgen. I. Chemie, vorkommen und faktoren, welche die bildung dieser polyene beeinflussen, *Arch. Mikrobiol.*, **61**, 81–102.

CZYGAN, F.C., 1970, Untersuchungen über die Bedeutung der Biosyntheses von secundar-carotenoiden als artmerkmale bei grünalgen, *Arch. Mikrobiol.*, **74**, 77–81.

DAVIES, B.H., 1973, Carotene biosynthesis in fungi, *Pure Appl. Chem.*, **35**, 1–28.

DEASON, T.R., CZYGAN, F.C. and SOEDER, C.J., 1977, Taxonomic significance of secondary carotenoid formation in *Neospongiococcus* (Chlorococcales), *J. Phycol.*, **13**, 176–180.

DERSCH, G., 1960, Mineralsalzmangel und secundar-carotenoide in grünalgen, *Flora*, **149**, 566–603.

DI MASCIO, P., KAISER, S. and SIES, H., 1989, Lycopene as the most efficient biological carotenoid singlet oxygen quencher, *Arch. Biochem. Biophys.*, **274**, 532–538.

DING, S.Y. and LEE, Y.K., 1994, Growth of entrapped *H. lacustris* in alginate beads in a fluidised bed air-lift bioreactor, in Phang, S.M., Lee, Y.K., Borowitzka, M. and Whitton, B. (Eds), *Algal Biotechnology in Asia-Pacific Region*, pp. 130–134. Kuala Lumpur: University of Malaya.

DONKIN, P., 1976, Ketocarotenoid biosynthesis by *H. pluvialis*, *Phytochemistry*, **15**, 711–715.

DROOP, M.R., 1955, Carotenogenesis in *Haematococcus pluvialis*, *Nature*, **175**, 42.

DROOP, M.R., 1961, *Haematococcus pluvialis* and its allies III. Organic nutrition, *Rev. Algology*, **3**, 247–259.

ELIOT, A.M., 1934, Morphology and life history of *Haematococcus pluvialis*, *Arch. Protistenk.*, **82**, 250–272.

EUGSTER, C.H., 1979, Characterisation, chemistry and stereochemistry of carotenoids, *Pure Appl. Chem.*, **51**, 463–506.

GIRARD, P., FALCONNIER, B., BRICOUT, J. and VLADESCU, B., 1994, Beta-carotene producing mutants of *Phaffia rhodozyma*, *Appl. Microbiol. Biotechnol.*, **41**, 183–191.

GOOD, B.H. and CHAPMAN, R.L., 1979, A comparison between the walls of *Haematococcus* cysts and the loricas of *Dysmorphococcus*, and the possible presence of sporopollenin, *J. Phycol.* (suppl.) **15**, 7.

GOODWIN, T.W., 1952a, Carotenogenesis III. Identification of minor polyene compounds of the *Phycomyces blakesleeanus* and a study under various cultural conditions, *Biochem. J.*, **50**, 550–558.

GOODWIN, T.W., 1952b, *The Comparative Biochemistry of the Carotenoids*, p. 356, London: Chapman and Hall.

GOODWIN, T.W., 1971, Biosynthesis, in Isler, O. (Ed.), *Carotenoids*, pp. 577–636, Basel: Birkhauser.

GOODWIN, T.W., 1980, *The Biochemistry of Carotenoids*, 2nd edn, Vol.1, *Plants*, London: Chapman and Hall.

GOODWIN, T.W. and JAMIKORN, M., 1954, Studies in carotenogenesis 2. Carotenoid synthesis in the alga *H. pluvialis*, *Biochem. J.*, **57**, 376–381.

GREEN, J., 1963, The occurrence of astaxanthin in the Euglenoid *Trachelomonas volvocina*, *Comp., Biochem. Physiol.*, **9**, 313–316.

GRUNG, M., BJERKENG, B., BOROWITZKA, M.A., SKULBERG, O. and LIAAEN-JENSEN, S., 1990, Alternative source of astaxanthin, including secondary carotenoids of microalgae. Abstract, 9th IUPAC symposium on carotenoids (Kyoto) p. 116.

GRUNG, M., D'SOUZA, F.M.L., BOROWITZKA, M. and LIAAEN-JENSEN, S., 1992, Algal carotenoids 51. Secondary carotenoids 2, *Haematococcus pluvialis* aplanospores as a source of (3S, 3'S)-astaxanthin esters, *J. Appl. Phycol.*, **4**, 165–191.

HAGEN, C., BRAUNE, W. and GREULICH, F., 1993, Functional aspects of secondary carotenoids in *Haematococcus lacustris* (Girad) Rostafinski (Volvocales) IV. Protection from photodynamic damage, *J. Photochem. Photobiol.*, **B20**, 153–160.

HAGEN, C., BRAUNE, W. and BJÖRN, L.O., 1994, Functional aspects of secondary carotenoids in *Haematococcus lacustris* (Volvocales) III. Action as a sunshade, *J. Phycol.*, **30**, 241–248.

HARKER, M., TSAVALOS, A.J. and YOUNG, A.J., 1995, Use of response surface methodology to optimise carotenogenesis in the microalga *H. pluvialis*, *J. Appl. Phycol.*, **7**, 399–406.

HARKER, M. TSAVALOS, A.J. and YOUNG, A.J., 1996a, Factors responsible for astaxanthin formation in the Chlorophyte *Haematococcus pluvialis*, *Bioresource Technol.*, **55**, 207–214.

HARKER, M., TSAVALOS, A.J. and YOUNG, A.J., 1996b, Autotrophic growth and carotenoid production of *Haematococcus pluvialis* in a 30 litre airlift photobioreactor, *J. Ferment. Bioeng.*, **82**, 113–118.

HUNDLE, B.S., O'BRIEN, D.A., BEYER, P., KLEINIG, H. and HEART, J.E., 1993, *In vitro* expression and activity of lycopene cyclase and β-carotene hydroxylase from *Erwinia herbicola*, *FEBS Lett.*, **315**, 329–334.

JOHNSON, E.A. and AN, G.H, 1991, Astaxanthin from microbial sources, *Crit. Rev. Biotechnol.*, **11**, 297–326.

JOHNSON, F.A. and LEWIS, M.J., 1979, Astaxanthin formation by the yeast *Phaffia rhodozyma*, *J. Gen. Microbiol.*, **115**, 173–183.

JYONOUCHI, H., ZHANG, L. and TOMITA, Y., 1993, Study of immuno-modulating actions of carotenoids II. Astaxanthin enhances *in vivo* antibody production to T-dependent antigens without facilitating polyclonal B-cell activation, *Nutr. Cancer*, **19**, 269–280.

KAJIWARA, S., KAKIZONO, T., SAITO, T., KONDO, K., OHTANI, T., NISHIO, N., NAGAI, S. and MISAWA, N., 1995, Isolation and function identification of a novel cDNA for astaxanthin biosynthesis from *H. pluvialis* and astaxanthin synthesis in *E. coli*, *Plant Molecular Biology*, **29**, 343–352.

KAKIZONO, T., KOBAYASHI, M. and NAGAI, S., 1992, Effect of carbon/nitrogen ratio on encystment accompanied with astaxanthin formation in a green alga, *Haematococcus pluvialis*, *J. Ferment. Bioeng.*, **74**, 403–405.

KESSLER, E. and CZYGAN, F., 1965, *Chlorella zofingiensis* Donz, Isolierung neuer stamme und ihre physiologisch-biochemischen eigenschaften, *Berichte der Deutschen Botanischen Gesellschaft*, **78**, 342–347.

KOBAYASHI, M., KAKIZONO, T. and NAGAI, S., 1991, Astaxanthin production by a green alga, *Haematococcus pluvialis* accompanied with morphological changes in an acetate medium, *J. Ferment. Bioeng.*, **71**, 335–339.

KOBAYASHI, M., KAKIZONO, T., NISHIO, N. and NAGAI, S., 1992a, Effect of light intensity, light quality and illumination cycle on astaxanthin formation in a green alga, *Haematococcus pluvialis*, *J. Ferment. Bioeng.*, **74**, 61–63.

KOBAYASHI, M., KAKIZONO, T., YAMAGUCHI, K., NISHIO, N. and NAGAI, S., 1992b, Growth and astaxanthin formation of *Haematococcus pluvialis* in heterotrophic and mixotrophic conditions, *J. Ferment. Bioeng.*, **74**, 17–20.

KOBAYASHI, M., KAKIZONO, T. and NAGAI, S., 1993, Enhanced carotenoid biosynthesis by oxidative stress in acetate-induced cyst cells of a green unicellular alga, *Haematococcus pluvialis*, *Appl. Environ. Microbiol.*, **59**, 867–873.

LANG, H.P., COGDELL, R.J., GARDINER, A.T. and HUNTER, K.N., 1994, Early steps in carotenoid biosynthesis, sequence and transcriptional analysis of the Crt I and Crt B genes of *Rhodobacter sphaeroides* and overexpression and reactivation of Crt I in *E. coli* and *R. sphaeroides*, *J. Bacteriol.*, **176**, 3859–3869.

LEE, Y.K. and DING, S.Y., 1994, Cell cycle and accumulation of astaxanthin in *Haematococcus lacustris* (Chlorophyta), *J. Phycol.*, **30**, 445–449.

LEE, Y.K. and DING, S.Y., 1995, Effect of dissolved oxygen partial pressure on the accumulation of astaxanthin in chemostat cultures of *H. lacustris* (Chlorophyta), *J. Phycol.*, **31**, 922–924.

LEE, Y.K. and SOH, C.W., 1991, Accumulation of astaxanthin in *Haematococcus lacustris* (Chlorophyta), *J. Phycol.*, **27**, 575–577.

LINDEN, H., MISAWA, N., SAITO, T. and SANDMANN, G., 1994, A novel carotenoid biosynthesis gene coding for δ-carotene desaturase, functional expression, sequence and phylogenetic origin, *Plant Mol. Biol.*, **24**, 369–379.

LOPEZ-DIAZ, I. and CERDA-OLMEDO, E., 1980, Relationship of photocarotenogenesis to other behavioral and regulatory response in *Phycomyces*, *Planta*, **150**, 134–139.

LOTAN, T. and HIRSCHBERG, J., 1995, Cloning and expression in *E. coli* of the gene encoding β-C-4-oxygenase, that converts β-carotene to the ketocarotenoid canthaxanthin in *Haematococcus pluvialis*, *FEBS Letters*, **364**, 125–128.

LU, F., VONSHAK, A. and BOUSSIBA, S., 1994, Effect of temperature and irradiance on growth of *H. pluvialis*, *J. Phycol.*, **30**, 829–833.

LU, F., VONSHAK, A., GABBAY, R., HIRSHBERG, J., COHEN, Z. and BOUSSIBA, S., 1995, The biosynthetic pathway of astaxanthin in a green alga, *Haematococcus pluvialis* as indicated by inhibition with diphenylamine, *Plant Cell Physiol.*, **36**, 1519–1524.

LU, F., VONSHAK, A. and BOUSSIBA, S., 1996, Does astaxanthin play a photoprotective role in *Haematococcus pluvialis* (Chlorophyceae)? *J. Appl. Phycol.*, **8**, 445.

MCLEAN, R.J., 1967, Primary and secondary carotenoids of *Spongiochloris typica*, *Physiol. Plantarum*, **20**, 41–47.

MCLEAN, R.J., 1968, Ultrastructures of *Spongiochloris typica* during senescence, *J. Phycol.*, **4**, 277–283.

MIKI, W., 1991, Biological function and activity of animal carotenoids, *Pure Appl. Chem.*, **63**, 141–146.

MISAWA, N., NAKAGAWA, M., KOBAYASHI, K., YAMANO, S., IZAWA, Y. and NAKAMURA, K., 1990, Elucidation of the *Erwinia uredovora* carotenoids biosynthetic pathway by the functional analysis of gene products expressed in *E. coli*, *J. Bacteriol.*, **172**, 6704–6712.

MISAWA, N., KAJIWARA, S., KONDO, K., YOKOYAMA, A., SATOMI, Y., SAITO, T., MIKI, W. and OHTANI, T., 1995a, Canthaxanthin biosynthesis by the conversion of methylene to keto groups in the hydrocarbon β-carotene by single gene, *Biochem. Biophys. Res. Com.*, **209**, 867–876.

MISAWA, N., SATOMI, Y., KONDO, K., YOKOYAMA, A., KAJIWARA, S., SAITO, T., OHTANI, T. and MIKI, W., 1995b, Structure and functional analysis of a marine bacterial carotenoid biosynthesis gene cluster and astaxanthin biosynthetic pathway proposed at gene level, *J. Bacteriol.*, **177**, 6575–6584.

MURILLO, F.J., 1976, Regulation of carotene synthesis in *Phycomyces*, *Mol. Gen. Genet.*, **148**, 19–24.

MURILLO, F.J., 1980, Effect of CPTA on carotenogenesis by *Phycomyces* car A mutants, *Plant Sci. Lett.*, **17**, 201–205.

NELIS, H.J. and DE LEENHEER, A.P.D., 1991, Microbial sources of carotenoid pigments used in food and feeds, *J. Appl. Bact.*, **70**, 181–191.

NINET, L., RENAUT, J. and TISSIER, R., 1969, Activation of the biosynthesis of carotenoids by *Blakesleea trispora*, *Biotechnol. Bioeng.*, **11**, 1115–1120.

OOTAK, T., 1973, Complementation between mutants of *Phycomyces* deficient with respect to carotenogenesis, *Mol. Gen. Genet.*, **121**, 57–70.

PALOZZA, P. and KRINSKY, N. I., 1992, Astaxanthin and canthaxanthin are potent antioxidants in membrane model, *Arch. Biochem. Biophys.*, **297**, 291–295.

PARN, P. and SEVIOUR, R.J., 1974, Pigments induced by organo mercurial compounds in *Cephalosporium diospyros*, *J. Gen. Microbiol.*, **85**, 228–236.

POCKOCK, M.A., 1961, *Haematococcus* in southern Africa, *Trans. of the Royal Soc. of South Africa*, **36**, 5–55.

PROCTOR, V.W., 1957, Preferential assimilation of nitrate by *Haematococcus pluvialis*, *Am. J. Bot.*, **44**, 141–143.

RAO, S. and MODI, V.V., 1977, Carotenogenesis, possible mechanism of action of trisporic acid in *Blakeslea trispora*, *Experientia*, **33**, 31–33

RAU, W., 1976, Photoregulation of carotenoid biosynthesis in plants, *Pure Appl. Chem.*, **47**, 237–243.

RENSTROM, B. and LIAAEN-JENSEN, S., 1981, Fatty acid composition of some esterified caro-tenols, *Comp. Biochem. Physiol. B. Comp. Biochem.*, **69**, 625–627.

RICKETTS, T.R., 1970, The pigments of the prasinophyceae and related organisms, *Phytochemistry*, **9**, 1835–1842.

RISE, M., COHEN, E., VISHKAUTSAN, M., COJOCARU, M., GOTTLIEB, H.E. and ARAD, S., 1994, Accumulation of secondary carotenoids in *Chlorella zofingiensis*, *J. Plant Physiol.*, **144**, 287–292.

ROMAN, K.S., 1989, *Photosynthetic Pigments of the Algae*. Cambridge, Cambridge University Press.

RUDDAT, M. and GARDER, E.D., 1983, Biochemistry, physiology and genetics of carotenogen-esis, in Benneta, J.W. and Ciegler, A. (Eds), *Secondary Metabolism and Differentiation in Fungi*, New York: Marcel Dekker.

SANDMANN, G., 1980, The inhibitory mode of action of the pyridazinone herbicide norflurazon on the cell-free carotenogenic enzyme system, *Pesticide Biochem. Physiol.*, **14**, 185–191.

SANDMANN, G. and MISAWA, N., 1992, New functional assignment of the carotenogenic genes Crt B and Crt E with constructs of these genes from *Erwinia* species, *FEMS Microbiol. Lett.*, **90**, 253–258.

SANTAMARIA, L., 1988, Chemoprevention of indirect and direct chemical carcinogenesis by carotenoids as oxygen radical quenchers, *Ann. N. Y. Acad. Sci.*, **534**, 584–596.

SANTOS, M.F. and MESQUITA, J.F., 1984, Ultrastructural study of *Haematococcus lacustris* (Girod.) Rotafinski (volvocales), 1. Some aspects of carotenogenesis, *Cytologia*, **49**, 215–228.

SCHROEDER, W.A. and JOHNSON, E.A., 1995, Singlet oxygen and peroxyl radicals regulate carotenoid biosynthesis in *Phaffia rhodozyma*, *J. Biol. Chem.*, **270**, 18374–18379.

SHI, T., FRANCIS, X., CUNNINGHAM, Jr, YOUMANS, M., GRABOWSKI, B., SUN, Z. and GANTT, E., 1995, Cytochrome f loss in astaxanthin-accumulating red cells of *Haematococcus pluvialis* (Chlorophyceae): comparison of photosynthetic activity, photosynthetic enzymes and thylakoid membrane polypeptides in red and green cells, *J. Phycol.*, **51**, 897–905.

SIEFERMANN-HARMS, D., 1987, The light-harvesting and protective function of carotenoids in photosynthetic membranes, *Physiol. Plant*, **69**, 561–568.

SKREDE, G. and STOREBAKKEN, T., 1987, Characteristic of colour in raw, baked and smoked wild and open-reared Atlantic salmon, *J. Food Sci.*, **51**, 804–808.

SOMMER, T.R., POTTS, W.T. and MORRISSY, N.M., 1991, Utilisation of microalgal astaxanthin by rainbow trout (*Oncorhynchus mykiss*), *Aquaculture*, **94**, 79–88.

SOMMER, T.R., D'SOUZA, F.M.L. and MORRISSY, N.M., 1992, Pigmentation of adult rainbow trout *Oncorhynchus mykiss*, using the green alga *Haematococcus*, *Aquaculture*, **106**, 63–74.

SPENCER, K.G., 1989, Pigmentation supplements for animal feed composition, US. Patent No. 4871551.

SPURGEON, S.L. and PORTER, J.W., 1983, Biosynthesis of carotenoids, in Porter, J.W. and Spurgeon, S.L. (Eds), *Biosynthesis of Isoprenoid Compounds*, Vol. 2, pp. 1–122, New York: Wiley.

STOREBAKKEN, T., FOSS, P., SCHIEDT, K., AUSTRENG, E., LIAAEN-JENSEN, S. and MANZ, V., 1987, Carotenoids in diets for salmonids IV. Pigmentation of Atlantic salmon with astaxanthin, astaxanthin dipalmitate and canthaxathin, *Aquaculture*, **65**, 279–292.

STROSS, R.G., 1963, Nitrate preference in *H. pluvialis* is controlled by strain, age of inoculum and pH of the medium, *Can. J. Microbiol.*, **9**, 37–40.

SUTTER, R.P., 1970, Effects of light on beta-carotene accumulation in *Blakesleea trispora*, *J. Gen. Microbiol.*, **64**, 215–221.

TANAKA, T., MORISHITA, Y., SUZUI, M., KOJIMA, T., OKUMURA, A. and MORI, A., 1994, Chemoprevention of mouse urinary bladder carcinogenesis by the naturally occurring carotenoid astaxanthin, *Carcinogenesis*, **15**, 15–19.

TERAO, J., 1989, Antioxidant activity of beta-carotene related carotenoids in solution, *Lipids*, **24**, 659–661.

TJAHJONO, A.E., KAKIZONO, T., YARYAMA, Y. and NAGAI, S., 1993, Formation and regeneration of protoplast from a unicellular green alga, *H. pluvialis*, *J. Ferment. Bioeng.*, **75**, 196–200.

TJAHJONO, A.E., HAYAMA, Y., KAKIZONO, T., TERADA, Y., NISHIO, N. and NAGAI, S., 1994a, Hyper-accumulation of astaxanthin in a green alga *Haematococcus pluvialis* at elevated temperature, *Biotechnol. Lett.*, **16**, 133–138.

TJAHJONO, A.E., KAKIZONO, T., YAVAMA, Y., WISHIO, N. and NAGAI, S., 1994b, Isolation of resistant mutants against carotenoid biosynthesis inhibitors for a green alga *Haematococcus pluvialis* and those hybrid formation by protoplast fusion for breeding of higher astaxanthin producers, *J. Ferment. Bioeng.*, **77**(4), 352–357.

TO, K.E., LAI, E., LEE, L., LIN, T., HUNG, C., CHEN, C., CHANG, Y. and LIN, S., 1994, Analysis of the gene cluster encoding carotenoid biosynthesis in *Erwinia herbicola* Eho13, *Microbiology*, **140**, 331–339.

TORRES-MARTINEZ, S., MURILLO, F.J. and CERDA-OLMEDO, E., 1980, Generics of lycopene cyclization and substrate transfer in beta-carotene biosynthesis in *Phycomyces*, *Genet. Res.*, **36**, 299–309.

TORRISSEN, O.J., 1986, Pigmentation of salmonids, a comparison of astaxanthin and cantha-xanthin as pigment sources for rainbow trout, *Aquaculture*, **53**, 271–278.

TSAVALOS, A.T., HARKAER, M., DANIELS, M. and YOUNG, A.J., 1992, Secondary carotenoids synthesis in microalgae, in Murata, N. (Ed.), *Research in Photosynthesis*, Vol.III. pp. 47–51, Dordrecht, Kluwer.

VECHTEL, B., EICHENBERGER, W. and RUPPEL, H.G., 1992a, Lipid bodies in *Eremosphaera viridis* De Bary (Chlorophyceae), *Plant Cell Physiol.*, **31**, 41–48.

VECHTEL, B., PISTORIOUS, E.K. and RUPPEL, H.G., 1992b, Occurrence of secondary carotenoids in PSI complex isolated from *Eremosphaera viridis* De Bary (Chlorophyceae), *Z. Naturforsch.*, **47C**, 51–56.

VIALA. G., 1966, L'astaxanthin chez le *Chlamydomonas nivalis* Wille, *C. R. Acad. Sci. Paris. D*, **263**, 1383–1386.

WEISS, R.L., 1983, Fine structure of snow algae (*Chlamydomonas nivalis*) and associated bacteria, *J. Phycol.*, **19**, 200–204.

YOKOYAMA, A. and MIKI, W., 1995, Composition and presumed biosynthetic pathway of carotenoids in the astaxanthin-producing bacterium, *Agrobacterium aurantiacum*, *FEMS Microbiol. Lett.*, **128**, 139–144.

YOKOYAMA, A., IZUMIDA, H. and MIKI, W., 1994, Production of astaxanthin and 4-keto-zeaxanthin by the marine bacterium, *Agrobacterium aurantiacum*, *Biosci. Biotech. Biochem.*, **58**, 1842–1844.

YONG, Y.Y.R. and LEE, Y.K., 1991, Do carotenoids play a photoprotective role in the cyto-plasm of *Haematococcus lacustris* (Chlorophyta)?, *Phycologia*, **30**, 257–261.

ZHANG, D.H., LEE, Y.K., NG, M.L. and PHANG, S.M., 1997, Composition and accumulation of secondary carotenoids in *Chlorococcum* sp, *J. Appl. Phycol.*, **9**, 147–155.

ZLOTNIK, I., SUKENIK, A. and DUBINSKY, Z., 1993, Physiological and photosynthetic changes during the formation of red aplanospores in the Chlorophyte *Haematococcus pluvialis*, *J. Phycol.*, **29**, 463–469.

9

Production of β-carotene from *Dunaliella*

AMI BEN-AMOTZ

Biology and halotolerance

The biflagellated alga *Dunaliella* is classified under Chlorophyceae, Volvocales, which includes a variety of ill-defined marine and fresh water unicellular species (Avron and Ben-Amotz, 1992). *Dunaliella*, like *Chlamydomonas*, is characterized by an ovoid cell volume in the range of 2–8 μm wide × 5–15 μm length, usually in the shape of pear, wider at the basal side and narrow at the anterior flagella top (Figure 9.1). The cellular organization of *Dunaliella* is no different than that of other members of the Volvocales, presenting one big chloroplast with single-centred starch surrounded pyrenoid, a few vacuoles, nucleus and nucleolus. However, unlike other green algae, *Dunaliella* lacks a rigid polysaccharide cell wall, and the cell is enclosed by a thin elastic plasma membrane covered by a mucous surface coat. The lack of a rigid cell wall permits rapid cell volume changes in response to extracellular changes in osmotic pressure; thus, commonly observed *Dunaliella* vary from round to eight shapes. Under extreme salt concentrations of above 4M *Dunaliella* loses its flagella and the surrounding mucus, and the cell rounds with a build-up of a thick surrounding wall to form a dehydration resistant cyst.

Dunaliella occurs in a wide range of marine habitats such as oceans, brine lakes, salt marshes and salt water ditches near the sea, predominantly in water bodies containing more than 2M salt and a high level of magnesium. The effect of magnesium on the distribution of *Dunaliella* is not clear, but in many 'bittern' habitats of marine salt producers, *Dunaliella* usually flourishes. The phenomenon of orange–red algal bloom in such marine environments is usually related to combined sequential growth of *Dunaliella* first, and then halophilic bacteria, and is commonly observed in concentrated saline lakes such as the Dead Sea in Israel, the Pink Lake in Western Australia, the Great Salt Lake in Utah, USA, and in many other places around the globe. *Dunaliella* is the most halotolerant eukaryotic organism known, showing a remarkable degree of adaptation to a variety of salt concentrations, from as low as 0.1M to salt saturation.

Dunaliella uses a special mechanism of osmoregulation to adapt to the various salt concentrations in the surrounding media by varying the intracellular concentration

Figure 9.1 Scanning electron micrograph of *Dunaliella bardawil,* an ovoid biflagellate, 15 μm long, 5 μm wide.

of glycerol in response to the extracellular osmotic pressure. The intracellular concentration of glycerol is directly proportional to the extracellular salt concentration and is sufficient to account for all the required cytoplasmic and most of the chloroplastic osmotic pressures. Glycerol biosynthesis and elimination occur in the light or dark by a novel glycerol cycle which involves several specific enzymes (Avron, 1992).

β-Carotene production

Among the many known species of *Dunaliella*, only a few subspecies of *D. salina* Teod. have been shown to produce and accumulate large amounts of β-carotene. In hypersaline lakes, which are generally low in available nitrogen and exposed to high solar irradiation, these β-carotene producing strains of *Dunaliella* predominate to a seasonal bloom of about 0.1 mg β-carotene per litre. Under such stressful environmental conditions more than 12 per cent of the algal dry weight is β-carotene, usually associated with a sharp decline in the thylakoids' chlorophyll. The β-carotene in

Dunaliella accumulates within distinctive oily globules in the interthylakoid spaces of the chloroplast periphery. Analysis of the globules showed that the β-carotene of *Dunaliella* is composed mainly of two stereoisomers – all-*trans* and 9-*cis* – with the rest a few other mono-*cis* and di-*cis* stereoisomers of the β-carotene. Both the amount of the accumulated β-carotene and the 9-*cis* to all-*trans* ratio depend on light intensity and on the algal division time as affected by the growth conditions. Thus, any growth stress which will slow down the rate of cell division under light will in turn increase β-carotene production in *Dunaliella*. In fact, high light and many environmental stress conditions such as high salt, low temperature, extreme pH, nutrient deficiency, and others affect the content of β-carotene in *Dunaliella*. It was previously suggested that the equation of the amount of light absorbed by the cell during one division cycle integrates the effect of all growth variables on the content and isomeric ratio of β-carotene in *Dunaliella* (Ben-Amotz and Avron, 1990). The exceptions to this integration are nitrogen deficiency and low growth temperatures; both induce exceptionally extreme intracellular accumulation of β-carotene under any light intensity. Taking into account that nitrogen starvation inhibits chlorophyll production in algae as part of its inhibitory effect on protein biosynthesis, the prolonged nitrogen independent carotenoid biosynthesis will then protect the chlorophyll-low cells against the lethal damage of light. The effect of low temperatures on *Dunaliella* is analyzed by specific stimulation effect of chilling on the biosynthesis of 9-*cis* β-carotene. The physicochemical properties of 9-*cis* β-carotene differ from those of all-*trans* β-carotene; *all-trans* β-carotene is practically insoluble in oil and is easily crystallized at low temperatures, while 9-*cis* β-carotene is much more soluble in hydrophobic–lipophilic solvents, very difficult to crystallize and generally oily in its concentrated form. To avoid cellular crystallization of all-*trans* β-carotene and to survive on growth at low temperatures, *Dunaliella* produces a higher ratio of 9-*cis* to all-*trans* β-carotene, where the 9-*cis* stereoisomer functions *in vivo* as an oily matrix to all-*trans* β-carotene (Ben-Amotz, 1996).

Carotenogenesis in *Dunaliella* follows the same biosynthetic pathway of carotenoids, with the same substrates and same intermediates as found in other eukaryotic organisms and plants. Mevalonic acid through isopentenyl diphosphate, geranyl diphosphate, and geranylgeranyl diphosphate forms phytoene which undergoes a few desaturation steps and cyclization to β-carotene (Goodwin, 1988). The isomerization reaction, which produces 9-*cis* β-carotene, occurs most probably early in the pathway of carotene biosynthesis at or before the formation of phytoene. Thereafter, all the intermediates occur in two isomeric forms. The induction of β-carotene in *Dunaliella* as described above was successfully applied to accumulate any choice of stereoisomeric intermediates in the carotenogenesis pathway, yielding the twin stereoisomers, all-*trans* and 9-*cis*: phytoene, phytofluene, neurosporene, ξ-carotene, β-zeacarotene, γ-carotene and β-carotene (Shaish *et al.*, 1990, 1991). The induced enzyme, enzymes or enzyme complex responsible for the high accumulation of β-carotene in *Dunaliella* have not been identified as yet.

A few speculative hypotheses have been suggested for the function of the β-carotene globules in *Dunaliella*. One suggested that β-carotene accumulation is related to a protecting effect against chlorophyll catalyzed singlet oxygen production under high intensity. However, the electron micrographs of *Dunaliella bardawil* (Ben-Amotz and Avron, 1990) which indicate that the massive accumulation of β-carotene occurs in globules located in the interthylakoid space of the chloroplast, is in disagreement with this hypothesis. The large distance between the β-carotene globules and the thylakoid located chlorophyll would not allow efficient quenching of singlet oxygen, or any other chlorophyll generated product. Other hypotheses on the possible function of the globules such as carbon storage for use under limiting growth conditions were eliminated by the observation that β-carotene rich *Dunaliella* cannot utilize or metabolize the β-carotene in the dark, nor under CO_2-less conditions (Ben-Amotz *et al.*, 1989a). The most accepted hypothesis suggests that the β-carotene globules protect the cell against injury by high intensity irradiation under limiting growth conditions by acting as a screen to absorb excess irradiation. The harmful effect of the blue region of the spectrum is screened by the algal peripheral located globules, thus preventing any cellular damage. Strains of *Dunaliella* and other algae unable to accumulate β-carotene die when exposed to lethal irradiation while the β-carotene-rich *Dunaliella* flourishes. Photosynthetic active radiation (PAR) absorbed by the algal chlorophylls triggers the massive synthesis of β-carotene in *Dunaliella*. The question of the involvement of chlorophyll generated active oxygen species was studied kinetically, by adding promoters of oxygen radicals and by azide, an inhibiter of catalase and superoxide dismutase, to *Dunaliella* during the selective induction period of β-carotene (Shaish *et al.*, 1993). The actual action of active oxygen species in triggering the biosynthesis of β-carotene remains open for further confirmation.

Biotechnology of β-carotene production by *Dunaliella*

Dunaliella is the most suitable organism for mass cultivation outdoors in open ponds. The ability to thrive in media with high sodium, magnesium, calcium and the respective anions, chloride and sulphate, in desert high solar irradiated land with access to brackish water or sea water at extreme temperatures from around −5°C to above 40°C, makes *Dunaliella* most attractive for the biotechnologists and venture capitalists. In fact, since 1980 a few firms, government authorities, and industries have invested capital in applied *Dunaliella* for the production of natural β-carotene. The large-scale *Dunaliella* is based on autotrophic growth in media containing inorganic nutrients with carbon dioxide as an exclusive carbon source. Attempts to develop heterotrophic strains or mutants of *Dunaliella* for growth on glucose or acetate such as *Chlorella* or *Chlamydomonas*, respectively, were not successful. Due to the demand for high light intensity for maximal β-carotene production beyond that required for normal growth, production facilities are located in areas where solar output is maximal and cloudiness is minimal. Most of the present *Dunaliella* plants are located close to available sources of salt water, for example, sea-salt industries usually in warm and sunny areas where the rate of water evaporation is high and non-agricultural land is abundant.

Four modes of cultivation have been used in large-scale production of *Dunaliella*. One, termed extensive cultivation, uses no mixing and minimal control of the environment. To decrease attacks by zooplanktonic predators, such as certain types of ciliates, amoebae, or brine shrimps, the extensive growers employ very high salt concentrations. *Dunaliella* grows slowly in shallow lagoons in nearly saturated brine and predators are largely eliminated. The naturally selected strain of *Dunaliella* is well adapted to nearly salt saturation: it partially loses its flagella and produces a thick cell wall on the transformation to a cyst form. Extensive cultivation productivity is low, and the area needed for commercial production is very large;

however, the low operating costs of such facilities led to the development of two commercial plants in Australia and China (Figure 9.2). The second, termed intensive cultivation, uses high biotechnology to control all factors affecting cell growth and chemistry. The ponds are usually oblong lined constructed raceways, varying in size up to a production surface area of approximately 3000 m^2. The use of long arms and slow revolution paddle wheels is currently common in the large-scale facilities in Israel, USA, China, Chile, and Portugal. One production shallow water pond of 20 cm on an area of 3000 m^2 (600 m^3) containing 15 g β-carotene/m^3 yields 9 kg β-carotene on total harvest. The current large scale production of β-carotene under intensive cultivation is around 200 mg β-carotene/m^2/day on a yearly average; thus, a modern intensive plant of 50 000 m^2 produces 10 kg β-carotene/day. Thirdly, between the extensive and intensive modes there are examples in Australia and China of a semi-intensive mode, where the ponds are enlarged ten times to about 50 000 m^2 each with partial control and no mixing. The fourth mode is highly intensive cultivation in closed photobioreactors. Many trials have been initiated in the last decade to grow *Dunaliella* in different models of closed photobioreactors (Figure 9.3) with attempts to design the best sunlight harvesting unit for β-carotene optimization. The different designs include narrow, very long plastic tubes, plastic bags, trays and others. However, as of today none of these trials has been scaled up above the laboratory, or small pilot plant, scale, mainly due to economic limitations and non-feasible large scale development. The few industrial ventures of high intensive-closed photobioreactors went insolvent and none exists today self dependent. Generally, large-scale optimization of β-carotene production is achieved in all modes of cultivation by high salt stress and by nitrogen deficiency, but practically, the first

Figure 9.2   A typical commercial growth pond of *Dunaliella*. Extensive mode of cultivation. Plant location: Betatene Ltd, Adelaide, Australia

Figure 9.3 Pilot large-scale tubular closed bioreactor for cultivation of *Dunaliella*. Highly intensive mode of cultivation. Notice the construction of long clear plastic tube to provide exposure to a daily cycle of 270° solar radiation. Plant location: PBL Ltd, Murcia, Spain.

is applied in the extensive mode, while nitrogen starvation controls the intensive mode.

Separation and harvesting of *Dunaliella* out of the growth medium from around 0.1 per cent biomass to a concentrated algal paste of more than 15 per cent should take into consideration the specific features of this alga. Industrial extensive producers of *Dunaliella* are presently using flocculation and cell surface absorption, while the intensive growers employ centrifugation and flocculation. The resulting paste is then dehydrated to a β-carotene-rich powder of *Dunaliella*, or is extracted to a few purified stereoisomeric forms of β-carotene. The detailed technologies of harvesting and drying *Dunaliella* are described in Ben-Amotz *et al.* (1989a).

Natural versus synthetic β-carotene

β-Carotene has been used for many years as a food colouring agent, as pro-vitamin A in food and animal feed, as an additive to cosmetics, in multivitamin preparations, and in the last decade as a health food product under the claim antioxidant Many epidemiological and oncological studies suggest that humans fed on a diet high in carotenoid-rich vegetables and fruits, who maintain higher than average levels of serum carotenoids, have a lower incidence of several types of cancer and cardiovascular disease. Recently, the Alpha-Tocopherol, β-Carotene (ATBC) Cancer Prevention Study Group, and the β-Carotene and Retinol Efficacy Trial (CARET) clearly showed that not only did β-carotene fail to reduce the incidence of cancer and cardiovascular disease, but in fact it increased it in smokers and workers exposed to asbestos (Hennekens et al., 1996; Omen *et al.*, 1996). These two USA National Cancer Institute supported studies and numerous earlier trials which dealt with the question of the protective role of β-carotene against chronic diseases, used only synthetically formed all-*trans* β-carotene.

Despite the fact that the most convincing reports support a direct connection between high intake of fruit and vegetables and low incidence of cancer and cardio-vascular disease, the literature lacks any information on the possible medical con-tribution of natural carotenoids, isomers of carotenoids, and carotenoid fatty acid esters. Experimental nutrition and medical studies with natural carotenoids originat-ing from different plants, fruits, vegetables and algae have been very limited and such research is in its infancy.

β-Carotene is present in most plants and algae in small amounts of approx. 0.2 per cent of the dry weight with approximately one-third as 9-*cis* β-carotene. The obser-vations of the high β-carotene containing more than 50 per cent 9-*cis* β-carotene in *Dunaliella* gave impetus to studies on the metabolism, storage and function of 9-*cis* β-carotene in animals and humans with emphasis on the possible role of the 9-*cis* stereoisomer in scavenging reactive oxygen species. Low doses of the algal β-caro-tene are as potent as doses of synthetic all-*trans* β-carotene in providing retinol in rats and chicks (Ben-Amotz *et al.*, 1989b). Preferential and selective uptake of either all-*trans* or 9-*cis* β-carotene was noted in different animals and humans. 9-*Cis* β-carotene is not detected in serum of chicks and rats nor in humans fed a diet rich in *Dunaliella* (Ben-Amotz and Levy, 1996) or oil extract of *Dunaliella* (Stahl *et al.*, 1993, Gaziano *et al.*, 1995). The β-carotene detected in any of these sera was all-*trans* β-carotene. Rats and chicks that generally do not accumulate β-carotene in their tissues showed uptake of β-carotene into the tissues when the diet was supplemented with β-carotene-rich *Dunaliella*. The lack of 9-*cis* β-carotene in mammals' serum and the different response to CNS toxicity (Bitterman and Ben-Amotz, 1994) and to *in vitro* oxidation (Levin and Mokady, 1994) led to the hypothesis that the isomeric structure of 9-*cis* β-carotene acts as a quencher of singlet oxygen and other free radicals, being an effective *in vivo* antioxidant. Generally, carotenoids exert an anti-oxidative effect by a mechanism that results in the formation of new products that are more stable and more polar. Different carotenoids exhibit different antioxida-tive–antiperoxidative capacities (Krinsky, 1989). Opening of the β-ionone ring, the addition of a chemical group on the ring, or replacement of the ring by various groups can modify the antioxidative activity. Therefore, structural variables other than the length of the polyene chain may direct the scavenging properties of the carotenoids. Insertion of any *cis* along the conjugated double bond chain may lead to modified antioxidative properties. Assuming that the *cis* conformational change leads to a higher steric interference between the two parts of the carotene molecule, the *cis* polyenic chain will be less stable and more susceptible to low oxygen tension in *in vivo* oxidation. Recent studies strongly supported this hypothesis by measuring different antioxidative activity (Levy *et al.*, 1995, 1996) and much lower concentration of low-ultraviolet dienes in the serum of humans supplemented with β-carotene rich-*Dunaliella* compared with all-*trans* β-carotene (Ben-Amotz and Levy, 1996). The possibility of synergistic effects and the possible beneficial potency of the different plant nutrients and carotenoids are still obscure and warrant further research.

Commercial producers

At present, the following companies are actively engaged in cultivating *Dunaliella* for commercial purposes. These are listed below, with their modes of cultivation:

1 Betatene Ltd, Cheltenham, Victoria 3192, Australia, a division of Henkel Co., Germany. Extensive mode.
2 Cyanotech Corp., Kailua-Kona, HI 96740, USA. Intensive mode.
3 Inner Mongolia Biological Eng. Co., Inner Mongolia, 750333, PR China. Semi-intensive mode.
4 Nature Beta Technologies (NBT) Ltd, Eilat 88106, Israel, a subsidiary of Nikken Sohonsha Co., Gifu, Japan. Intensive mode.
5 Nutrilite, Calipatria, CA 92233, USA, a division of Amway Corporation, USA. Intensive mode.
6 Photo Bioreactors, PBL Ltd, Murcia 30003, Spain. Highly intensive mode, closed photobioreactors
7 Tianjin Lantai Biotechnology, Inc., Nankai, Tianjin, in collaboration with the Salt Scientific Research Institute of Light Industry Ministry, PR China. Intensive mode.
8 Western Biotechnology Ltd, Bayswater, WA 6053, Australia, a subsidiary of Coogee Chemicals Pty Ltd, Australia. Semi-intensive mode.

References

AVRON, M., 1992, Osmoregulation, in Avron, M. and Ben-Amotz, A. (Eds), *Dunaliella: Physiology, Biochemistry, and Biotechnology*, pp. 135–164, Boca Raton: CRC Press.

AVRON, M. and BEN-AMOTZ, A. (Eds), 1992, *Dunaliella: Physiology, Biochemistry, and Biotechnology*, Boca Raton: CRC Press.

BEN-AMOTZ, A., 1996, Effect of low temperature on the stereoisomer composition of β-carotene in the halotolerant alga *Dunaliella bardawil* (Chlorophyta) *J. Phycol.*, **32**, 272–275.

BEN-AMOTZ, A. and AVRON, M., 1990, The biotechnology of cultivating the halotolerant alga *Dunaliella, Trends Biotechnol.*, **8**, 121–126.

BEN-AMOTZ, A. and LEVY, I., 1996, Bioavailability of a natural isomer mixture compared with synthetic *all-trans* β-carotene in human serum, *Am. J. Clin. Nutr.*, **63**, 729–734.

BEN-AMOTZ, A., MOKADY, S., EDELSTEIN, S. and AVRON, M., 1989a, Bioavailability of natural isomer mixture as compared with synthetic all-*trans* β-carotenein rats and chicks. *J. Nutr.*, **119**, 1013–1019.

BEN-AMOTZ, A., SHAISH, A. and AVRON, M., 1989b, Mode of action of the massively accumulated β-carotene of *Dunaliella bardawil* in protecting the alga against damage by excess radiation, *Plant Physiol.*, **91**, 1040–1043.

BITTERMAN, N. and BEN-AMOTZ, A., 1994, β-Carotene and CNS toxicity in rats, *J. Appl. Physiol.*, **76**, 1073–1076.

GAZIANO, J.M., JOHNSON, E.J., RUSSEL, E.M., *et al.*, 1995, Discrimination in absorbance or transport of β-carotene isomers after oral supplementation with either all-*trans*- or 9-*cis* β-carotene, *Am. J. Clin. Nutr.*, **61**, 1248–1252.

GOODWIN, T.W. (Ed.), 1988, *Plant Pigments*, London: Academic Press.

HENNEKENS, C.H., *et al.*, 1996, Lack of effect of long-term supplementation with beta-carotene on the incidence of malignant neoplasms and cardiovascular disease, *N. Eng. J. Med.*, **334**, 1145–1149.

KRINSKY, N.I., 1989, Antioxidant functions of carotenoids, *Free Radic. Bio. Med.*, **7**, 617–635.

LEVIN, G. and MOKADY, S., 1994, Antioxidant activity of 9-*cis* compared to all-*trans* β-carotene *in vitro, Free Radic. Biol. Med.*, **17**, 77–82.

LEVY, Y., BEN-AMOTZ, A. and AVIRAM, M., 1995, Effect of dietary supplementation of different β-carotene isomers on lipoprotein oxidative modification, *J. Nutr. Environm. Med.*, **5**, 13–22.

LEVY, Y., KAPLAN, M., BEN-AMOTZ, A. and AVIRAM, M., 1996, Effect of dietary supplementation of β-carotene on human monocyte-macrophage-mediated oxidation of low density lipoprotein, *Isr. J. Med. Sci.* **32**, 473–478.

OMEN, G.S., *et al.*, 1996, Effects of combination of beta-carotene and vitamin A on lung cancer and cardiovascular disease, *N. Eng. J. Med.*, **334**, 1150–1155.

SHAISH, A., AVRON, M. and BEN-AMOTZ, A., 1990, Effect of inhibitors on the formation of stereoisomers in the biosynthesis of β-carotene in *Dunaliella bardawil*, *Plant Cell Physiol.*, **31**, 689–696.

SHAISH, A., BEN-AMOTZ, A. and AVRON, M., 1991, Production and selection of high β-carotene mutants of *Dunaliella bardawil* (Chlorophyta), *J. Phycol.*, **27**, 652–656.

SHAISH, A., AVRON, M., PICK, U. and BEN-AMOTZ, A., 1993, Are active oxygen species involved in induction of β-carotene in *Dunaliella bardawil?*, *Planta*, **190**, 363–368.

STAHL, W., SCHWARTZ, W. and SIES, H., 1993, Human serum concentrations of all-*trans* and α-carotene but not 9-*cis* β-carotene increase upon ingestion of a natural mixture obtained from *Dunaliella salina* (Betatene), *J. Nutr.*, **123**, 847–851.

10

Chemicals of *Botryococcus braunii*

P. METZGER AND C. LARGEAU

INTRODUCTION

Botryococcus braunii (Kützing, 1849) is a colonial member of *Chlorophyceae*, characterized by an original organization of colonies and a considerable production of lipids. Widely distributed in freshwater lakes, reservoirs or ponds, it also inhabits brackish waters and ephemeral lakes. On light microscopy observations (Figure 10.1), the colonies, variable in size from 30 μm up to 2 mm, appear to consist of single or multiple cell clusters united by transparent and refringent strands. The cells, generally pyriform (7 × 13 μm), are arranged around the periphery of the cluster and embedded within cups of a matrix impregnated by oil. Ultrastructural investigations show that the matrix consists of outer walls originating from successive cellular divisions (Figure 10.1); more or less closely pressed to each other, these outer walls ensure colony cohesion.

In many water bodies, *B. braunii* can be the dominant species of the algal flora and forms spectacular blooms (Aaronson *et al.*, 1983), occasionally enduring, and rubbery deposits on the shores (Wake and Hillen, 1981). Such an ability associated with the oil richness of this microorganism – an oil content of 86 per cent of dry weight was reported by Brown *et al.* (1969) from a bloom in a freshwater lake – has drawn the attention of scientists since the end of the last century.

The pioneering studies on *Botryococcus* were concerned with the implication of this microalga in the formation of some fossil materials ranging in age from Ordovician to present (Cane, 1969, 1977; Cane and Albion, 1973) and exhibiting high petroleum potentials. Later, fossil chemical compounds, specifically derived from *B. braunii*, were found in high abundances in some crude oils (Moldowan and Seifert, 1980; Brassell *et al.*, 1986; McKirdy *et al.*, 1986), or in some recent sediments (Huang *et al.*, 1996).

Interest in modern *B. braunii* arose from the discovery in the 1960s, that the living algae are able to synthesize high amounts of hydrocarbons (Swain and Gilby, 1964; Maxwell *et al.*, 1968), while it was considered until then that the oil impregnating the matrix of the colonies consisted essentially of fatty acids (Zalessky, 1926; Blackburn, 1936).

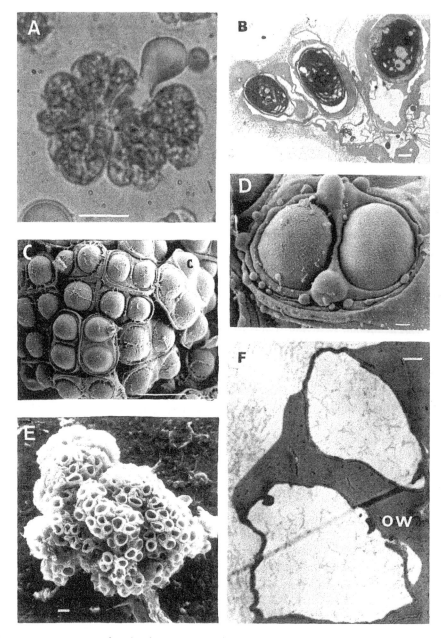

Figure 10.1 Micrographs of *B. braunii*. **(A)** Light microscopy of the Cambridge strain (A race); a refringent globule of lipids is excreted from the colony by pressure on the cover glass. **(B)** Transmission electron microscopy of the Yamoussoukro strain (L race); the section shows some intracellular lipid inclusions and starch granules; several successive outer walls surround each cell. **(C,D)** Scanning electron microscopy of a Martinique strain (B race); the basal part of the cells is embedded in thick outer walls when the apical part is covered by a thin cap (c); note some oil droplets near the cells in D. **(E)** Scanning electron microscopy of a *B. braunii* colony in a recent (40 000 years) sediment core from East Africa; only the outer walls are well preserved. **(F)** Transmission electron microscopy of the residual resistant material obtained after chemical treatments. All the cell contents have been removed; the empty areas are surrounded by the algaenan-composed outer walls (ow). Scale bars: A, C and E: 10 μm; B, D and F: 1 μm.

Hydrocarbon synthesis in *B. braunii* was subsequently thought to be rather complex, when two very different types of hydrocarbons were described, apparently in two different physiological states of the alga. Indeed, *n*-alkadienes and trienes, odd numbered from C_{23} to C_{33}, were considered to be synthesized by colonies in the active growth phase, while C_{30}–C_{37} triterpenes, called botryococcenes, were supposed to be produced by algae in a resting state characterized by an intense orange colour due to the accumulation of carotenoids (Knights *et al.*, 1970). Intensive analyses of samples collected in lakes of various geographical zones, as well as the isolation of numerous strains from wild samples and subsequent analyses of their hydrocarbons at different stages of growth, led us to refute this statement (Metzger *et al.*, 1985a). Indeed, each strain produces a single type of hydrocarbon, whatever its physiological state may be. Furthermore, in the course of screenings on *B. braunii*, a third type of hydrocarbon was discovered; this was mainly dominated by one tetra-terpene, named lycopadiene, synthesized by strains originating from some freshwater tropical lakes (Metzger and Casadevall, 1987).

Due to the rather small morphological differences existing between strains producing the three types of hydrocarbons, i.e. the presence of a denser matrix in algae yielding botryococcenes or lycopadiene, a somewhat less conspicuous pyrenoid and the absence of oil droplets surrounding the colonies in the lycopadiene-synthesizing algae, which also exhibit smaller cells, we chose to classify the strains into three chemical races: A, B and L by reference to alkadienes, botryococcenes and lycopadiene, respectively (Metzger *et al.*, 1985a, 1990). In Table 10.1 are summarized the geographical sources from which *B. braunii* samples have been collected and for some of them strains isolated, in relation to the hydrocarbon type.

If chemists and geochemists generally agree with this point of view, there exists however a controversy on *B. braunii* taxonomy. Indeed, some phycologists consider that many algae of the genus *Botryococcus* would be wrongly referred to as the morphotype *B. braunii*. Thus, on the basis of the above mentioned morphological differences, Komárek and Marvan (1992) have proposed a new taxonomy for *Botryococcus*, including 13 species. In this context we have recently found (Metzger *et al.*, 1997) that two strains we classify in the L chemical race are also considered as being *Botryococcus neglectus* (Jeeji Bai, 1996). Considering the great number of papers referring to the traditional name, *B. braunii*, and considering the difficulties encountered to appreciate morphological changes, not always stable under different culture conditions (Plain *et al.*, 1993), we prefer however to maintain this species name, paired with the appropriate epithet of the chemical race, A, B or L.

Although especially known as a hydrocarbon-rich microorganism, *B. braunii* also produces some usual lipids commonly found in algal oil and non-classical lipids, sometimes in very high yields. Furthermore, the structural elements building up the matrix of the colonies in the three races appear to consist of insoluble and chemically resistant biopolymers arising from an extended cross-linkage of macromolecular lipids. In this review, attention is focused on the structure and biosynthesis of all these metabolites, which for the most part are unique to *B. braunii*. Polysaccharides are also often important by-products of the culture of these algae and they will be briefly discussed. The extensive biotechnological studies carried out on the production of lipids, especially hydrocarbons, will be considered as well. Finally a comparative overview of the abundances of the main classes of lipidic compounds in the three chemical races will be given.

Table 10.1 Hydrocarbon types, content and origin of *B. braunii* samples.

Hydrocarbon types	Origin[1] and hydrocarbon content[2]		References
n-Alkadienes and trienes, odd from C_{23} to C_{33}; *n*-alkenes and *n*-alkatetraenes, minor, rarely found	Australia:	Tarago (30), Sorrento (27–30), Jillamatong (5–15)	Wake and Hillen (1981) Metzger *et al.* (1997)
	Bolivia:	Challapata (6), Overjuyo (0.4–2), Pota Khota (52), Titicaca (13), Icha Khota (0.9)	Metzger *et al.* (1989) Metzger *et al.* (1996)
	England:	Maddingley Brick Pits (9), Grasmere (28)	Knights *et al.* (1970) Metzger *et al.* (1997)
	France:	Chaumeçon (61), Crescent (40), Lingoult (32), Grosbois, Coat ar Herno (1.6)	Metzger *et al.* (1985a) Metzger *et al.* (1986) Metzger *et al.* (1989)
	Japan:	Yamanoka (16)	Okada *et al.* (1995)
	Morocco:	Oukaimden (19–40)	Metzger *et al.* (1985a)
	Portugal:	Amieiro	Metzger *et al.* (1997)
Botryococcenes C_{30} to C_{37}	Australia:	Darwin River Reservoir (29–33), Devilbend (40), Green Lake (27), McCay's Reservoir (32), Coolamatong	Wake and Hillen (1981) Summons and Capon (1991)
	Bolivia:	Overjuyo	Metzger *et al.* (1988)
	England:	Oak Mere (76)	Maxwell *et al.* (1968)
	France:	Sanguinet (34), Pareloup, Vioreau	Metzger *et al.* (1985a) Metzger *et al.* (1988)
	Ivory Coast:	Ayame (27)[3], Taabo (20)[3], Yamoussoukro	Metzger *et al.* (1988)
	Japan:	Yayoi (33), Kawaguchi (10,19)	Oakada *et al.* (1995)

208

Table 10.1 Hydrocarbon types, content and origin of *B. braunii* samples (contd.).

Hydrocarbon types	Origin[1] and hydrocarbon content[2]		References
	Philippines:	Katugday	Metzger *et al.* (1988)
	Thailand:	Khao Kho Hong (8)	Metzger *et al.* (1988)
	USA:	Michigan, Norman,	Wolf and Cox (1981)
		Berkeley (24–39)	Wolf *et al.* (1985a)
	Venezuela:	Caroni River	Jaffé *et al.* (1995)
	West Indies		
	Martinique:	La Manzo (32–43),	Metzger *et al.* (1985a)
		Paquemar (28–38)	
	Guadeloupe:	Chateaubrun (37)	Metzger *et al.* (1985a)
Lycopadiene C$_{40}$	India:	Kulavai[4] (0.1)	Metzger *et al.* (1997)
	Ivory Coast:	Kossou, Yamoussoukro (3)	Metzger *et al.* (1990)
	Thailand:	Khao Kho Hong (2–8)	Metzger and Cadadevall (1987)

[1]Freshwaters, Titicaca (brackish water) and Jillamatong (ephemeral lake) excepted.
[2]Content: values in parentheses, in % of dry wt.
[3]Unpublished values.
[4]Strains from this lake are also considered as being *B. neglectus*.

Isolation and methodology for the study of chemicals from *B. braunii*

Studies on the three chemical races have revealed that the bulk of *B. braunii* lipid material is located in the outer walls which build up the matrix of the colonies. Thus in the case of the A race, sequential extractions with solvents of increasing polarity and *in vivo* examination of the algae by Raman microprobe established that 95 per cent of hydrocarbons are stored in the outer walls (Largeau *et al.*, 1980b). A short extraction of the dry biomass with an apolar solvent such as hexane allows recovery of the external lipids as an oil coloured by carotenoids. A longer extraction of the remaining biomass with $CHCl_3$ furnishes a rather elastic material, often refractory to redissolution after complete elimination of the solvent. Subsequently (see below) this rubbery material was considered to be an additional part of the external constituents. Finally, the internal lipids present in the membranes and as cytoplasmic inclusions are obtained by prolonged extraction with $CHCl_3$–MeOH.

Other procedures for lipid extraction have been developed. It must be stated, however, that in the absence of extraction with $CHCl_3$ or a similar solvent, before extraction of internal lipids, the entire recovery of the rubbery material may be impossible (Gelin *et al.*, 1994a).

Application of chromatographic techniques, including normal and reversed phase HPLC leads to the isolation from the extracts, along with variable amounts of hydrocarbons and of some usual lipids such as sterols, triacylglycerols and carotenoids, of non-classical lipids. Depending on the considered race and also on the strain, several series of α-branched aldehydes, alkenylphenol derivatives, epoxides and ether lipids were identified. Moreover the 'rubbers' isolated from the $CHCl_3$ extracts appeared as polymers of an entirely new type: they can be considered as macromolecular lipids derived from the condensation of fatty aldehydes. IR, mass spectrometry (EI, CI and FAB), 1D and 2D ^1H and ^{13}C NMR spectroscopies were used extensively for structure determination. Hydrogenation, hydrolysis, chemical degradation by hydrogenolysis, ozonolysis and ether cleavage, contributed to structure elucidation. Biosynthetic experiments were carried out by feeding the algae with ^2H, ^3H, ^{13}C and ^{14}C labelled precursors; incubations with cell free extracts were also performed. Radioactive labellings were examined by means of radio-GC, scintillation counting and radio-thin layer chromatography. NMR spectroscopy allowed the determination of the labelling patterns obtained from precursors labelled with stable isotopes.

Insoluble and non-hydrolyzable biopolymers were shown to form the chemically resistant part of outer walls in the three races. These non-polysaccharidic materials which have been shown to occur in some microalgal cell walls and termed algaenans, are isolated from the lipid-free biomass by means of basic and acid treatments. Their structures were investigated using FTIR, solid ^{13}C NMR and analysis of the pyrolysis products.

Extracellular polysaccharides, produced in high amounts by some strains, are usually isolated by lyophilization of the non-dialysable filtrates obtained from the culture media. The mass ranges were determined by gel chromatography and the compositions established after hydrolysis and analysis of the derived sugars.

The procedure used for the isolation of the lipids, 'rubbers', algaenans and polysaccharides is summarized in Figure 10.2.

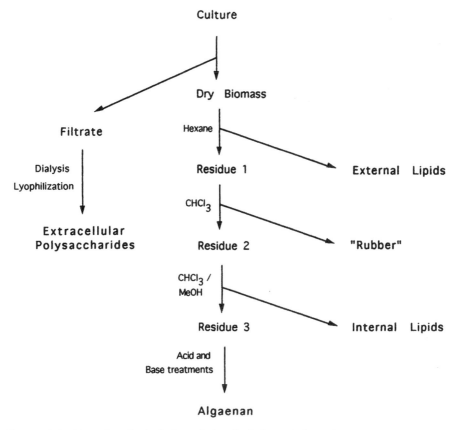

Figure 10.2   Procedure for isolation of chemicals from *B. braunii*.

Hydrocarbons

Hydrocarbons are easily separated from the crude lipid extracts by SiO_2 or Al_2O_3 chromatography, using hexane as eluent. They are mainly recovered from the external lipids extracted from the biomass by hexane.

n-Long chain hydrocarbons (A race)

Some 50 hydrocarbons exhibiting normal chains have been detected so far from several algal samples collected in nature or from strains grown in the laboratory (Table 10.1); some 30 structures have been determined.

All the A race algae analyzed exclusively showed the presence of odd compounds, with the exception of one strain which synthesizes low amounts of C_{26} and C_{28} hydrocarbons (Metzger, 1993). On the whole, the hydrocarbon range is C_{23}–C_{33}, each compound exhibiting a terminal double bond (Figure 10.3).

When present, C_{23}–C_{27} monoenes **1** are always minor components (Metzger, 1993; Metzger *et al.*, 1989, 1997), while dienes **2** often predominate; these latter always comprise a mid-chain ω9 unsaturation, of *cis* or *trans* configuration (Knights *et al.*,

$CH_3-(CH_2)_x-CH=CH_2$ **1**

x=20, 22, 24

$CH_3-(CH_2)_7-CH=CH-(CH_2)_x-CH=CH_2$ **2**
ω9

cis ; x=odd, 11-21
trans ; x=odd,11-19

$CH_3-(CH_2)_5-CH=CH-CH=CH-(CH_2)_x-CH=CH_2$ **3**
ω7   ω9

cis, cis ; x=odd,15-19
cis, trans ; x=17, 19

         E or Z   E or Z
$CH_3-(CH_2)_7-CH=CH-(CH_2)_{13}-CH=CH-CH=CH_2$ **4**
         18   13   3

trans, trans
trans, cis
cis, trans
cis, cis

         E or Z   E   E or Z
$CH_3-(CH_2)_7-CH=CH-(CH_2)_{11}-CH=CH-CH=CH-CH=CH_2$ **5**
         18   11   5   3

cis, trans, cis
cis, trans, trans
trans, trans, trans
trans, trans, cis

Figure 10.3 Alkenes, dienes, trienes and tetraenes of the A race

1970; Metzger *et al.*, 1986). Two types of trienes have been identified so far; both comprise two conjugated double bonds. In the first, type **3**, the most commonly found, the conjugated system occurs in mid-chain positions at ω7 and ω9; *cis, cis* and *cis, trans* isomers have been isolated (Metzger *et al.*, 1986, 1997; Villarreal-Rosales *et al.*, 1992). In the second, type **4**, only formed of C_{27} hydrocarbons, the conjugation occurs in the terminal position; four isomers have been identified in relation to the *cis* or *trans* configuration of the double bonds at positions 3 and 18 (Metzger, 1993). Only one type of tetraenes, **5** has been clearly identified, with three conjugated double bonds in the terminal position (Metzger, 1993); *cis* and *trans* isomerisms occur at C-3 and C-18, as in **4**, while the configuration at C-5 is exclusively *trans*.

A very wide hydrocarbon variability, both in yield and composition, was observed from the analyses of A strains cultivated under identical conditions (Table 10.2). Thus, while the strain originating from the French lake of Chaumeçon synthesizes hydrocarbons up to 61 per cent of its dry weight (Metzger *et al.*, 1985a), the content of another French strain, Coat ar Herno, is only 1.6 per cent (Metzger *et al.*, 1989) and the one of a Bolivian strain 0.4 per cent (Metzger, 1993). However, intermediate levels corresponding to substantial values are generally observed; for example an English strain and an Australian one (Metzger *et al.*, 1997), listed in Table 10.2, accumulate hydrocarbons up to 28 and 8 per cent of their dry weight, respectively. Data listed in Table 10.2 also illustrate the variation in hydrocarbon composition: C_{27}, C_{29} or C_{31} prevalence, presence of alkenes, *trans*-dienes and(or) trienes.

The structures of these hydrocarbons suggested they are derived from fatty acids. Some biosynthetic experiments with [16-^{14}C] palmitic, [18-^{14}C] stearic and [10-^{14}C] oleic acids confirmed this assumption (Templier *et al.*, 1984). In addition, the existence of a *cis* -ω9 unsaturation both in oleic acid and in most of the hydrocarbons synthesized by the tested strains of A race, as well as the substantially higher incorporation levels noticed with labelled oleic acid, were indicative of a direct precursor role for this acid in hydrocarbon biosynthesis. Moreover feeding experiments with [9, 10-^{3}H, 1-^{14}C] and [9, 10-^{3}H, 10-^{14}C] oleic acid showed a total conservation of the initial ^{3}H/^{14}C ratio, suggesting that hydrocarbon biosynthesis takes place via an elongation mechanism by successive addition of C_2 units, very likely via malonate. This mechanism was confirmed by the inhibition of both hydrocarbon and very long chain fatty acid biosynthesis with trichloroacetic acid, an inhibitor of the elongation process (Templier *et al.*, 1987).

Furthermore, the extensive inhibition of hydrocarbon synthesis by dithioerythritol, an inhibitor of decarboxylation, suggested the involvement of a decarboxylation step in the formation of the terminal unsaturation. Regarding this possible mechanism requiring high energy, it was proposed that very long chain fatty acid derivatives, activated by a β-substituent would be implicated in the decarboxylation process. The incorporation of the ^{14}C-labelled β-hydroxy C_{28} fatty acid into C_{27} diene was consistent with such a mechanism (Chan Yong *et al.*, 1986) (Figure 10.4). All these observations therefore indicated that the n-alkenes of the A race of *B. braunii*

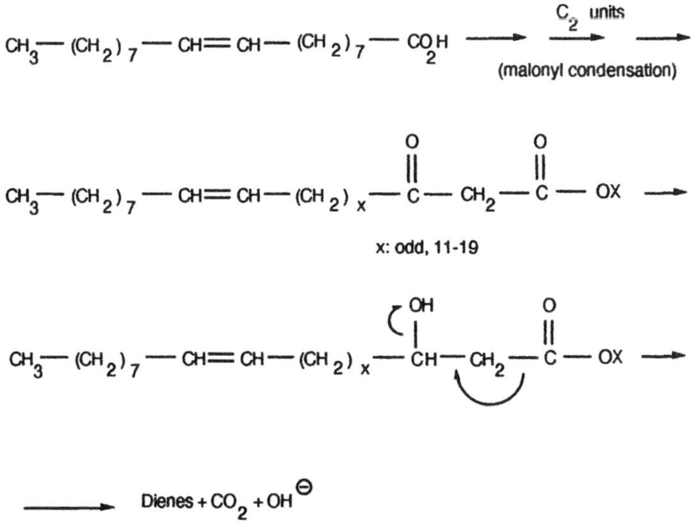

Figure 10.4 Biosynthesis of alkadienes from oleic acid via the elongation–decarboxylation mechanism.

Table 10.2 Hydrocarbon composition and yield in some strains of the A race.[1]

Hydocarbons	Location and stereochemistry of the double bonds	Australia[2] Jillamatong (strain no. 4)	Bolivia[3] Overjuyo (strain no. 7)	England[2] Grasmere	France Chaumeçon[4]	France Coat ar Herno[5]
23:1	1	–	1.5	–	–	3.5
23:2	1,14 (Z)	tr.	1	2.1	0.7	1
23:2	1,14 (E)	–	0.5	–	–	–
25:1	1	–	1.5	–	–	2.5
25:2	1,16 (Z)	tr.	5.5	3.7	2.6	2.5
25:2	1,16 (E)		1.5	1.4	–	–
27:1	1	–	–	–	–	2.5
27:2	1,18 (Z)	3.6	12	17.8	19.5	31.0
27:2	1,18 (E)	–	4	5.9	–	–
27:3	1,3,18	–	39	–	–	–
27:3	1,18 (Z), 20 (Z)	–	–	–	–	25.2
27:4	1,3,5,18	–	12	–	–	–
29:2	1,20 (Z)	16.3	5	62.7	61.8	11.4
29:2	1,20 (E)	–	2	–	–	–
29:3	1,20 (Z), 22 (Z)	12.9	–	–	–	2.3
29:3	1,20 (E), 22 (Z)	1.6	–	–	–	–
31:2	1,22 (Z)	26.7	–	3.8	9.6	–
31:3	1,22 (Z), 24 (Z)	2.6	–	–	–	–
31:3	1,22 (E), 24 (Z)	28.6	–	–	–	–
33:4		0.6	–	–	–	–
33:5		2.1	–	–	–	–
Others		5.0	14.5	2.6	5.8	18.1
% dry weight		8.3	0.4	27.6	61.0	1.6

[1] Algae were grown under the same conditions and harvested when they were in the stationary phase of growth (air-lift cultures supplied by 1% CO_2, permanent illumination).
[2] From Metzger et al. (1997); [3] from Metzger (1993); [4] from Metzger et al. (1985a); [5] from Metzger et al. (1989) with kind permission from Elsevier Science Ltd

tr.: race amounts.

214

shall be formed via an elongation–decarboxylation pathway similar to the one observed for the cuticular hydrocarbons of some higher plants (Kolattukudy *et al.*, 1976).

More recently an alternative pathway was proposed on the basis of incubations of radio-labelled precursors with cell free preparations (Dennis and Kolattukudy, 1991). In these experiments, very long chain fatty aldehydes, derived from the corresponding acids by an enzymatic reduction (Wang and Kolattukudy, 1995), were decarbonylated; however the hydrocarbons so released did not show any terminal unsaturation. Thus further evidence is needed before the intervention of this mechanism in the synthesis of hydrocarbons in the A race can be considered.

Concerning the biosynthesis of *trans*-dienes, feeding experiments with ^{14}C-labelled elaidic and oleic acids demonstrated that the presence of an enzymatic system capable of converting oleic acid into its *trans* isomer, is necessary. Indeed, when algae naturally devoid of *trans*-dienes are fed with elaidic acid they synthesize these hydrocarbons (Templier *et al.*, 1991a).

With regard to the biosynthesis of trienes, **3**, the results of incubation experiments with ^{14}C-labelled dienes and [1-^{14}C] linoleic acid, the only 18:2 fatty acid synthesized by *B. braunii*, show that $\omega 7$, $\omega 9$ trienes do not derive from these putative precursors (Templier *et al.*, 1991b). It was suggested that the additional $\omega 7$ double bond is introduced at an early stage of the elongation–decarboxylation pathway either via direct desaturation of oleic acid into an $\omega 7$, $\omega 9$ acid, or via desaturation of very long chain fatty acyl derivatives.

Botryococcenes and squalene-related hydrocarbons (B race)

Botryococcenes, the typical hydrocarbons of the B race, form a wide family of triterpenoid hydrocarbons, comprising both acyclic (Figure 10.5) and cyclic (Figure 10.6) compounds. To date more than 50 compounds of general formula C_nH_{2n-10}, with n = 30–37, have been detected through GC-EIMS analyses, and 16 structures have been determined. The main difficulties encountered in their structural determination arise from the purification of such compounds from often complex hydrocarbon mixtures. In some cases, application of reversed-phase HPLC, using C_{18} columns provided pure compounds or enriched mixtures. To avoid the polymerization of these hydrocarbons, reported to occur both with pure fractions or even in the biomass kept in air (Douglas *et al.*, 1969; Dubreuil *et al.*, 1989), botryococcenes must be stored in hexane solution and under an inert atmosphere at $-25°C$.

The first described hydrocarbon of this family was a C_{34} compound, named 'botryococcene', a term which has been further used as generic name. It was isolated from algae collected in Oakmere lake in England (Maxwell *et al.*, 1968). Its structure, **6**, (Figure 10.5) was established from ^{13}C NMR data and oxidative degradation using permanganate–periodate cleavages by Cox *et al.* (1973). The authors considered that this irregular triterpene probably arised from a tail-to-tail linkage of two dimethylated C_{15} units. The absolute stereochemistry was later established via oxidative degradation and comparison of the resulting enantiomers with optically pure synthons (White *et al.*, 1986, 1992): the stereocentres at C-3, C-7, C-10, C-16 and C-20 were found to be *S*, while the *R* configuration was ascribed to C-13. Moreover, a total synthesis confirmed these findings (White *et al.*, 1988).

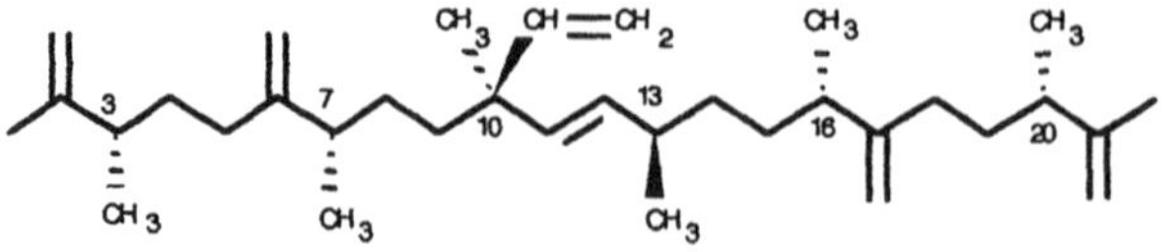

Figure 10.5 Acyclic botryococcenes of the B race.

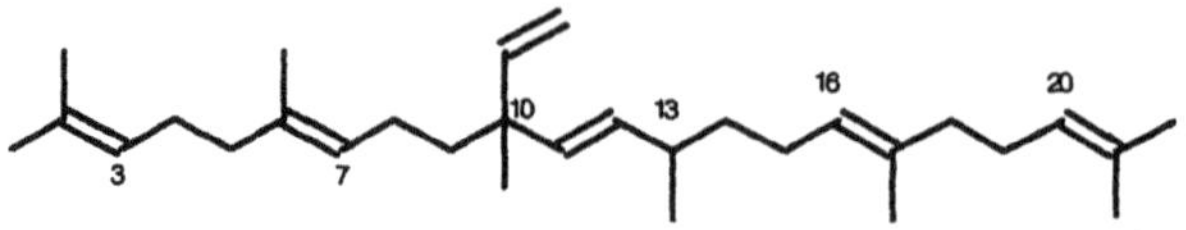

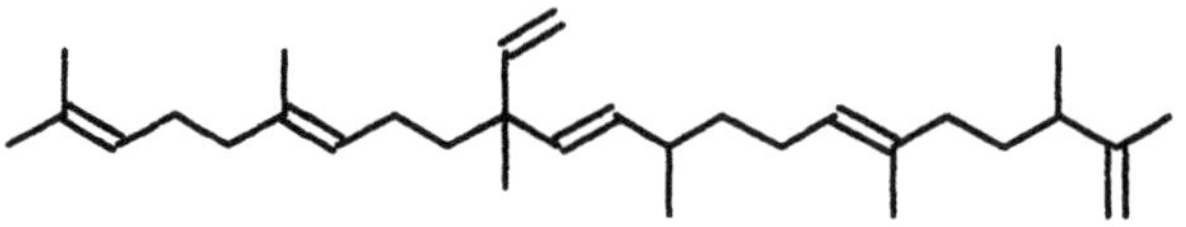

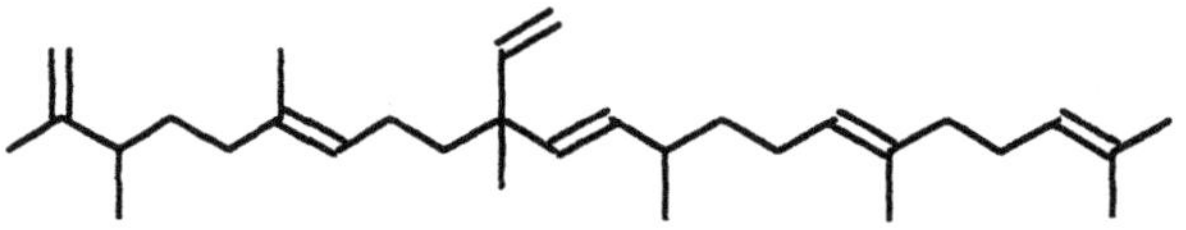

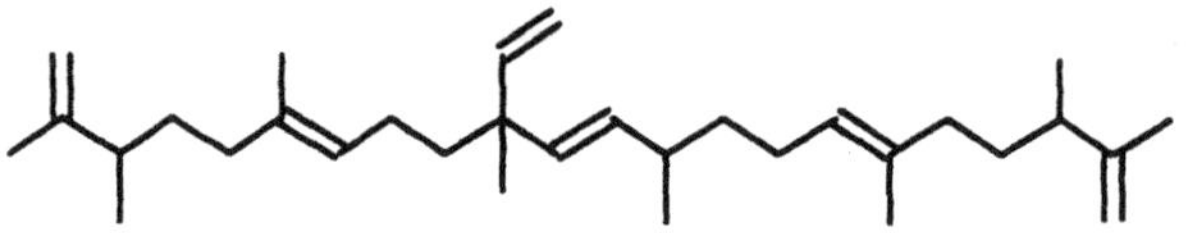

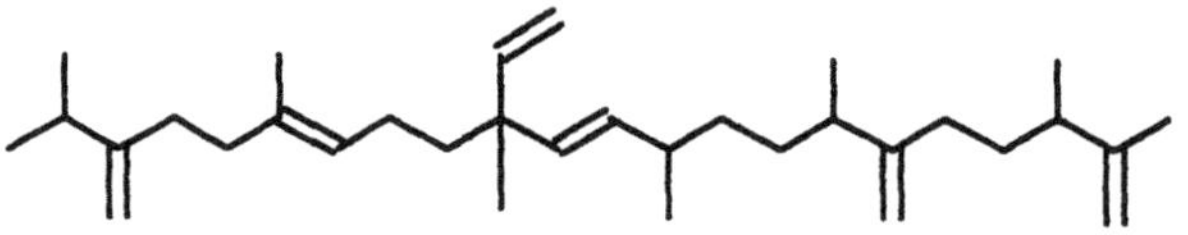

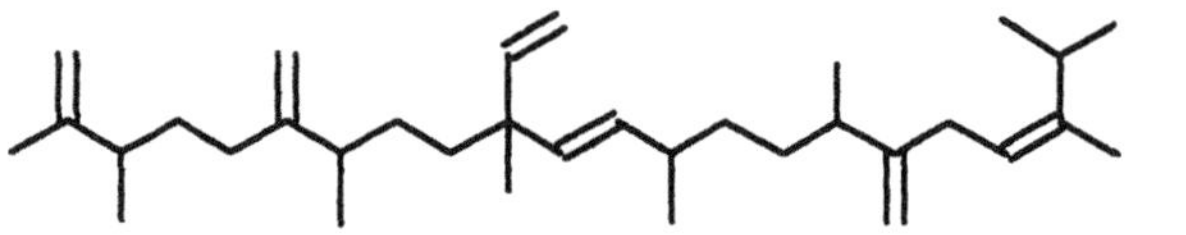

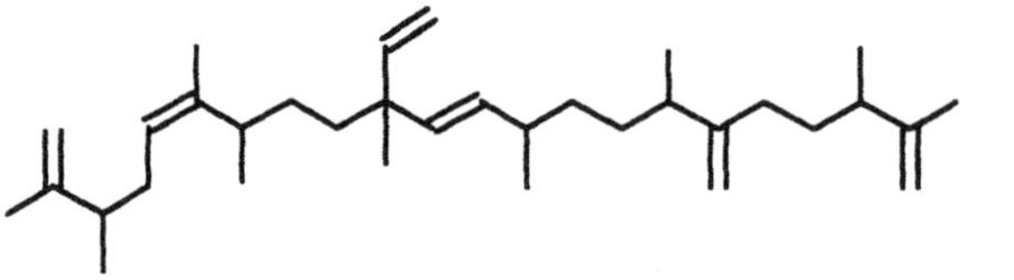

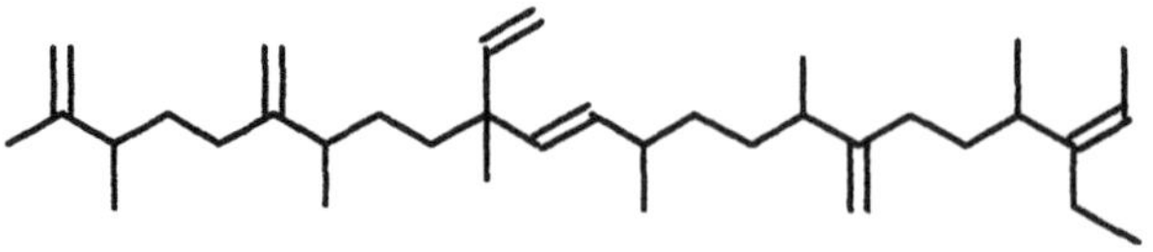

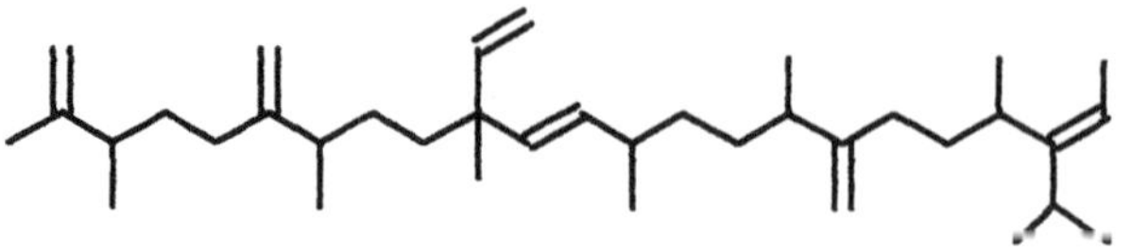

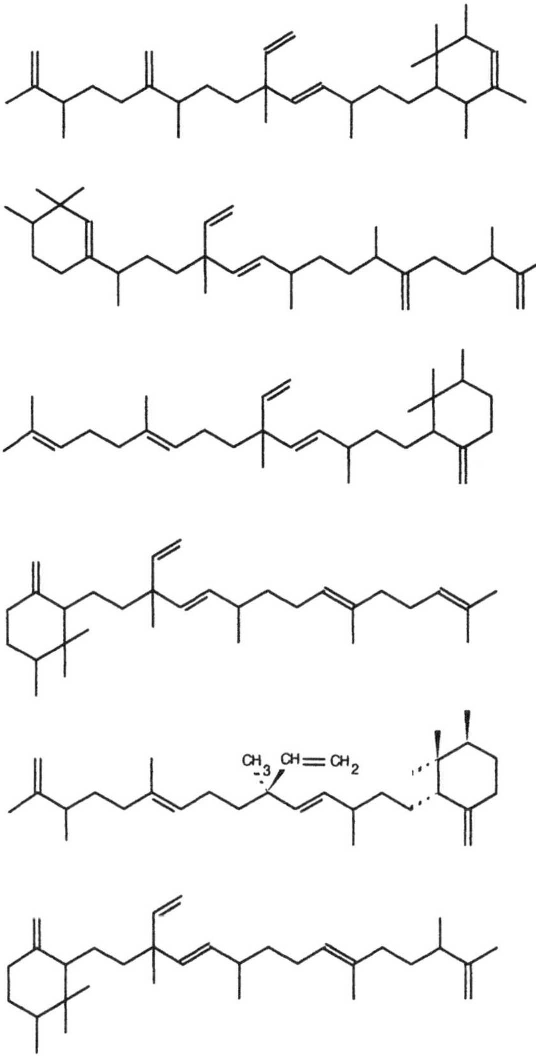

Figure 10.6 Cyclobotryococcenes of the B race.

The structural elucidation of some other botryococcenes, together with some preliminary biosynthetic studies, then suggested that botryococcene C_{30}, **7**, formerly isolated from a French West Indian strain, fills a key position in this terpene family (Metzger and Casadevall, 1983; Casadevall *et al.*, 1984; Wolf *et al.*, 1985a, 1985b). All higher botryococcenes derive from this simpler compound by successive methylations at C-3, C-7, C-16 and C-20, in a manner similar to the one operating in the alkylation of the side chain of sterols. An important feature of this biogenetic pathway is that the resulting carbocation, generated by the alkylation mechanism, can give rise to non-cyclized higher homologues or can be cyclized to cyclobotryococcenes.

In the group of acyclic botryococcenes, a single methylation at C-20 or C-3 leads to the C_{31} compounds **8** or **9**, respectively, while two to four successive methylations

occur in the synthesis of **6, 10–13**; furthermore, three and four successive methylations on a single C_5 unit are involved in the formation of C_{36}, **14** and C_{37}, **15**
(Galbraith *et al.*, 1983; Metzger *et al.*, 1985b; Huang and Poulter, 1989c).

The first cyclic botryococcene to be described was the C_{34} hydrocarbon **16** (Figure
10.6), isolated from algae collected in the Darwin River Reservoir in Australia
(Metzger *et al.*, 1985b). Spectroscopic analyses showed that it contains a penta-
methylated cyclohexenyl ring. Another cyclo C_{34}, **17**, was found as a major constituent (14 per cent of dry weight) of an Ivorian strain, originating from a lake in
Yamoussoukro. It comprises a trimethylated cyclohexenyl ring (David *et al.*, 1988).
From the Berkeley strain, four hydrocarbons **18–21** appeared to contain the same
trimethylated methylenecyclohexyl ring (Huang and Poulter, 1988a, 1988b; Huang *et
al.*, 1988; Murakami *et al.*, 1988); the C_{32}, **20**, was also isolated from a Bolivian strain
(David *et al.*, 1988). Unexpectedly, 1D and 2D ^1H NMR experiments gave evidence
of an axial configuration for the largest substituent (a $C_{22}H_{37}$ group) of the cyclo-
hexyl ring in **20**; that configuration, however, suppresses important steric hindrances
existing for other conformations.

Squalene co-occurs with botryococcenes in all the hydrocarbon extracts of algae of
the B race, and two methylated squalenes were formally identified: monomethyl-
squalene, **22** in a Martinique strain (Metzger and Casadevall, 1983) and tetramethylsqualene, **23** in the Berkeley strain (Huang and Poulter, 1989b). Recently a
trimethylsqualene was tentatively identified in the culture of a Japanese strain
(Okada *et al.*, 1995), along with relatively high amount of **23** (ca 1.3 per cent of
dry weight). In the Berkeley strain, naturally poor in tetramethylsqualene, Huang
and Poulter (1989b) observed that the level of **23** can be increased by the addition of
methionine to the culture medium.

As for the A race, a wide chemical variability is observed for the hydrocarbons on
comparison of B race strains grown under the same conditions (Table 10.3). The
hydrocarbon content, although generally high (ca 30 per cent of dry weight), can
range from 8 per cent (Okada *et al.*, 1995) to nearly 40 per cent of dry weight
(Metzger *et al.*, 1985a). Very different botryococcene profiles can be observed
from GCMS analyses: presence of numerous isomers; accumulations of lower
(C_{30}–C_{32}) or higher ($\geq C_{34}$) botryococcenes; predominance of one hydrocarbon, generally C_{34}, **6**, as in algae from Songkla Nakarin lake, Khao Kho Hong in Thailand
(Metzger *et al.*, 1988).

The biosynthetic pathway of botryococcenes was extensively investigated using
several possible precursors labelled with stable or radioactive isotopes: CO_2,
NaHCO$_3$, sodium acetate, L-leucine, sodium mevalonate, farnesol, farnesal, isopentenyl diphosphate and *S*-methyl methionine. Botryococcene biosynthesis raised
essentially two questions: how is the linkage of the two C_{15} units leading to a $1'–3$

Table 10.3 Botryococcene composition and yield in some strains of the B race.[1]

Botryococcenes	Bolivia Overjuyo (strain no. 4)	France Pareloup	Ivory Coast Ayame	Thailand Khao Kho Hong
$C_{30}H_{50}$ **7**	–	5.8	–	–
$C_{31}H_{52}$ **8**	–	8.6	–	–
$C_{32}H_{54}$ **10**	47.9	11.2	–	–
$C_{32}H_{54}$ **20**	6.6	–	–	–
$C_{34}H_{58}$ **6**	32.8	27.0	8.9	90.3
$C_{34}H_{58}$ **13**	3.3	7.2	1.8	7.0
$C_{34}H_{58}$ **17**	–	–	53.2	–
$C_{36}H_{62}$ **14**	–	4.2	–	–
Others	9.4	36.0[2]	36.1[3]	2.7
% dry weight	35.9	n.d.	26.7	7.6

[1]From Metzger et al. (1988) with kind permission from Elsevier Science Ltd; unshaken cultures, no CO_2 added, 14 h–10 h light–dark cycle.
[2]Mainly C_{36} of unknown structures.
[3]Mainly C_{33} of unknown structures.
n.d., not determined.

fused triterpene, the botryococcene C_{30}, 7, made and which is the biogenesis of the higher compounds?

Structural investigations of botryococcenes C_{34}, 6, and C_{32}, 20 showed that C-10 has the same absolute configuration as the cyclopropyl C-3 of presqualene pyrophosphate, 24 (Figure 10.7), which suggested a biosynthetic pathway via this close precursor of squalene (White *et al.*, 1986; Huang *et al.*, 1988). The efficient incorporation of [5-^{3}H, 1-^{14}C] farnesol into botryococcenes, without degradation, confirmed this assumption (Huang and Poulter, 1989a). Moreover the results of studies using (R)- and (S)-[1-^{2}H] farnesol and (S)-[1-^{2}H, 1-^{13}C] farnesol were consistent with the existence of an intermediate common to squalene and botryococcene synthesis. From presqualene pyrophosphate, 24, the rearrangement of the cyclopropyl cation (path a), followed by hydride attack from NADPH leads to squalene,

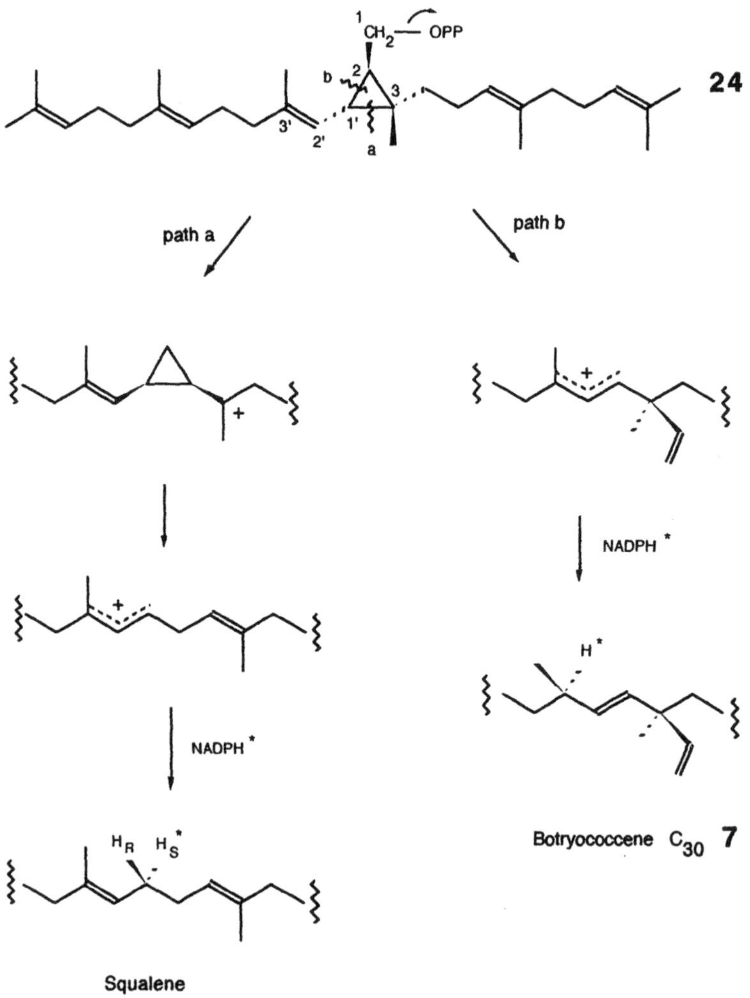

Figure 10.7 Proposed mechanisms for the biosynthesis of squalene and botryococcene 7 from presqualene pyrophosphate 24 (redrawn from White *et al.* (1992) with kind permission from the American Chemical Society).

while the opening of the cyclopropyl ring (path b), also followed by reaction with NADPH would result in botryococcene C_{30}, **7** (Huang and Poulter, 1989a; White *et al.*, 1992); Figure 10.7 illustrates these parallel pathways. Moreover, the authors suggested that only a small shift of the cofactor relative to the C_{30} substrate, near C-3$'$ in the botryococcene synthetase, near C-1$'$ in the squalene synthetase could suffice to explain the regioselective formation of **7** or of squalene in the B race of *B. braunii*, and that squalene biosynthesis could be 'the result of an evolutionary adaptation of the simpler botryococcenoid pathway' (Huang and Poulter, 1989a; White *et al.*, 1992).

Direct evidence for the formation of botryococcene, **7**, from farnesyl pyrophosphate has not been yet obtained. Using cell free extracts, Inoue *et al.* (1994a) only observed the conversion of this precursor into squalene, but not into botryococcene, **7**. The properties of the farnesol phosphokinase from *B. braunii* B race and the formation of farnesal from farnesol by protoplasts and of farnesyl fatty acid esters from farnesol by cell free extracts of the Berkeley strain were studied (Inoue *et al.*, 1993, 1994b, 1995). Furthermore, these authors proposed farnesal, 3-hydroxy-2,3-dihydrofarnesal or farnesyl phosphate as possible candidate precursors for botryococcene biosynthesis; no experiment has yet verified these assumptions.

With regard to the alkylation process operating in the synthesis of higher botryococcenes, the very high incorporation level of L -[Me-^{14}C]-methionine, nearly 65 per cent (David, 1985), suggested that this amino acid acts as the methyl donor. Further experiments conducted with L-[Me-^{13}C]-methionine clearly demonstrated that ^{13}C enrichment occurred for methyl groups at positions 3, 7, 16 and 20 (Metzger *et al.*, 1987). Moreover, a pulse-chase experiment performed with ^{14}CO$_2$ and chase experiments with ^{14}C-labelled acetate and methionine demonstrated that each botryococcene acts as precursor for its direct higher homologue (Wolf *et al.*, 1985b; Metzger *et al.*, 1987). Considering the absolute configuration of the non-terpenoid methyls in **6**, it was also deduced that the methyl groups were transferred at the *si* face of the trisubstituted double bonds (White *et al.*, 1986). Based on the structures of cyclobotryococcene, **16**, we proposed that methylation could be the starter for cyclization (Metzger *et al.*, 1985b), as illustrated in Figure 10.8 (concerted pathway). An alternative mechanism (sequential pathway) is the initiation of cyclization by protonation of a methylated botryococcene (Huang and Poulter, 1988a). This mechanism (Figure 10.8) is suggested by the same absolute stereochemistry found at C_{20} in **20** and in **6**. Further work is needed to determine the exact mechanism by which cyclization occurs.

Lycopadiene (L race)

This compound of general formula $C_{40}H_{78}$ was isolated as the exclusive or main hydrocarbon of nine strains of *B. braunii* originating from some tropical freshwater lakes (Table 10.1). It was identified as 2,6(*R*), 10(*R*), 14, 19, 23(*R*), 27(*R*), 31-octamethyldotriaconta-14(*E*), 18(*E*) diene, **25**, on the basis of spectroscopic studies, hydrogenation and oxidative cleavage. The absolute configuration of the four stereocentres was deduced from the comparison of the C_{18}-ketone obtained by ozonolysis with the 6(*R*), 10(*R*), 14-trimethylpentadecanone derived from natural phytol by the same method (Metzger and Casadevall, 1987); this suggests that

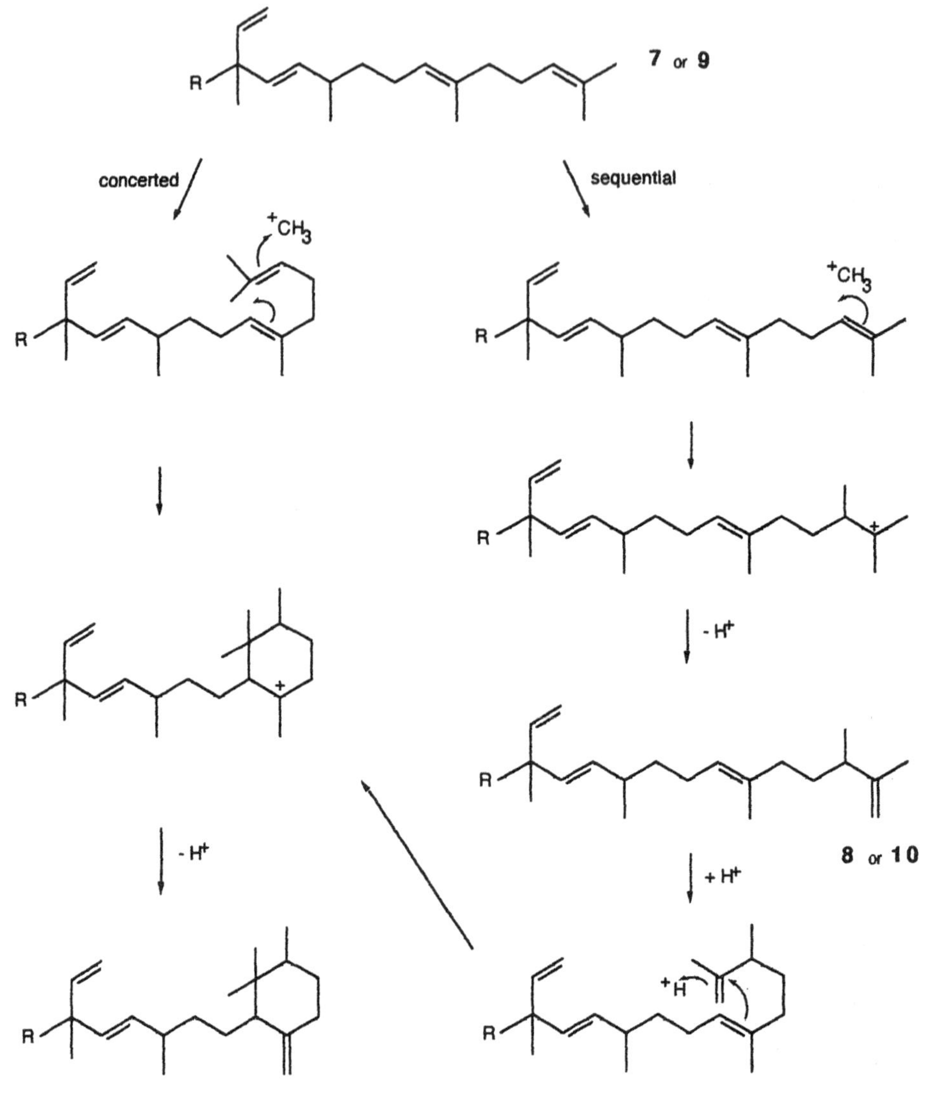

Figure 10.8 Proposed mechanisms for alkylation and cyclization of botryococcenes (R = $C_{11}H_{19}$ for **7** or $C_{12}H_{21}$ for **9** (redrawn from Huang and Poulter (1988a), with kind permission from the American Chemical Society).

synthesis of lycopadiene would occur via dimerization of phytol or of some derivatives. Alternatively, it can be considered that lycopadiene derives from reduction of an acyclic carotenoid such as lycopene.

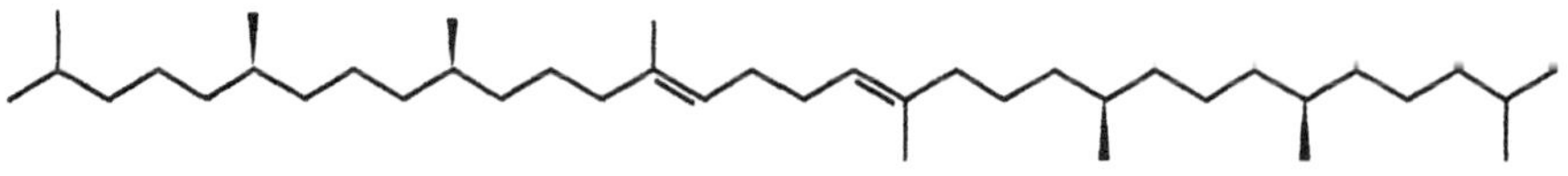

In addition, another $C_{40}H_{78}$ compound was identified, in trace amounts, in the Ivorian and Thai strains. More recently this compound was shown to account for 8 to 20 per cent of the hydrocarbon fraction from the Indian strains originating from Kulavai lake (Metzger *et al.*, 1997). The latter strains were also described as *B. neglectus* (Jeeji Bai, 1996).

Although always low, the lycopadiene content varies with the strain origin: less than 0.1 per cent in the Indian strains, and up to 8 per cent in the Thai strain from Khao Kho Hong lake. Furthermore, it was observed that lycopadiene content is maximum at the end of the active growth phase (Metzger *et al.*, 1990); thereafter, the slow decrease observed could be related to the involvement of this hydrocarbon in the biogenesis of epoxides and related compounds.

Usual lipids

Fatty acids and triacylglycerols

Fatty acids were studied in the three races where they account for 5 to 14 per cent of dry biomass (Metzger *et al.*, 1990). Obtained via saponification of the external and internal lipid extracts, these even acids are in the range C_{14}–C_{32}. Similar distributions were observed for the internal and external pools. In the three races, the main acids are palmitic, oleic and octacosenoic (28:1) acids. Oleic acid predominates in the A race (more than 80 per cent of the total fatty acids), which is certainly related to the precursor role of this compound in the biosynthesis of *n*-very long chain lipids.

Fatty acids of the French strain of Coat ar Herno lake (A race) were shown to contain a minor series of trimethylated fatty acids, **26**, so far not identified in the other strains of the three races. They are odd saturated compounds, ranging from C_{23} to C_{29} (Metzger *et al.*, 1991) and accounting for ca 8 per cent of total fatty acids. Their methyl esters derivatives exhibit in GCMS analysis a base peak at m/z 88, indicative of an α-methyl substitution. The location of the two other methyl substituents was deduced from GC EIMS data of their derived alkanes obtained by chemical treatments. Structural considerations suggest that methyl-branched acids arise from the elongation pathway of fatty acids with the branching carbons derived from condensation with methylmalonate instead of malonate (Figure 10.9). However, a second pathway with the branching originating from the transfer of the *S*-methyl group of methionine on unsaturated intermediate cannot be excluded.

$$CH_3 - (CH_2)_5 - \overset{\overset{\displaystyle CH_3}{|}}{CH} - (CH_2)_5 - \overset{\overset{\displaystyle CH_3}{|}}{CH} - (CH_2)_x - \overset{\overset{\displaystyle CH_3}{|}}{CH} - CO_2H \qquad \textbf{26}$$

x=odd, 5-11

Generally found in minute amounts in a free form, fatty acids occur essentially as triacylglycerols and, as indicated later, as the acylating agents of the alcohol functions present in ether lipids. Triacylglycerols were detected in the hexane extracts of all strains of the A race. In a Bolivian strain originating from Challapata lake, they constitute, besides hydrocarbons, the dominant class of lipids (Metzger *et al.*, 1989). Analysis by reversed-phase HPLC on a C_{18} column showed that triolein is a major

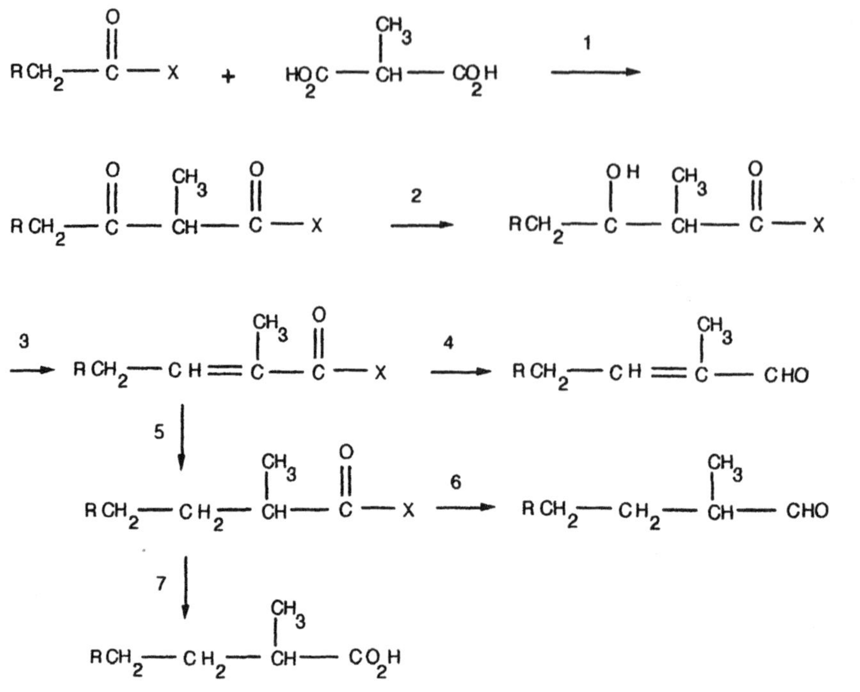

Figure 10.9 Proposed mechanisms for the biosynthesis of methyl-branched fatty acids and fatty aldehydes in *B. braunii* (A race); 1: CO_2 elimination; 2: β-keto acyl reduction; 3: dehydration; 4: acyl reduction; 5: α,β reduction; 6: acyl reduction; 7: free fatty acid release (redrawn from Metzger *et al.* (1991) with kind permission from Elsevier Science Ltd).

constituent; molecular species involving fatty acyl groups with chains longer than C_{18} were also detected.

Sterols

24-Methylcholest-5-en-3β-ol and its 24-ethyl homologue are the main sterols occurring in the three races of *B. braunii* (Metzger *et al.*, 1990). They are accompanied by minor amounts of cholesterol, and in the case of the L race of some unidentified sterols. On the whole, sterols account for 0.1 to 0.2 per cent of the dry biomass and are primarily recovered in the internal lipids.

Carotenoids

Whereas algae of the A race of *B. braunii* turn from green to pale green or pale yellow when they enter the stationary phase of growth, those of the B and L races become orange to red in relation to accumulation of carotenoids (0.2 to 0.4 per cent of dry weight). Chromatographic and spectroscopic analyses have shown that the B race (La Manzo strain from Martinique) and the L race (Yamoussoukro strain from

Ivory Coast) contain the β,ε-carotene derivatives (6'R) β,ε-carotene **27**, (3R, 3'R, 6'R)-lutein **28**, and (3R, 3'R, 6'R)-loroxanthin **29**, and the β,β-carotene derivatives β,β-carotene **30**, (3R, 3'R)- zeaxanthin, **31**, (3R, 5R, 6S, 3'S, 5'R, 6'S)-violaxanthin, **32**, (3S, 5R, 6R, 3'S, 5'R, 6'S)-neoxanthin, **33**, echinenone, **34**, (3'R)-3'-hydroxye-chinenone, **35**, and canthaxanthin, **36** (Grung *et al.*, 1989) (Figure 10.10). In addition (3S)-3-hydroxyechinenone, **37**, (3S, 3'S)-astaxanthin, **38** and (3S, 3'R)-adoni-xanthin, **39** were identified in a second strain from Yamoussoukro (Grung *et al.*, 1994a).

In the linear phase of growth, lutein, **28** is the dominant carotenoid, while cantha-xanthin, **36** and to a lesser extend echinenone, **34** predominate in the stationary phase (Grung *et al.*, 1989). Quantitative analyses performed at the exponential, linear and stationary phases of growth, on whole colonies, on matrices which ensure the colony cohesion and on free cells showed that ketocarotenoids **34** and **36** are produced by the cells during the linear and stationary phases and continuously transferred to the matrices (Grung *et al.*, 1994b); this suggests that they exert a protective role against light, when the lipid-rich colonies float at the water surface. Moreover, the initiation of canthaxanthin, **36** and echinenone, **34** synthesis seems to be concomitant with the onset of nitrogen deficiency in the culture medium. From these analyses and on structural grounds, a biosynthetic pathway was proposed for *B. braunii* carotenoids, starting from an acyclic precursor such as lycopene (Figure 10.11).

Very recently, the isolation of botryaxanthin A, **40** (Figure 10.10), a member of a new class of carotenoids, was reported from the Berkeley strain (Okada *et al.*, 1996). This pigment is a ketal probably originating from the condensation of echinenone, **34** with a dihydroxy-tetramethylsqualene (Figure 10.11); this diol is probably derived from an epoxide whose structure will be discussed later.

Non-classical lipids

Since the end of the 1980s, we have extensively investigated strains of the A race for isolation of non-classical lipids. Recent work shows that algae of the L race and to a lesser degree those of the B race also synthesize lipids of unprecedented types.

α-Branched fatty aldehydes

Botryals are long chain, α-branched fatty aldehydes occurring in all strains of the A race. Their carbon skeletons comprise an even number of carbons, from 52 to 64 and three unsaturations: two *cis* double bonds in ω9 and ω9' and another conjugated with the aldehyde function. Two series of botryals, **41** and **42** (Figure 10.12) were obtained by preparative TLC, differing by the *E* and *Z* configurations of the con-jugated double bond, respectively (Metzger and Casadevall, 1989). The structures were resolved through utilization of spectroscopic methods, the position of the mid-chain unsaturations being determined by means of ozonolysis. Isolated at first in a very high yield (18 per cent of dry weight) from a Titicaca strain, botryals were found in all the tested strains (Metzger and Casadevall, 1989), with a level ranging from 1 to 11 per cent of the dry biomass; botryals, **42** predominated in all strains analyzed.

A feeding experiment with $[1,2-^{13}C]$ acetate demonstrated the repeated condensa-tion of C_2 units, very likely via malonate, leading to even C_{26} to C_{32} *n*-fatty alde-

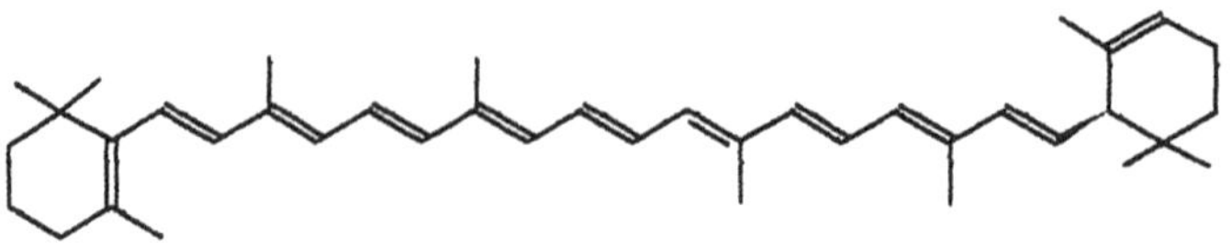

Figure 10.10 Carotenoids of the B and/or L races.

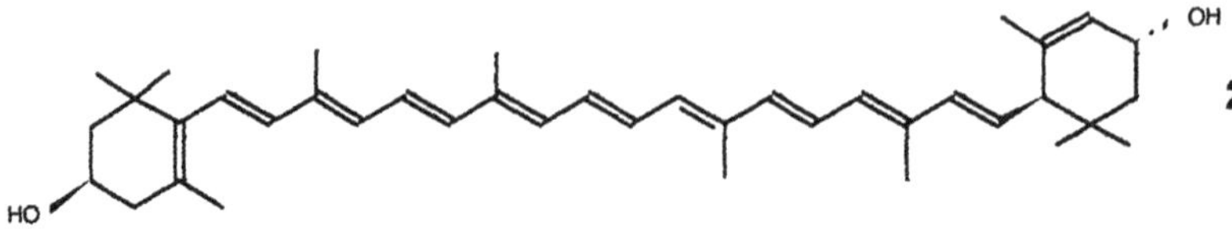

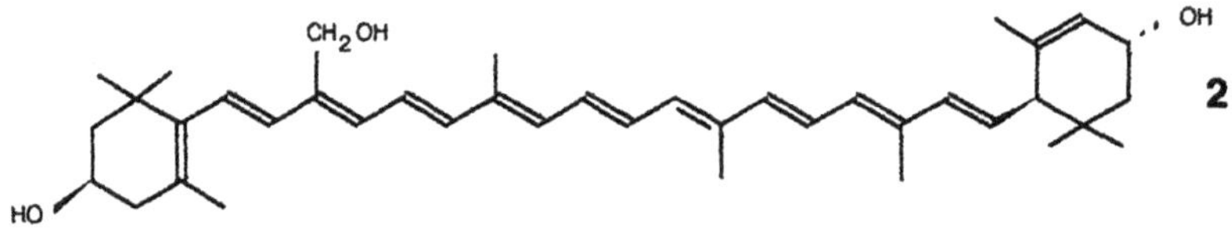

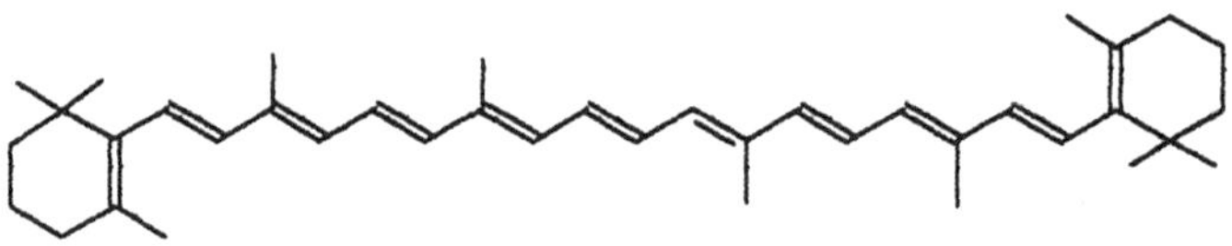

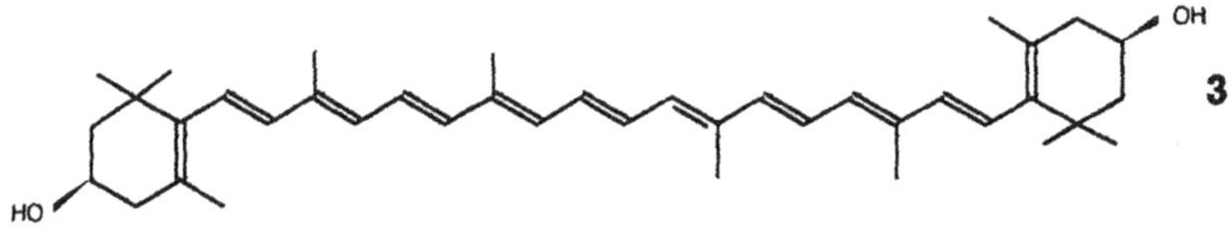

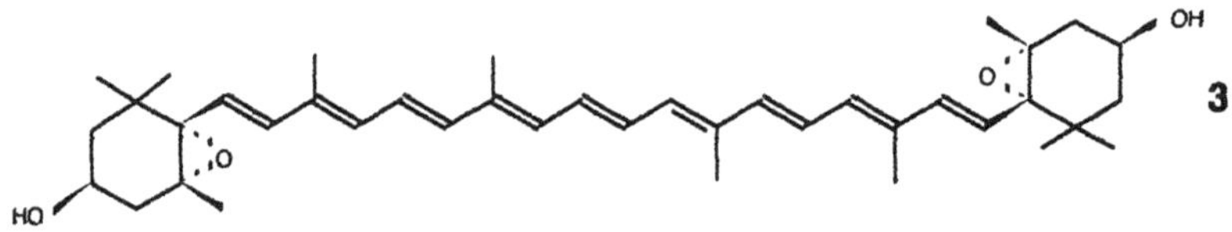

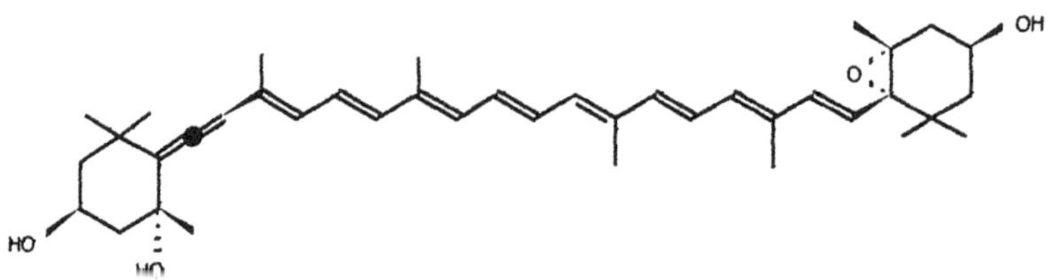

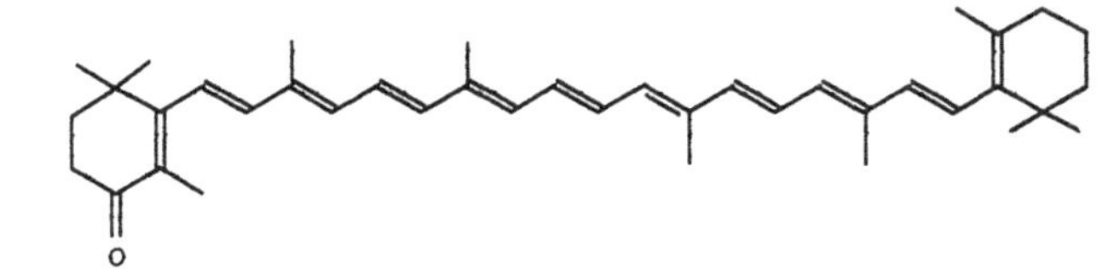

34

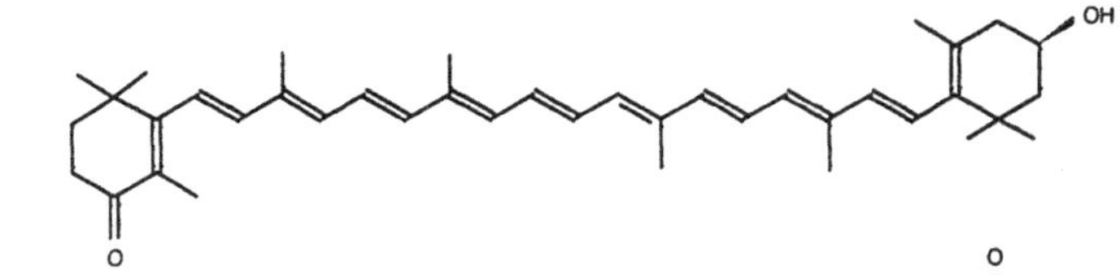

35

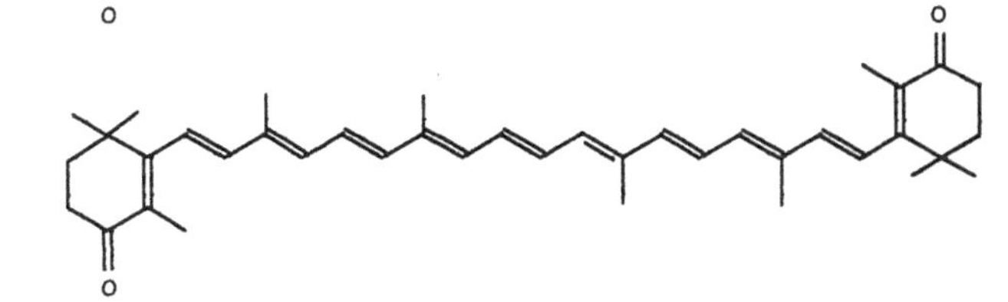

36

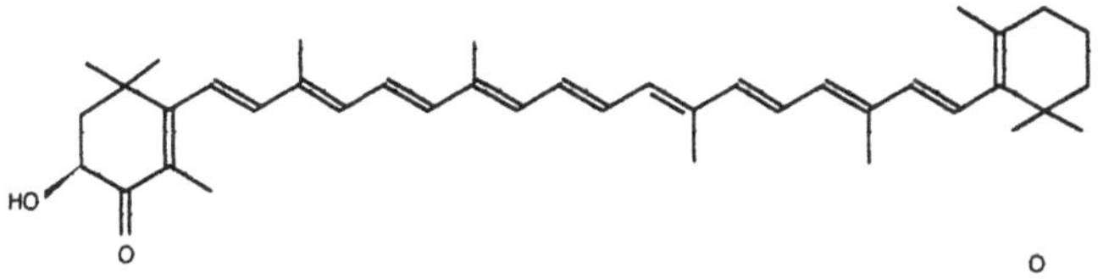

37

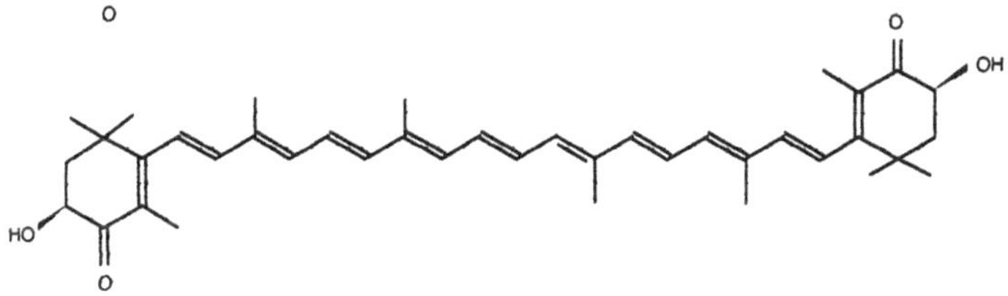

38

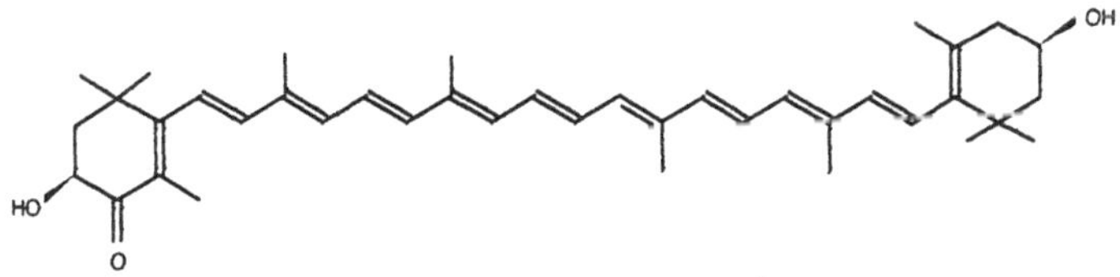

39

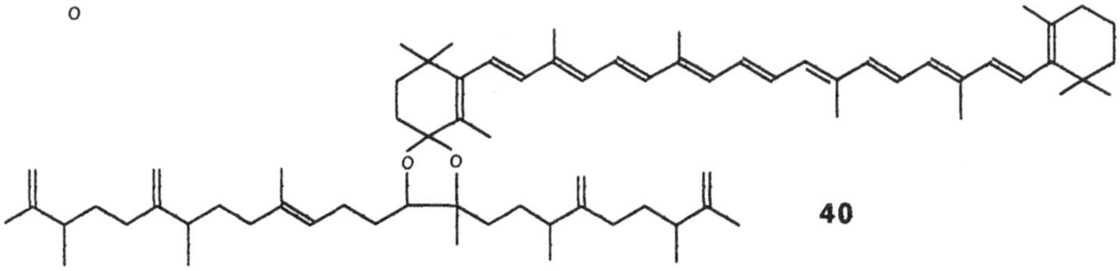

40

Figure 10.10 Continued.

hydes. The condensation of two of these aldehydes generated aldols which were dehydrated to botryals, **41** and **42** (Figure 10.13) (Metzger and Casadevall, 1989).

To the best of our knowledge, aldehydes structurally related to botryals have only been isolated, in minute amounts, from some red algae (de Rosa *et al.*, 1995). They differ from botryals, however, by shorter chains, whose carbon numbers preclude

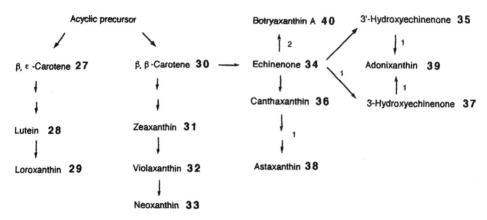

Figure 10.11 Proposed biogenesis of carotenoids in the B and L races of *B. braunii*. 1: only in the L race; 2: only in the B race (partially redrawn from Grung *et al.* (1989, 1994a), with kind permission from Elsevier Science Ltd).

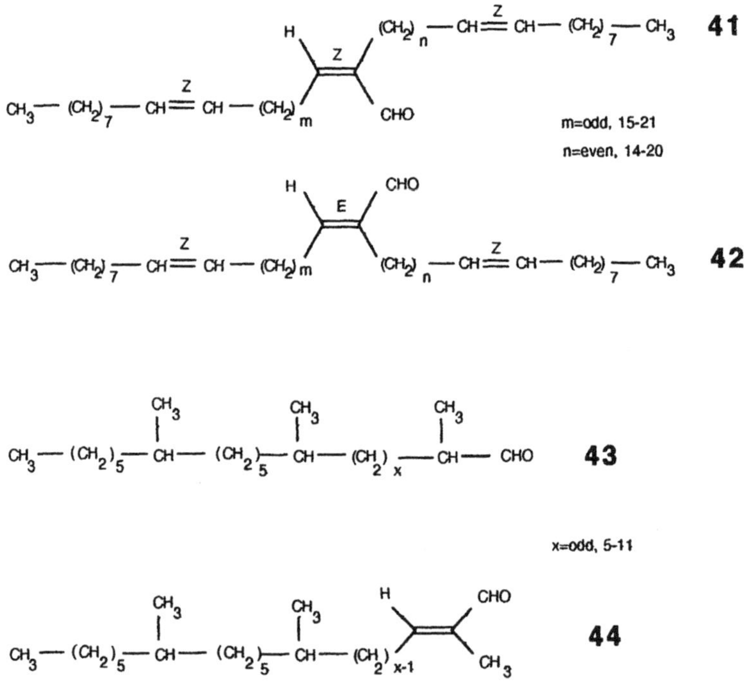

Figure 10.12 α-Branched fatty aldehydes of the A race.

that they are formed by condensation of even fatty aldehydes. Interestingly, the structure and biogenesis of botryals are rather close to those of mycolic acids. These important constituents of the cell walls of Corynebacteria, Nocardia and Mycobacteria, are α-branched, β-hydroxy acids synthesized by condensation of fatty acids (Brennan, 1988).

Besides botryals, the strain originating from Coat ar Herno lake, synthesizes trimethyl fatty aldehydes, **43** and the α-unsaturated homologous compounds, **44**

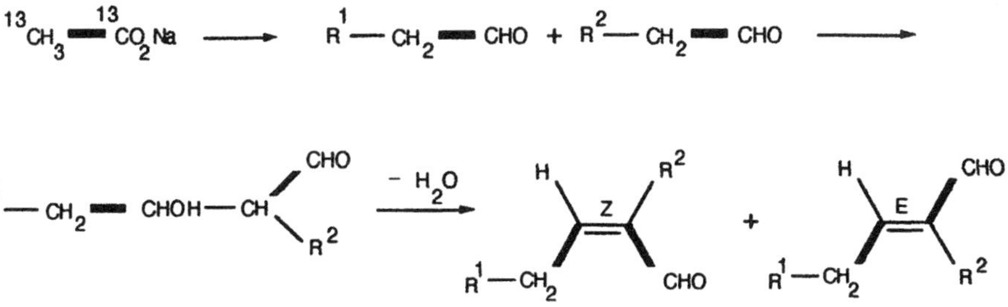

Figure 10.13 Proposed biosynthesis of botryals from [1,2-^{13}C]acetate (redrawn from Metzger and Casadevall (1989) with kind permission from Elsevier Science Ltd).

(Figure 10.12), in a yield of 2.7 and 2.3 per cent of dry weight, respectively (Metzger *et al.*, 1991). These unusual fatty aldehydes, comprising odd carbon numbers from C_{23} to C_{29}, exhibit the same gross structure as the trimethyl fatty acids, **26**, identified in this same strain. These structural features suggest a close biogenetic pathway via common intermediates: the α-unsaturated fatty acyl derivatives. Reduction of the acyl group would give the unsaturated aldehydes, **44**, while reduction of the carbon–carbon double bond with subsequent reduction of the acyl group or hydrolysis of this latter would produce the saturated aldehydes, **43** or the acids, **26**, respectively (Figure 10.9).

n-Alkenylphenols

The first *n*-alkenylphenols identified in *B. braunii* were the *n*-alkenylhydroquinol dimethyl ethers, **45** (Figure 10.14). These compounds exhibiting C_{27}, C_{29} and C_{31} hydrocarbon chains were isolated (1.5 per cent of dry weight) from the Austin Collection strain (Metzger and Casadevall, 1989). Later, from this strain, and from the one of Coat ar Herno, we succeeded in separating a *n*-alkyl and a series of *n*-alkenylpyrogallol dimethyl ethers, **46** and **47**, respectively, accounting for 0.85 per cent of dry weight (Metzger and Pouet, 1995). A systematic investigation of algae of the A race, showed us that these two types of phenols are present in all strains of this chemical race, with the exception of one originating from Overjuyo lake in Bolivia. This last strain contains a series of *n*-alkenylresorcinol dimethyl ethers, **48** (0.44 per cent of dry weight), with C_{25}, C_{27} and C_{29} hydrocarbon chains (Metzger, 1994; Metzger and Largeau, 1994). All these phenolic compounds exhibit strong UV absorptions at 289, 222 and 215 nm for **45**, 215 nm for **46** and **47** and 203 nm for **48**. The structures were mainly elucidated by NMR methods and by chemical degradation.

On the basis of a feeding experiment with [1,2-^{13}C] acetate, hydroquinol derivatives, **45** were assumed to have a tetraketide origin in *B. braunii* (Metzger and Casadevall, 1989). Indeed, the ^{13}C NMR spectrum of the ^{13}C-enriched metabolites revealed that the bond between the benzylic carbon and the aromatic C-6 is derived from an intact acetate unit. The ring would be formed by a standard aldol condensation, followed by dehydration to form the carboxylic intermediate, **49** (Figure 10.15).

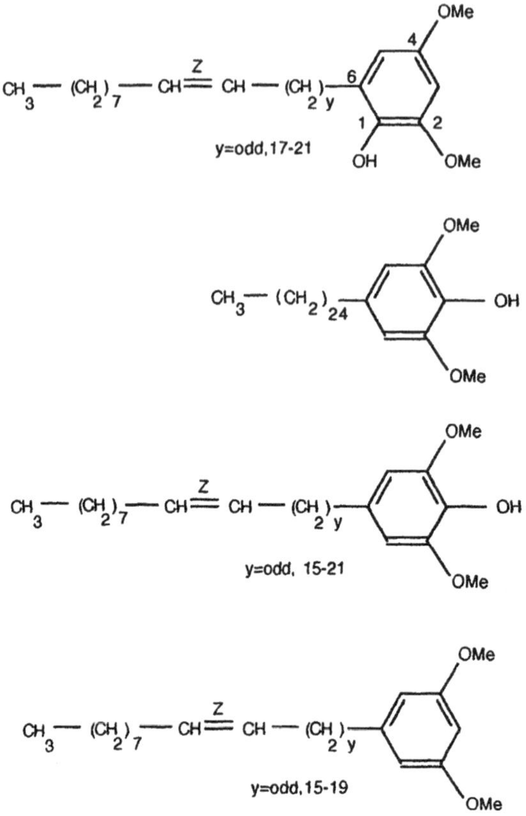

Figure 10.14   *n*-Alkyl and alkenyl phenols of the A race.

Furthermore the labelling patterns suggested that the hydroxyl at C-1 is very likely introduced after decarboxylation of **49**. Moreover, hydroxylation at the para position relative to the aliphatic chain, would lead to the *n*-alkenylpyrogallol derivatives, **47**. In contrast with all the other strains examined so far, the one from Overjuyo would be unable to hydroxylate the aromatic nucleus and thus resorcinol dimethyl ethers, **48** would accumulate.

Phenolic lipids are widely distributed in nature, for instance in marine brown algae (Amico, 1995), higher plants such as *Anacardium occidentale* (Tyman, 1979) and also in some marine sponges (Barrow and Capon, 1991). These compounds generally display antioxidant properties, reinforced, in lipidic regions of these organisms, by the presence of aliphatic chains. Based on the localization of the alkenylphenols in the outer walls of *B. braunii*, as deduced from their quick extraction by hexane, these compounds could protect the external lipids against oxidative degradation.

Epoxides

Our investigations into the lipids of *B. braunii* resulted in the isolation of non-isoprenoid epoxides in the A race (Figure 10.16) and of isoprenoid epoxides in the B and L races (Figure 10.17).

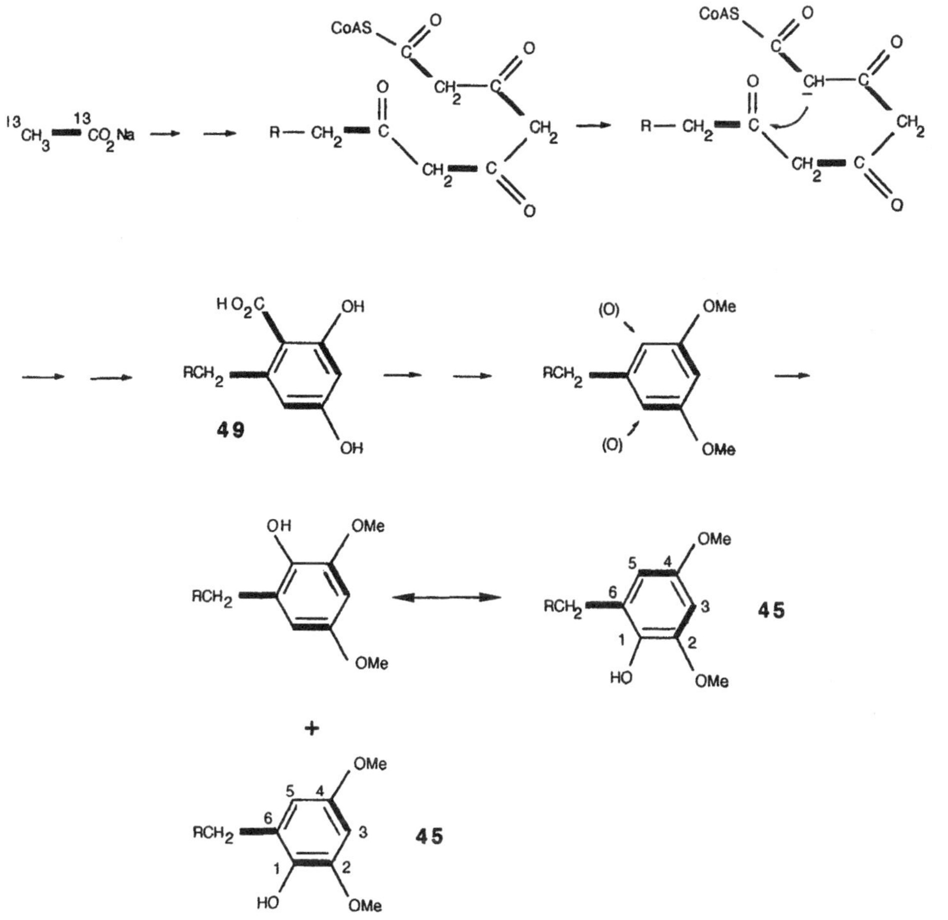

Figure 10.15 Proposed mechanism for the biosynthesis of *n*-alkenylhydroquinol dimethyl ethers, **45** (redrawn from Metzger and Casadevall (1989) with kind permission from Elsevier Science Ltd).

Three series of non-isoprenoid epoxides were recovered from the Austin strain (A race), accounting on the whole for 4 per cent of dry biomass. They were epoxyalkenes, **50**, epoxybotryals, **51** and epoxyalkylhydroquinol dimethyl ethers, **52**. These *cis* epoxides derive from parent compounds: alkadienes, **2**, botryals, **42** and alkenylhydroquinol dimethyl ethers, **45**, respectively, by epoxidation of the $C\omega9-C\omega10$ double bond. (Metzger and Casadevall, 1989).

From a Martinique strain of the B race (La Manzo), a series of five triterpenoid epoxides was isolated in a yield of 0.8 per cent of dry biomass (Delahais and Metzger, 1997). EIMS, ^1H and ^{13}C NMR data showed that they are 10,11-epoxy-squalene derivatives exhibiting two to four non-terpenoid additional methyl groups. Four structures were established, one C_{32}, **53**, two C_{33}, **54** and **55** and one C_{34}, **56**; the non-isoprenoid methyls are located at positions C-3, C-7, C-18 and/or C-22. These epoxides can be considered as arising from methylated squalenes, such as tetramethylsqualene, **23** in the case of **56**.

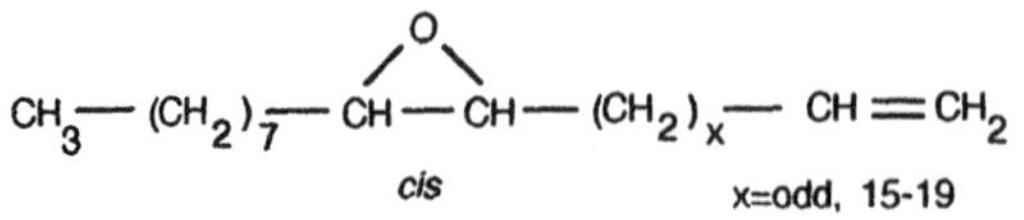

$CH_3-(CH_2)_7-CH-CH-(CH_2)_x-CH=CH_2$ **50**

cis

x=odd, 15-19

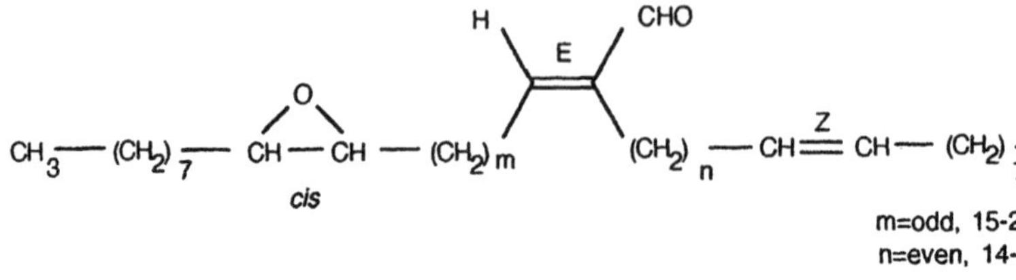

51

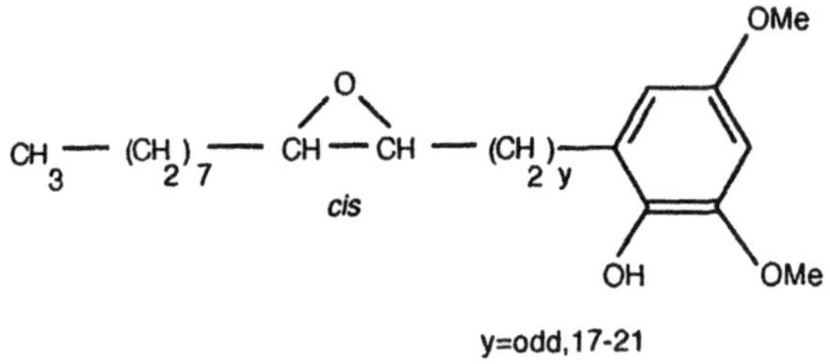

$CH_3-(CH_2)_7-CH-CH-(CH_2)_m$ ... $(CH_2)_n-CH=CH-(CH_2)_7-CH_3$

cis

m=odd, 15-21
n=even, 14-20

52

$CH_3-(CH_2)_7-CH-CH-(CH_2)_{2y}$

cis

y=odd,17-21

Figure 10.16 Epoxides of the A race.

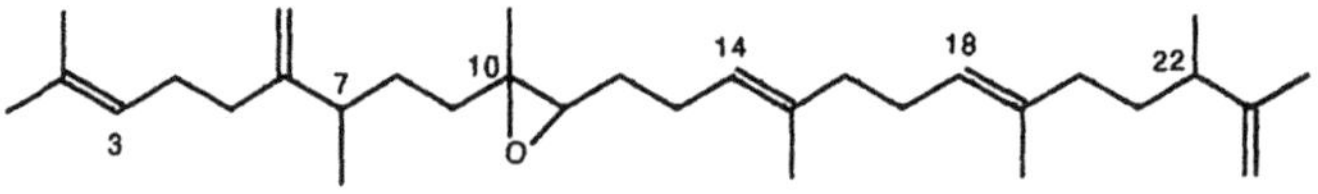

53

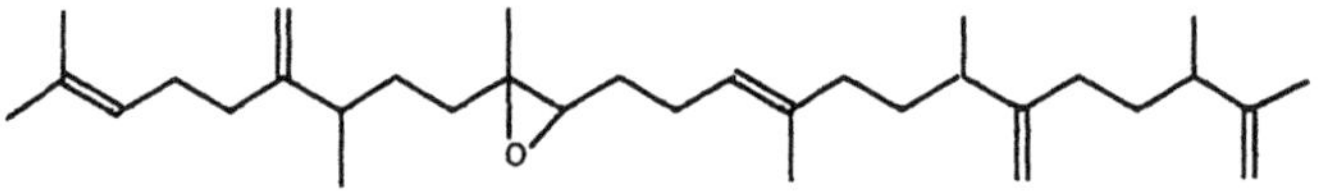

54

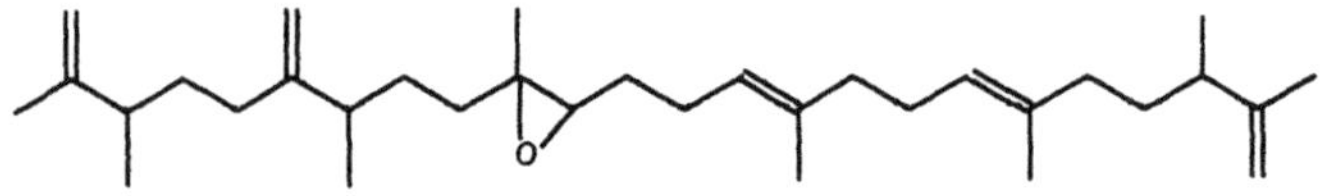

55

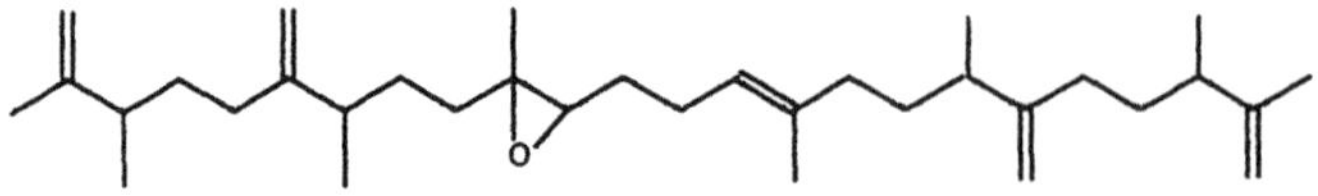

56

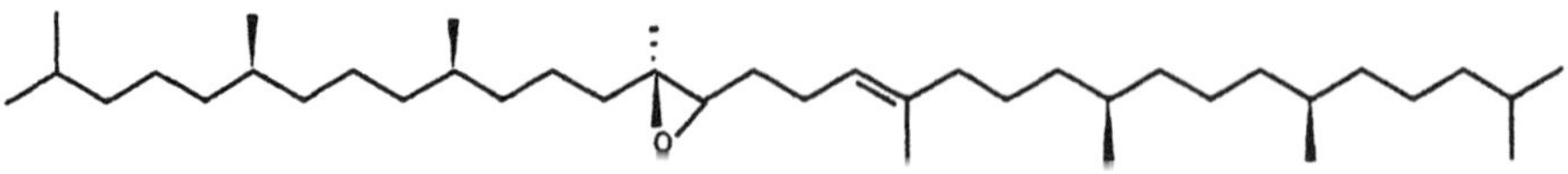

57

Figure 10.17 Epoxides of the B (**53–56**) and L races (**57**).

trans-trans-Epoxylycopaene, **57** accounts for 0.3 per cent of dry biomass of Yamoussoukro strain (L race) (Delahais and Metzger, 1997). The absolute stereochemistry (6*R*, 10*R*, 14*R*, 15*R*, 23*R*, 27*R*) was established via chemical degradation and comparison with 6*R*, 10*R*, 14-trimethyl pentadecanone and application of Mosher ester methodology on the derived diol, **58**. This diol, **58** was also found to occur naturally in the alga.

As will be shown later all these epoxides are key intermediates in the synthesis of more complex lipids; in the case of tetramethyl-epoxy squalene, **56** we suggested that it is very likely implicated in the formation of botryaxanthin A, **40**.

Ether lipids

In 1991 we reported the isolation and structural determination of two series of unusual ether lipids of a new type from the Coat ar Herno strain (Metzger and Casadevall, 1991). These compounds exhibit oxygen bridges between very long hydrocarbon chains and, therefore, differ from ether lipids occurring widely in nature, the alkylglycerolipids which comprise normal or terpenoid chains ether linked to glycerol (Mangold and Paltauf, 1983). Since this report more than 20 series of ether lipids have been described: many have been studied in the A race, two series were identified in the L race, and to date none in the B one.

Ether lipids of the A race

On structural grounds ether lipids of the A race can be classified into: alkoxy ether lipids, based on long straight chains linked by O-alkyl ether bridges, on mid-chain or terminal carbons; and phenoxy ether lipids, characterized by involvement of the phenolic moiety of *n*-alkenylresorcinols in ether bridge formation.

Alkyl-O-alkadienyl ethers, **59** (Figure 10.18) are the simplest alkoxy ether lipids. They were isolated from the external lipids of a French strain (Coat ar Herno) and of a Bolivian one (Overjuyo, strain 6), in 1.1 per cent of dry biomass (Metzger and Casadevall, 1991). TLC and HPLC separation of the oil extracted with hexane afforded a solid mixture of two compounds, **59a** and **59b** in an almost 1:1 ratio. EIMS and ^{1}H NMR data showed that these two compounds comprise a C_{27} alkadienyl chain exhibiting a terminal unsaturation and a *cis* double bond at C-7. This chain is linked, at C-9 by an ether bridge to the terminal carbon of a saturated C_{24} (**59a**) or C_{26} (**59b**) normal chain; one hydroxyl group is also present, α to the ether bridge, on the C_{27} chain.

From the two above mentioned strains, the alkadienyl-O-alkatrienyl ether, **60** (C_{27}–O–C_{27}) was also isolated in a very high yield (26 per cent of dry weight). Its structure was elucidated by spectroscopic techniques and via hydrogenolysis and ozonolysis. The stereochemistry of the double bonds was deduced from the magnitude of the coupling constants, J_{7-8}, $J_{8'-9'}$ and $J_{10'-11'}$. More recently, a novel ether,

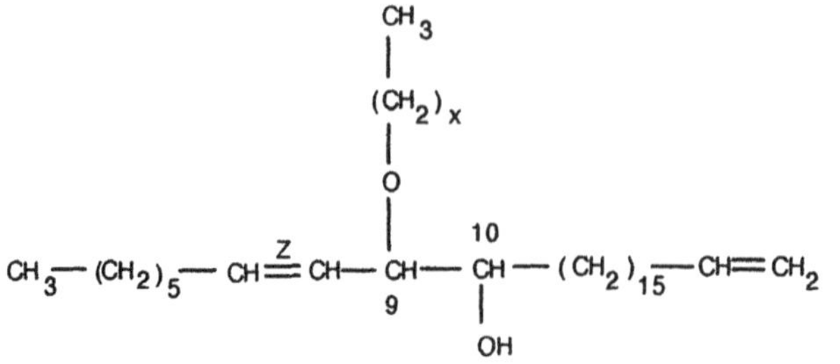

59a x=23

59b x=25

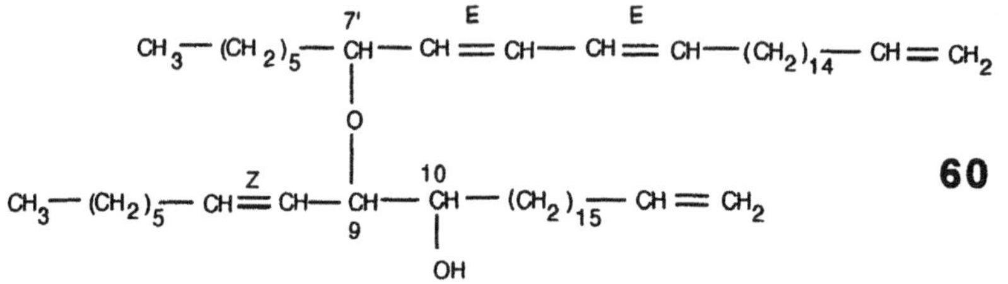

60

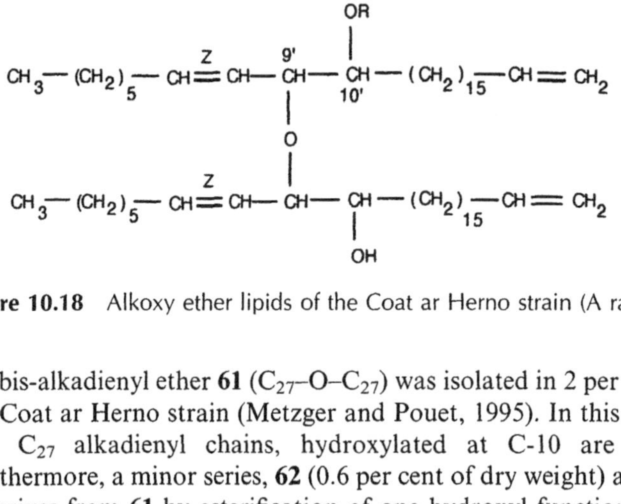

61 R=H

62 R=fatty acyl

Figure 10.18 Alkoxy ether lipids of the Coat ar Herno strain (A race).

the bis-alkadienyl ether **61** (C_{27}–O–C_{27}) was isolated in 2 per cent of dry weight from the Coat ar Herno strain (Metzger and Pouet, 1995). In this compound two equivalent C_{27} alkadienyl chains, hydroxylated at C-10 are ether linked at C-9. Furthermore, a minor series, **62** (0.6 per cent of dry weight) also occurs in this strain; it derives from **61** by esterification of one hydroxyl function by fatty acids ranging from C_{16} to C_{32} ($C_{30:1}$ dominant).

On the whole, ether lipids identified in the Coat ar Herno strain account for ca 30 per cent of dry biomass. A common structural feature of these compounds is the presence of the hydroxylated C_{27} alkadienyl chain. Structural considerations suggested a close biosynthetic relationship with a lower metabolite identified in the hydrocarbon fraction: the heptacosa-1,18(*Z*),20(*Z*) triene, **3** (Villarreal-Rosales *et al.*, 1992). Indeed an inverse relationship was observed along the algal growth between alkadienyl-O-alkatrienyl ether, **60** and *cis*, *cis* C_{27} triene **3**: the abundance of the triene was lower as ether lipid biosynthesis was high. Moreover a feeding experiment showed the incorporation of radioactivity from this ^{14}C-labelled C_{27} triene into **60**. On the basis of these results and on structural grounds, triene C_{27}, **3** would be first oxidized into two monoepoxides, **63**, and **64** (Figure 10.19). Initiation of the coupling reaction would be the protonation of epoxide **63**, promot-

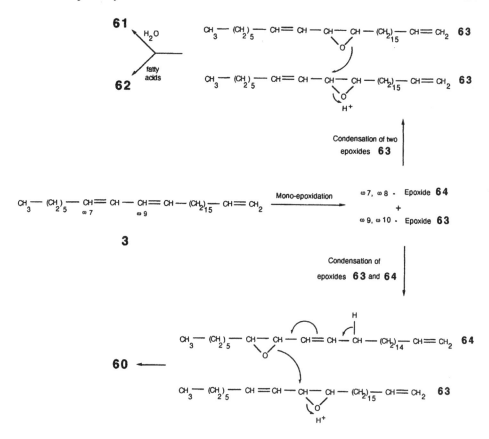

Figure 10.19 Proposed mechanisms for the biosynthesis of ether lipids **60-62**, in the Coat ar Herno strain (A race) (redrawn from Villarreal-Rosales *et al.* (1992) and from Metzger and Pouet (1995) with kind permission from Elsevier Science Ltd).

ing the cleavage of the C9–O bond. In a second step, the ether bridge would be formed by reaction of the resulting carbocation with epoxide **64**. The concomitant migration of the C9′–C10′ double bond and the loss of an allylic proton at C11′ would finally result in the *trans, trans* conjugated system. The involvement of two molecules of epoxide **63** was proposed for the synthesis of **61** and **62** (Metzger and Pouet, 1995); the hypothetical mechanism requires however the formation of a carbocation at an unfavourable position (Figure 10.19).

The collection strain of Austin contains a lower amount of alkoxy ether lipids: 4.5 per cent of dry weight (Metzger and Casadevall, 1992). The simplest ethers isolated from this strain are formed by a series of di-O-alkenyl ethers, **65** (Figure 10.20), accounting for 0.3 per cent, in which two hydrocarbon chains, very likely derived from C_{27}, C_{29} and C_{31} alkadienes, **2**, are ether linked at C9′ and C10 and hydroxylated at C10′ and C9. Their mono fatty acyl ester derivatives, **66**, were isolated in a higher yield (2 per cent). They predominantly comprise an oleyl moiety, and also saturated and monounsaturated C_{24} to C_{32} acyl chains. Four other series of minor alkoxy ether lipids were also isolated from the Austin strain: an unseparable mixture of **67** and **68** which are alkenyl-O-alkylhydroquinol ethers (0.2 per cent), a series of

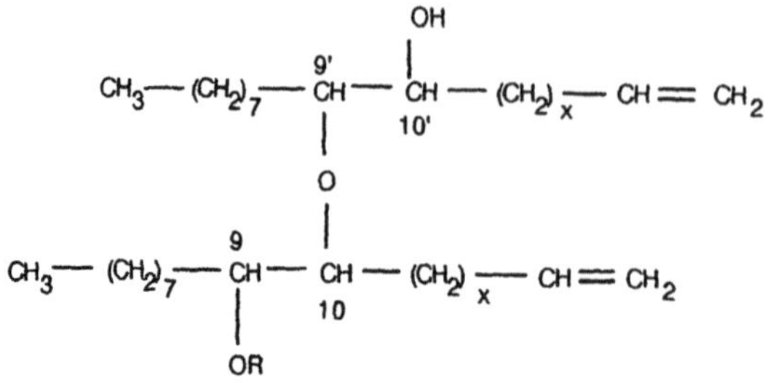

65 R=H

66 R=fatty acyl

x=odd, 15-19

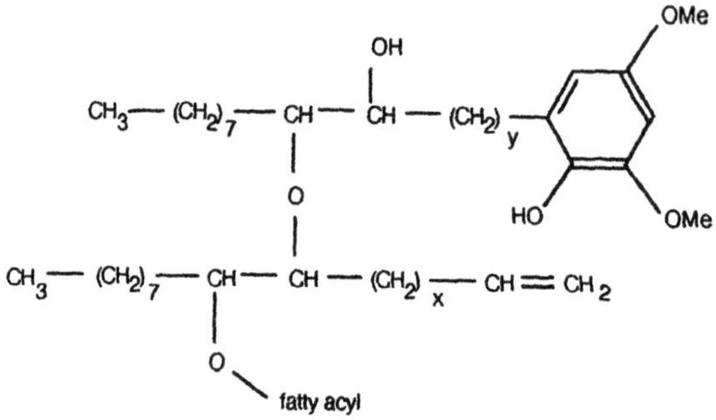

67

x=odd, 15-19

y=odd, 17-21

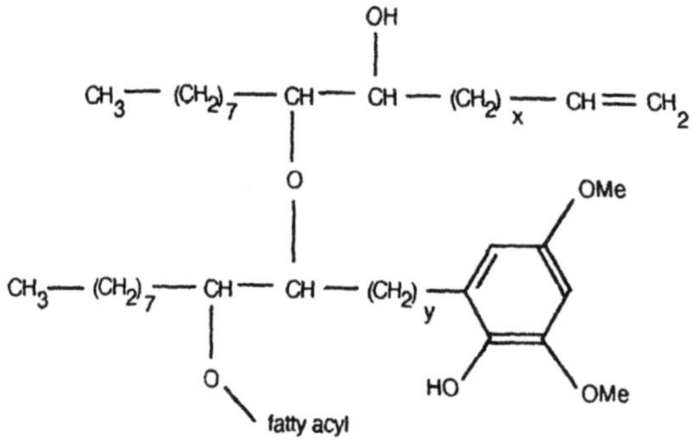

68

x, y as in **67**

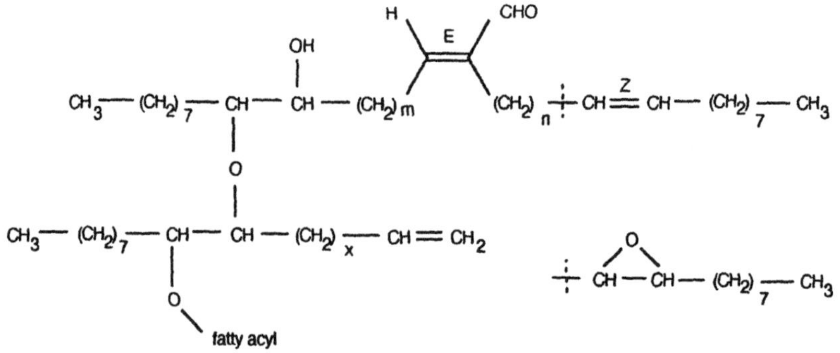

69

70

m=odd, 15-21

n=even, 14-20

x=odd, 15-19

Figure 10.20 Alkoxy ether lipids of the Austin strain (A race).

alkenyl-O-botryalyl ethers, **69** (1 per cent) and an epoxidized series, **70** derived from this latter (0.7 per cent).

All these compounds, **65–70**, exhibit oxygen functions at C9, C10, C9' and C10', suggesting that they could be synthesized by coupling of two epoxides: epoxyalkenes **50**, epoxybotryals, **51** and/or epoxyalkylhydroquinols, **52**. This was confirmed by feeding the algae with ^{14}C-labelled 9,10-epoxynonacos-28-ene: ca 10 per cent of the radioactivity was recovered in ether lipids, **65–70**. As illustrated in Figure 10.21, ring opening from a protonated epoxide function would occur in a first step and the C9–O and C9'–O bonds would be specifically cleaved. Neutralization of the carbocationic species either by water or by a fatty acyl derivative would thus generate ether lipids **65** or **66–70**, respectively.

Phenoxy ether lipids were first discovered in another strain from Overjuyo lake, in which they account for 35 per cent of dry weight (Metzger, 1994). Seven series, **71–77** (Figure 10.22), were isolated from the external lipids by chromatographic techniques. In all these compounds, an *n*-alkenylresorcinol, with a C_{25}, C_{27} (major) or C_{29} monounsaturated chain, is involved in the formation of phenoxy bonds with aliphatic chains derived from alkadienes (as in **71, 74–77**) and/or tetraenes (as in **72, 73** and **76**). In turn hydrocarbon chains can be ether linked, in a manner similar to that

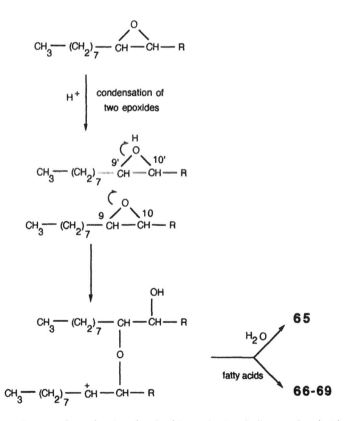

Figure 10.21 Proposed mechanism for the biosynthesis of alkoxy ether lipids **65-69** in the Austin strain (A race). R: chain portion of **2, 42** or **45** (redrawn from Metzger and Casadevall (1992) with kind permission from Elsevier Science Ltd).

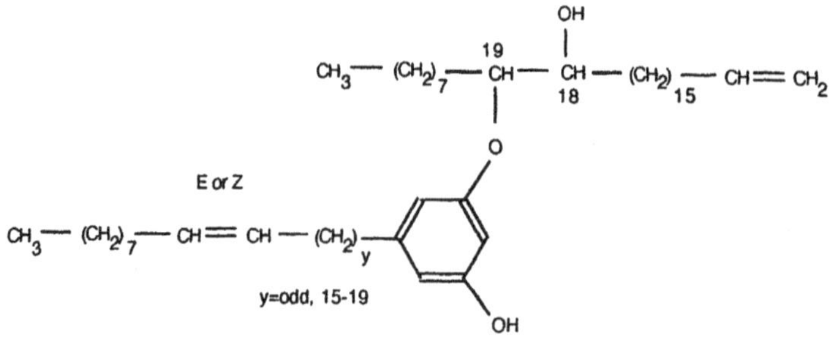

71

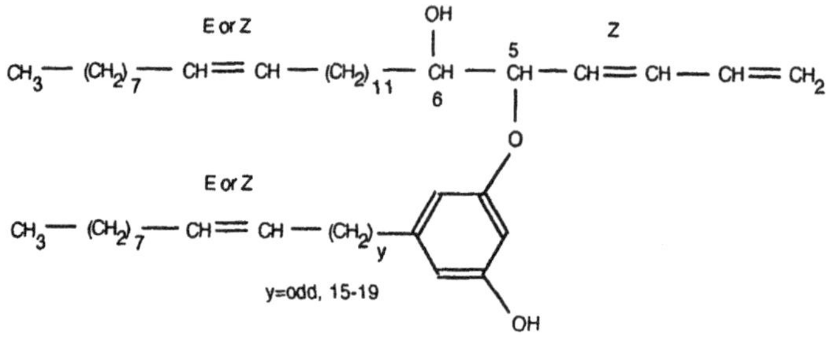

72

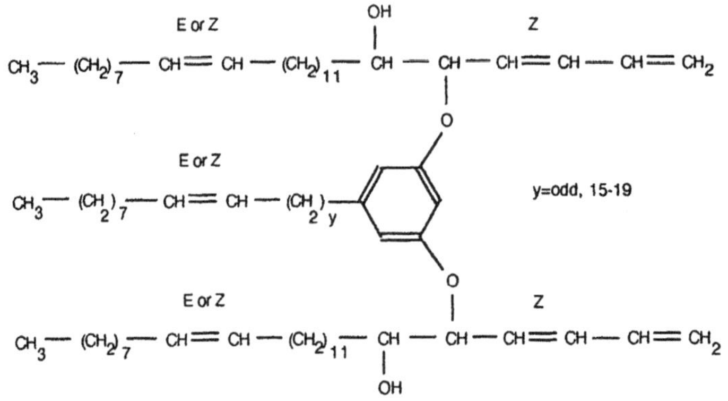

73

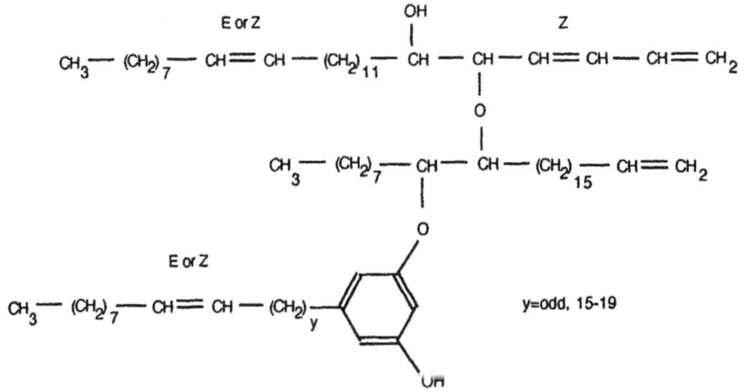

74

Figure 10.22 Phenoxy ether lipids of an Overjuyo isolate, strain 7 (A race).

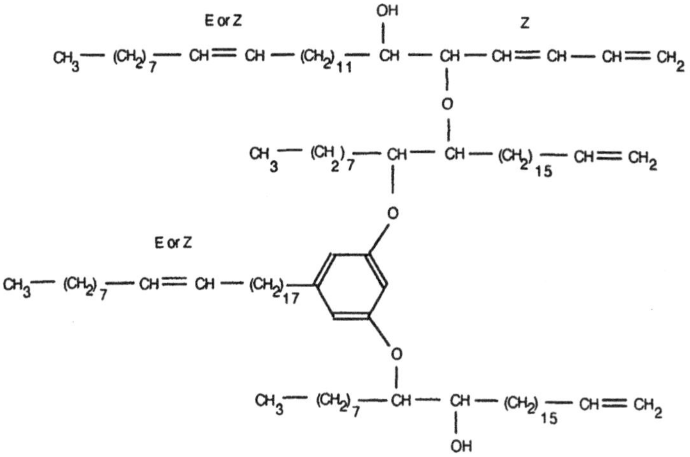

Figure 10.22 Continued

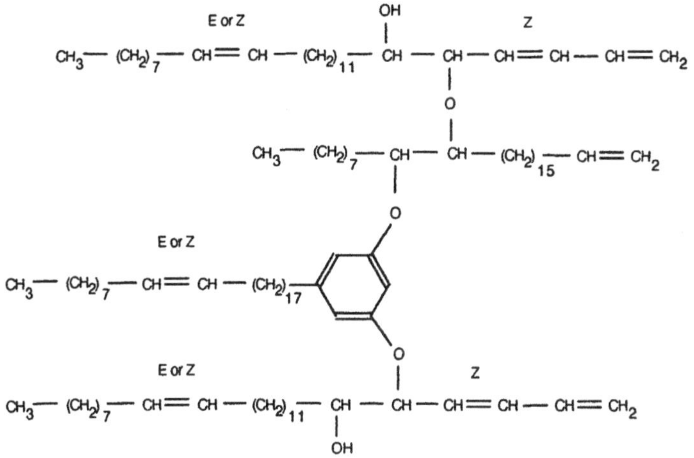

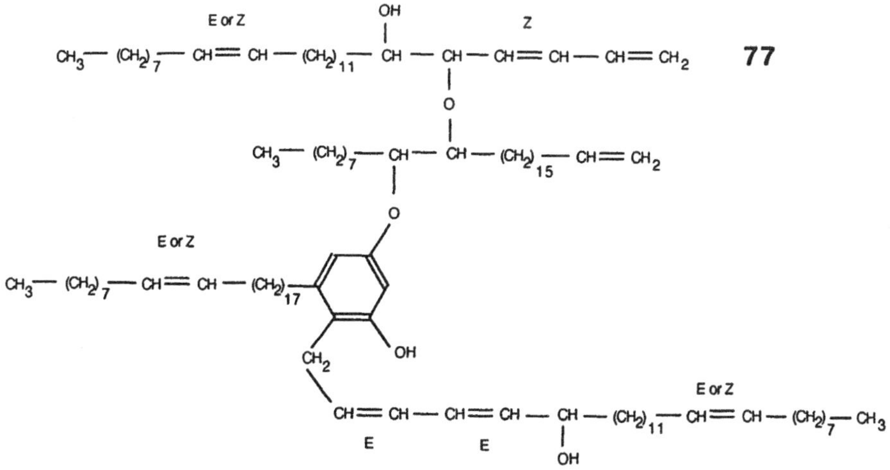

observed in alkoxy ether lipids, to another hydrocarbon chain (as in **74–77**). Moreover one phenol group (**71, 72, 74** and **77**) or two (**73, 75, 76**) can be involved in the formation of the ether bridges. In each series, *cis* and *trans* isomerisms occur at the $\omega 9$ unsaturation (*cis* configurations predominate), and all the non-phenolic moieties exhibit C_{27} hydrocarbon chains.

Structural considerations suggest that phenoxy bonds arise from the reaction of phenol functions with epoxides: either a 9, 10-epoxyalk-26-ene or a 5,6-epoxyalkatri-1,3,18-ene. These two putative precursors would be derived from the corresponding C_{27} dienes and tetraenes identified in the hydrocarbon fraction (Metzger, 1993). Moreover the accumulation of heavy ether lipids **75–77**, to the detriment of lower ones is in favour of a multistep pathway rather than a cascade mechanism in their biosynthesis. Furthermore, in the synthesis of **77** which is alkenylresorcinol-O-[alkenyl-O-alkatrienyl]-C-[alkatrienyl] diether it seems likely that C-alkylation of the resorcinol ring occurs, via a nucleophilic substitution as illustrated in Figure 10.23.

Subsequent chemical investigations of the collection strain of Austin led to isolation of four other series of phenoxy ether lipids (ca 0.3 per cent). Each series, **78–81** (Figure 10.24), contains an alkenylresorcinol moiety, ether linked by phenoxy bonds

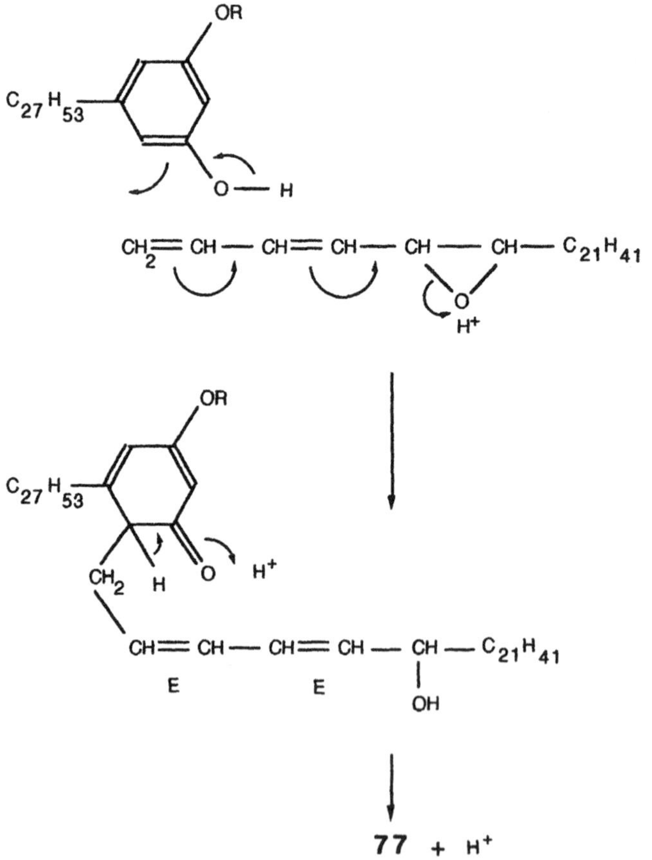

Figure 10.23 Proposed mechanism for the C-alkylation leading to **77**; R: alkenyl-O-alkatrienyl subunit (redrawn from Metzger (1994) with kind permission from Elsevier Science Ltd).

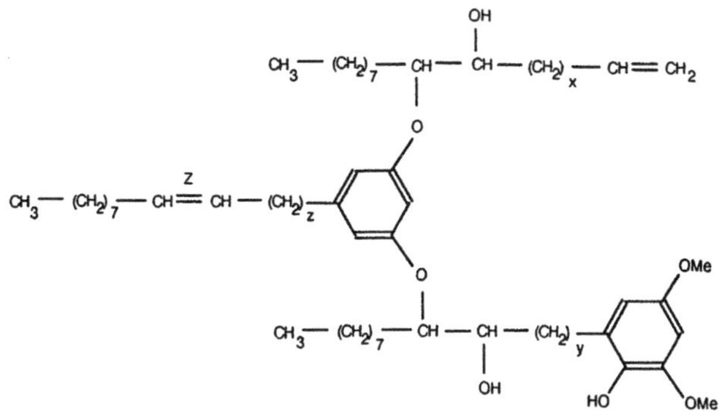

x=odd, 15-19
y=odd, 17-21
z=odd, 15-19

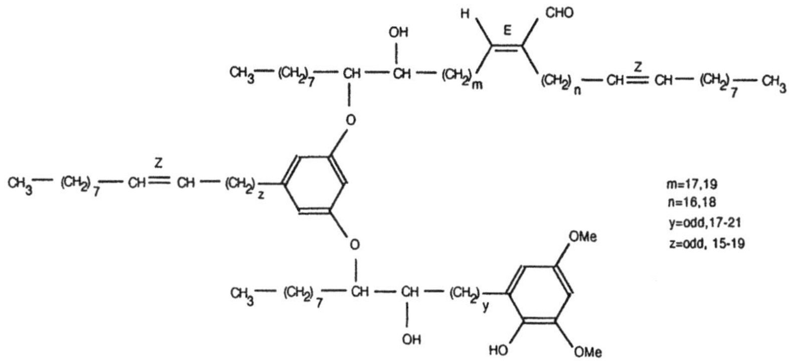

78

m=17,19
n=16,18
y=odd,17-21
z=odd, 15-19

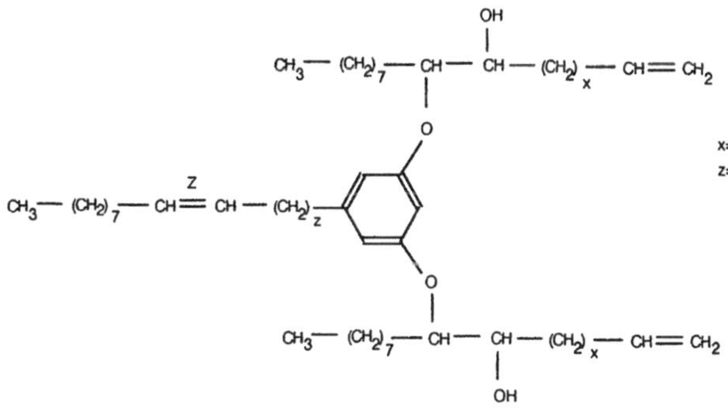

79

x=odd, 13-19
z=odd, 15-19

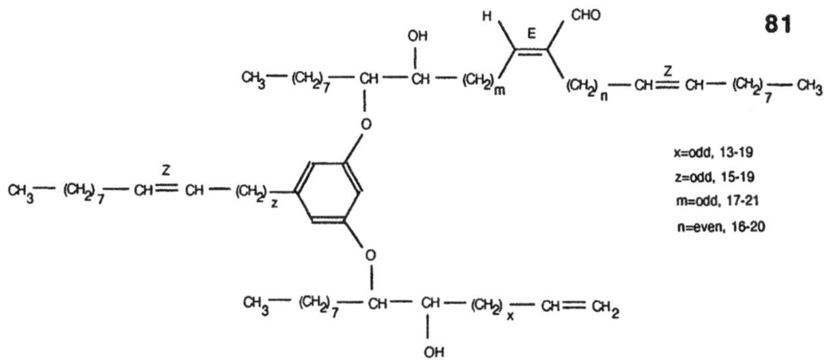

80

x=odd, 13-19
z=odd, 15-19
m=odd, 17-21
n=even, 16-20

81

Figure 10.24 Phenoxy ether lipids of the Austin strain (A race).

to two alkyl chains derived from alkadienes (**78, 80** and **81**), alkenylhydroquinols (**78** and **79**) and botryals (**79** and **81**) (Metzger and Pouet, 1995).

Ether lipids of the L race

Lycopanerols (Figure 10.25) form another class of high molecular weight ether lipids recently identified in all strains of the L race. These molecules are hydroxylated polyethers comprising two tetraterpene skeletons linked by an ether bridge.

Lycopanerols A, **82** (Metzger and Aumelas, 1997), and B, **83** (Metzger, unpublished results), are the two structurally complex members of this family isolated and identified so far; they were recovered from external lipids by CC and TLC techniques with a yield of 12.6 per cent and 0.6 per cent of dry weight, respectively. Their molecular formulae were determined by FABMS. The planar structures were established by means of ^{1}H and ^{13}C NMR, ^{1}H-^{1}H COSY, ^{1}H-^{13}C HETCOSY, EIMS and EIMS of the trimethylsilyl derivatives. The structure of lycopanerols, **82** was also investigated by chemical methods: hydrogenation, hydrogenolysis, ether cleavage and ozonolysis. Compounds of both series **82** and **83** exhibit a tetrahydrofuran (THF) ring in one tetraterpene moiety and a tetrahydropyran (THP) ring in the other; the THF moiety is α-hydroxylated and the THP ring bears, via ether linkage, an even *n*-long chain, C_{28}, C_{30} or C_{32} (major), containing one ω9,10 double bond in **82** and one epoxide at this same position in **83**.

Polyethers from natural sources constitute a wide family of compounds; representative examples are the dinoflagellate toxins such as brevetoxins (Lin *et al.*, 1981), the THF-acetogenins from *Annonaceae* (Zafra-Polo *et al.*, 1996) or the THF, THP-squalene derivatives from the red seaweeds of the genus *Laurencia* (Suzuki *et al.*, 1985; Sakemi *et al.*, 1986). All the biogenetic pathways proposed for these ethers involve epoxides as precursors (Lee *et al.*, 1989; Rupprecht *et al.*, 1990; Suzuki *et al.*, 1993). In the present case, a diepoxylycopane derived from lycopadiene, **25** could be

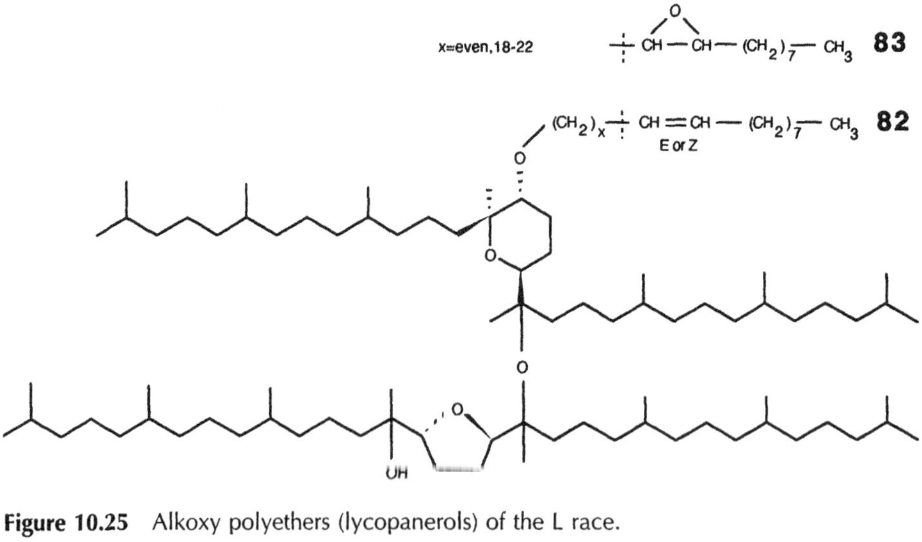

Figure 10.25 Alkoxy polyethers (lycopanerols) of the L race.

the precursor. The formation of C_{40}-THF and C_{40}-THP compounds by hydrolysis of diepoxylycopane catalyzed by perchloric acid supports this assumption.

Miscellaneous

Monooleyl esters of aliphatic diols, **84** and **85**, are minor lipids of the strain of Coat ar Herno (A race), accounting for 1.2 and 0.2 per cent of dry weight, respectively (Metzger and Pouet, 1995). The former are structurally related to C_{27} and C_{29} alkadienes, **2**, while the second are derived from C_{56}–C_{62} botryals, **42**, with the corresponding epoxides as probable precursors.

Botryococcenone, **86** is the sole ketone, the ketocarotenoids excepted, identified in *B. braunii* so far. It was isolated from an algal sample collected in the Australian lake of Coolamatong (Summons and Capon, 1991). Structurally related to a C_{33} botryococcene, it comprises however only five carbon–carbon double bonds.

Macromolecular lipids and algaenans

Macromolecular lipids

Lipid polymers of very high molecular weights, extractable by some organic solvents, occur in all strains of the three chemical races of *B. braunii*. These polymers of a rubbery consistency are isolated from freshly dried biomass by extraction with solvents of low polarity such as chloroform or dichloromethane. Addition of methanol or acetone to the extracts led to the formation of elastic precipitates which are purified by repeated solution in CHCl$_3$ and subsequent addition of MeOH. Through analysis by size exclusion HPLC on ultrastyragel columns, using THF as eluent, they show broad peaks corresponding to wide mass distributions ranging from 10^4 to 10^7 daltons with maxima of 1.4×10^5, 2.7×10^5 and 2.4×10^5 daltons for the A, B and L races, respectively.

The polymer of the A race was the first investigated (Metzger *et al.*, 1993); further it appeared as the simplest. The combination of data obtained from elemental and spectroscopic analyses and ozonolysis showed the existence of a repetitive C$_{32}$ monomer unit. The structure of the polymer, **87,** is based on an α-branched, α-unsaturated aldehyde similar to that observed in botryals, but differing by the presence of two *cis* double bonds (Figure 10.26). With regard to the biosynthesis of this polymer, named aliphatic polyaldehyde, the results of feeding experiments with [1-^{14}C] and [10-^{14}C] oleic acid showed that oleic acid is a precursor. Moreover the data of the analyses of the diacids recovered after ozonolysis of the ^{14}C-labelled polymer suggested, as illustrated in Figure 10.27, that the biosynthesis would proceed via elongation of oleic acid to an *n*-C$_{32}$ acid, which then would undergo an ω-oxidation. Afterwards

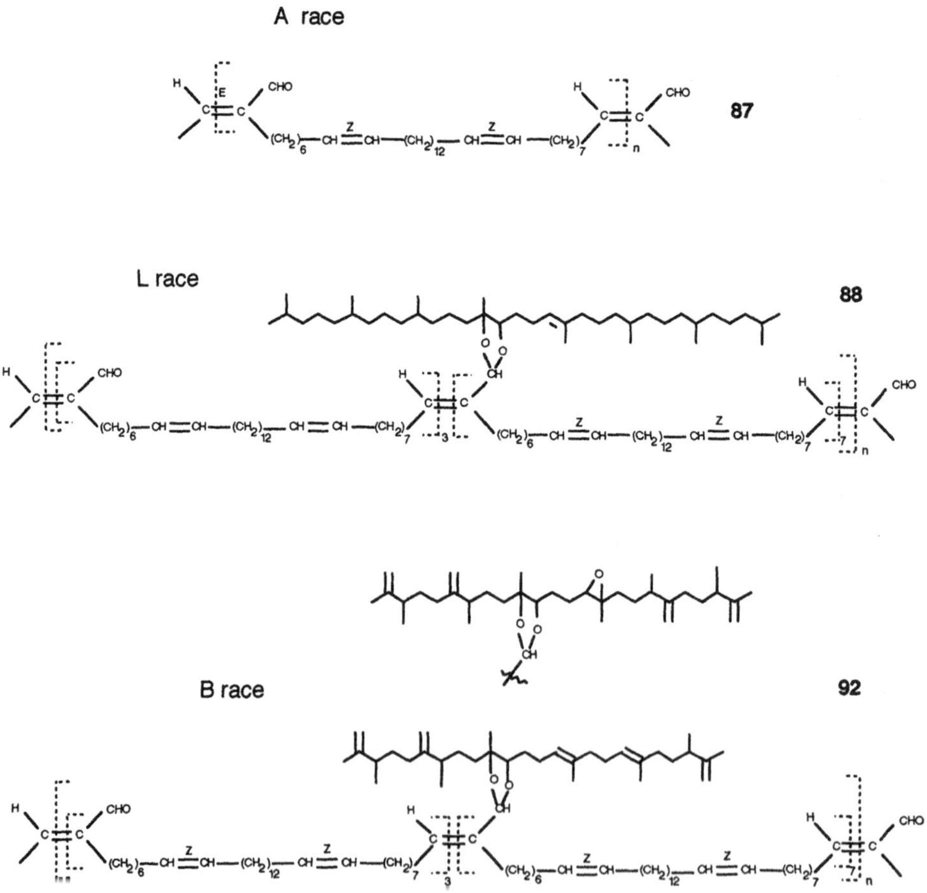

Figure 10.26 Structures of the macromolecular lipids of the three races of *B. braunii.*

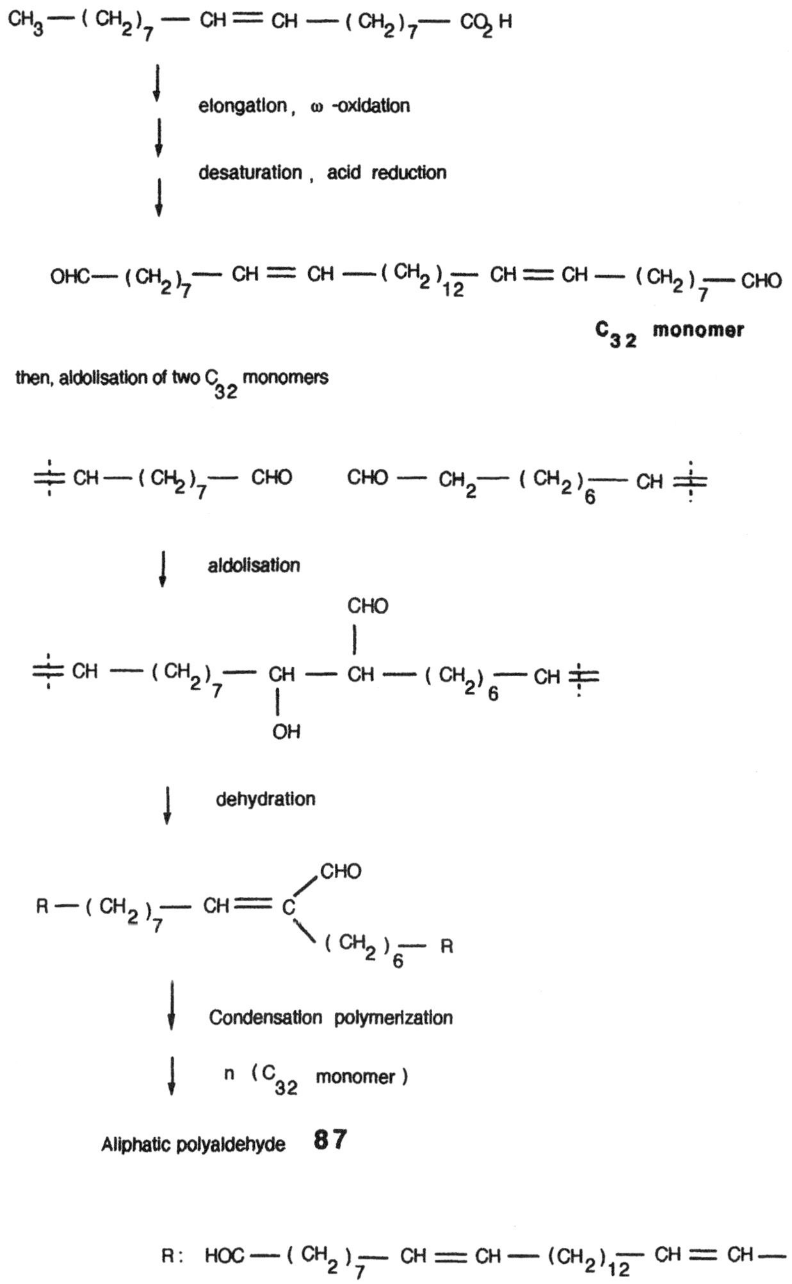

$$CH_3 - (CH_2)_7 - CH = CH - (CH_2)_7 - CO_2H$$

↓ elongation, ω-oxidation

↓ desaturation, acid reduction

$$OHC - (CH_2)_7 - CH = CH - (CH_2)_{12} - CH = CH - (CH_2)_7 - CHO$$

C_{32} monomer

then, aldolisation of two C_{32} monomers

$$\doteqdot CH - (CH_2)_7 - CHO \qquad CHO - CH_2 - (CH_2)_6 - CH \doteqdot$$

↓ aldolisation

$$\doteqdot CH - (CH_2)_7 - CH - CH - (CH_2)_6 - CH \doteqdot$$

with CHO on the CH and OH below

↓ dehydration

$$R - (CH_2)_7 - CH = C \begin{cases} CHO \\ (CH_2)_6 - R \end{cases}$$

↓ Condensation polymerization

↓ n (C_{32} monomer)

Aliphatic polyaldehyde **87**

$$R: \quad HOC - (CH_2)_7 - CH = CH - (CH_2)_{12} - CH = CH -$$

Figure 10.27 Proposed biosynthetic pathway of the aliphatic polyaldehyde, **87**, of *B. braunii* (redrawn from Metzger *et al.* (1993) with kind permission from Elsevier Science Ltd).

the repeated condensation of the derived C_{32} dialdehyde would result in aldols which would be dehydrated.

The structures of the polymers extracted from the B and L races derive from **87** by acetalization of approx. seven aldehyde functions out of ten with terpene diols

(Berthéas, 1994; Delahais, 1995). Indication for the presence of acetals arose from the observation in the ^{13}C NMR spectrum, of carbon signals typical of these functions at ca δ 105. Confirmation was given by acid hydrolysis of these two polymers in THF/aqueous HCl. In both cases, on the one hand a rubbery material was recovered which only exhibited spectroscopic properties identical to those observed with the aliphatic polyaldehyde of the A race, **87**, and, on the other hand, terpene diols were isolated. In the hydrolysis of the polymer of the L race the diol was found to be an acyclic tetraterpene, the dihydroxy-lycopaene **58**, that suggested structure **88** for the polymer (Figure 10.26). Several triterpenes were released from the hydrolysis of the polymer of the B race: all are structurally related to methylated squalenes and derive from 10,11-epoxides **53–56**. The structures of two major triterpenes were elucidated: the 10,11-dihydroxy, 3,7,22-trimethyl squalene, **89** and, surprisingly, the chlorinated tetramethylsqualene triol, **90**. Based on the presence in the NMR spectra of the native polymer of signals characteristic of epoxide functions, this last compound **90** was assumed to derive very likely from the epoxy-dihydroxy tetramethyl squalene, **91** via opening of the epoxide function by HCl during hydrolysis. From these data, structure **92** was proposed for the complex polymer extracted from the B race (Figure 10.26). The existence in this polymer of a terpene moiety bearing an epoxide is of importance for a possible cross-linkage with some free aldehyde function, via hydrolysis and then acetalization. Thus we can easily imagine a possible extension and/or reticulation of the polymer network. Moreover such cross-linkage may also take place in the polymer of the L race, via epoxidation of the tetraterpene moiety.

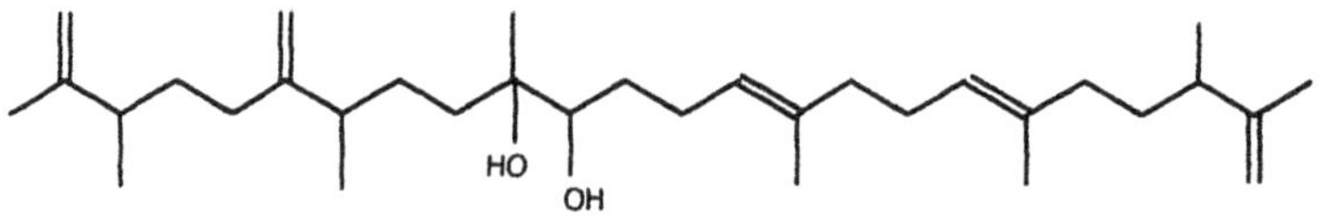

89

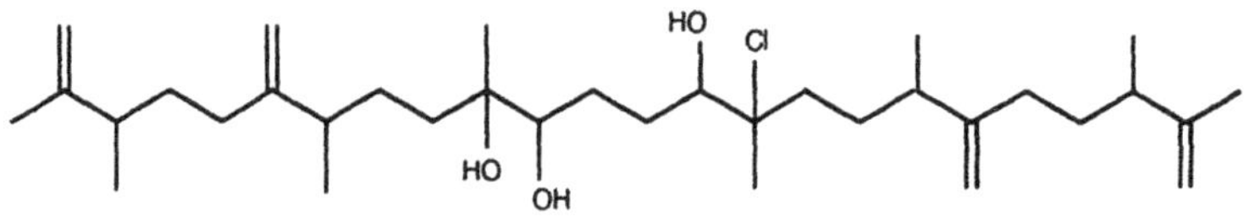

90

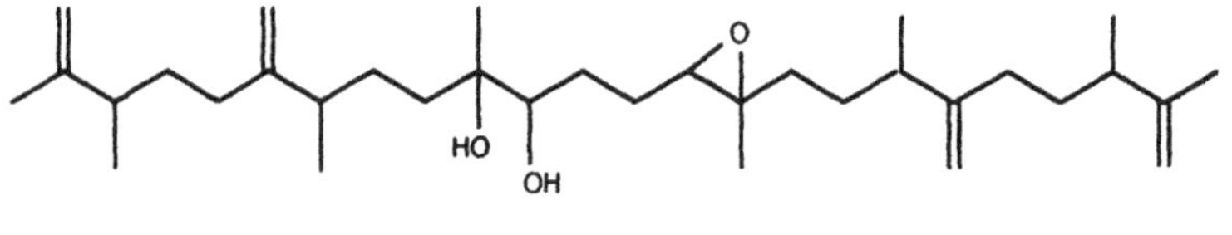

91

Algaenans

After lipid extraction, drastic treatment of *B. braunii* biomass, under basic and then acid conditions, furnishes an organic residual material. Examination by electron microscopy showed that the cell content of the algae is degraded, while the outer walls exhibit an unchanged organization (Figure 10.1). The occurrence of such chemically resistant and insoluble biopolymers in some microalgae is now well-known; these materials have been termed algaenans (Tegelaar *et al.*, 1989).

The occurrence of a chemically resistant polymer in *B. braunii* was first established in the A race (Berkaloff *et al.*, 1983), using KOH–aqueous MeOH and then concentrated H_3PO_4 in order to eliminate hydrolyzable constituents. The gross structural features of algaenan A were identified using FTIR and high resolution solid state ^{13}C NMR spectroscopy. The observation in the IR spectrum of a substantial absorption at 720 cm^{-1} and in the ^{13}C NMR spectrum of a predominant peak at δ29 was indicative of the presence of long methylenic chains. The similarity of the chemical structures of insoluble algaenan A and of soluble aliphatic polyaldehyde of the A race, **87**, was revealed by analytical pyrolysis (Gelin *et al.*, 1994a). Both materials yield identical series of pyrolysis products, mainly *n*-alkanes, *n*-alkenes and *n*-alkyl-cyclohexanes, with similar distributions. Accordingly it was concluded that the molecular structure of algaenan A is probably a more condensed and/or a reticulated form of the soluble aliphatic polyaldehyde **87**.

From elemental analysis, FTIR, solid state ^{13}C NMR spectroscopy and GCMS analysis of pyrolysis products, the chemically resistant biopolymer isolated from the B race exhibited very similar structural features when compared to the one issued from the A race (Kadouri *et al.*, 1988). Only a slightly higher level of methyl-branching was observed that could suggest the presence of terpenoid compounds in the polymeric network. Recently, a modified procedure was used for the isolation of algaenan B: it comprises the hydrolysis of the remaining biomass by aqueous trifluoroacetic acid and then concentrated HCl in THF solution, instead of phosphoric acid; these treatments were found to allow a better cleavage of hydrolyzable functions hindered by aliphatic chains (Allard *et al.*, 1997). Spectroscopic investigations of algaenan B obtained by this procedure clearly indicate a similarity with aliphatic polyaldehyde **87** (Allard, unpublished results).

The resistant polymer isolated from the L race appeared at first of a very different nature (Derenne *et al.*, 1989): spectroscopic analyses showed an important contribution of methyl groups relative to the polymethylenic chains. Moreover GCMS analysis of the pyrolysis products allowed identification of numerous terpenoid compounds present in high amounts, among them C_{40} isoprenoid hydrocarbons and ketones (Derenne *et al.*, 1990). According to their structures, these tetraterpenoids were considered as directly related to lycopadiene **25**, and it was assumed that C_{40} ketones would result from the thermal cleavage of C–O–C links. Thus it was hypothesized that the resistant polymer of the L race would comprise mainly C_{40} units, probably linked by ether bridges (Derenne *et al.*, 1990; Gelin *et al.*, 1994b). Application of the new isolation process leads however to a reconsideration of this structural statment (Berthéas *et al.*, 1997). Indeed, the FTIR spectrum of algaenan L recently isolated by this way, exhibits spectroscopic features similar to those observed with algaenans A and B: occurrence of long $(CH_2)_n$ alkyl chains (absorption at 720 cm^{-1}) and of α-unsaturated aldehyde functions (absorptions at 2706 and 1690 cm^{-1}). The only difference lies in small absorptions due to methyl branching. GCMS analyses of the pyrolysis products of this algaenan L and of their derivatives, obtained by catalytic hydrogenation, corroborate these observations. Thermal cleavage predominantly releases the same compounds as those identified after pyrolysis of algaenan A; furthermore the contribution of C_{40} hydrocarbons and ketones was lower than that observed by Derenne *et al.* (1990) (Metzger *et al.*, unpublished results). These results strongly suggest a close relationship between algaenan L and the soluble polymer **88**: algaenan L very likely arises from the hydrolysis of acetal functions present in a more cross-linked form of polymer **88**. Identification of dihydroxy-lycopaene, **58** as the dominant component in the products recovered from acid treatments supports this assumption. The observation of C_{40} hydrocarbons and ketones in the pyrolysis products could be related to an incomplete hydrolysis of some acetal functions during the isolation process of algaenan L used in the first studies.

Furthermore, we can speculate that algaenan B probably corresponds to a more cross-linked or reticulated form of the soluble polymer **92** isolated from the B race. Thus it appears that the algaenans from the three races of *B. braunii* exhibit the same basic structure: an aliphatic polyaldehydic network.

Structural elements of the outer walls of *B. braunii*, i.e. the algaenans, could also act as 'sponges' towards the external lipids. Indeed we observed that dried *Botryococcus* algaenans can strongly increase in volume when immersed in some solvents of low polarity such as $CHCl_3$. The absorption of lipids into the cross-linked polymers would be at the origin of the hydrophobic matrix which ensures the colony cohesion. Moreover the chemical properties of *B. braunii* algaenans are responsible, on the one hand, for the preservation of the morphology of the alga during fossilization and, on the other hand, for the very high oil potential of the derived algal coals (Largeau *et al.*, 1984, 1986).

Polysaccharides

In comparison with lipids and algaenans, little is known about *B. braunii* polysaccharides even though very high amounts are released into the growth medium. A first indication for the production of polysaccharides arose from increasing viscosity of the medium during growth of a strain of the A race (Casadevall *et al.*, 1985). Since then, five strains were investigated, two of both A and B races, and one of the L race: the production of extracellular polysaccharides ranges from 0.25 to 1 g per litre of medium (Allard and Casadevall, 1990). The molecular weights were determined by gel filtration over a Sephacryl column. Most of these polymers are highly polydispersed; thus the extracellular polysaccharides of the Austin strain exhibit *Mr* at maximum concentrations of 2×10^6, 3×10^5 and 6×10^4 daltons.

The crude polysaccharides were subjected to complete acid hydrolysis and the resulting sugars examined, as suitable derivatives, by GCMS analysis. The sugar composition of the polysaccharides of four strains of *B. braunii* is summarized in Table 10.4. Galactose is the main component in the three races, and all the other identified monosaccharides often occur in the extracellular polysaccharides of green microalgae, the O-methylated sugars excepted. Distinct features are: the co-occurrence, in high amounts, of glucose and galactose in the polysaccharides of the L race, only reported in *Ankistrodesmus densus* (Vieira and Myklestad, 1986); and the presence of 3-OMe fucose in the B race, never identified in algae and of 3-OMe rhamnose in the A and B races, only observed in the polysaccharides of blue–green algae (Madhavi Shekharam *et al.*, 1987).

Biotechnological studies

As described in the previous sections, *B. braunii* is characterized by a conspicuous production of long chain lipidic compounds exhibiting a high diversity in chemical

Table 10.4 Sugar composition of the extracellular polysaccharides of *B. braunii* (relative percentages).

	A race		B race La Manzo	L race Yamoussoukro
	Austin	Coat ar Herno		
Xylose	1.1	4.5	5.2	tr.
Arabinose	4.9	9.1	7.6	4.1
Rhamnose	2.2	1.8	19.5	tr.
3-OMe rhamnose	1.5	12.7	4.9	–
Fucose	13.6	32.4	9.1	31.7
3-OMe fucose	–	–	13.7	–
Manose	tr.	tr.	tr.	tr.
Glucose	tr.	tr.	tr.	17.7
Galactose	72.1	36.3	37.1	38.7
Glucuronic acid	2.2	tr.	tr.	6.1
n.d.	2.2	3.2	3.1	1.7

Reprinted from Allard and Casadevall (1990) with kind permission from Elsevier Science Ltd. n.d., not determined; tr., trace amounts.

structure. A number of studies were therefore carried out so as to examine the various factors controlling the production of such compounds, especially hydrocarbons. Most of these studies were carried out on the A race.

The first considered topic was the relationship between the formation of hydrocarbons and the physiological state of *B. braunii* cells. It is well documented that some green microalgae like *Chlorella* sp. (Iwamoto and Sugimoto, 1955) can accumulate very large amounts of fatty acid derivatives, mostly as triacylglycerols: up to 80 per cent of total algal dry biomass in stationary cultures. Such high accumulations are observed when cell divisions are inhibited due to nutrient depletion, especially upon nitrogen starvation. Under these conditions, the bulk of the carbon fixed by photosynthesis is diverted towards fatty acid production. As already stressed, the biosyntheses of *B. braunii* fatty acids and hydrocarbons are related in the A race. Accordingly, it was important to determine whether the very large hydrocarbon contents noted in some *B. braunii* cultures reflected the same process as fatty acid accumulations. The level of hydrocarbons in *B. braunii* (as percentage of total biomass) and their productivity (amount of hydrocarbons produced per litre of culture and per day) were therefore determined, through batch experiments, for physiological states ranging from exponentially growing cultures to late stationary ones (Casadevall *et al.*, 1985). Hydrocarbons were shown to be produced at any physiological state. Moreover, hydrocarbon productivity was at a maximum during exponential growth phase. It thus appeared that a large production of hydrocarbons does not reflect a diversion taking place when the other metabolic pathways are strongly inhibited. Indeed, such a large production is a normal feature of *B. braunii* metabolism in healthy cells. As indicated in a previous section, the bulk of *B. braunii* hydrocarbons is stored in outer walls, so that the intracellular contents can remain relatively low even for hydrocarbon-rich colonies. Moreover, studies using labelled precursors (Largeau *et al.*, 1980a) indicated that this species seems unable to catabolize its own hydrocarbons. Taken together, all the above features are consistent with the ability of *B. braunii* to accumulate extremely large amouts of hydrocarbons, although a part of these compounds is further transformed into high molecular weight lipids as dicussed later on.

Parallel studies, aimed at improving the growth rate of *B. braunii*, were also carried out. The pioneering works on *B. braunii* cultures (Chu, 1942; Belcher, 1968) showed a slow growth with mean biomass doubling times of around one week. A marked improvement of growth rate was achieved from continuously illuminated 'air-lift' cultures, where stirring and aeration are provided by CO_2-enriched air, 1 per cent v/v (Largeau *et al.*, 1980c; Casadevall *et al.*, 1985). Under such conditions, a mean doubling time of 2.3 days was obtained both for batch cultures in the exponential state and for continuous cultures. In the batch cultures the final biomass was about 6 g/l (dry weight) and the hydrocarbon content around 30 per cent. A number of other microalgae can exhibit much faster growth rates when compared to *B. braunii*, with mean doubling times of 6–8 hours. It should be noted, however, that *B. braunii* provides a biomass rich in hydrocarbons, i.e. rich in energetically expensive compounds, whereas the biomass in all these fast growing microalgae is dominated by compounds, especially proteins, which require less energy for their biosynthesis. Accordingly, doubling times of a few hours are probably impossible to be obtained in the case of *B. braunii* and the values observed in the air lift experiments are probably close to the maximum growth rate of that species.

A rapid alteration of the ultrastructure of *B. braunii* cells was noted, along the growth curve, for the batch cultures carried out under such conditions (Casadevall *et al.*, 1985). This alteration appears as early as the beginning of the progressive deceleration stage. Such an early degeneration results in cytoplasm and chloroplast disorganization, accumulation of starch grains between thylakoids and bleaching. A complete degeneration appears rather rapidly in the late stationary stage. In sharp contrast, a high resistance of *B. braunii* to stress (anaerobic conditions, continuous illumination and prolonged darkness) was noted in other experiments. However the latter were carried out under conditions (Chu, 1942; Belcher, 1968) where the growth of *B. braunii* is much slower when compared to the air-lift cultures. This degeneration process is very rapidly reverted when the cells, at the beginning of the stationary stage, are inoculated into fresh medium. Most of the subsequent studies were therefore based on these air-lift 'standard' conditions. Their purpose was to derive information on the influence of various culture parameters on *B. braunii* total growth and hydrocarbon production.

Studies on nutrients showed that phosphate is not limiting (Casadevall *et al.*, 1985) since increases in initial phosphate concentration have no effect on growth. However, the hydrocarbon contents are substantially increased (ca 25 per cent) in a phosphate-rich medium, although the nature and the relative abundance of the hydrocarbons do not change. This higher hydrocarbon production reflects the increase in the phosphate/nitrate ratio in the growth medium since such a ratio is known to influence lipid content in a number of microalgae. However, as observed for the influence of nitrogen starvation, the changes observed in the latter species were mainly concerned with fatty acid derivatives whereas hydrocarbons are increased in the case of *B. braunii*.

A higher initial concentration of nitrate in the growth medium resulted in a longer exponential phase (Brenckmann *et al.*, 1985a) and hence in a higher final biomass, thus indicating that nitrogen was limiting in the 'standard' medium. Due to the decrease in the phosphate/nitrate ratio, the biomass obtained from these nitrate-rich cultures exhibits a lower hydrocarbon content. However, the increase in total biomass outweighs this variation in content so that the hydrocarbon production is substantially improved (ca 25 per cent). On the contrary, ammonia is not a suitable nitrogen source for *B. braunii*. In fact, the addition of 1M NH_4Cl to the medium results in an almost complete inhibition of hydrocarbon production, along with marked decreases in respiration and photosynthetic activity (Ohmori *et al.*, 1984).

CO_2 limitation was shown to exert large detrimental effects, still more pronounced for the hydrocarbons than for the total biomass (Chirac *et al.*, 1985). Thus, when air enriched with 1 per cent CO_2 is replaced by non-enriched air, in standard air-lift cultures, a fivefold decrease is observed for hydrocarbon production against a four-fold one for total biomass. *B. braunii* was also grown with various sources of organic carbon added to the growth medium (Weetall, 1985). A limited number of the tested compounds, such as glucose, were shown to improve biomass and hydrocarbon production significantly in illuminated cultures. Acetate appears to stimulate the carbon flux to lipid formation (Tenaud *et al.*, 1989).

Studies on light intensity showed that a broad range of irradiances (40–180 $W.m^{-2}$) is acceptable for *B. braunii* (Cepák and Lukavsky, 1994). It was also observed that irradiance influences the composition of the biomass. Thus the formation of starch, of hydrocarbons and of exopolysaccharides is favoured by a low, a medium and a high irradiance, respectively. Comparisons of batch cultures grown under 'standard' air-lift conditions, continuously illuminated with different intensities, showed the same optimum irradiance for both total biomass and hydrocarbon production (Brenckmann *et al.*, 1985b). The detrimental effects induced by a high irradiance (photoinhibition) still appear more pronounced for hydrocarbons than for total biomass.

Finally, the influence of biotic parameters was examined from cultures carried out under standard air-lift conditions (Chirac *et al.*, 1985). Comparisons of parallel cultures showed important variations in both total biomass and hydrocarbon production between various axenic strains of the A race. A tenfold difference in the amount of formed hydrocarbons was thus noted between the least and the most productive strain. The influence of selected microorganisms on *B. braunii* was also examined. To this end, associated cultures were performed by inoculating, with the alga, bacteria or yeasts commonly occurring in aquatic environments. Sharply different influences, ranging from a pronounced inhibition (observed for most of the tested species, e.g. with *Pseudomonas oleovorans*) to a large beneficial influence (obtained with *Flavobacterium aquatile*), were thus noted both on total biomass and hydrocarbons. It is extremely difficult to keep large scale cultures of microalgae axenic and most of the contaminant microbial species should exhibit, as shown above, marked detrimental effects on hydrocarbon production. Nevertheless, the presence of an associated microorganism should afford an efficient protection against such extraneous contaminations. Accordingly, associated cultures with *B. braunii* and a suitable bacteria, like *F. aquatile*, should provide both an improved hydrocarbon production and a strong reduction of contaminations by antagonistic species. Such associations should thus help to solve, in part, one of the major problems encountered in large scale cultures of *B. braunii*.

Growth of *B. braunii* was also carried out under natural illumination (Gudin and Chaumont, 1981, 1984). Tubular photobioreactors with sunlight trapping areas of 1 and 6 m^2, corresponding to culture volumes up to 200 l, were used in such experiments. In these continuous, fully automated cultures all the parameters (temperature, CO_2 input, pH and nutrient concentration) are controlled and light is the limiting factor. These experiments allowed the feasibility of hydrocarbon production to be tested via large scale cultures of *B. braunii*. The ability of *B. braunii* to grow in saline media was also demonstrated (Aaronson *et al.*, 1980). Finally, the feasibility of using secondarily treated sewage for *B. braunii* cultures was recently examined (Sawayama *et al.*, 1994). Such a medium was thus shown to provide suitable conditions for continuous growth of *B. braunii*, along with removal of nitrate and phosphate from waste-water.

In addition to the above experiments, concerned with free cultures, parallel studies were carried out on batch cultures of immobilized cells of *B. braunii*. Entrapment in calcium alginate beads (Bailliez *et al.*, 1985, 1986) was shown to result in: (1) an increase in chlorophyll content of the algal biomass along with an increase in chlorophyll photosynthetic activity; (2) a delayed degeneration, probably reflecting the protection of the entrapped cells, against photoinhibition, by gel shadowing; (3) a decrease in the rate of biomass production, probably due to steric constraints hampering colony growth in the alginate gel; and (4) a substantial improvement (around 20 per cent) in hydrocarbon production relative to free control cultures (*B. braunii* metabolism is still more strongly diverted towards hydrocarbon biosynthesis than in controls, as a result of two of the features just mentioned, i.e. a lower production of total biomass whereas photosynthetic CO_2 fixation is higher). Immobilization of *B. braunii* with polyurethane foams (Bailliez *et al.*, 1988) revealed some drawbacks related to foam toxicity and cell release in the liquid medium.

Simple and efficient recovery of hydrocarbons is also required for the development of a large scale production. As already indicated, the bulk of *B. braunii* hydrocarbons is stored in outer walls along with other lipids. The oily content of such walls can be partly recovered by filter press (Weetall, 1985). Recovery by biocompatible solvents was also considered (Frenz *et al.*, 1989a, 1989b). For free cells, the highest yield was achieved via short contact of hexane with algae concentrated by filtration. A large fraction of the hydrocarbons (ca 70 per cent) can thus be recovered, without impairing cell viability, even after repeated extractions. When applied to cells immobilized in calcium alginate beads or polyurethane foams, contact with hexane results in recovery yields higher than those obtained with free cells.

Concluding remarks

A wide variety of chemicals, mainly lipids, have been isolated and identified from a large number of strains or wild samples of *B. braunii*. Differences in lipid composition have been observed not only between the three chemical races but also between strains belonging to the same race. A comparative overview of the relative abundances of various chemicals is given in Table 10.5 for three strains of the A race and for one strain of each of the B and L races. So far many of these lipids appear not only species-specific but also unique to a chemical race: botryococcenes, methylated squalenes, botryals, botryaxanthin, alkoxy and phenoxy ether lipids or acetalized forms of polyaldehydes.

The formation of large amounts of algaenan, building up the outer walls and originating from extended cross-linking of aliphatic polyaldehydes, accounts for the geochemical importance of *B. braunii*. Owing to their extremely high resistance to degradation, algaenans remain virtually unaffected during fossilization so that the typical organization of *B. braunii* colonies can be retained and deposits with a very

Table 10.5 Abundances of the main types of lipids and of algaenans in some strains of *B. braunii* (as % of dry weight).

Compounds	A race			B race	L race
	Austin	Coat ar Herno	Overjuyo (strain no. 7)	La Manzo	Yamoussoukro
Total lipids	63	71	62	53	35
Hydrocarbons	8.8	1.6	0.4	32	3
Triacylglycerols	6.3	1.9	1.7	n.d.	n.d.
Carotenoids	n.d.	n.d.	n.d.	0.3	0.4
Epoxides	3.8	–	–	0.8	0.3
Aldehydes	11.1	6.5	n.d.	n.d.	n.d.
Phenols	1.6	0.3	0.1	tr.	tr.
Alkoxy ether lipids	4.5	30	–	n.d.	13.2[1]
Phenoxy ether lipids	0.4	–	35	–	–
Sterols	0.1	n.d.	n.d.	0.2	0.2
Extractable polymer	5.1	2.4	3.1	1.3	15.2
Algaenans	5	n.d.	n.d.	5	12

n.d., not determined; tr., trace amounts; –, not identified so far.
[1]Lycopanerols.

high oil potential are formed. In addition, botryococcanes, the fully hydrogenated derivatives of botryococcenes, are specific biomarkers of the B race of *B. braunii* in petroleums. Indeed, such long-chain isoprenoid compounds are the only species-specific biomarkers so far identified in crude oils.

All the above features explain why *B. braunii* is probably the microalga in which lipidic composition has been the most extensively studied. The feasibility of the production of chemicals via large scale cultures of this species has also been considered and extensive biotechnological studies have been carried out for this purpose.

REFERENCES

AARONSON, S., BERNER, T. and DUBINSKY, Z., 1980, Microalgae as a source of chemicals and natural products, in Shelef, G. and Soeder C.J. (Eds), *Algae Biomass*, pp. 575–601, Elsevier/North-Holland Biomedical Press.

AARONSON, S., BERNER, T., GOLD, K., KUSHNER, L., PATNI, N.J., REPAK, A. and RUBIN, D., 1983, Some observations on the green planktonic alga *Botryococcus braunii* and its bloom form, *Journal of Plankton Research*, **5**, 693–700.

ALLARD, B. and CASADEVALL, E., 1990, Carbohydrate composition and characterization of sugars from the green microalga *Botryococcus braunii*, *Phytochemistry*, **29**, 1875–1878.

ALLARD, B., TEMPLIER, J. and LARGEAU, C., 1997, Artifactual origin of mycobacterial bacterans. Formation of melanoidin-like artifact macromolecular material during the usual isolation process, *Organic Geochemistry*, **26**, 691–703.

AMICO, V., 1995, Marine brown algae of family *Cystaseiraceae*: chemistry and chemotaxonomy, *Phytochemistry*, **39**, 1257–1279.

BAILLIEZ, C., LARGEAU, C. and CASADEVALL, E., 1985, Growth and hydrocarbon production of *Botryococcus braunii* immobilized in calcium alginate gel, *Applied Microbiology and Biotechnology*, **23**, 99–105.

BAILLIEZ, C., LARGEAU, C., BERKALOFF, C. and CASADEVALL, E., 1986, Immobilization of *Botryococcus braunii* in alginate: influence on chlorophyll content, photosynthetic activity and degeneration during batch cultures, *Applied Microbiology and Biotechnology*, **23**, 361–366.

BAILLIEZ, C., LARGEAU, C., CASADEVALL, E., YANG, L.W. and BERKALOFF, C., 1988, Photosynthesis, growth and hydrocarbon production of *Botryococcus braunii* immobilized by entrapment and adsorption in polyurethane foams, *Applied Microbiology and Biotechnology*, **29**, 141–147.

BARROW, R.A. and CAPON, J., 1991, Alkyl and alkenylresorcinols from an Australian marine sponge, *Haliclona* sp. (Haplosclerida: Haliclonidae), *Australian Journal of Chemistry*, **44**, 1393–1405.

BELCHER, J.H., 1968, Notes on the physiology of *Botryococcus braunii* Kützing, *Archiv für Mikrobiologie*, **61**, 335–346.

BERKALOFF, C., CASADEVALL, E., LARGEAU, C., METZGER, P., PERACCA, S. and VIRLET, J., 1983, The resistant polymer of the walls of the hydrocarbon-rich alga *Botryococcus braunii*, *Phytochemistry*, **22**, 389–397.

BERTHÉAS, O., 1994, Etude d'un nouveau type de polymère à haut poids moléculaire isolé de l'algue *Botryococcus braunii* (race L), unpublished DEA, Université Pierre et Marie Curie, Paris.

BERTHÉAS O., DELAHAIS, V., METZGER, P. and LARGEAU, C., 1997, Acetal-containing macromolecular lipids in *Botryococcus braunii* (B and L races). Chemical structure and relationship with algaenans, in *Organic Geochemistry*, EAOG meeting, part II, pp. 855–856, Forschungzentrum Jülich, Germany.

BLACKBURN, K.B., 1936, A reinvestigation of the alga *Botryococcus braunii* Kützing, *Transactions of the Royal Society of Edinburgh*, **58**, 845–854.

BRASSELL, S.C., EGLINTON, G. and FU JIA MO, 1986, Biological marker compounds as indicators of the depositional history of the Maoming oil shale, *Organic Geochemistry*, **10**, 927–941.

BRENCKMANN, F., LARGEAU, C., CASADEVALL, E. and BERKALOFF, C., 1985a, Influence de la nutrition azotée sur la croissance et la production d'hydrocarbures de l'algue unicellulaire *Botryococcus braunii*, in Palz, W., Coombs, J. and Hall, D.O. (Eds), *Energy from Biomass*, pp. 717–721, London: Elsevier Applied Science Publishers.

BRENCKMANN, F., LARGEAU, C., CASADEVALL, E., CORRE, B. and BERKALOFF, C., 1985b, Influence of light intensity on hydrocarbon and total biomass production of *Botryococcus braunii*. Relationships with photosynthetic characteristics, in Palz, W., Coombs, J. and Hall, D.O. (Eds), *Energy from Biomass*, pp. 722–726, London: Elsevier Applied Science Publishers.

BRENNAN, P.J., 1988, Mycobacterium and other actinomycetes, in Ratledge C. and Wilkinson, S.G. (Eds), *Microbial Lipids*, Vol. 1, pp. 203–298, London: Academic Press.

BROWN, A.C., KNIGHTS, B.A. and CONWAY, E., 1969, Hydrocarbon content and its relationship to physiological state in the green alga *Botryococcus braunii*, *Phytochemistry*, **8**, 543–547.

CANE, R.F., 1969, Coorongite and the genesis of oil shale, *Geochimica et Cosmochimica Acta*, **33**, 257–265.

CANE, R.F., 1977, Cooroongite, Balkashite and related substances. An annoted bibliography, *Transactions of the Royal Society of South Australia*, **101**, 153–164.

CANE, R.F. and ALBION, P.R., 1973, The organic geochemistry of Torbanite precursors, *Geochimica et Cosmochimica Acta*, **37**, 1543–1549.

CASADEVALL, E., METZGER, P. and PUECH, M-P., 1984, Biosynthesis of triterpenoid hydrocarbons in the alga *Botryococcus braunii*, *Tetrahedron Letters*, **25**, 4123–4126.

CASADEVALL, E., DIF, D., LARGEAU, C., GUDIN, C., CHAUMONT, D. and DESANTI, O., 1985, Studies on batch and continuous cultures of *Botryococcus braunii*: hydrocarbon production in relation to physiological state, cell ultrastructure and phosphate nutrition, *Biotechnology and Bioengineering*, **27**, 286–295.

CEPÁK, V. and LUKAVSKY, J., 1994, The effect of high irradiances on growth, biosynthetic activities and the ultrastructure of the green alga *Botryococcus braunii* strain Droop 1950/ 807–1, *Archiv für Hydrobiologie*, Supplement-Band, **101**, 1–17.

CHAN YONG, T-P., LARGEAU, C. and CASADEVALL, E., 1986, Biosynthesis of non-isoprenoid hydrocarbons by the microalga *Botryococcus braunii*; evidence for an elongation decarboxylation mechanism; activation of decarboxylation, *Nouveau Journal de Chimie*, **10**, 701–707.

CHIRAC, C., CASADEVALL, E., LARGEAU, C. and METZGER, P., 1985, Bacterial influence upon growth and hydrocarbon production of the green alga *Botryococcus braunii*, *Journal of Phycology*, **21**, 380–387.

CHU, S.P., 1942, The influence of the mineral composition of the medium on the growth of planktonic algae. I. Methods and culture media, *Journal of Ecology*, **30**, 284–304.

COX, R.E., BURLINGAME, A.L., WILSON, D.M., EGLINTON, G. and MAXWELL, J.R., 1973, Botryococcene – a tetramethylated acyclic triterpenoid of algal origin, *Journal of Chemical Society Chemical Communications*, **12**, 284–285.

DAVID, M., 1985, Contribution à l'étude du mécanisme de biosynthèse des hydrocarbures terpéniques chez l'algue verte *Botryococcus braunii*, unpublished DEA, University Pierre et Marie Curie, Paris.

DAVID, M., METZGER, P. and CASADEVALL, E., 1988, Two cyclobotryococcenes from the B race of the green alga *Botryococcus braunii*, *Phytochemistry*, **27**, 2863–2867.

DELAHAIS, V., 1995, Etude d'un nouveau type de biopolymère d'origine algaire. Structure des composants triterpéniques et de leurs précurseurs, unpublished DEA, Université Pierre et Marie Curie, Paris.

DELAHAIS, V. and METZGER, P., 1997, Four polymethylsqualene epoxides and one acyclic tetraterpene epoxide from *Botryococcus braunii*, *Phytochemistry*, **44**, 671–678.

DENNIS, M.W. and KOLATTUKUDY, P.E., 1991, Alkane biosynthesis by decarbonylation of aldehyde catalyzed by a microsomal preparation from *Botryococcus braunii*, *Archives of Biochemistry and Biophysics*, **287**, 268–275.

DERENNE, S., LARGEAU, C., CASADEVALL, E. and BERKALOFF, C., 1989, Occurrence of a resistant biopolymer in the L race of *Botryococcus braunii*, *Phytochemistry*, **28**, 1137–1142.

DERENNE, S., LARGEAU, C., CASADEVALL, E. and SELLIER, N., 1990, Direct relationship between the resistant biopolymer and the tetraterpenic hydrocarbon in the lycopadiene-race of *Botryococcus brauni*, *Phytochemistry*, **29**, 2187–2192.

DE ROSA, S., DE GIUILIO, A., IODICE, C., ALCARAZ, M.J. and PAYA, M., 1995, Long-chain aldehydes from the red alga, *Corallina mediterranea*, *Phytochemistry*, **40**, 995–996.

DOUGLAS, A.G., EGLINTON, G. and MAXWELL, J.R., 1969, The hydrocarbons of Coorongite, *Geochimica et Cosmochimica Acta*, **33**, 569–577.

DUBREUIL, C., DERENNE, S., LARGEAU, C., BERKALOFF, C. and ROUSSEAU, B., 1989, Mechanism of formation and chemical structure of Coorongite – I. Role of the resistant biopolymer and of the hydrocarbons of *Botryococcus braunii*. Ultrastructure of Coorongite and its relationship with Torbanite, *Organic Geochemistry*, **14**, 543–553.

FRENZ, J., LARGEAU, C. and CASADEVALL, E., 1989a, Hydrocarbon recovery by extraction with a biocompatible solvent from free and immobilized cultures of *Botryococcus braunii*, *Enzyme and Microbial Technology*, **11**, 717–724.

FRENZ, J., LARGEAU, C., CASADEVALL, E., KOLLERUP, F. and DAUGULIS, A.J., 1989b, Hydrocarbon recovery and biocompatibility of solvents for extraction from cultures of *Botryococcus braunii*, *Biotechnology and Bioengineering*, **34**, 755–762.

GALBRAITH, M.N., HILLEN, L.W. and WAKE, L.V., 1983, Darwinene: a branched hydrocarbon from a green form of *Botryococcus braunii*, *Phytochemistry*, **22**, 1441–1443.

GELIN, F., LEEUW, J.W. DE, SINNINGHE DAMSTÉ, J.S., DERENNE, S., LARGEAU, C. and METZGER, P., 1994a, The similarity of chemical structures of soluble aliphatic polyaldehyde and insoluble algaenan in the green microalga *Botryococcus braunii* race A as revealed by analytical pyrolysis, *Organic Geochemistry*, **21**, 423–435.

GELIN, F., LEEUW, J.W. DE., SINNINGHE DAMSTÉ, J.S., DERENNE, S., LARGEAU, C. and METZGER, P., 1994b, Scope and limitations of flash pyrolysis-gas chromatography-mass spectrometry as revealed by the thermal behaviour of high-molecular-weight lipids derived from the green microalga *Botryococcus braunii*, *Journal of Analytical and Applied Pyrolysis*, **28**, 183–204.

GRUNG, M., METZGER, P. and LIAAEN-JENSEN, S., 1989, Primary and secondary carotenoids in two races of the green alga *Botryococcus braunii*, *Biochemical Systematics and Ecology*, **17**, 263–269.

GRUNG, M., METZGER, P. and LIAAEN-JENSEN, S., 1994a, Secondary carotenoids in the green alga *Botryococcus braunii*, race L, new strain, *Biochemical Systematics and Ecology*, **22**, 25–29.

GRUNG, M., METZGER, P., BERKALOFF, C. and LIAAEN-JENSEN, S., 1994b, Studies on the formation and localization of primary and secondary carotenoids in the green alga *Botryococcus braunii*, including the regreening process, *Comparative Biochemistry and Physiology. B. Comparative Biochemistry*, **107B**, 265–272.

GUDIN, C. and CHAUMONT, D., 1981, For a solar biotechnology based on microalgae, in Chartier, P. and Palz, W. (Eds), *Energy from Biomass*, serie E, Vol. 1, pp. 81–84, D. Reidel Publishing Company.

GUDIN, C. and CHAUMONT, D., 1984, Solar biotechnology study and development of tubular solar receptors for controlled production of photosynthetic cellular biomass for methane production and specific exocellular biomass, in Palz, W. and Pirrwitz, D. (Eds), *Energy from Biomass*, serie E, Vol. 5, pp. 184–193, D. Reidel Publishing Company.

HUANG, Z. and POULTER, C.D., 1988a, Braunicene. Absolute stereochemistry of the cyclohexane ring, *Journal of Organic Chemistry*, **53**, 4089–4094.

HUANG, Z. and POULTER, C.D., 1988b, Isobraunicene, wolficene and isowolficene. New cyclic 1'-3 fused isoprenoids from *Botryococcus braunii*, *Journal of Organic Chemistry*, **53**, 5390–5392.

HUANG, Z. and POULTER, C.D., 1989a, Stereochemical studies of botryococcene biosynthesis: analogies between 1'-1 and 1'-3 condensations in isoprenoid pathway, *Journal of the American Chemical Society*, **111**, 2713–2715.

HUANG, Z. and POULTER, C.D., 1989b, Tetramethylsqualene, a triterpene from *Botryococcus braunii* var. Showa, *Phytochemistry*, **28**, 1467–1470.

HUANG, Z. and POULTER, C.D., 1989c, Isoshowacene, a C_{31} hydrocarbon from *Botryococcus braunii* var. Showa, *Phytochemistry*, **28**, 3043–3046.

HUANG, Z., POULTER, C.D., WOLF, F.R., SOMERS, T.C. and WHITE, J.D., 1988, Braunicene, a novel cyclic C_{32} isoprenoid from *Botryococcus braunii*, *Journal of the American Chemical Society*, **110**, 3959–3964.

HUANG, Y., MURRAY, M., METZGER, P. and EGLINTON, G., 1996, Novel unsaturated triterpenoid hydrocarbons from sediments of Sacred Lake, Mont Kenya-Kenya, *Tetrahedron*, **52**, 6973–6982.

INOUE, H., KORENAGA, T., SAGAMI, H., KOYAMA, T. and OGURA, K., 1994, Phosphorylation of farnesol by a cell-free system from *Botryococcus braunii*, *Biochemical and Biophysical Research Communications*, **200**, 1036–1041.

INOUE, H., KORENAGA, T., SAGAMI, H., KOYAMA, T., SUGIYAMA, H. and OGURA, K., 1993, Formation of farnesal and 3-hydroxy-2,3-dihydrofarnesal from farnesol by protoplasts of *Botryococcus braunii*, *Biochemical and Biophysical Research Communications*, **196**, 1401–1405.

INOUE, H., KORENAGA, T., SAGAMI, H., KOYAMA, T., SUGIYAMA, H. and OGURA, K., 1994, Formation of farnesyl oleate and three other farnesyl fatty acid esters by cell-free extracts from *Botryococcus braunii* B race, *Phytochemistry*, **36**, 1203–1207.

INOUE, H., SAGAMI, H., KOYAMA, T. and OGURA, K., 1995, Properties of farnesol phosphokinase of *Botryococcus braunii*, *Phytochemistry*, **40**, 377–381.

IWAMOTO, H. and SUGIMOTO, H., 1955, Fat synthesis in unicellular algae, *Bulletin Agricultural Chemical Society of Japan*, **19**, 247–252.

JAFFÉ, R., WOLF, G.A., CABRERA, A.C. and CARVAJAL-CHITTY, H., 1995, The biogeochemistry of lipids in rivers of the Orinoco Basin, *Geochimica et Cosmochimica Acta*, **59**, 4507–4522.

JEEJI BAI, N., 1996, Interesting developmental and morphological features of *Botryococcus* (Chlorococcales, Chlorophyceae) in culture, *Nova Hedwigia*, **112**, 489–498.

KADOURI, A., DERENNE, S., LARGEAU, C., CASADEVALL, E. and BERKALOFF, C. 1988, Resistant biopolymer in the outer walls of *Botryococcus braunii*, B Race, *Phytochemistry*, **27**, 551–557.

KNIGHTS, B.A., BROWN, A.C., CONWAY, E. and MIDDLEDITCH, B.S., 1970, Hydrocarbons from the green form of the freshwater alga *Botryococcus braunii*, *Phytochemistry*, **9**, 1317–1324.

KOLATTUKUDY, P.E., CROTEAU, R., and BUCKNER, J.S., 1976, Biochemistry of plant waxes, in Kolattukudy, P.E. (Ed.), *Chemistry and Biochemistry of Natural Waxes*, pp. 290–349, Elsevier.

KOMÁREK, J. and MARVAN, P., 1992, Morphological differences in natural populations of the genus *Botryococcus* (Chlorophyceae), *Archiv für Protistenkunde*, **141**, 65–100.

KÜTZING, H., 1849, *Species Algarum*, Jena.

LARGEAU, C., CASADEVALL, E. and BERKALOFF, C., 1980a, The biosynthesis of long-chain hydrocarbons in the green alga *Botryococcus braunii*, *Phytochemistry*, **19**, 1081–1085.

LARGEAU, C., CASADEVALL, E., BERKALOFF, C. and DHAMELINCOURT, P., 1980b, Sites of accumulation and composition of hydrocarbons in *Botryococcus braunii*, *Phytochemistry*, **19**, 1043–1051.

LARGEAU, C., CASADEVALL, E., DIF, D. and BERKALOFF, C., 1980c, Renewable hydrocarbon production from the alga *Botryococcus braunii*, in Palz, W., Chartier, P. and Hall, D.O. (Eds), *Energy from Biomass*, pp. 653–658, London: Applied Science Publishers.

LARGEAU, C., CASADEVALL, E., KADOURI, A. and METZGER, P., 1984, Formation of *Botryococcus braunii* kerogens. Comparative study of immature Torbanite and of the extant alga *Botryococcus braunii*, *Organic Geochemistry*, **6**, 327–332.

LARGEAU, C., DERENNE, S., CASADEVALL, E., KADOURI, A. and SELLIER, N., 1986, Pyrolysis of immature Torbanite and of the resistant biopolymer (PRBA) isolated from extant alga *Botryococcus braunii*. Mechanism of formation and structure of Torbanite, *Organic Geochemistry*, **10**, 1023–1032.

LEE, M.S., QIN, G.-W., NAKANISHI, K. and ZAGORSKI, M.G., 1989, Biosynthetic studies of brevetoxins, potent neurotoxins produced by the dinoflagellate *Gymnodinium breve*, *Journal of the American Chemical Society*, **111**, 6234–6241.

LIN, Y.-Y., RISK, M., RAY, S.M., VAN ENGEN, D., CLARDY, J., GOLIK, J., JAMES, J.C. and NAKANISHI, K., 1981, Isolation and structure of Brevetoxin B from the red tide dinoflagellate *Ptychodiscus brevis* (Gymnodinium breve), *Journal of the American Chemical Society*, **103**, 6773–6775.

MADHAVI SHEKHARAM, K., VENKATARAMAN, L.V. and Salimath, P.V., 1987, Carbohydrate composition and characterisation of two unusual sugars from the blue green alga, Spirulina platensis, *Phytochemistry*, **26**, 2267–2269.

MANGOLD, H.K. and PALTAUF, F., 1983, *Ether lipids, Biochemical and Biomedical Aspects*, New-York: Academic Press.

MAXWELL, J.R., DOUGLAS, A.G., EGLINTON, G. and MCCORMICK, A., 1968, The Botryococenes. Hydrocarbons of novel structure from the alga *Botryococcus braunii* Kützing, *Phytochemistry*, **7**, 2157–2171.

MCKIRDY, D.M., COX, R.E., VOLKMAN, J.K. and HOWELL, V.J., 1986, Botryococcane in a new class of Australian non-marine crude oils, *Nature*, **320**, 57–59.

METZGER, P., 1993, *n*-Heptacosatrienes and tetraenes from a Bolivian strain of *Botryococcus braunii*, *Phytochemistry*, **33**, 1125–1128.

METZGER, P., 1994, Phenolic ether lipids with an *n*-alkenylresorcinol moiety from a Bolivian strain of *Botryococcus braunii* (A race), *Phytochemistry*, **36**, 195–212.

METZGER, P. and AUMELAS, A., 1997, Lycopanerols A, di-tetraterpenoid tetraether derivatives from the green microalga *Botryococcus braunii*, L strain, *Tetrahedron Letters*, **38**, 2977–2980.

METZGER, P. and CASADEVALL, E., 1983, Structure de trois nouveaux botryococcènes synthétisés par une souche de *Botryococcus braunii* cultivée en laboratoire, *Tetrahedron Letters*, **24**, 4013–4016.

METZGER, P. and CASADEVALL, E., 1987, Lycopadiene, a tetraterpenoid hydrocarbon from new strains of the green alga *Botryococcus braunii*, *Tetrahedron Letters*, **28**, 3931–3934.

METZGER, P. and CASADEVALL, E., 1989, Aldehydes, very long chain alkenylphenols, epoxides and other lipids from an alkadiene-producing strain of *Botryococcus braunii*, *Phytochemistry*, **28**, 2097–2104.

METZGER, P. and CASADEVALL, E., 1991, Botryococcoid ethers, ether lipids from *Botryococcus braunii*, *Phytochemistry*, **30**, 1439–1444.

METZGER, P. and CASADEVALL, E., 1992, Ether lipids from *Botryococcus braunii* and their biosynthesis, *Phytochemistry*, **31**, 2341–2349.

METZGER, P. and LARGEAU, C., 1994, A new type of ether lipid comprising phenolic moieties in *Botryococcus braunii*. Chemical structure and abundance, and geochemical implications, *Organic Geochemistry*, **22**, 801–814.

METZGER, P. and POUET, Y., 1995, *n*-Alkenylpyrogallol dimethyl ethers, aliphatic diol monoesters and some minor ether lipids from *Botryococcus braunii*, A race, *Phytochemistry*, **40**, 543–554.

METZGER, P., BERKALOFF, C., CASADEVALL, E. and COUTE, A., 1985a, Alkadiene- and botryococcene-producing races of wild strains of *Botryococcus braunii*, *Phytochemistry*, **24**, 2305–2312.

METZGER, P., CASADEVALL, E., POUET, M.J. and POUET, Y., 1985b, Structures of some botryococcenes: branched hydrocarbons from the B race of the green alga *Botryococcus braunii*, *Phytochemistry*, **24**, 2995–3002.

METZGER, P., TEMPLIER, J., LARGEAU, C. and CASADEVALL, E., 1986, An *n*-alkatriene and some *n*-alkadienes from the A race of the green alga *Botryococcus braunii*, *Phytochemistry*, **25**, 1869–1872.

METZGER, P., DAVID, M. and CASADEVALL, E., 1987, Biosynthesis of triterpenoid hydrocarbons in the B-Race of the green alga *Botryococcus braunii*. Sites of production and nature of the methylating agent, *Phytochemistry*, **26**, 129–134.

METZGER, P., CASADEVALL, E. and COUTE, A., 1988, Botryococcene distribution in strains of the green alga *Botryococcus braunii*, *Phytochemistry*, **27**, 1383–1388.

METZGER, P., VILLARREAL-ROSALES, E., CASADEVALL, E. and COUTÉ, A., 1989, Hydrocarbons, aldehydes and triacylglycerols in some strains of the A race of the green alga *Botryococcus braunii*, *Phytochemistry*, **28**, 2349–2353.

METZGER, P., ALLARD, B., CASADEVALL, E., BERKALOFF, C. and COUTÉ, A., 1990, Structure and chemistry of a new chemical race of *Botryococcus braunii* (Chlorophyceae) that produces lycopadiene, a tetraterpenoid hydrocarbon, *Journal of Phycology*, **26**, 258–266.

METZGER, P., VILLARREAL-ROSALES, E. and CASADEVALL, E., 1991, Methyl-branched fatty aldehydes and fatty acids in *Botryococcus braunii*, *Phytochemistry*, **30**, 185–191.

METZGER, P., POUET, Y., BISCHOFF, R., and CASADEVALL, E., 1993, An aliphatic polyaldehyde from *Botryococcus braunii* (A race), *Phytochemistry*, **32**, 875–883.

METZGER, P., POUET, Y. and SUMMONS, R., 1997, Chemotaxonomic evidence for the similarity between *Botryococcus braunii*, L race and *Botryococcus* neglectus, *Phytochemistry*, **44**, 1071–1075.

MOLDOWAN, J.M. and SEIFERT, W.K., 1980, First discovery of botryococcene in petroleum, *Journal of Chemical Society Chemical Communications*, **19**, 912–914.

MURAKAMI, M., NAKANO, H., YAMAGUCHI, K., KONOSU, S., NAKAYAMA, O., MATSUMOTO, Y. and IWAMOTO, H., 1988, Meijicoccene, a new cyclic hydrocarbon from *Botryococcus braunii*, *Phytochemistry*, **27**, 455–457.

OHMORI, M., WOLF, F.R. and BASSHAM, J.A., 1984, *Botryococcus braunii* carbon/nitrogen metabolism as affected by ammonia addition, *Archives of Microbiology*, **140**, 101–106.

OKADA, S., MURAKAMI, M. and YAMAGUCHI, K., 1995, Hydrocarbon composition of newly isolated strains of the green microalga *Botryococcus braunii*, *Journal of Applied Phycology*, **7**, 555–559.

OKADA, S., MATSUDA, H., MURAKAMI, M. and YAMAGUCHI, K., 1996, Botryaxanthin A, a member of a new class of carotenoids from the green microalga *Botryococcus braunii*, Berkeley, *Tetrahedron Letters*, **37**, 1065–1068.

PLAIN, N., LARGEAU, C., DERENNE, S. and COUTÉ, A., 1993, Variabilité morphologique de *Botryococcus braunii* (Chlorococcales, Chlorophyta): corrélations avec les conditions de croissance et la teneur en lipides, *Phycologia*, **32**, 259–265.

RUPPRECHT, J.K., HUI, Y.-H. and MCLAUGHLIN, J.L., 1990, Annonaceous acetogenins: a review, *Journal of Natural Products*, **53**, 237–278.

SAKEMI, S., HIGA, T., JEFFORD, C.W. and BERNARDINELLI, G., 1986, Venustatriol, a new anti-viral triterpene tetracyclic ether from *Laurencia venusta*, *Tetrahedron Letters*, **27**, 4287–4290.

SAWAYAMA, S., INOUE, S. and YOKOYAMA, S., 1994, Continuous culture of hydrocarbon-rich microalga *Botryococcus braunii* in secondarily treated sewage, *Applied Microbiology and Biotechnology*, **41**, 729–731.

SUMMONS, R.E. and CAPON, R.J., 1991, Botryococcenone, an oxygenated botryococcene from *Botryococcus braunii*, *Australian Journal of Chemistry*, **44**, 313–322.

SUZUKI, T., SUZUKI, M., FURUSAKI, A., MATSUMOTO, T., KATO, A., IMANAKA, Y. and KUROSAWA, E., 1985, Teurilene and thyrsiferyl 23-acetate, meso and remarkably cytotoxic compounds from the marine red alga *Laurencia obtusa* (Hudson) Lamouroux, *Tetrahedron Letters*, **26**, 1329–1332.

SUZUKI, M., MATSUO, Y., TAKEDA, S. and SUZUKI, T., 1993, Intricatetraol, a halogenated triterpene alcohol from the red alga *Laurencia intricata*, *Phytochemistry*, **33**, 651–656.

SWAIN, F.M. and GILBY, J.M., 1964, Ecology and taxonomy of Ostracoda and an alga from lake Nicaragua, *Publicazioni della Stazione Zoologica di Napoli*, **33**(suppl.), 361–371.

TEGELAAR, E.W., DE LEEUW, J.W., DERENNE, S. and LARGEAU, C., 1989, A reappraisal of kerogen formation, *Geochimica et Cosmochimica Acta*, **53**, 3103–3106.

TEMPLIER, J., LARGEAU, C. and CASADEVALL, E., 1984, Mechanism of non-isoprenoid hydrocarbon biosynthesis in *Botryococcus braunii*, *Phytochemistry*, **23**, 1017–1028.

TEMPLIER, J., LARGEAU, C. and CASADEVALL, E., 1987, Effect of various inhibitors on biosynthesis of non-isoprenoid hydrocarbons in *Botryococcus braunii*, *Phytochemistry*, **26**, 377–383.

TEMPLIER, J., LARGEAU, C. and CASADEVALL, E., 1991a, Non-specific elongation-decarboxylation in biosynthesis of *cis*- and *trans*-alkadienes by *Botryococcus braunii*, *Phytochemistry*, **30**, 175–183.

TEMPLIER, J., LARGEAU, C. and CASADEVALL, E., 1991b, Biosynthesis of *n*-alkatrienes in *Botryococcus braunii*, *Phytochemistry*, **30**, 2209–2215.

TENAUD, M., OHMORI, M. and MIYACHI, S., 1989, Inorganic carbon and acetate assimilation in *Botryococcus braunii* (Chlorophyta), *Journal of Phycology*, **25**, 662–667.

TYMAN, J.H.P., 1979, Non-isoprenoid long chain phenols, *Chemical Society Reviews*, **8**, 499–537.

VIEIRA, A.A. and MYKLESTAD, S., 1986, Production of extracellular carbohydrate in cultures of *Ankistrodesmus densus* Kors. (Chlorophyceae), *Journal of Plankton Research*, **8**, 985–994.

VILLARREAL-ROSALES, E., METZGER, P. and CASADEVALL, E., 1992, Ether lipid production in relation to growth in *Botryococcus braunii*, *Phytochemistry*, **31**, 3021–3027.

WAKE, L.V. and HILLEN, L.W., 1981, Nature and hydrocarbon content of blooms of the alga *Botryococcus braunii* occurring in Australian freshwater lakes, *Australian Journal of Marine and Freshwater Research*, **32**, 353–367.

WANG, X. and KOLATTUKUDY, P.E., 1995, Solubilization and purification of aldehyde-generating fatty acyl-CoA reductase from green alga *Botryococcus braunii*, *FEBS Letters*, **370**, 15–18.

WEETALL, H.H., 1985, Studies on the nutritional requirements of the oil-producing alga *Botryococcus braunii*, *Applied Biochemistry and Biotechnology*, **11**, 377–391.

WHITE, J.D., SOMERS, T.C. and REDDY, G.N., 1986, Absolute configuration of (−) botryococcene, *Journal of the American Chemical Society*, **108**, 5352–5353.

WHITE, J.D., REDDY, G.N. and SPESSARD, G.O., 1988, Total synthesis of (−) botryococcene, *Journal of the American Chemical Society*, **110**, 1624–1626.

WHITE, J.D., SOMERS, T.C. and REDDY, G.N., 1992, Degradation and absolute configurational assignment to C_{34}-botryococcene, *Journal of Organic Chemistry*, **57**, 4991–4998.

WOLF, F.R. and COX, R.E., 1981, Ultrastructure of active and resting colonies of *Botryococcus braunii* (Chlorophyceae), *Journal of Phycology*, **17**, 395–405.

WOLF, F.R., NONOMURA, A.M. and BASSHAM, J.A., 1985a, Growth and branched hydrocarbon production in a strain of *Botryococcus braunii* (Chlorophyta), *Journal of Phycology*, **21**, 388–396.

WOLF, F.R., NEMETHY, E.K., BLANDING, J.H. and BASSHAM, J.A., 1985b, Biosynthesis of unusual acyclic isoprenoids in the alga *Botryococcus braunii*, *Phytochemistry*, **24**, 733–737.

ZAFRA-POLO, C., GONZÁLEZ, C., ESTORNELL, E., SAHPAZ, S. and CORTES, D., 1996, Acetogenins from Annonaceae, inhibitors of mitochondrial complex I, *Phytochemistry*, **42**, 253–271.

ZALESSKY, M.M.D., 1926, Sur les nouvelles algues découvertes dans le sapropélogène du lac Beloe et sur une algue sapropélogène *Botryococcus braunii*. Kützing, *Revue Générale de Botanique*, **38**, 31–48.

11

Phycobiliproteins

ALEXANDER N. GLAZER

Introduction

Phycobiliproteins (*phykos* (Gr.), seaweed) are a family of proteins with covalently attached linear tetrapyrrole prosthetic groups. These prosthetic groups are called *bilins* because of their close structural relationship to the well-known bile pigments of humans, biliverdin and bilirubin (Lemberg and Legge, 1949). The conformational and steric constraints imposed upon the bilins and their environment within the native phycobiliproteins endow these tetrapyrroles with special spectroscopic properties which are manifested by the brilliant colours and vivid fluorescence of these proteins.

The phycobiliproteins are found in cyanobacteria (formerly called blue–green algae) (Stanier *et al.*, 1971; Rippka *et al.*, 1979), in the chloroplasts of the Rhodophyta (the red algae) (Glazer *et al.*, 1982; Honsell *et al.*, 1984; Mörschel and Rhiel, 1987), and in those of Cryptophyceae, a class of biflagellate unicellular eukaryotic algae (cryptomonads) (Allen *et al.*, 1959; Haxo and Fork, 1959; Ó hEocha and Raftery, 1959; Gantt, 1979; Glazer and Wedemayer, 1995; Wedemayer *et al.*, 1996). In all of these organisms, the phycobiliproteins function as photosynthetic accessory proteins. They absorb light over a wide range of wavelengths in the visible part of the spectrum and transfer the excitation energy by radiationless processes to the reaction centres in the photosynthetic membranes for conversion to chemical energy (Gantt, 1979, 1980; Glazer, 1976, 1981, 1984, 1985, 1986, 1989).

Phycobiliprotein-containing organisms are ubiquitous in the biosphere. The cyanobacteria are an extraordinarily diverse group of prokaryotes (Rippka *et al.*, 1979; Rippka, 1988). They grow in a wide range of habitats: arid deserts, hot springs up to about 75°C, both temperate freshwater lakes and strongly saline ones, ice-cold lakes of the Antarctic, the open ocean, soils, and rock surfaces. Since the majority of cyanobacterial species are obligately phototrophic, they are restricted to lighted environments, although in some of their habitats, such as caves, the light levels are very low (Fogg *et al.*, 1973).

Methods for the isolation and culture of cyanobacteria are well described (Rippka *et al.*, 1979; Rippka, 1988). Axenic cultures of many different cyanobacteria are available from such sources as the Pasteur Collection of Cyanobacteria (PCC; Institut Pasteur, Paris), the American Type Culture Collection (ATCC; Rockville, Maryland), or the Culture Collection of Algae at the University of Texas (UTEX). Cyanobacteria can be readily grown in mass culture to densities of several grams per litre (wet weight), as long as provision is made for agitation (or circulation) of the culture, adequate illumination, and appropriate supply of carbon dioxide. The highest amounts of phycobiliproteins (as much as 24 per cent of the dry weight) occur in organisms cultured in white light of low intensity (Myers and Kratz, 1955).

The red algae are an extremely diverse group with over 400 genera encompassing some 4000 species. Morphologically, the red algae range from unicellular organisms to large plants. The red algae are predominantly marine and are found in both tropical and colder waters. Some are found in freshwater streams, others are terrestrial (Cole and Sneath, 1990). Unicellular red algae, such as *Porphyridium cruentum* (Lee and Bazin, 1990; Dermoun *et al.*, 1992; Chaumont, 1993) or *Rhodella reticulata* (Mihova *et al.*, 1996), can be readily grown in pure culture in large amounts in the laboratory. Many multicellular red algae are also maintained in pure culture in the laboratory, but in general, they grow very slowly and field material is commonly used for biochemical work.

The cryptomonads are found in both marine and freshwater habitats (Gantt, 1979). Many strains are available in pure culture and can be readily grown on a large scale.

Phycobiliproteins from cyanobacteria or red algae

The value of a particular phycobiliprotein as a reagent, whether as a fluorescent label in cell analyses or immunoassays, assay of reactive oxygen species, or histochemistry, is critically dependent on its: (a) subunit composition; (b) bilin composition and content; (c) oligomerization state; (d) monodispersity of the oligomer; (e) visible absorbance spectrum and absorbance coefficient(s) at the maxima; and (f) fluorescence quantum yield. Let us address each of these points in turn for cyanobacterial and red algal phycobiliproteins and then, separately, for the very different phycobiliproteins of the cryptomonads.

Subunit composition

Cyanobacterial and rhodophytan phycobiliproteins can be divided into three classes on the basis of their amino acid sequences and their major absorbance maxima: phycoerythrins (λ_{max} ~550–565 nm), phycocyanins (λ_{max} ~610–625 nm), and allophycocyanins (λ_{max} 650 nm) (Glazer, 1981, 1985). By eye, the phycoerythrins appear red, the phycocyanins range from purple (phycoerythrocyanin, R-phycocyanin) to deep blue (C-phycocyanin), and the allophycocyanins are blue with a hint of green.

For each of these different types of proteins, the building block is an $\alpha\beta$ heterodimer. The mass of the α and β subunits ranges between 18 000 and 22 000 daltons. The heterodimer is a very stable unit which resists dissociation at protein concentrations as low as 10^{-12}M, or in the presence of chaotropic agents (MacColl *et al.*, 1981) such as ammonium thiocyanate.

Bilin composition and content

Four isomeric bilins, phycourobilin (PUB), phycobiliviolin (PXB), phycoerythrobilin (PEB), and phycocyanobilin (PCB) are found attached to cyanobacterial and rhodophytan phycobiliproteins (Glazer, 1985; Ong and Glazer, 1991). These bilins are linked to the polypeptide through thioether bonds to cysteinyl residues. The structures of the polypeptide-bound bilins are shown in Figure 11.1 (see colour section). In native phycobiliproteins, PUB gives rise to an absorbance peak at ~495 nm, PEB to absorbance peaks at 540–565 nm, PXB at 568 nm, and PCB at 590–625 nm.

The sequences of many phycobiliproteins have been determined either by amino acid sequence analysis, or by the determination of DNA sequences of the genes encoding these proteins (Wilbanks and Glazer, 1993a,b; Sidler, 1994). Extensive studies have established the locations of the cysteinyl residues that serve as sites of bilin attachment and the identity of the bilin at each of these locations (Figure 11.2). Allophycocyanin carries one bilin on each α and β subunit. Phycocyanins carry one bilin on the α subunit and two on the β subunit. Phycoerythrins carry either two or three bilins on the α subunit and three on the β subunit.

Oligomerization state

Within cyanobacterial cells and red algal chloroplasts, phycobiliproteins are components of large complexes, called phycobilisomes (Gantt, 1980; Glazer, 1985, 1988a, 1989; Wilbanks and Glazer, 1993b). Phycobilisomes are attached to the cytoplasmic surface of the photosynthetic lamellae. The phycobilisome is made up of disk-shaped trimeric, $(\alpha\beta)_3$, and hexameric, $(\alpha\beta)_6$, complexes of phycobiliproteins. The hexameric complexes are formed by face-to-face interaction of the trimers. The trimers are about 12 nm in diameter and 3 nm thick. In the phycobilisome, the assembly of the trimers and hexamers is mediated by position- and phycobiliprotein-specific linker polypeptides. These linker polypeptides also stabilize the assembled complexes.

The stability of the trimers and hexamers varies markedly with the organismal source of the phycobiliprotein, but these oligomers are less stable than the heterodimer. When phycobiliproteins are released by cell breakage and subjected to the conventional procedures for protein purification, the linker polypeptides are stripped away in the majority of instances, and the oligomerization state of the $\alpha\beta$ heterodimers depends strongly on the conditions of purification (Glazer, 1988b). Only certain phycoerythrins from unicellular marine cyanobacteria and red algae survive purification as monodisperse double-disk $(\alpha\beta)_6\gamma$ complexes of ~250 000 daltons, where γ is a linker polypeptide (Glazer and Hixson, 1977; Yu *et al.*, 1981).

Single phycobiliproteins are therefore artifacts of purification; the oligomers that are isolated share many structural features in common with the portions of the phycobilisome from which they are derived. However, they generally lack the interactions with the linker polypeptides and with the nearest-neighbour phycobiliprotein complexes that are characteristic of the intact phycobilisome.

PHYCOCYANINS

	α-84 (α-1)	β-82 (β-1)	β-155 (β-2)
C-Phycocyanin	PCB	PCB	PCB
R-Phycocyanin	PCB	PCB	PCB
Phycoerythrocyanin	PXB	PCB	PEB
R-Phycocyanin II	PEB	PCB	PEB

PHYCOERYTHRINS

	α-75 (α-3)	α-83 (α-1)	α-140 (α-2)	β-50,61 (β-3)	β-82 (β-1)	β-159 (β-2)
C-PE		PEB	PEB	PEB	PEB	PEB
B-PE		PEB	PEB	PEB	PEB	PEB
R-PE		PEB	PEB	PUB	PEB	PEB
WH8020 PE(I)		PEB	PEB	PEB	PEB	PEB
WH8020 PE(II)	PUB	PEB	PEB	PUB	PEB	PEB
WH8103 PE(I)		PEB	PUB	PUB	PEB	PEB
WH8103 PE(II)	PUB	PUB	PUB	PUB	PEB	PEB
WH8501 PE(I)	PUB	PUB	PUB	PUB	PEB	PUB

Figure 11.2 Bilin types and sites of attachment in various phycocyanins and phycoerythrins. The bold face designations, such as '*α–84*', indicate the phycobiliprotein subunit and position of the cysteinyl residue that functions as the site of bilin attachment. The designations in parentheses, such as '(α-1)', represent older nomenclature for particular bilin-bearing peptides obtained by degradation of the phycobiliproteins, used before complete amino acid sequences were known. Bilins shown in outline font absorb light energy in the intact phycobiliprotein and transfer the excitation energy to other bilins. The terminal energy acceptor bilin is believed to be at β-82 in all of these cyanobacterial and red algal phycobiliproteins. The abbreviations for bilins are defined in the legend to Figure 11.1 which shows their structures as bound to the polypeptide. For references to the original data sources, see Ong and Glazer (1991). The marine unicellular cyanobacterial strains WH8020, WH8103, and WH8501 are described in Waterbury and Rippka (1989).

Monodispersity of the oligomer

If a particular phycobiliprotein oligomer is to be used as a fluorescent tag, it is of considerable importance that it is monodisperse at very low protein concentration (Glazer, 1984, 1985). This condition is met by B- and R-phycoerythrins, where the $(\alpha\beta)_6\gamma$ complexes do not dissociate even at 10^{-12}M. Whereas many phycocyanins form trimers or hexamers at concentrations above 10^{-5}M–10^{-6}M, at lower concentrations, they dissociate in a concentration-dependent manner to the $\alpha\beta$ heterodimer (Neufeld and Riggs, 1969; MacColl *et al.*, 1971; Glazer *et al.*, 1973). The monomer with only three bilins is a much less effective fluorescent tag than the hexamer with 18 bilins. No effective technique has been described to stabilize phycocyanin trimers or hexamers, and this doubtless accounts for the limited use this class of phycobiliproteins has found as fluorescent tags.

Dissociation of allophycocyanin to the monomer poses even more severe problems (Ong and Glazer, 1985). At near neutral pH, at concentrations of $\sim 10^{-5}$M and higher, allophycocyanin exists as a trimer, $(\alpha\beta)_3$, with λ_{max} at 650 nm and a fluorescence emission maximum at 660 nm. However, upon dilution to lower concentrations allophycocyanin dissociates to the $\alpha\beta$ heterodimer, which has one third the molar absorbance, a shift in the λ_{max} to 615 nm, and a drop in the fluorescence quantum yield (Ong and Glazer, 1985; Yeh *et al.*, 1986). Fortunately, the trimer can be stabilized by site-specific cross-linking with 1-ethyl-3-(3-dimethylaminopropyl) carbodiimide (Ong and Glazer, 1985). The spectroscopic properties of the cross-linked trimer are virtually the same as those of the unmodified trimer. However, unlike the unmodified trimer, it is fully stable to dilution to $< 10^{-12}$M, or to chaotropic salts. The cross-linked trimer also shows much higher thermostability than its unmodified counterpart (Yeh *et al.*, 1987). Because of the availability of the cross-linked trimer, allophycocyanin has found wide use as a near-red emitting fluorescent tag.

It is common to isolate two different phycoerythrins from a single red alga, one of which is polydisperse, the other a monodisperse, highly stable hexamer. Another caution is that it cannot be assumed that the same phycobiliprotein isolated from different organisms will have the same oligomerization state or oligomer stability.

Visible absorbance spectrum and absorbance coefficient(s) at the maxima

The absorbance maxima of the phycobiliproteins are given in Table 11.1. In part because of the presence of multiple bilins (as depicted in Figure 11.3 (see colour section) for example), the absorbance coefficients of phycobiliproteins greatly exceed those of the most strongly absorbing dyes (Glazer, 1984, 1985, 1988b). The highest absorbance coefficients reported for dyes in the visible region of the spectrum are $\sim 250\,000$ M^{-1} cm^{-1}.

High resolution crystal structure determinations have been published for several phycobiliproteins: allophycocyanin (Brejc *et al.*, 1995), C-phycocyanin (Duerring *et al.*, 1991), phycoerythrocyanin (Duerring *et al.*, 1990), and B-phycoerythrin (Ficner *et al.*, 1992; Ficner and Huber, 1993). Within all of these native phycobiliproteins, the bilins are held in extended configurations (as depicted in Figure 11.4 for example). In this configuration, bilins exhibit the strongest absorbance at their long wavelength maxima (Scheer and Kufer, 1977; Glazer, 1984). When phycobiliproteins

Table 11.1 Spectroscopic properties of representative phycobiliproteins[a]

Protein (subunit composition)	Molecular weight	λ_{max}^{Abs} nm	ε_M $M^{-1}\,cm^{-1}$	λ_{max}^{F} nm	Φ_F
Allophycocyanin [$(\alpha\beta)_3$]	100 000	650	696 000	660	0.68
C-phycocyanin [$(\alpha\beta)_{n=1-6}$]	36 700/$\alpha\beta$	620	281 000/$\alpha\beta$	642	0.51
B-phycoerythrin [$(\alpha\beta)_6\gamma$]	240 000	545	2 410 000	575	0.98
R-phycoerythrin [$(\alpha\beta)_6\gamma$]	240 000	565	1 960 000	578	0.82

[a]References to primary data sources are given in Oi *et al.* (1982).

are denatured by urea, sodium dodecyl sulphate, or at extremes of pH, the bilins, released from three-dimensional constraints, assume cyclohelical (lock-washer) conformations with a concomitant large decrease in their long wavelength absorbance and complete loss of fluorescence emission (Scheer and Kufer, 1977; Glazer, 1984, 1986; see below).

Fluorescence quantum yield

The bilins in the phycobiliproteins are held rigidly and are largely protected from the solvent (Figure 11.5 (see colour section)). Excitation of phycobiliproteins leads to fluorescence emission with high quantum yields (Table 11.1; Figure 11.6 (see colour section)). Radiationless decay due to motion or collision with quencher molecules is minimized. In contrast, free bilins or denatured phycobiliproteins are virtually non-fluorescent (Figure 11.7 (see colour section)).

Cryptomonad phycobiliproteins

The manner in which the cryptomonad phycobiliproteins are organized in living cells is not known. Purified cryptomonad phycobiliproteins, independent of bilin composition, share certain properties in common (Glazer and Wedemayer, 1995). These properties are detailed below.

Subunit composition

The subunit composition of cryptomonad biliproteins is $\alpha\alpha'\beta_2$, where the α subunits are approximately 8000–9000 daltons and the β subunits are ~20 000 daltons (Mörschel and Wehrmeyer, 1975, 1977; Sidler *et al.*, 1985, 1990). The $\alpha\alpha'\beta_2$ tetramer is thus ~50 000–60 000 daltons. Like the $\alpha\beta$ heterodimeric building block of the cyanobacterial and red algal phycobiliproteins, the $\alpha\alpha'\beta_2$ tetramer is very stable.

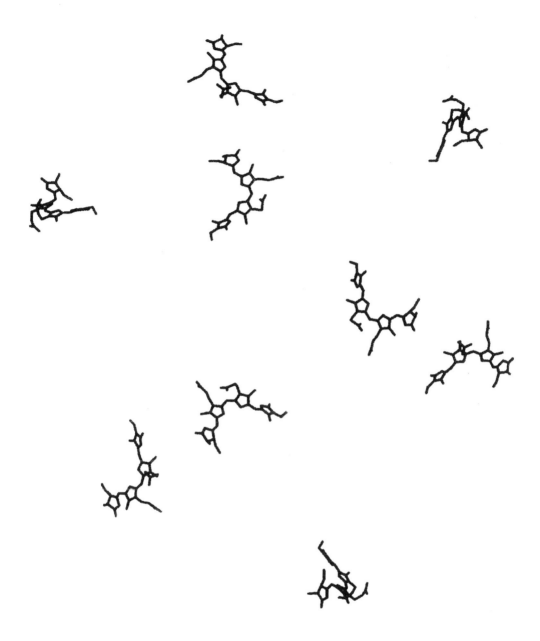

Figure 11.4 Locations and conformations of the the nine bilins in the crystal structure of the *Fremyella diplosiphon* C-phycocyanin trimer (from Duerring *et al.*, 1991).

Bilin composition and content

Whereas only four different bilins (PUB, PEB, PXB, and PCB) are found attached to cyanobacterial and red algal phycobiliproteins, six different bilins are found attached to cryptomonad phycobiliproteins. These are PEB, PCB, dihydrobiliverdin (DBV), bilin 584, bilin 618, and bilin 684 (for structures, see Figure 11.1). Because of the greater diversity of bilin prosthetic groups, the cryptomonad phycobiliproteins frequently have complex absorbance spectra (Glazer and Wedemayer, 1995; Wedemayer *et al.*, 1996).

Spectroscopic properties

Each α or α' subunit carries only a thioether-linked bilin and each β subunit carries three bilins (Wedemayer *et al.*, 1996). The types and locations of bilins on different cryptomonad phycobiliproteins are given in Table 11.2.

The spectroscopic properties of the cryptomonad phycobiliproteins are given in Table 11.3. Since $\alpha\alpha'\beta_2$ tetramer carries only eight bilins (frequently of more than one kind), these proteins do not have as favourable spectroscopic properties as the hexameric R-phycoerythrin with its 34 bilins, or cross-linked allophycocyanin trimer with its six PCBs and sharp long wavelength (650 nm) absorbance maximum. However, because of their high stability and diversity of spectroscopic properties, greater consideration should be given in the future to the use of these proteins as fluorescent tags.

Applications of the phycobiliproteins

Properties of phycobiliproteins relevant to their use as fluorescent tags in cell sorting and cell analyses

Phycobiliproteins find their widest use in fluorescence-activated cell sorting (FACS) and flow cytometry analyses of cell populations (Oi *et al.*, 1982; Glazer and Stryer, 1984, 1990; Glazer, 1994). Phycobiliprotein conjugates are used routinely in such high throughput clinical analyses worldwide. Hundreds of research studies exploit these reagents annually. Since their introduction in 1982, the applications of phycobiliproteins as fluorescent tags have continued to expand. This longstanding popularity rests on a number of unique properties of these proteins.

Stable oligomers

As noted above, phycoerythrin hexamers with subunit composition $(\alpha\beta)_6\gamma$ are extremely stable down to very low protein concentrations. Chemically crosslinked allophycocyanin trimers $(\alpha\beta)_3$ (Ong and Glazer, 1985) likewise do not dissociate on dilution to below 10^{-9}M.

Table 11.2 Positions of the main absorption and fluorescence emission maxima of cryptophycean biliproteins at visible wavelengths.[a]

Biliprotein	Absorption[b,c] (nm)	Emission[c,d] (nm)
Cr-PE 545	538–551	580–587
Cr-PE 555	553–556	578
Cr-PE 566	563–568	600–619
Cr-PC 569	568–569	650, 656
Cr-PC 612[e]	612–615	634–641
Cr-PC 630	625–630	648, 649
Cr-PC 645	641–650	654–662

[a]A widely used nomenclature for cryptomonad phycobiliproteins employs a prefix 'Cr' to designate the cryptophycean origin of the protein, followed by 'PE' or 'PC', depending on the appearance of the protein solution; red-coloured proteins are designated as PE, whereas blue ones are designated PC. This table is modified from Wedemayer *et al.* (1992). References to primary data sources are given in Wedemayer *et al.* (1992).

[b]This column gives the range of published values of the major absorption maxima for the particular biliprotein.

[c]Single values in the columns represent those reported in a single study. Values separated by commas were those reported by two different studies, and those separated by dashes represent the extremes reported by three or more studies.

[d]This column gives the range of published values of the fluorescence emission maxima for the particular biliprotein.

[e]This biliprotein has also been referred to as (Cr)-PC 615 by some authors.

Molar absorbance coefficients and fluorescence quantum yields

The phycobiliproteins contain multiple rigidly held bilin chromophores and, as a consequence, have absorbance coefficients and fluorescence quantum yields that are superior to those of the best available organic dye fluorophores (Haugland, 1996). The fluorescence quantum yields of the phycobiliproteins are very high ($\Phi_f = 0.3$–0.9; Grabowski and Gantt, 1978). The bilin chromophores are buried deep within the proteins (Duerring *et al.*, 1990, 1991; Ficner *et al.*, 1992; Ficner and Huber, 1993; Brejc *et al.*, 1995). As long as the phycobiliproteins are native, the fluorescence is not quenched by other molecules present in the solution. The fluorescence emission of the phycobiliproteins is unaffected by pH over a broad range (from pH 4 to 9 for B- and R-phycoerythrin, for example), nor does it vary markedly with temperature (e.g., Yeh *et al.*, 1987).

In B-phycoerythrin, 32 PEBs and 2 PUBs are contained within a disk-shaped hexamer 12 nm in diameter and 6 nm in height. This density of chromophores corresponds to a concentration of 83 mM and gives rise to a molar absorbance coefficient (ε_M) of $2.41 \times 10^6 \, M^{-1} \, cm^{-1}$ at 545 nm (Glazer and Hixson, 1977) and a fluorescence quantum yield of 0.98 (Grabowski and Gantt, 1978). For comparison, fluorescein, the most widely used 'green' fluorophore, has an ε_M at pH 9 of $8.8 \times 10^4 \, M^{-1} \, cm^{-1}$ at 490 nm and a Φ_f of 0.93 (Haugland, 1996). The ε_M of the

Table 11.3 Bilin types and locations on cryptomonad biliproteins.

Strain and biliprotein	α-Cys-18[a]	β-DiCys-50,61	β-Cys-82	β-Cys-158
PrS:PE 545	Cys-DBV (562)	DiCys-PEB (550)	Cys-PEB (550)	Cys-PEB (550)
IVF$_2$:FE 555	Cys-PEB (550)	DiCys-DBV (562)	Cys-PEB (550)	Cys-PEB (550)
Ber:PE 566	Cys-Bilin 584 (584)	DiCys-Bilin 584 (584)	Cys-PEB (550)	Cys-Bilin 584 (584)
CBD:FE 566	Cys-Bilin 618 (618)	DiCys-Bilin 584 (584)	Cys-PEB (550)	Cys-Bilin 584 (584)
FaD:PC 569	Cys-PCB (643)	DiCys-Bilin 584 (584)	Cys-PCB (643)	Cys-Bilin 584 (584)
HP9001:PC612	Cys-PCB (643)	DiCys-DBV (562)	Cys-PCB (643)	Cys-PCB (643)
UW374:PC645	Cys-MBV (684)	DiCys-DBV (562)	Cys-PCB (643)	Cys-PCB (643)

Abbreviations used: DBV, 15,16-dihydrobiliverdin; PEB, phycoerythrobilin; PCB, phycocyanobilin; MBV, mesobiliverdin. Numbers in parentheses indicate long wavelength absorption maxima of solutions of bilin peptides in 10 mM trifluoroacetic acid. See Figure 11.1 for structures of all these bilins. From Wedemayer *et al.* (1991, 1992, 1996) and Wemmer *et al.* (1993).

[a] These designations specify the subunit and residue numbers of the cysteine residue(s) that serve as site(s) of bilin attachment. For example, β-DiCys-50,61 designates the attachment site on the β subunit where the bilin is held by two thioether linkages, one to Cys-50 the other to Cys-61.

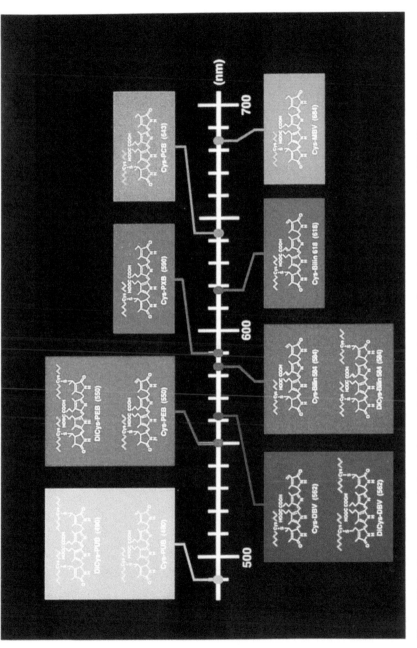

Figure 11.1 Structures and absorption maxima (given in parentheses in nanometers, in 10 mM aqueous trifluoroacetic acid) of peptide-bound bilins. Phycocyanobilin (PCB), singly- and doubly-linked phycoerythrobilin (PEB) and phycourobilin (PUB), are present in both cyanobacterial and red algal phycobiliproteins, and phycobiliviolin (PXB) only in cyanobacterial phycobiliproteins. PCB, PEB, mesobiliverdin (MBV), bilin 618, singly-and doubly-linked bilin 584, and singly- and doubly-linked bilin 562 are found on cryptomonad phycobiliproteins.

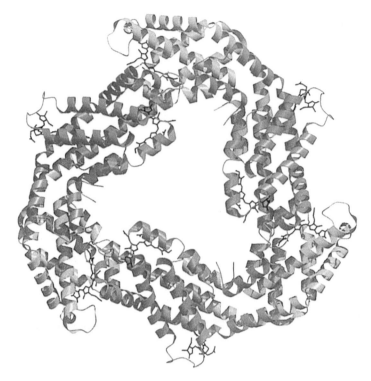

Figure 11.3 Face view of the crystal structure of the *Fremyella diplosiphon* C-phycocyanin trimer. The α subunits are shown in green and the β subunits in gold color. The bilins are shown in red color. Data from Duerring *et al.*, 1991.

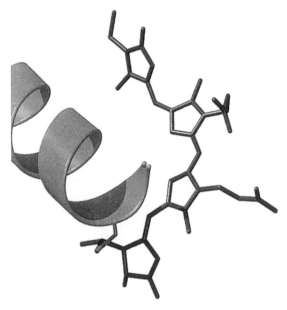

Figure 11 5 Detail taken from the structure of the *Fremyella diplosiphon* C-phycocyanin trimer showing the bilin at β-82. Note that the bilin is held in an extended conformation (from Duerring *et al.*, 1991).

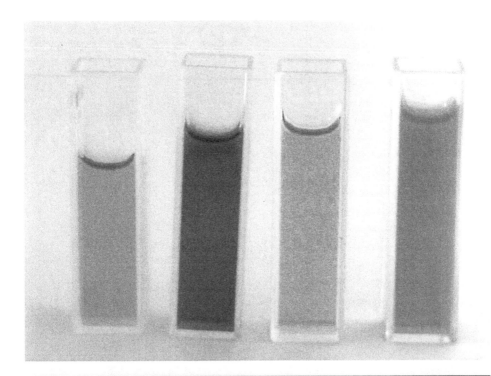

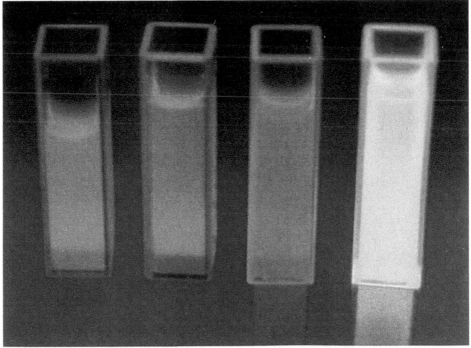

Figure 11.6 The upper photograph shows the colours of aqueous solutions at neutral pH of allophycocyanin, C-phycocyanin, phycoerythrocyanin, and B-phycoerythrin (from left to right). The lower photograph shows the fluorescence emission from these solutions with irradiatiation with near-ultraviolet light (courtesy of Dr Gary J. Wedemayer).

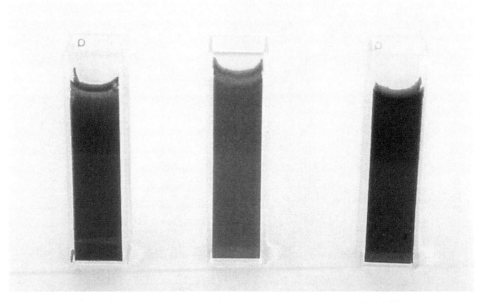

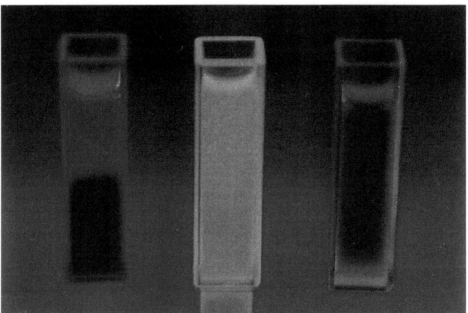

Figure 11.7 The upper photograph shows cryptomonad phycoerythrin 566 at pH 2.5, 7.0, and 12 (from left to right). The lower photograph shows the fluorescence emission from these solutions with irradiatiation with near-ultraviolet light. Note that upon exposure to strongly acid or strongly alkaline pH, the protein becomes non-fluorescent (courtesy of Dr Gary I. Wedemayer)

allophycocyanin trimer $(\alpha\beta)_3$, is $6.96 \times 10^5 \, \text{M}^{-1} \, \text{cm}^{-1}$ at 650 nm (Cohen-Bazire *et al.*, 1977) and its Φ_f is 0.68 (Grabowski and Gantt, 1978). Again, for comparison, a widely used, strongly absorbing, red-emitting cyanine dye, Cy5.18 in phosphate-buffered saline, has an ε_M of $2.5 \times 10^5 \, \text{M}^{-1} \, \text{cm}^{-1}$ at 650 nm, a $\lambda^F_{\max}$ at 667 nm, and a Φ_f of 0.27 (Mujumdar *et al.*, 1993), and the sulphobenzindocyanine dye Cy5.205 has an ε_M of $1.9 \times 10^5 \, \text{M}^{-1} \, \text{cm}^{-1}$ at 674 nm, a $\lambda^F_{\max}$ at 704 nm, and a Φ_f of 0.23 (Mujumdar *et al.*, 1996).

Stokes shift

When a solution of a fluorophore is excited, a fluorescence emission spectrum of exactly the same shape is observed irrespective of the choice of excitation wavelength. The fluorescence emission lies at a longer wavelength than the absorbance spectrum. The red shift of the emission band is explained as follows. When a photon of energy $h\nu_{exc}$ is absorbed by the fluorophore in its ground state (S_o), the fluorophore is promoted to an electronic excited state S_1'. This excited state decays very rapidly (within 10^{-12} seconds) to the lowest singlet excited state S_1. This decay is very fast relative to the lifetime of the excited state, ca. 10^{-9} sec. Transition from S_1 to S_o is accompanied by the emission of a photon of energy $h\nu_{em}$. The dissipation of energy acquired in absorption over that of the S_1–S_o transition is brought about by the rapid exchange of vibrational energy between the excited fluorophore and the surrounding solvent molecules or even by partition among the degrees of freedom within the polyatomic excited molecule (Weber and Teale, 1965). The difference in energy or wavelength represented by $\Delta E = [h\nu_{exc} - h\nu_{em}]$ is called the Stokes shift. It is the Stokes shift that allows the discrimination between the emitted photons and the exciting photons.

The larger the Stokes shift, the higher the sensitivity of detection of the fluorophore. With a large Stokes shift, the emission photons are well separated from the excitation photons, from Rayleigh scattering, and from background fluorescence from materials such as plastics, or from photons emitted by 'contaminating' fluorophores such as flavins and porphyrins, commonly found in samples of biological origin. As described in detail below, phycobiliprotein fluorescent tags provide uniquely large Stokes shifts.

The multiple chromophores carried by a single phycobiliprotein are spectroscopically distinguishable and together give rise to broad visible absorbance bands. In other words, within the native multimeric phycobiliproteins, the energy levels of the bilins are not equivalent. Consequently, whereas all of the bilins in a given phycobiliprotein absorb excitation energy, the fluorescence of the protein originates from the bilins with the longest wavelength absorbance bands. Bilins which absorb light energy and then transfer the excitation energy to other bilins are called donors. Those bilins that both absorb excitation energy and fluoresce are called acceptors. The intramolecular energy transfer within the phycobiliprotein trimers and hexamers is very fast, and in consequence the steady state fluorescence emission originates almost exclusively from the acceptors (Glazer, 1989; Ong and Glazer, 1991). The broad absorbance bands and the highly efficient intramolecular energy transfer allow choice of excitation wavelengths that lead to large Stokes shifts. For example, B- and R-phycoerythrin can be excited efficiently at 488 nm leading to maximal emission at 576 nm. This represents a Stokes shift of 88 nm, three times the Stokes shift of fluorescein. The intense absorbance and large Stokes shift of the fluorescence emission allow the detection of single molecules of B-phycoerythrin in solution (Peck *et al.*, 1989) and on cell surfaces (Wilson *et al.*, 1996).

Photobleaching

Photostability is an important parameter in assessing the value of a fluorophore as a label. The phycobiliproteins show excellent photostability. White and Stryer (1987) reported the photodestruction quantum yield (ϕ_d) at 514.5 nm for *Gastroclonium coulteri* R-phycoerythrin, and *Anabaena variabilis* C-phycocyanin and allophyco-cyanin as 1.1×10^{-5}, 2.5×10^{-6} and 4.5×10^{-6}, respectively. Lao and Glazer (1996) reported photodestruction quantum yields for C-phycocyanin for visible photons in the range of $1.6–2.7 \times 10^{-7}$, under a different set of experimental conditions. These photodestruction quantum yields are similar to those of fluorescein ($\phi_d 2.7 \times 10^{-5}$; Mathies and Stryer, 1986) and rhodamine dyes ($\phi_d 5.6 \times 10^{-6}$–$5.5 \times 10^{-5}$; Soper *et al.*, 1993) in aqueous solution.

Isoelectric points

The isoelectric points of the phycobiliproteins lie between pH 4.7 and 5.5. Consequently, these proteins are negatively charged at physiological pH. This characteristic is of particular importance when phycobiliprotein conjugates are used as fluorescent tags for cell surface markers in flow cytometry or in fluorescence-activated cell sorting. Generally, cells surfaces have a net negative charge and do not bind phycobiliprotein conjugates non-specifically.

Phycobiliprotein tandem conjugates

Fluorescent phycobiliprotein labels with new spectroscopic properties have been generated by exploiting intermolecular fluorescence energy transfer. As the first example of such labels, Glazer and Stryer (1983) prepared a 1:1 tandem conjugate of hexameric B-phycoerythrin with trimeric allophycocyanin using N-succinimidyl 3-(2-pyridylthio)-propionate as a cross-linking agent. Excitation of the B-phyco-erythrin in this conjugate at 545 nm led to emission from allophycocyanin at 660 nm. Intermolecular energy transfer has also been exploited by synthesizing conjugates of B- and R-phycoerythrin and allophycocyanin with sulphoindocyanine dyes such as Cy5.18 and bisCy7 (N,N'-biscarboxypentyl-5,5'-disulphonatoindotri-carbocyanine). The conjugates are prepared by coupling the protein with the reactive bis-succinimidyl esters of the cyanine dyes. In conjugates containing three molecules or more of Cy5 per R-phycoerythrin hexamer, energy transfer to the cyanine dye, with an emission maximum at 667 nm, proceeds with an efficiency exceeding 90 per cent (Lansdorp *et al.*, 1991; Shih *et al.*, 1993; Waggoner *et al.*, 1993). Excitation of phycoerythrin in the phycoerythrin-Cy7 conjugate at 488 nm, or of allophycocyanin in the allophycocyanin-Cy7 conjugate at 633 nm, leads to emission from Cy7 at 780 nm (Beavis and Pennline, 1996; Roederer *et al.*, 1996). The tandem conjugates linked to a biospecific molecule (antibody, lectin, etc.) facilitate multiparameter analyses. For example, a set of labels tagged with fluorescein, phycoerythrin, phycoerythrin-Cy5, and phycoerythrin-Cy7 can all be efficiently excited at the argon ion laser line at 488 nm. The resulting strong and well-separated fluorescence emissions lie at 530, 580, 667, and 780 nm, respectively.

Biospecific phycobiliprotein conjugates

Phycobiliprotein conjugates with biological specificity are generally prepared by coupling the molecule that confers specificity (biotin, avidin, antibody, lectin, etc.) to the phycobiliproteins through the ε-amino groups of lysyl residues (see Hardy, 1986; Monji, 1988; Glazer and Stryer, 1990; for experimental protocols). At a modest degree of conjugation (1–2 other molecules coupled per phycobiliprotein $\alpha\beta$ heterodimer), the spectroscopic properties of the phycobiliproteins are not modified.

Some examples of the applications of phycobiliprotein fluorescent tags

Phycobiliprotein antibody conjugates are widely used in routine multiparameter flow cytometry analyses of peripheral blood lymphocytes. However, they have found many other uses as well as illustrated by the few examples that follow.

Imaging cell surface markers

Phycobiliprotein conjugates are used in various multicolour imaging applications. For example, red fluorescence-emitting phycoerythrin conjugates with monoclonal antibodies specific to muscarinic acetylcholine receptors, and green fluorescence-emitting fluorescein-labeled monoclonal antibodies to nicotinic acetylcholine receptor proteins were used in a double-labelling study to examine human cortical neurons in biopsy and autopsy specimens. This study demonstrated the presence and localization of a subpopulation of human cortical neurons that displayed immunoreactivity for both receptor types (Schroeder *et al.*, 1989). Cell populations can be transfected with a retroviral vector encoding a fusion protein with the green fluorescent protein (Chalfie *et al.*, 1994; Cubitt *et al.*, 1995). The green fluorescent protein fusion emission lies in the green (< 520 nm). Consequently, the transformed cells can be stained with a cell surface marker specific phycoerythrin-Cy5 conjugate (fluorescence emission at 670 nm) to allow multiparameter analysis (Lybarger *et al.*, 1996).

Phycoerythrin conjugates with monoclonal antibodies directed to specific cell surface antigens were used in conjunction with confocal laser scanning microscopy to image cellular differention markers in human muscle biopsy specimens (Schubert, 1991). Phycoerythrin-interleukin 8 conjugates allowed the imaging of the spatial distribution of a cell surface receptor on living human neutrophils during phagocytosis (Kindzelskii *et al.*, 1994). Similarly, phycoerythrin-interleukin-1β and the sandwich phycoerythrin-avidin/biotin-labelled transforming factor β-1 were used to follow cytokine receptor alterations during HIV infection in the human nervous system (Masliah *et al.*, 1994).

Lectin histochemistry of embryonal carcinoma cells exposed to potentially toxic agents may be a useful method for predicting embryotoxicity of heavy metals and organic compounds (Hooghe and Ooms, 1995). Phycoerythrin–lectin (wheat germ agglutinin) conjugates were used to follow changes in the glycosylation pattern of cell surface glycoconjugates in murine embryonal carcinoma cells after exposure *in vitro* to xenobiotics for 1–3 days. FACS analysis provided rapid quantitation of bound conjugate.

Discriminating between different cell types in a heterogeneous population in conjunction with the measurement of intracellular events

Sei and Arora (1991) showed that a phycoerythrin conjugate with a cell surface-specific antibody could be used in conjunction with the dye fluo-3 (sensitive to cytoplasmic free $[Ca^{2+}]$) to relate patterns of calcium mobilization to specific cell types after stimulation of a heterogeneous population of immune cells with mitogen.

Detection of biotinylated oligonucleotide probes by second-step staining with streptavidin-conjugated phycoerythrin

These labelling procedures involve the formation of an oligonucleotide–biotin/strep-tavidin–phycoerythrin sandwich. Biotinylated deoxynucleotide triphosphates are readily incorporated into oligonucleotides by DNA polymerases, or through the use of biotinylated primers during amplification by the polymerase chain reaction, or by direct labelling by photocross-linking a psoralen derivative of biotin directly to the nucleic acid fragments. After hybridization of the probe to the complementary target oligonucleotide, fluorescent labelling is achieved by the addition of streptavi-din-conjugated phycoerythrin (Chee *et al.*, 1996). Davis *et al.* (1996) described an interesting application of this labelling strategy. A high affinity oligonucleotide ligand for human neutrophil elastase was modified at either the 5′- or 3′-end by the addition of a polyoxyethylene tether, consisting of six ethylene glycol units, and carrying a biotin at the end distal to the oligonucleotide. To introduce the fluorescent label for flow cytometric assay, the elastase was coated on 3.3μ polystyrene beads exposed to the biotinylated DNA ligand, washed, exposed to the streptavidin-conjugated phycoerythrin, and washed again.

Phycoerythrin-based assays for reactive oxygen species

The quantum yield of the fluorescence emission of phycoerythin is a sensitive measure of the physical and chemical integrity of this macromolecule. This property can be exploited to detect physiologically important antioxidant molecules that protect cells and tissues from damage by various reactive oxygen species.

Glazer (1988c) showed that reaction of phycoerythrin with peroxy radicals pro-duced at a constant rate by the thermal decomposition of the water-soluble free radical initiator 2,2′-azobis(2-amidinopropane) hydrochloride results in a linear decay in phycoerythrin fluorescence. Other reactive oxygen species produced by a mixture of Cu^{2+} and ascorbate produce a first order decay in phycoerythrin fluor-escence consistent with site-specific damage mediated by Cu^{2+} bound to the protein. This assay has been used to assess the free radical scavenging capacity of human plasma, of proteins, of DNA and of numerous small molecules (Glazer, 1988c; DeLange and Glazer, 1989, 1990; Cao *et al.*, 1995; Miller *et al.*, 1996). Molecules that react with peroxy radicals much more rapidly than does phycoerythrin prevent loss of phycoerythrin fluorescence until they are completely destroyed. Examples of such molecules are the water-soluble vitamin E analogue, Trolox, glutathione, and ergothioneine. The stoichiometry of the reaction between the free radical species and the small molecule can be calculated from the known rate of peroxy radical generation and the period of complete protection of phycoerythrin. Assay of free-radical damage by the Cu^{2+}-ascorbate system is particularly useful in screening for compounds that protect macromolecules by chelating metal ions necessary for the site-specific formation of the reactive oxygen species. Such a protective role was demonstrated for creatinine (Glazer, 1988c).

The phycoerythrin assay has been used to screen for tissue-specific protective agents which might yield distinctive oxidation products upon reaction with peroxy radicals. Such products are potential markers for diseases or processes associated with high levels of oxidative stress. Protective agents uncovered by this assay which fit the above criteria are homovanillic acid in the brain (DeLange and Glazer, 1989), and bile acids in the gastrointestinal tract (DeLange and Glazer, 1990).

References

ALLEN, M.B., DOUGHERTY, E.C. and McLAUGHLIN, J.J.A., 1959, Chromoprotein pigments of some cryptomonad flagellates, *Nature*, **184**, 1047–1049.

BEAVIS, A.J. and PENNLINE, K.J., 1996, ALLO-7: A new fluorescent tandem dye for use in flow cytometry, *Cytometry* 24, 390–394.

BREJC, K., FICNER, R., HUBER, R. and STEINBACHER, S., 1995, Isolation, crystallization, crystal structure analysis and refinement of allophycocyanin from the cyanobacterium *Spirulina platensis* at 2.3 Å resolution, *Journal of Molecular Biology*, **249**, 424–440.

CAO, G., VERDON, C.P., WU, A.H.B., WANG, H. and PRIOR, R.L., 1995, Automated assay of oxygen radical absorbance capacity with the COBAS FARA II, *Clinical Chemistry*, **41**, 1738–1744.

CHALFIE, M., TU, Y., EUSKIRCHEN, G., WARD, W.W. and PRASHER, D.C., 1994, Green fluorescent protein as a marker for gene expression, *Science*, **263**, 802–805.

CHAUMONT, D., 1993, Biotechnology of algal biomass production: A review of systems for outdoor mass culture. *Journal of Applied Phycology*, 5, 593–604.

CHEE, M., YANG, R., HUBBELL, E., BERNO, A., HUANG, X.C., STERN, D., *et al.*, 1996, Accessing genetic information with high-density DNA arrays, *Science*, **274**, 610–614.

COHEN-BAZIRE, G., BÉGUIN, S., RIMON, S., GLAZER, A.N. and BROWN, D.M., 1977, Physicochemical and immunological properties of allophycocyanin, *Archives of Microbiology*, **111**, 225–238.

COLE, K.M. and SNEATH, R.G. (Eds.), 1990, *Biology of the Red Algae*, Cambridge, UK: Cambridge University Press.

CUBITT, A.B., HEIM, R., ADAMS, S.R., BOYD, A.E., GROSS, L.A. and TSIEN, R.Y., 1995, Understanding, using and improving green fluorescent protein, *Trends in Biochemical Sciences*, **20**, 448–455.

DAVIS, K.A., ABRAMS, B., LIN, Y., JASAYENA, S.D., 1996, Use of a high affinity DNA ligand in flow cytometry, *Nucleic Acids Research*, **24**, 702–706.

DELANGE, R.J. and GLAZER, A.N., 1989, Phycoerythrin fluorescence-based assay for peroxy radicals: a screen for biologically relevant protective agents, *Analytical Biochemistry*, **177**, 300–305.

DELANGE, R.J. and GLAZER, A.N., 1990, Bile acids: antioxidants or enhancers of peroxidation depending on lipid concentration, *Archives of Biochemistry and Biophysics*, **276**, 19–25.

DERMOUN, D., CHAUMONT, D., THEBAULT, J-M. and DAUTA, A., 1992, Modelling of growth of *Porphyridium cruentum* in connection with two interdependent factors: light and temperature, *Bioresource Technology*, **42**, 113–117.

DUERRING, M., HUBER, R., BODE, W., RUEMBELI, R. and ZUBER, H., 1990, Refined three-dimensional structure of phycoerythrocyanin from the cyanobacterium *Mastigocladus laminosus* at 2.7 Å, *Journal of Molecular Biology*, **211**, 633–644.

DUERRING, M., SCHMIDT, G.B. and HUBER, R., 1991, Isolation, crystallization, crystal structure analysis and refinement of constitutive C-phycocyanin from the chromatically adapting cyanobacterium *Fremyella diplosiphon* at 1.66Å resolution, *Journal of Molecular Biology*, **217**, 577–592.

FICNER, R. and HUBER, R., 1993, Refined crystal structure of phycoerythrin from *Porphyridium cruentum* at 0.23-nm resolution and localization of the γ subunit. *European Journal of Biochemistry*, **218**, 103–106.

FICNER, R., LOBECK, K., SCHMIDT, G. and HUBER, R., 1992, Isolation, crystallization, crystal structure analysis and refinement of B phycoerythrin from the red alga *Porphyridium sordidum* at 2.2 Å resolution, *Journal of Molecular Biology*, **228**, 935–950.

FOGG, G.E., STEWART, W.D.P., FAY, P. and WALSBY, A.E., 1973, *The Blue–Green Algae*, London: Academic Press.

GANTT, E., 1979, Phycobiliproteins of cryptophyceae, in Levandowsky, M. and Huttner, S.H. (Eds), *Biochemistry and Physiology of Protozoa*, Vol. 1, 2nd edn, pp. 121–137, New York: Academic Press.

GANTT, E., 1980, Structure and function of phycobilisomes: Light harvesting complexes in red and blue-green algae, *International Review of Cytology*, **66**, 45–80.

GLAZER, A.N., 1976, Phycocyanins, structure and function, in Smith, K.C. (Ed.), *Photochemical and Photobiological Reviews*, Vol. 1, pp. 71–115, New York: Plenum Press.

GLAZER, A.N., 1981, Photosynthetic accessory proteins with bilin prosthetic groups, in Hatch, M.D. and Boardman, N.K. (Eds), *The Biochemistry of Plants*, Vol. 8, *Photosynthesis*, pp. 51–96, New York: Academic Press.

GLAZER, A.N., 1984, Phycobilisome. A macromolecular complex optimized for energy transfer, *Biochimica and Biophysica Acta*, **768**, 29–51.

GLAZER, A.N., 1985, Light harvesting by phycobilisomes, *Annual Review of Biophysics and Biophysical Chemistry*, **14**, 47–77.

GLAZER, A.N., 1986, Phycobilisomes: Relation of structure to energy flow dynamics, in Youvan, D.C. and Daldal, F. (Eds), *Microbial Energy Transduction. Genetics, Structure and Function of Membrane Proteins*, pp. 31–36, Current Communications in Molecular Biology, New York: Cold Spring Harbor Laboratory.

GLAZER, A.N., 1988a, Phycobilisomes, *Methods in Enzymology* **167**, 304–312.

GLAZER, A.N., 1988b, Phycobiliproteins, *Methods in Enzymology* **167**, 291–303.

GLAZER, A.N., 1988c, Fluorescence-based assay for reactive oxygen species: a protective role for creatinine, *FASEB Journal*, **2**, 2487–2491.

GLAZER, A.N., 1989, Light guides. Directional energy transfer in a photosynthetic antenna, *Journal of Biological Chemistry*, **264**, 1–4.

GLAZER, A.N., 1994, Phycobiliproteins – a family of valuable, widely used fluorophores, *Journal of Applied Phycology*, **6**, 105–112.

GLAZER, A.N. and HIXSON, C.S., 1975, Characterization of R-phycocyanin. Chromophore content of R-phycocyanin and C-phycoerythrin, *Journal of Biological Chemistry*, **250**, 5487–5495.

GLAZER, A.N. and HIXSON, C.S., 1977, Subunit structure and chromophore composition of rhodophytan phycoerythrins. *Porphyridium cruentum* B-phycoerythrin and b-phycoerythrin, *Journal of Biological Chemistry*, **250**, 5487–5495.

GLAZER, A.N. and STRYER, L., 1983, Fluorescent tandem phycobiliprotein conjugates. Emission wavelength shifting by energy transfer, *Biophysics Journal*, **43**, 383–386.

GLAZER, A.N. and STRYER, L., 1984, Phycofluor probes, *Trends Biochem. Sci.*, **9**, 423–427.

GLAZER, A.N. and STRYER, L., 1990, Phycobiliprotein-avidin and phycobiliprotein-biotin conjugates. *Methods in Enzymology*, **184**, 188–194.

GLAZER, A.N. and WEDEMAYER, G.J., 1995, Cryptomonad biliproteins – an evolutionary perspective, *Photosynthesis Research*, **46**, 93–105.

GLAZER, A.N. and WEDEMAYER, G.J., 1996, Cryptomonad biliproteins: bilin types and locations, *Photosynthesis Research*, **48**, 163–170.

GLAZER, A.N., FANG, S. and BROWN, D.M., 1973, Spectroscopic properties of C-phycocyanin and of its α and β subunits, *Journal of Biological Chemistry*, **248**, 5679–5685.

GLAZER, A.N., WEST, J.A. and CHAN, C., 1982, Phycoerythrins as chemotaxonomic markers in red algae: a survey. *Biochemical Systematics and Ecology*, **10**, 203–215.

GRABOWSKI, J. and GANTT, E., 1978, Photophysical properties of phycobiliproteins from phycobilisomes: fluorescence lifetimes, quantum yields, and polarization spectra, *Photochemistry and Photobiology* **28**, 39–45.

HARDY, R.R.,1986, Purification and coupling of fluorescent proteins for use in flow cytometry, in Weir, D.A., Herzenberg, L.A., Blackwell, C.C. and Herzenberg L.A. (Eds), *Handbook of Experimental Immunology*, 4th edn, p. 31.1. Edinburgh: Blackwell.

HAUGLAND, R.P., 1996, *Handbook of Fluorescent Probes and Research Chemicals*, 6th edn, pp. 19–20, Eugene, OR: Molecular Probes, Inc.

HAXO, F.T. and FORK, D.C., 1959, Photosynthetically active accessory pigments of cryptomonads, *Nature*, **184**, 1051–1052.

HONSELL, E., KOSOVEL, V. and TALARICO, L., 1984, Phycobiliprotein distribution in Rhodophyta: studies and interpretation on the basis of their absorption spectra, *Botanica Marina*, **27**, 1–16.

HOOGHE, R.J. and OOMS, D., 1995, Use of the fluorescence-activated cell sorter (FACS) for in vitro assays of developmental toxicity, *Toxicology In Vitro*, **9**, 349–354.

KINDZELSKII, A.L., XUE, W., TODD, R.F. III AND PETTY, H.R., 1994, Imaging the spatial distribution of membrane receptors during neutrophil phagocytosis, *Journal of Structural Biology*, **113**, 191–198.

LANSDORP, P.M., SMITH, C., SAFFORD, M., TERSTAPPEN, L.W. and THOMAS, T.E., 1991, Single laser three color immunofluorescence staining procedures based on energy transfer between phycoerythrin and cyanine 5, *Cytometry*, **12**, 723–730.

LAO, K. and GLAZER, A.N., 1996, Ultraviolet-B photodestruction of a light-harvesting complex, *Proceedings of the National Academy of Sciences USA*, **93**, 5258–5263.

LEE, E.T.Y. and BAZIN, M.J., 1990, A laboratory scale air-lift helical photobioreactor to increase biomass output of photosynthetic algal cultures, *New Phytologist*, **116**, 331–335.

LEMBERG, R. and LEGGE, J.W., 1949, *Hematin Compounds and Bile Pigments*, New York: Interscience Publishers.

LYBARGER, L., DEMPSEY, D., FRANEK, K.J. and CHERVENAK, R., 1996, Rapid generation and flow cytometric analysis of stable GFP-expressing cells, *Cytometry*, **25**, 211–220.

MACCOLL, R., LEE, L.J. and BERNS, D.S., 1971, Protein aggregation in C-phycocyanin. Studies at very low concentrations with the photoelectric scanner of the ultracentrifuge, *Biochemical Journal*, **122**, 421–426.

MACCOLL, R., CSATORDAY, K., BERNS, D.S. and TRAEGER, E., 1981, The relationship of the quaternary structure of allophycocyanin to its spectrum, *Archives of Biochemistry and Biophysics*, **208**, 42–48.

MASLIAH, E., GE, N., ACHIM, C.L. and WILEY, C.A., 1994, Cytokine receptor alterations during HIV infection in the human central nervous system, *Brain Research*, **663**, 1–6

MATHIES, R.A. and STRYER, L., 1986, Single-molecule fluorescence detection: a feasibility study using phycoerythrin, in Taylor, D.L., Waggoner, A.S., Lanni, F., Murphy, R.F. and Birge, R. (Eds), *Applications of Fluorescence in Biomedical Sciences*, pp. 129–140, New York: Alan R. Liss.

MIHOVA, S.G., GEORGIEV, D.I., MINKOVA, K.M. and TCHERNOV, A.A., 1996 Phycobiliproteins in *Rhodella reticulata* and photoregulatory effects on their content, *Journal of Biotechnology*, **48**, 251–257.

MILLER, J.W., SELHUB, J. and JOSEPH, J.A., 1996, Oxidative damage caused by free radicals produced during catecholamine autoxidation; protective effects of O-methylation and melatonin, *Free Radical Biology and Medicine*, **21**, 241–249.

MONJI, N., 1988, Conjugation of haptens and macromolecules to phycobiliprotein for application in fluorescence immunoassay, in Pal, S.B. (Ed.), *Reviews on Immunoassay Technology*, Vol. 1, pp. 95–110, New York: Chapman and Hall.

MÖRSCHEL, E. and RHIEL, E., 1987, Phycobilisomes and thylakoids: the light-harvesting system of cyanobacteria and red algae, in Harris, J.R. and Horne, R.W. (Eds), *Membraneous Structures*, pp. 210–254, London: Academic Press.

MÖRSCHEL, E. and WEHRMEYER, W., 1975, Cryptomonad biliprotein: phycocyanin-645 from a *Chroomonas* species, *Archives of Microbiology*, **105**, 153–158.

MÖRSCHEL, E. and WEHRMEYER, W., 1977, Multiple forms of phycoerythrin-545 from *Cryptomonas maculata*, *Archives of Microbiology*, **113**, 83–89.

MUJUMDAR, R.B., ERNST, L.A., MUJUMDAR, S.R., LEWIS, C.J. and WAGGONER, A.S., 1993, Cyanine labeling reagents: sulfoindocyanine succinimidyl esters, *Bioconjugate Chemistry*, **4**, 105–111.

MUJUMDAR, S.R., MUJUMDAR, R.B., GRANT, C.M. and WAGGONER, A.S., 1996, Cyanine labeling reagents: sulfobenzindocyanine succinimidyl esters, *Bioconjugate Chemistry*, **7**, 356–362.

MYERS, J. and KRATZ, W.A., 1955, Relations between pigment content and photosynthetic characteristics in a blue-green alga, *Journal of General Physiology*, **39**, 11–22.

NEUFELD, G.J. and RIGGS, A., 1969, Aggregation properties of C-phycocyanin from *Anacystis nidulans*, *Biochimica et Biophysica Acta*, **181**, 234–243.

Ó hEOCHA, C. and RAFTERY, M., 1959, Phycoerythrins and phycocyanins of cryptomonads, *Nature*, **184**, 1049–1051.

OI, V.T., GLAZER, A.N. and STRYER, L., 1982, Fluorescent phycobiliprotein conjugates for analyses of cells and molecules, *Journal of Cell Biology*, **93**, 981–986.

ONG, L.J. and GLAZER, A.N., 1985, Crosslinking of allophycocyanin, *Physiologie Végetale*, **23**, 777–787.

ONG, L.J. and GLAZER, A.N., 1991, Phycoerythrins of marine unicellular cyanobacteria. I. Bilin types and locations and energy transfer pathways in *Synechococcus* sp. phycoerythrins, *Journal of Biological Chemistry*, **266**, 9515–9527.

PECK, K., STRYER, L., GLAZER, A.N. and MATHIES, R.A., 1989, Single-molecule fluorescence detection: autocorrelation criterion and experimental realization with phycoerythrin, *Proceedings of the National Academy of Sciences USA*, **86**, 4087–4091.

RIPPKA, R., 1988, Recognition and identification of cyanobacteria, *Methods in Enzymology*, **167**, 28–67.

RIPPKA, R., DERUELLES, J., WATERBURY, J.B., HERDMAN, M AND STANIER, R.Y., 1979, *Journal of General Microbiology*, **111**, 1–61.

ROEDERER, M., KANTOR, A.B., PARKS, D.R. and HERZENBERG, L.A., 1996, Cy7PE and Cy7APC: Bright new probes for immunofluorescence, *Cytometry*, **24**, 191–197.

SCHEER, H. and KUFER, W., 1977, Conformational studies on C-phycocyanin from *Spirulina platensis*, *Zeitschrift für Naturforschung*, **32c**, 513–519.

SCHROEDER, H., ZILLES, K., LUITEN, P.G.M., STROSBERG, A.D. and AGHCHI, A., 1989, Human cortical neurons contain both nicotinic and muscarinic acetylcholine receptors: an immunochemical double-labeling study, *Synapse*, **4**, 319–326.

SCHUBERT, W., 1991, Triple immunofluorescence confocal laser scanning microscopy: spatial correlation of novel cellular differentiation markers in human muscle biopsies, *European Journal of Cell Biology*, **55**, 272–285.

SEI, Y. and ARORA, P.K., 1991, Quantitative analysis of calcium mobilization after stimulation with mitogens or anti-CD3 antibodies: simultaneous fluo-3 and immunofluorescence flow cytometry, *Journal of Immunological Methods*, **137**, 237–244.

SHIH, C.C., BOLTON, G., SEHY, D., LAY, G., CAMPBELL, D. and HUANG, C.M., 1993, A novel dye that facilitates three-color analysis of PMBC by flow cytometry, *Annals of the New York Academy of Science*, **677**, 389–395.

SIDLER, W.A., 1994, Phycobilisome and phycobiliprotein structures, in Bryant, D.A. (Ed.), *The Molecular Biology of Cyanobacteria*, pp. 139–216, Dordrecht: Kluwer Academic Publishers.

SIDLER, W., KUMPF, B., SUTER, F., MORISSET, W., WEHRMEYER, W. and ZUBER, H.,1985, Structural studies on cryptomonad biliprotein subunits. Two different α subunits in *Chroomonas* phycocyanin-645 and *Cryptomonas* phycoerythrin-545. *Biol. Chem. Hoppe-Seyler*, **366**, 233–244.

SIDLER, W., NUTT, H., KUMPF, B., FRANK, G., BRENZEL, A., WEHRMEYER, W. and ZUBER, H., 1990, The complete amino acid sequence and the phylogenetic origin of phycocyanin-645 from the cryptophytan alga *Chroomonas* sp., *Biol. Chem. Hoppe-Seyler*, **371**, 537–547.

SOPER, S.A., NUTTER, H.L., KELLER, R.A., DAVIS, L.M. and SHERA, E.B., 1993, The photophysical constants of several fluorescent dyes pertaining to ultrasensitive fluorescence spectroscopy, *Photochemistry and Photobiology*, **57**, 972–977.

STANIER, R.Y., KUNISAWA, R., MANDEL, M. and COHEN-BAZIRE, G., 1971, Purification and properties of unicellular blue-green algae (Order *Chroococcales*), *Bacteriological Reviews*, **35**, 171–205.

WAGGONER, A.S., ERNST, L.A., CHEN, C.H. and RECKTENWALD, D.J., 1993, PE-CY5. A new fluorescent antibody label for three-color flow cytometry with a single laser, *Annals of the New York Academy of Science*, **677**, 185–193.

WATERBURY, J.B. and RIPPKA, R., 1989, The order Chroococcales, in N.R. Krieg and J.G. Holt, (Eds), *Bergey's Manual of Systematic Bacteriology*, Vol. 3, pp. 1746–1770, Baltimore: Williams and Wilkins.

WEBER, G. and TEALE, F.W.J., 1965, Interaction of proteins with radiation, in Neurath, H. (Ed.) *The Proteins*, 2nd edn, pp. 445–521, New York: Academic Press.

WEDEMAYER, G.J., WEMMER, D.E. and GLAZER, A.N., 1991, Phycobilins of cryptophycean algae. Structures of novel bilins with acryloyl substituents from phycoerythrin 566, *Journal of Biological Chemistry*, **266**, 4731–4741.

WEDEMAYER, G.J., KIDD, D.G., WEMMER, D.E. and GLAZER, A.N., 1992, Phycobilins of cryptophycean algae. Occurrence of dihydrobiliverdin and mesobiliverdin in cryptomonad biliproteins, *Journal of Biological Chemistry*, **267**, 7315–7331.

WEDEMAYER, G.J., KIDD, D.G. and GLAZER, A.N., 1996, Cryptomonad biliproteins: bilin types and locations, *Photosynthesis Research*, **48**, 163–170.

WEMMER, D.E., WEDEMAYER, G.J. and GLAZER, A.N., 1993, Phycobilins of cryptophycean algae. Novel linkage of dihydrobiliverdin in a phycoerythrin 555 and a phycocyanin 645. *Journal of Biological Chemistry*, **268**, 1658–1669.

WHITE, J.C. and STRYER, L., 1987, Photostability studies of phycobiliprotein fluorescent labels, *Analytical Biochemistry*, **161**, 442–452.

WILBANKS, S.M. and GLAZER, A.N., 1993a, Rod structure of a phycoerythrin II-containing phycobilisome. I. Organization and sequence of the gene cluster encoding the major phycobiliprotein rod components in the genome of marine *Synechococcus* sp. WH8020, *Journal of Biological Chemistry*, **268**, 1226–1235.

WILBANKS, S.M. and GLAZER, A.N., 1993b, Rod structure of a phycoerythrin II-containing phycobilisome. II. Complete sequence and bilin attachment site of a phycoerythrin γ subunit, *Journal of Biological Chemistry*, **268**, 1226–1235.

WILSON, K.M., MORRISON, I.E.G., SMITH, P.R., FERNANDEZ, N. and CHERRY, R.J., 1996, Single particle tracking of cell-surface HLA-DR molecules using R-phycoerythrin labeled monoclonal antibodies and fluorescence digital imaging, *Journal of Cell Science*, **109**, 2101–2109.

YEH, S.W., GLAZER, A.N. and CLARK, J.R., 1986, Control of bilin transition dipole moment direction by macromolecular assembly: energy transfer in allophycocyanin, *Journal of Physical Chemistry*, **90**, 4578–4580.

YEH, S.W., ONG, L.J., CLARK, J.H. and GLAZER, A.N., 1987, Fluorescence properties of allophycocyanin and a crosslinked allophycocyanin trimer, *Cytometry*, **8**, 91–95.

YU, M-H., GLAZER, A.N., SPENCER, K.G. and WEST, J.A., 1981, Phycoerythrins of the red alga *Callithamnion*. Variation in phycoerythrobilin and phycourobilin content, *Plant Physiology*, **68**, 482–488.

12

Polysaccharides of red microalgae

SHOSHANA (MALIS) ARAD

Introduction

The biotechnology for the production of valuable biochemicals from microalgae is no longer a dream, but a firm reality, as may be seen in the successful commercial production of β-carotene as a health food from the green microalga *Dunaliella* and phycoerythrin as a diagnostic marker (phycofluor) from red microalgae. At this stage, it is essential that other successful products be developed if industry is to regard microalgae as a reliable source of valuable biochemicals. This bioindustrial area is, in fact, developing rapidly, the driving force being the tendency of the market to switch from synthetic to natural products. Of special interest are the red microalgae as a source of a wide variety of biochemicals, particularly sulphated polysaccharides.

Sulphated polysaccharides are used as gelling agents, stabilizers, thickeners and emulsifiers, mainly in footstuffs but also in paints, photographic films and pharmaceuticals and in tertiary oil recovery. The best known of these sulphated polysaccharides are agar–agar and carrageenan. The conventional sources of sulphated polysaccharides are seaweeds, especially the red macroalgae, which are usually harvested from their natural habitats (Gellenbeck and Chapman, 1983). This source is, however, not constant, because of intensive harvesting and ecological damage. The increasing demand for polysaccharides can thus not be met by traditional seaweed sources. Industry has therefore initiated a search for new sources of a more stable nature, e.g., xanthan gum from bacteria (Baird *et al.*, 1983). A potential new source of great promise is to be found in certain species of red microalgae (Ramus, 1986; Arad, 1987), the cells of which are surrounded by gel polysaccharides, as are the cells of the red macroalgae.

Red microalgae

The red microalgae belong to the division Rhodophyta, order Porphyridiales, which includes about eight genera, the best known of which are *Porphyridium*, *Rhodella* and *Rhodosorus*. Morphologically, the red microalgae are the simplest of all the rhodophytes, and reproduction is asexual. In most species, the algal cells are brown due to the presence of chlorophyll and phycoerythrin (Gantt and Lipschultz, 1972). Many strains grow in fresh water, while others grow in seawater, either as unicells or massed together into irregular colonies held in a mucilage. A great deal of work has been devoted to the genus *Porphyridium*, which grows naturally in a diverse spectrum of habitats. The ultrastructure of *Porphyridium* cells and the conditions for growth are well documented (Hoogenhaut, 1963; Brown and Richardson, 1968; Ramus, 1972).

The cells of the red microalgae are encapsulated within a sulphated polysaccharide in the form of a gel. The capsules are thinnest during the logarithmic phase of growth and thickest during the stationary phase, since the rate of production is higher than the rate of solubilization into the external medium (Ramus, 1972, 1986). During growth in a liquid medium, the viscosity of the medium increases due to the release of polysaccharide from the cell surface. Although it is believed that the capsule is homogeneous, it seems possible that the inner layer differs from the outer layer, as it does in the red macroalgae.

The main functions of the capsular polysaccharide are to protect the cells from desiccation (Ramus, 1981) and to enable the algae to grow in a marine environment (Kloareg and Quatrano, 1988). The polysaccharide may also serve as an ion exchanger (Ritchie and Larkum, 1982; Mariani *et al.*, 1985) or an ion reservoir; it may afford mechanical protection to the cell; or it may create a buffer layer around the cells to protect them from extreme environmental conditions (Arad, 1988). It has indeed been shown that the cells of *Porphyridium* are relatively stable over a wide range of pH values, temperatures and salinities (Ucko *et al.*, 1989).

Cell-wall polysaccharides of the red microalgae

Knowledge concerning the chemical composition and structure of red microalgal polysaccharides is limited, both becuase of their complexity and because of the lack – until a short time ago – of degrading enzymes capable of splitting the complex molecule. The polysaccharides of the different species of red microoaglae are heteropolymers, which have different chemical compositions and varying amounts of sulphate (Evans *et al.*, 1974; Heaney-Kieras *et al.*, 1977; Medcalf *et al.*, 1985; Geresh and Arad, 1991; Dubinsky *et al.*, 1992; Simon *et al.*, 1992). The main species being studied belong to the genera *Porphyridium* (Ramus, 1972, 1974, 1986; Ramus and Groves, 1972, 1974; Ramus and Robins, 1975) and *Rhodella* (Evans *et al.*, 1974; Evans and Callow, 1976; Kroen and Ramus, 1990; Dubinsky *et al.*, 1992). In all the species studied so far, the main sugars in the polymers are xylose, glucose and galactose in various ratios. Additional minor sugars, i.e., rhamnose, ribose, arabinose, mannose, 4-*O*-methyl galactose, and 3-*O*-methyl pentose, have also been detected. All the polysaccharides have glucuronic acid and half-sulphate ester groups, which confer acidic properties on the molecules. It was previously reported that the polysaccharide also contains protein (Heaney-Kieras *et al.*, 1977). Recently, we have resolved non-covalently bound proteins – the main one being a 66-kdalton glycoprotein moiety – that were found to be specific to *Porphyridium* sp. (Shrestha *et al.*, 1995). Studies on the predator–prey relationship between *Porphyridium* sp. and its predator, the dinoflagellate *Crypthecodinium cohnii*, revealed that the

66-kdalton glycoprotein on the cell wall is the biorecognition site for the dinoflagellate (Ucko, 1996).

The polysaccharides of *Porphyridium* sp., *P. cruentum*, *P. aerugineum* and *R. reticulata* were found to contain a basic disaccharide building block, which is an aldobiuronic acid, 3-*O*-(α-D-glucopyranosyluronic acid)-L-galactopyranose (Geresh *et al.*, 1990), which has sulphate groups attached to glucose and galactose in the 6 or 3 position (Lupescu *et al.*, 1991). The polysaccharides of *Porphyridium* are known to contain various fractions. Two fractions with differing amounts of sulphate were shown by electrophoresis and by anion-exchange chromatography (Evans *et al.*, 1974; Medcalf *et al.*, 1975; Heaney-Kieras and Chapman, 1976; Heaney-Kieras *et al.*, 1976, 1977). In *R. reticulata* and *Porphyridium* sp. it was also shown that the polysaccharides are composed of two fractions that differ in their charge (low and high) and in their chemical composition (Dubinsky, 1993; Geresh *et al.*, 1992). The molecular mass of polysaccharides from various red microalgae has been estimated to be $2-7 \times 10^6$ daltons (Heaney-Kieras and Chapman, 1976; Heaney-Kieras *et al.*, 1976, 1977; Arad, 1988; Geresh and Arad, 1991; Simon *et al.*, 1992; Arad *et al.*, 1993a).

In concentrated solutions (1–2 g/l), the polysaccharide was found to be stable (as reflected by its viscosity) when exposed to wide range of pH values (2–9), temperatures (30–120°C) and salinities. From the dependence of the intrinsic viscosity on ionic strength, it was estimated that the stiffness of the *Porphyridium* sp. polysaccharide chains is in the same range as that of xanthan and DNA (Eteshola *et al.*, 1996). It was thus hypothesized (from the overall dilute solution features) that the biopolymer chain molecules adopt an ordered conformation in solution.

Until very recently, there were no known enzymes – carbohydrolases – capable of cleaving red microalgal polysaccharides. However, two enzymatic activities (that have not yet been fully defined) that specifically degrade the polysaccharide have been found – a dinoflagellate, which preys specifically on *Porphyridium* sp. and degrades its polysaccharide (Ucko *et al.*, 1989), and a mixture of soil bacteria (Arad *et al.*, 1993b). Although the degradation products had relatively high molecular weights (similar to that of the native polysaccharide), their viscosities were significalty lower than the viscosity of the native polysaccharide (Arad *et al.*, 1993b; Simon *et al.*, 1992, 1993). The bacterial enzymes probably degrade a protein bridge in the molecule, causing a significant decrease in viscosity.

Ambient environmental conditions influence polysaccharide production; for example, nitrate and sulphate starvation of *Porphyridium* not only enhance production and solubilization of the polysaccharide (Köst *et al.*, 1984; Wanner and Köst, 1984; Thepenier *et al.*, 1985; Adda *et al.*, 1986; Arad *et al.*, 1988, 1992), but also influence its chemical composition (Ucko *et al.*, 1994). Preliminary results in our laboratory indicate that CO_2 concentration affects both algal growth and production of cell wall polysaccharide, as well as influencing the distribution between bound and dissolved polysaccharide fractions and modifying polysaccharide composition.

Kroen and Ramus (1990) suggested that polysaccharide production is controlled at the carbon fixation level. However, findings from our laboratory indicate the exocellular polysaccharide production is controlled at the level of carbon partitioning rather than by total photosynthetic loading (Friedman *et al.*, 1991; Arad *et al.*, 1992).

Cell wall formation

In studies of the kinetics of polysaccharide release from *Porphyridium* cells, it was shown that the uptake of [^{35}S]sulphate by *P. aerugineum* cells is light dependent (Ramus, 1972; Ramus and Groves, 1972). Pulse-chase experiments revealed rapid labelling of the solubilized capsular polysaccharide recovered from the medium, indicating that the polysaccharide is sulphated intracellularly (Ramus and Groves, 1972) and then secreted (Ramus, 1974). Micrographs of *P. aerugineum* show membrane-bound vesicles containing fibrillar material identical to the capsular polysaccharide (Ramus, 1972; Ramus and Groves, 1972). These vesicles originate in the Golgi apparatus, migrate to the cell membrane, and release their contents to the cell surface by exocytosis. In the log phase, the Golgi apparatuses are large and more numerous, and the rate of polysaccharide synthesis is higher than in the stationary phase (Ramus and Robins, 1975). In *R. reticulata*, the polysaccharide mucilage is packed in the Golgi bodies and passes to the plasmalemma in large vesicles. Sulphation of the mucilage occurs in the Golgi cisternae (Evans *et al.*, 1974; Evans and Callow, 1976; Callow *et al.*, 1978; Triemer, 1980). The involvement of Golgi complex in selective transport and exocytosis has also been visualized using freeze-fracture (Tsekos and Reiss, 1988). By means of ^{14}C and ^{35}S labelling, combined with nitrogen and sulphate starvation and inhibition of carbon polymerization, our group has shown that the process of sulphation of the polysaccharide is not coupled to carbon polymerization but is regulated separately (Dubinsky, 1993).

One of the ways to study cell wall formation is using inhibitors of cell-wall biosynthesis. In our studies, a herbicide known to inhibit cellulose biosynthesis without affecting protein or nucleic acid synthesis – 2,6-dichlorobenzonitrile (DCB) (Delmer, 1991) – was found to inhibit growth and polysaccharide production in red microalgae (Arad *et al.*, 1993a, 1994). This finding is especially interesting, since no cellulose has been found in the cell wall of the red microalgae. In addition, a spontaneous DCB-resistant mutant of *R. reticulata* with a totally modified cell wall has been isolated: instead of ten sugars, one sugar – methyl galactose – predominates in the polysaccharide cell wall of the mutant, but the sulphate and protein contents of the mutant are like those of the wild type. In *Porphyridium* sp., it has been shown that DCB inhibits cell wall regeneration in protoplasts, and a DCB-resistant mutant of *Poryphyridium* sp. has been isolated (Arad *et al.*, 1994). In addition, another four mutants resistant to DCB were isolated spontaneously from *Porphyridium* sp., all with modified cell wall compositions that differ from one another. Various explanations have been suggested for the mode of inhibition by DCB: the herbicide may inhibit the bonding of sugars that are present in the polysaccharide (glucose/glucose; glucose/other sugars; or various sugars among themselves) or DCB may not be specific to glucose bonding.

Another approach used to study cell wall formation is following the cell cycle. To apply this approach, we synchronized cultures of *Porphyridium* sp. and *R. reticulata* by using alternating dark–light regimes; synchronization was confirmed by cell number and DNA replication (Cohen-Katz and Arad, 1998; Simon-Bercovitch *et al.*, 1999). By following the appearance of the various cell-wall constituents (sugars, proteins and sulphate) during the cycle, the sequence of events leading towards cell-wall formation was unravelled. From our preliminary studies, it seems that a xylose polymer is first produced and that the other sugars and the sulphate are then attached to this polymer: An intermediate polymer of about 0.5×10^6 daltons (a polyxylan) was isolated outside the cells at the beginning of the cycle. It appears that this polyxylan then polymerizes further to produce a higher molecular weight intermediate, which finally produces the cell-wall polysaccharide. The enzymatic polymerization thus occurs outside the cell.

Production of new polysaccharides by protoplast fusion

Since the know-how for the genetic manipulation of red microalgae lags far behind that of plants and since red microalgae reproduce asexually, protoplast fusion seemed to be the potentially most promising tool for genetic manipulation. For this type of operation, three prerequisites are necessary: viable protoplasts, genetic markers and somatic cell hybridization.

Production of protoplasts from algae is a relatively new field. Since the cell-wall structure of red algae is complex, only a few successful cases of production of protoplasts from this group of algae have been reported (Polne-Fuller and Gibor, 1986; Polne-Fuller *et al.*, 1986; Cheney and Kurtzman, 1992). Mixtures of cell-wall degrading enzymes were not successful for the production of viable protoplasts of red microalgae. However, using the two different enzyme sources mentioned above – one from the dinoflagellate (Roth-Bejerano *et al.*, 1991) and the other from a mixture of soil bacteria (Sivan *et al.*, 1992) – we have succeeded in producing viable protoplasts in *Porphyridium* sp.

Complementation of selectable markers provides a powerful selection system for the enrichment of somatic hybrids from protoplast fusion mixtures. A long-term programme to produce genetic markers in red microalgae has yielded mutants resistant to the herbicide sulphometuron methyl (SMM) (van Moppes *et al.*, 1989) and DCB (Arad *et al.*, 1993a, 1994) as well as phenotypic pigment mutants (Sivan and Arad, 1993).

Reports on protoplast fusion and regeneration of red algae are very limited. Some time ago, protoplasts of two species of *Porphyra* were fused (Matsumoto *et al.*, 1991). We have succeeded in fusing colour mutants of *Porphyridium* sp. (Sivan *et al.*, 1995), the first reported fusion in red microalgae. Recently, fusion of two herbicide-resistant mutants of *Porphyridium* sp. yielded a double mutant (Sivan and Arad, 1998). By fusion of various cell wall modified mutants, new families of polysaccharides can be produced. We are therefore not so far from producing specially tailored polysaccharides!

Applications of red microalgal polysaccharides

Red microalgal polysaccharides may be used in a wide field of activities from engineering through cosmetics to health foods. In tertiary oil recovery, the biopolymer of *P. aerugineum* has been applied as thickening agent for aqueous driving fluids to enhance recovery of petroleum trapped in the pore space of reservoir rocks (Savins, 1978; Ramus *et al.*, 1989). The biopolymer product was found to be superior, showing a fourfold higher apparent low shear rate viscosity yield than an equivalent weight of Kelzan (from *Xanthomonas campestris*).

The polysaccharides may also find application in human and animal health as dietary fibres and antiviral agents. It has been shown that serum cholesterol, triglyceride and very-low-density lipoprotein levels were considerably lower (about 25 per cent) in rodents receiving the biomass of *R. reticulata* or its polysaccharide in their diets than in control animals; no toxic effects were found. In glucose loading experiments, the levels of serum insulin and glucose were much lower in rats fed with algal biomass or polysaccharide than in control animals. These findings suggest that the polysaccharide acts as a dietary fibre (Dvir *et al.*, 1994; Yaron *et al.*, 1995). The sulphated polysaccharide of a red microalga has shown promising antiviral activity against *Herpes simplex* virus types 1 and 2 and *Varicella zoster* virus, with no cytotoxic effects on vero cells. It seems that the polysaccharide is able to inhibit viral infection either by preventing adsorption of virus into the host cells and/or by inhibiting the production of new viral particles inside the host cells (Huheihel *et al.*, 1995). These findings are currently being developed in our laboratories for pharmaceutical uses.

A variety of other potential applications have been evaluated in our laboratories. Recently, the polysaccharide of *R. reticulata* was cross-linked to form a chemically and mechanically stable system that could be used as an ion-exchanger for the removal of metal ions (e.g., Ni, Co, and Cd) (Geresh *et al.*, 1997). The stabilization properties of a red microalgal polysaccharide were used to prevent the degradation of aloe vera gel, as was shown by rheological studies (i.e., increased apparent viscosities that did not deteriorate during storage, apparent yield points, and, in some cases, hysteresis) (Yaron *et al.*, 1992). Applications of the heteropolymers as natural soil aggregating agents (biofertilizers), as matrixes for slow release (Kolani *et al.*, 1995), and for cosmetics are also being investigated and developed.

References

ADDA, M., MERCHUK, J.C. and ARAD (MALIS), S., 1986, Effect of nitrate on growth and production of cell wall polysaccharide by the unicellular red alga *Porphyridium*, *Biomass*, **10**, 131–140.

ARAD (MALIS), S., 1987, Biochemicals from unicellular red algae, *International Industrial Biotechnology*, June/July, 281–284.

ARAD (MALIS), S., 1988, Production of sulfated polysaccharides from red unicellular algae, in Stadler, T., Mollion, J., Verdus, M.C., Karamanos, Y., Morvan, H. and Christiaen, D. (Eds), *Algal Biotechnology*, pp. 65–87, London: Elsevier Applied Science.

ARAD (MALIS), S., FRIEDMAN (DAHAN), O. and ROTEM, A., 1988, Effect of nitrogen on polysaccharide production in *Porphyridium* sp., *Applied and Environmental Microbiology*, **54**, 2411–2414.

ARAD (MALIS), S., LERENTAL (BROWN), Y. and DUBINSKY, O., 1992, Effect of nitrate and sulfate starvation on polysaccharide production in *Rhodella reticulata*, *Bioresource Technology*, **42**, 141–148.

ARAD (MALIS), S., DUBINSKY, O. and SIMON, B., 1993a, A modified cell wall mutant of the red microalga *Rhodella reticulata* resistant to the herbicide DCB, *Journal of Phycology*, **29**, 309–313.

ARAD (MALIS), S. KEROSTIVSKY, G., SIMON, B., BARAK, Z. and GERESH, S., 1993b, Biodegradation of the sulphated polysaccharide of *Porphyridium* sp. by soil bacteria, *Phytochemistry*, **32**, 287–290.

ARAD (MALIS), S., KOLANI, R., SIMON-BERCOVITCH, B. and SIVAN, A., 1994, Inhibition by DCB of cell wall polysaccharide formation in the red microalga *Porphyridium* sp. (Rhodophyta), *Phycologia*, **33**, 158–162.

BAIRD, J.F., SANFORD, P.A. and COTTRELL, I.W., 1983, Industrial applications of some new microbial polysaccharides, *Bio/Technology*, **1**, 778–783.

BROWN, T.E. and RICHARDSON, F.L., 1968, The effect of growth environment on the physiology of algae: light intensity, *Journal of Phycology*, **4**, 38–42.

CALLOW, M.E., COUGHLAN, S.J. and EVANS, L.V.,1978, The role of Golgi bodies in polysacchride sulphation in *Fucus zygotes*, *Journal of Cell Science*, **32**, 337–356.

CHENEY, D.P. and KURTZMAN, L.A., 1992, Progress in protoplast fusion and gene transfer in red algae, in *Proceedings of the XIV International Seaweed Symposium*, Brittany, France, p. 2 (abstract).

COHEN-KATZ, A. and ARAD (MALIS), S., 1998, Biosynthesis of the cell-wall polysaccharide in the red microalga *Rhodella reticulata*, *Israel Journal of Plant Sciences*, **46**, 147–153.

DELMER, D.P., 1991, The biochemistry of cellulose synthesis, in Hoyd, C.W. (Ed.), *The Cytoskeleton Basis of Plant Growth and Form*, p. 101, London: Academic Press Ltd.

DUBINSKY, O., 1993, Sulfated Polysaccharide Production and Cell Wall Formation in the Unicellular Red Alga *Rhodella reticulata*, unpublished PhD thesis, Ben-Gurion University of the Negev, Beer-Sheva, Israel.

DUBINSKY, O., SIMON, B., KARAMANOS, Y., GERESH, S., BARAK, Z. and ARAD (MALIS), S., 1992, Composition of the cell wall polysaccharide produced by the unicellular red alga *Rhodella reticulata*, *Plant Physiology and Biochemistry*, **30**(4), 409–414.

DVIR, I., MAISLOS, M. and ARAD (MALIS), S., 1994, Feeding rodents with red microalgae, in *Dietary Fiber, Mechanisms of Action in Human Physiology and Metabolism*, June 20–22, France, Nantes, p. E6.

ETESHOLA, E., GOTTLIEB, M. and ARAD (MALIS), S., 1996, Dilute solution viscosity of red microalgae exopolysaccharide, *Chemical Engineering Science*, **51**(9), 1487–1494.

EVANS, L.V. and CALLOW, M.E., 1976, Secretory processes in seaweed, in Sunderland, N. (Ed.), *Perspectives in Experimental Biology*, Vol. 2, pp. 487–499, Oxford: Pergamon Press.

EVANS, L.V., CALLOW, M.E., PERCIVAL, E. and FAREED, V., 1974, Studies on the synthesis and composition of extracellular mucilage in the unicellular red alga *Rhodella*, *Journal of Cell Science*, **16**, 1–21.

FRIEDMAN, O., DUBINSKY, Z. and ARAD (MALIS), S., 1991, Effect of light intensity on growth and polysaccharide production in the red and blue-green Rhodophyta unicells, *Bioresource Technology*, **38**, 105–110.

GANTT, E. and LIPSCHULTZ, C.A., 1972, Phycobilisomes of *Porphyridium cruentum*, *Journal of Cell Biology*, **54**, 313–324.

GELLENBECK, K.W. and CHAPMAN, D.J., 1983, Seaweed uses: the outlook for mariculture, *Endeavour*, **7**, 377.

GERESH, S. and ARAD (MALIS), S., 1991, The extracellular polysaccharide of the red microalgae: chemistry and rheology, *Bioresource Technology*, **38**, 195–201.

GERESH, S., DUBINSKY, O., ARAD (MALIS), S., CHRISTIAEN, D. and GLASER, R., 1990, Structure of 3-*O*-(α-D-glucopyranosyluronic acid)-L-galactopyronase, an aldobiuronic acid isolated from polysaccharide of various unicellular red algae, *Carbohydrate Research*, **208**, 301–305.

GERESH, S., LUPESCU, N. and ARAD (MALIS), S., 1992, Fractionation and partial characterization of the sulfated polysaccharide of the red microagla *Porphyridium* sp., *Phytochemistry*, 4181–4186.

GERESH, S., ARAD (MALIS), S. and SHEFER, A., 1997, Chemically crossed linked polysaccharide of the red microalga *Rhodella reticulata* – an anion exchanger for toxic metal ions, *Journal of Carbohydrate Chemistry*, **16**(425), 703–708.

HEANEY-KIERAS, J. and CHAPMAN, D.J., 1976, Structural studies on the extracellular polysaccharide of the red alga *Porphyridium cruentum*, *Carbohydrate Research*, **52**, 169–177.

HEANEY-KIERAS, J., KIERAS, J.F. and BOWEN, D.V., 1976, 2-*O*-methyl-D-glucuronic acid, a new hexaronic acid of biological origin, *Biochemistry Journal*, **155**, 181–185.

HEANEY-KIERAS, J., RODEN, L. and CHAPMAN, D.J., 1977, The covalent linkage of protein to carbohydrate in the extracellular protein – polysaccharide from the red alga *Porphyridium cruentum*, *Biochemistry*, **165**, 1–9.

HOOGENHAUT, H., 1963, Synchronization of cell division in *Porphyridium aerugineum*, *Naturwissenschaften*, **50**, 456–457.

HUHEIHEL, M., SHANU, V.I., TAL, J. and ARAD (MALIS), S., 1995, Antiviral effect of red microalgal polysaccharide on *Herpes simplex* and *Varicella zoster* viruses, presentation at the Israel Society of Microbiology, Vol. 15, p. 103.

KLOAREG, B. and QUATRANO, R.S., 1988, Structure of the cell wall of marine algae and ecophysiological functions of the matrix polysaccharides, *Oceanography and Marine Biology an Annual Review*, **26**, 259–315.

KOLANI, R., ARAD (MALIS), S. and KOST, J., 1995, Red microalgal polysaccharide as an oral drug delivery matrix, presentation at International Symposium of Controlled Release Bioactive Materials, 22, Seattle, Washington, USA, August.

KÖST, H.P., SENSER, M. and WANNER, G., 1984, Effect of sulfate starvation on *Porphyridium cruentum* cells, *Zeitschrift für Pflanzenphysiologie*, **113S**, 231–249.

KROEN, K. and RAMUS, J., 1990, Viscous polysaccharide and starch synthesis in *Rhodella reticulata* (Porphyridiales, Rhodophyta), *Journal of Phytocology*, **26**, 226–271.

LUPESCU, N., GERESH, S., ARAD (MALIS), S., BERNSTEIN, M. and GLASER, R., 1991, Structure of some sulfated sugars isolated after acid hydrolysis of the extracellular polysaccharide of *Porphyridium* sp. unicellular red alga, *Carbohydrate Research*, **210**, 349–352.

MARIANI, P., TOLOMIO, C. and BRAGHETTA, P., 1985, An ultrastructural approach to the adaptive role of the cell wall in the intertidal alga *Fucus virsodides*, *Protoplasma*, **128**, 208–217.

MATSUMOTO, M., KAWASHIMA, Y., TOKUDA, H., FUKUI, E., AOYAMA, T. AND KAGEYAMA, H., 1991, Intrageneric protoplast fusion in *Porphyra*, *Journal of Phycology*, **27**, (Suppl), 48.

MEDCALF, D.G., SCOTT, J.R., BRANNON, J.H., HEMERICK, G.A., CUNNINGHAM, R.L., CHESSEN, J.H. and SHAH, J., 1975, Some structural features and viscometric properties of the extracellular polysaccharide from *Poryphyridium cruentum*, *Carbohydrate Research*, **44**, 87–96.

POLNE-FULLER, M. and GIBOR, A., 1986, Calluses, cells and protoplasts in the studies towards genetic improvement of seaweeds, *Aquaculture*, **57**, 117–123.

POLNE-FULLER, M., SAGA, N. and GIBOR, A., 1986, Algal cells, callus and tissue culture and selection of algal strains, in Barclay, W.R. and McIntosh, R.P. (Eds), *Algal Biomass Technologies*, pp. 30–36, Berlin Stuttgart: J. Cramer.

RAMUS, J., 1972, The production of extracellular polysaccharide by the unicellular red alga *Porphyridium aerugineum*, *Journal of Phycology*, **8**, 97–111.

RAMUS, J., 1974, *In-vivo* molybdate inhibition of sulfate transfer to *Porphyridium* capsular polysaccharide, Plant Physiology, **54**, 945–949.

RAMUS, J., 1981, The capture and transduction of light energy, in Lobban, C. S. and Wynne, M. J. (Eds), *The Biology of Seaweeds*, pp. 458–492, Boston: Blackwell Scientific Publications.

RAMUS, J., 1986, Rhodophyte unicells: biopolymer, physiology and production, in Barclay, W.R. and McIntosh, R.P. (Eds), *Algal Biomass Technologies*, pp. 51–55, Berlin-Stuttgart: J. Cramer.

RAMUS, J. and GROVES, S.T., 1972, Incorporation of sulfate into the capsular polysaccharide of the red alga *Porphyridium*, *Journal of Cell Biology*, **54**, 399–407.

RAMUS, J. and GROVES, S.T., 1974. Precursor-product relationships during sulfate incorporation into *Porphyridium* capsular polysaccharide, *Plant Physiology*, **53**, 434–439.

RAMUS, J. and ROBINS, D.M., 1975, The correlation of golgi activity and polysaccharide secretion in *Porphyridium*, *Journal of Phycology*, **11**, 70–74.

RAMUS, J., KENNEY, B. E. and SHAUGHNESSY, E.J., 1989, Drag reducing properties of micro-algal exopolymers, *Biotechnology Bioengineering*, **34**, 1203–1208.

RITCHIE, R.J. and LARKUM, D.V., 1982, Cation exchange properties of the cell-wall of *Enteromorpha intestinalis*, *Journal of Experimental Botany*, **33**, 125–139.

ROTH-BEJERANO, N., VAN MOPPES, D., SIVAN, A. and ARAD (MALIS), S., 1991, Potential production of protoplasts from *Porphyridium* sp. using an enzymatic extract of its pre-dator *Gymnodinium* sp., *Bioresource Technology*, **38**, 127–131.

SAVINS, J.G., 1978, Oil recovery process employing thickened aqueous driving fluid, Mobil Oil Corporation, NY, US patent 4,079,544.

SHRESTHA, R., BAR-ZVI, D. and ARAD (MALIS), S., 1995, Non-covalently bound cell wall proteins of the red microalga *Porphyridium* sp., in *7th Cell-wall Meeting*, p. 114, Santiago, Spain, Sept. 25–29.

SIMON, B., GERESH, S. and ARAD (MALIS), S., 1992, Degradation products of the cell wall polysaccharide of the red microalga *Porphyridium* sp. (Rhodophyta) by means of enzy-matic activity of its predator the dinoflagellate *Gymnodinium* sp. (Gymnodiales), *Journal of Phycology*, **28**, 460–465.

SIMON, B., GERESH, S. and ARAD (MALIS), S., 1993, Polysaccharide-degrading activities extracted during the growth of a dinoflagellate that preys on *Porphyridium* sp., *Plant Physiology and Biochemistry*, **31**, 387–393.

SIMON-BERCOVITCH, B., BAR-ZVI, D. and ARAD (MALIS), S. 1999, Cell wall formation during the cell cycle of *Porphyridium* sp. (UTEX 637) Rhodophyta, *Journal of Phycology* (in press).

SIVAN, A. and ARAD (MALIS), S., 1993, Induction and characterization of pigment mutants in the red microalga *Porphyridium* sp. (Rhodophyceae), *Phycologia*, **32**(1), 68–72.

SIVAN, A. and ARAD (MALIS), S., 1998, Intraspecific transfer of herbicide resistance in the red microalga *Porphyridium* sp. via protoplast fusion, *Journal of Phycology* **34**, 706–711.

SIVAN, A., VAN MOPPES, D. and ARAD (MALIS), S., 1992, Protoplast production from the unicellular red alga *Porphyridium* sp. (UTEX 637) by an extracellular bacterial lytic preparation, *Phycologia*, **31**, 253–261.

SIVAN, A., VAN MOPPES, D., THOMAS, J.C., DUBACQ, J.P. and ARAD (MALIS), S., 1995, Protoplast fusion and genetic complementation of pigment mutations in the red micro-alga *Porphyridium* sp. (UTEX 637), *Journal of Phycology*, **31**, 167–172.

THEPENIER, C., GUDIN, C. and THOMAS, D., 1985, Immobilization of *Porphyridium cruentum* in polyurethane foams for the production of polysaccharide, *Biomass*, **7**, 225–240.

TRIEMER, R.E., 1980, Role of golgi apparatus in mucilage production and cyst formation in *Euglena gracilis* (Euglenophyceae), *Journal of Phycology*, **16**, 46–52.

TSEKOS, I. and REISS, H.D., 1988, Occurrence and transport of particle 'tetrads' in the cell membrane of the unicellular red alga *Porphyridium* visualized by freeze-fracture, *Journal of Ultrastructure Molecular Structure Research*, **99**, 156–168.

UCKO, M., 1996, The dinoflagellate *Crypthecodinium cohnii*: description, life-cycle, food uptake and mode of recognition of its prey the red microalga *Porphyridium* sp., unpub-lished PhD thesis, Ben-Gurion University of the Negev, Beer-Sheva, Israel.

UCKO, M., COHEN, E., GORDIN, H. and ARAD (MALIS), S., 1989, Relationship between the unicellular red alga *Porphyridium* sp. and its predator the dinoflagellate *Gymnodinium* sp., *Applied and Environmental Microbiology*, **55**, 2990–2994.

UCKO, M., GERESH, S., SIMON-BERCOVITCH, B. and ARAD (MALIS), S., 1994, Predation by a dinoflagellate on a red microalga with cell wall modified by sulfate and nitrate starvation, *Marine Ecology Progress Series*, **104**(3), 293–298.

VAN MOPPES, D., BARAK, Z., CHIPMAN, D.M., GOLLOP, N. and ARAD (MALIS), S., 1989, A herbicide (sulfomeuron methyl) resistant mutant in *Porphyridium* (Rhodophyta), *Journal of Phycology*, **25**, 108–112.

WANNER, G. and KÖST, H.P., 1984, 'Membrane storage' of the red alga *Porphyridium cruentum* during nitrate and sulphate starvation, *Zeitschrift für Pflanzenphysiologie*, **113S**, 251–262.

YARON, A., COHEN, E. and ARAD (MALIS), S., 1992, Stabilization of *Aloe vera* gel by interaction with sulfated polysaccharides from red microalga and with xanthan gum, *Journal of Agriculture and Food Chemistry*, **40**, 1316–1320.

YARON, A., DVIR, I., MAISLOS, M., MOKADY, S. and ARAD (MALIS), S., 1995, The red microagla *Rhodella reticulata* as a source of the dietary ω-3 highly unsaturated fatty acid – eicosapentaenoic acid, in Charalambous, G. (Ed.), *Developments in Food Science. Food Flavors: Generation, Analysis and Process Influence*, Vol. 37, pp. 665–674, Amsterdam: Elsevier Science.

13

Polyhydroxyalkanoates

MASSIMO VINCENZINI AND ROBERTO DE PHILIPPIS

Introduction

Poly-hydroxyalkanoates (PHAs), for a long time thought to be exclusively represented by poly-β-hydroxybutyric acid [P(3HB)], the polyester first observed and characterized by Lemoigne at the Institute Pasteur (Lemoigne, 1923), now constitute a family of natural water-insoluble stereospecific polyesters of a wide range of different D(−)-hydroxyalkanoic acids (HAs), being characterized by the general chemical structure reported in Figure 1.

Polymers of HAs have been found in a wide variety of prokaryotes and also in many eukaryotic plant and animal cells, but only prokaryotes are able to accumulate high molecular weight PHAs in the form of cytoplasmic amorphous granules with almost no osmotic activity. Eukaryotic cells and some bacterial families, such as the Enterobacteriaceae, have been shown to possess a low molecular weight form of P(3HB), located in the cytoplasmic membrane and complexed to calcium ions and polyphosphate, possibly involved in genetic competence (Reusch and Sadoff, 1983, 1988; Reusch et al., 1986, 1987; Reusch, 1989).

The increasing interest in the development of biodegradable plastic products able to compete, at least to some extent, with the non-biodegradable, highly polluting, petrochemical polymers has given in recent years a tremendous impetus to researches on the high molecular weight PHAs of bacterial origin. This sharp scientific attention is documented by several recent reviews concerning either the general features and applications of PHAs (Byrom, 1987; Holmes, 1988; Lafferty et al., 1988; Brandl et al., 1990, 1995; Doi, 1990; Steinbüchel, 1991; Barham et al., 1992; Lenz et al., 1992; Mergaert et al., 1992; Doi et al., 1994; De Koning, 1995; Sasikala and Ramana, 1995; Lee, 1996) or the biology of PHA synthesizing bacteria (Anderson and Dawes, 1990; Doi, 1990; Steinbüchel, 1991; Steinbüchel and Schlegel, 1991; Stal, 1992; Steinbüchel et al., 1992; Fuller, 1995; Lee and Chang, 1995; Steinbüchel and Valentin, 1995). Surprisingly, not much research work has been done with cyanobacteria, in spite of the fact that their capability to synthesize P(3HB) was demonstrated 30 years ago (Carr, 1966) and that their photoautotrophic nature could lead to inexpensive PHA production from solar energy and carbon dioxide.

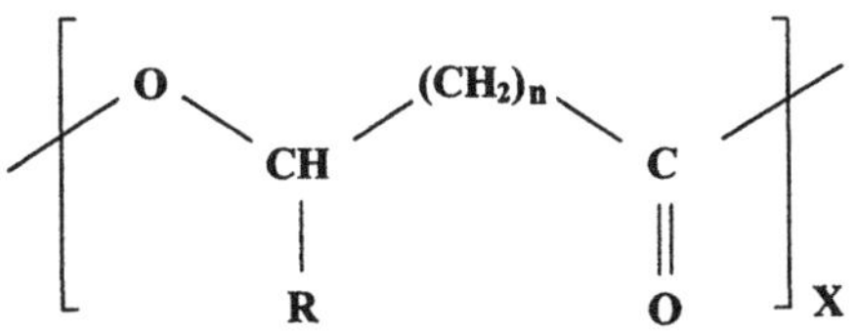

Figure 13.1 General structure of polyhydroxyalkanoates ('X' can reach values up to 30 000; the value of 'n' is generally 1 (in poly-3-hydroxyalkanoates), but in a few cases it has also assumed the value of 2 (in poly-4-hydroxyalkanoates) and 3 (in poly-5-hydroxyalkanoates) (Steinbüchel and Valentin, 1995; Lee, 1996); the R-pendant group includes the H-atom and a large variety of C-atom chains (Steinbüchel and Valentin, 1995)).

In this chapter, PHA research in cyanobacteria is emphasized and reviewed.

Properties of PHAs relevant to their use

PHAs are thermoplastic polymers with material properties ranging from brittle to flexible to rubbery, according to the presence of different kinds of hydroxyalkanoic acids. From a general point of view, the most attractive characteristics of these polymers are their material properties which are similar to those of conventional synthetic plastics, their biodegradability, their hydrophobicity and the possibility to use renewable resources for their production. On the other hand, their prices are still much higher than those of petrochemical derived plastics; e.g. US\$16/kg for BIOPOL, a copolymer of 3-hydroxybutyrate and 3-hydroxyvalerate [P(3HB-co-3HV)] manufactured in the UK by Zeneca BioProducts, in comparison with US\$1/kg for polypropylene (Lee, 1996), the synthetic polymer possessing material properties similar to those of PHAs.

A very important parameter to be considered in order to assess the quality of a polymer is its molecular mass, which generally affects the physical properties of all the polymers, including PHAs (Yeom and Yoo, 1995). Indeed, it has been reported that the mechanical properties of BIOPOL become unacceptable when the weight-average molecular mass is below 4×10^5 dalton (Taidi *et al.*, 1995). Anyway, many of the reported molecular weights of PHAs are above 11×10^6 dalton (Taidi *et al.* 1994), and can be modulated by either using different substrates and environmental conditions (Brandl *et al.*, 1988; Suzuki *et al.*, 1988; Daniel *et al.*, 1992; Kawaguchi and Doi, 1992; Taidi *et al.*, 1994, 1995) or changing the microbial strains (Ballard *et al.*, 1987; Doi, 1990; Bradel *et al.*, 1991), or by choosing appropriate procedures for the extraction of the polyesters from cells (Holmes, 1988; Berger *et al.*, 1989).

According to the terminology proposed by Anderson and Dawes (1990) and by Steinbüchel *et al.* (1992), PHAs can be divided into three classes, depending on the number of carbon atoms in the monomer units; short chain length (SCL), medium chain length (MCL) and long chain length polyhydroxyalkanoates (LCL-PHAs), composed by hydroxyacids with 3–5, 6–14 or more than 14 carbon atoms, respectively. At the present time, up to 91 different monomer units (but none with more than 14 C atoms) have been reported as constituents of PHAs, even if most of them are present only in a very limited number of cases and/or at very low concentrations (Steinbüchel and Valentin, 1995). Among SCL-PHAs, P(3HB) is the most commonly found in bacteria. It is a completely stereoregular polymer with 100 per cent of the asymmetric carbon atoms in the R stereochemical configuration (D (−) in traditional nomenclature). As a consequence, the polymer is isotactic, highly crystalline (crystallinity ranges between 55 and 80 per cent) and relatively stiff; its glass-to-rubber transition temperature (T_g) and its melting temperature (T_m) are comparable with isotactic polypropylene as well as some mechanical properties like Young's modulus and tensile strength (Table 13.1). Other properties of P(3HB) useful for specific applications are resistance to humidity, piezoelectricity and optical purity (Lee, 1996). On the other hand, the polymer possesses three main drawbacks: (1) a very low thermal stability (at temperatures around 170°C the molecular weight begins to decrease and at temperatures around 200°C the polymer decomposes, leading to the formation of crotonic acid); (2) values of the melt flow index (an industrial parameter used to assess the thermal processability of a polymer on industrial scale) well below acceptable levels (Uttley, 1986); and (3) a progressive embrittlement after the processing of the polymer, induced by a slow crystallization that constrains the amorphous phase between crystals (De Koning and Lemstra, 1993). These drawbacks can be overcomed by using P(3HB-co-3HV) copolymers, which are characterized by a random distribution of 3HB and 3HV units. The presence of 3HV units gives to the polymer better mechanical properties (Table 13.1) and a larger processability window, lowering T_m without affecting the value of the decomposition temperature (De Koning, 1995).

MCL-PHAs are fully isotactic, semicrystalline elastomers characterized by low T_m values, low tensile strength and high values of elongation to break (De Koning, 1995; Lee, 1996); one example of the characteristics of these polymers is reported in Table 13.1. This family of polyesters possesses two main drawbacks: they usually soften at temperatures around 40°C and they are characterized by very low crystallization rates. However, both these limitations can be overcome by cross-linking the polymers (De Koning, 1995).

As stated above, the most attractive features of these polymers are their biodegradability coupled with their hydrophobicity; indeed, even if cheaper biodegradable plastics are commercially available, only PHAs have a good moisture resistance, so that their use in applications for which these characteristics are needed (i.e. packaging of liquids and foods, paper coatings, etc.) can be envisaged. The possible applications of PHAs as plastic goods, related to their more significant properties, and their actual commercial uses are summarized in Table 13.2. However, in contrast with their great potential, a massive industrial production of these polyesters cannot yet be envisaged owing to the high prices of the substrates required for their production. In this connection, it was recently claimed that the use of vegetable oils for MCL-PHA production can reduce their manufacturing costs (Eggink et al., 1993).

Finally, in spite of the increasing number of studies on PHA biodegradability, only very few results on field experiments have been reported (Brandl and Püchner, 1992; Mergaert et al., 1992; Kaplan et al., 1994; Sawada, 1994), the bulk of papers referring to laboratory tests (recently reviewed by Brandl et al., 1995 and by Mukai and Doi, 1995). In any case, the biodegradation of PHAs has been observed in both anaerobic and aerobic environments (activated sludge, anaerobic sludge, estuarine sediment, soil, sea and lake water), and a large number of microbial species have been demonstrated to be capable of degrading PHAs.

Table 13.1 Material properties of some PHAs and commercial plastics (from Holmes, 1988; Lafferty et al., 1988; Brandl et al., 1990; Doi, 1990).

Polymer	T_g (°C)	T_m (°C)	Tensile strength (MPa)	Elongation to break (%)	Young's modulus (GPa)	Crystallinity (%)
P (3HB)	0–5	171–182	40	5	3.5	60–80
P (3HB-co-3HV)						
3 mol % 3HV	8	170	38	n.r.	2.9	n.r.
14 mol % 3HV	4	150	35	20[a]	1.5	n.r.
25 mol % 3HV	–6	137	30	100[b]	0.7	n.r.
P (3HHx-co-3HO)[c]	n.r.	61	10	300	n.r.	n.r.
Polypropylene	–10	176	38	400	1.7	50–70
Polyethylene	–40	141	8–35	400–800	0.2–1.4	n.r.
Polystyrene	n.r.	110	40–65	3–5	3.3	n.r.
Polyethyleneterephtalate	69	267	56	7300	2.2	30–50

T_g = glass-to-rubber transition temperature; T_m = melting temperature; n.r. = not reported; [a] Referred to a copolymer of 3-hydroxybutyrate (3HB) and 3-hydroxyvalerate (3HV) with 10 mol per cent 3HV; [b] referred to a copolymer with 20 mol per cent 3HV; [c] copolymer of 3-hydroxyhexanoate (3HHx) and 3-hydroxyoctanoate (3HO).

Table 13.2 Actual and potential industrial applications of PHAs (from Uttley, 1986; Lafferty et al., 1988; Doi, 1990; Brandl et al., 1990, 1995; Lee, 1996).

Application field	Uses	Properties useful for specific uses	Current status
Agriculture	Controlled release of pesticides, plant growth regulators, herbicides, fertilizers	Biodegradability, retarding properties	Research
	Covering foils	Biodegradability	Field experiments
	Seed encapsulation	Biodegradability	Conceptual
Chiral chromatography	Stationary phase for columns	Chiral properties	Conceptual
Chiral synthesis/cosmetics	Sources of chiral precursors	Stereoregularity, chiral properties	Research
Disposables	Razors, trays for food, utensils, etc.	Biodegradability, good mechanical properties	Commercial (Japan)
Hygiene products	Diapers, feminine hygiene products	Moisture resistance, good water barrier, biodegradability	Conceptual
Medical	Absorbable sutures, surgical pins, staples	Biocompatibility, biodegradability	Research/commercial
	Bone plates, films around bone fracture	Biocompatibility, piezoelectic properties[a]	Conceptual
Miscellaneous	Autoseparative air filters	Biodegradability[b]	Research
	Fibre-reinforced, biodegradable goods	Biodegradability, good mechanical properties	Research
Packaging	Bottles	Biodegradability, good mechanical properties, good liquid barrier, moisture resistance	Commercial (Europe, Japan, USA)
	Films for food packaging	Biodegradability, good liquid barrier, good mechanical properties, low O_2 permeability, moisture resistance	Research
Pharmaceutical	Paper coating	Biodegradability, moisture barrier	Conceptual
	Retarded drug release, drug carrier	Biocompatibility, biodegradability, retarding properties	Research

[a]Piezoelectricity stimulates bone growth; [b]biodegradation allows the autoseparation of particles from the filter.

Biosynthesis of PHAs

Most of the studies on PHA metabolism in bacterial cells deal with the most common poly-hydroxyalkanoate, P(3HB), whose biosynthesis and degradation usually occurs via a cyclic process, as first established in 1973 for *Azotobacter beijerinckii* and *Alcaligenes eutrophus* (Oeding and Schlegel, 1973; Senior and Dawes, 1973). In these microorganisms, as well as in many other bacteria when they are grown on carbohydrates or on acetyl-CoA yielding carbon sources, the biosynthetic route for the formation of P(3HB) starts from acetyl-CoA and includes three sequential, enzyme-mediated reactions: (1) a condensation of two acetyl-CoA molecules to yield acetoacetyl-CoA by the action of the 3-ketothiolase; (2) a reduction of acetoacetyl-CoA to D(−)-3-hydroxybutyryl-CoA catalyzed by an NADPH-dependent reductase; and (3) a PHA synthase-catalyzed polymerization of the 3-hydroxybutyrate monomer units. The first enzyme of the sequence, 3-ketothiolase, controls the synthesis of P(3HB) with free CoA acting as the key regulatory molecule (Oeding and Schlegel, 1973). Minor modifications to this biosynthetic pathway, like those occurring in *Rhodospirillum rubrum* which requires the additional action of two enoyl-CoA hydratases to form D(−)-3-hydroxybutyryl-CoA from acetoacetyl-CoA, were also established (Moskowitz and Merrick, 1969), but without changing the basic principles.

The overall cyclic nature of P(3HB) metabolism as it occurs in *Alcaligenes eutrophus* can be represented as shown in Figure 13.2.

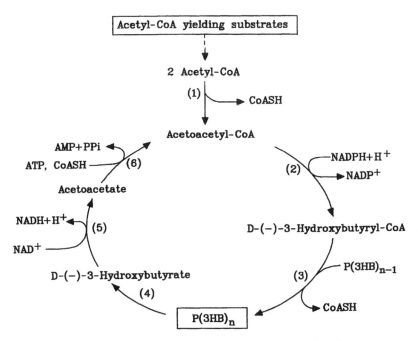

Figure 13.2 P(3HB) metabolism as it occurs in *Alcaligenes eutrophus* (from Doi, 1990; Doi *et al.*, 1990; Dawes, 1992). Enzymes: (1) 3-ketothiolase; (2) NADPH-linked acetoacetyl-CoA reductase; (3) PHA synthase; (4) PHA depolymerase; (5) 3-hydroxybutyrate dehydrogenase; (6) acetoacetyl-CoA synthase.

After the discovery of a copolyester (mainly 3HB-*co*-3HV) in activated sewage sludge, reported more than 20 years ago (Wallen and Rohwedder, 1974), and with the identification of an increasing number of different monomer units as constituents of PHAs from various bacteria (recently reviewed by Steinbüchel and Valentin, 1995), the picture of PHA biosynthesis in bacterial cells has become more complex. A further factor of complexity is derived from the discovery that hydroxyacid monomer units, usually originating from substrates having a carbon skeleton structurally related to the constituent monomer, can be also synthesized, as first observed by Dawes and collaborators (Anderson *et al.*, 1990), from structurally unrelated carbon sources. In this case, a central intermediate of the cell metabolism is directed to PHA biosynthesis so that complex polyesters, mainly consisting of MCL-3HAs, can be synthesized from simple sugars or other unrelated substrates (Haywood *et al.*, 1990, 1991; Timm and Steinbüchel, 1990; Anderson *et al.*, 1995; Kato *et al.*, 1996).

These findings have evidenced a great deal of flexibility in the functions of PHA biosynthetic key enzymes and in the reactions which convert metabolic intermediates into hydroxyacyl-CoA thioesters suitable for polymerization by PHA synthases. Indeed, evidence for at least two distinct types of both 3-ketothiolase and PHA synthase, exhibiting a broad substrate range but different substrate specificities, has been obtained from various bacteria (Haywood *et al.*, 1988a, 1988b, 1989, 1990).

The formation of the monomer units must always lead to hydroxyacyl-CoA thioesters with a D(−) configuration owing to the absolute stereospecificity of PHA synthases for the D(−) enantiomers (Anderson and Dawes, 1990). However, this can be accomplished in bacterial cells through few metabolic steps or after a complex sequence of reactions belonging to anabolic as well as catabolic pathways. In any case, the monomer composition of a biosynthetic PHA with constituents other than 3HB is generally dependent on the bacterial species and the carbon sources supplied. A general scheme for PHA biosynthesis in bacterial cells is given in Figure 13.3.

Occurrence of PHAs in cyanobacteria

To our knowledge, almost 90 cyanobacterial strains, belonging to more than 25 different genera included in four of the five subsections provided for cyanobacterial classification by the last edition of *Bergey's Manual of Systematic Bacteriology* (Vol. 3, 1989), have been tested for the presence of PHAs; no data are available for the cyanobacteria belonging to the II subsection (Pleurocapsales). Under photoautotrophic growth conditions, about 60 per cent of these strains were found to contain PHAs at concentrations ranging from 0.04 to 6.0 per cent of the cell dry weight (Table 13.3). However, it must be stressed that pure cultures were not always used and in many cases the analytical methods were only capable of identifying and quantifying the presence of P(3HB), as is the case for the methods based on the conversion of P(3HB) to crotonic acid and subsequent detection and quantification of this acid by UV spectroscopy (Law and Slepecky, 1961) or HPLC analysis (De Philippis *et al.*, 1992a). Thus, some of the strains that have been referred to as not containing P(3HB) could contain other PHAs, as has been observed in *Oscillatoria limosa* and *Spirulina subsalsa* by Stal *et al.* (1990); alternatively the PHA concentration could have been underestimated, if the strain has synthesized a copolymer. On the other hand, some of the papers that reported the occurrence of PHAs in cyanobacterial strains on the mere basis of microscopic observations not supported by

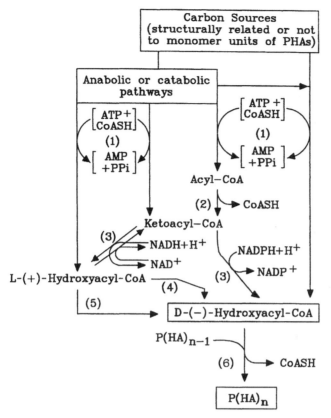

Figure 13.3 General scheme of PHA biosynthesis as it can occur in different bacterial strains (from Doi, 1990; Steinbüchel, 1991; Steinbüchel and Valentin, 1995). Enzymes: (1) Acyl-CoA synthetase; (2) 3-ketothiolase; (3) NAD(P)H-linked reductases; (4) Enoyl-CoA hydratases; (5) epimerases; (6) PHA synthases.

chemical analyses are difficult to interpret in order to establish the actual presence of the polyester, owing to the possibility of mistaking other cell inclusions for PHA granules, as stated by Stal in his recent review (1992). Finally, it should be noted that in spite of the increasing number of strains that are stated to contain PHAs, most of the papers deal with other aspects of cyanobacterial physiology and biochemistry, the presence of the polyester being reported without any further investigation of its metabolism. The only exceptions are the researches carried out with *Oscillatoria limosa* and *Gloeothece* sp. (Stal *et al.*, 1990; Stal, 1992), *Spirulina maxima*, now referred to as *Arthrospira maxima*, (De Philippis *et al.*, 1992a, 1992b), and *Anabaena cylindrica* (Lama *et al.*, 1996). However, from the data reported above, it appears that the presence of PHAs in cyanobacteria is much more widely distributed than was generally thought (Allen, 1984; Carr, 1988).

Since the first report on the stimulation induced by acetate on P(3HB) accumulation in cyanobacteria (Carr, 1966), many research groups have studied the effect of mixotrophic growth conditions with acetate on the amount of P(3HB) intracellularly accumulated. Among the approx. 40 strains tested (Table 13.3), more than 75 per cent showed a higher accumulation of PHAs in the presence of acetate with respect

Table 13.3 Cyanobacterial strains tested for PHA presence.

Species	PHA detection method [a]	PHA presence (P)[b] or absence (A)		References
		Photoautotrophy	Mixotrophy with acetate	
Subsection I: Chroococcales				
Aphanocapsa PCC 6308	CA-UV	A	A	Ihlenfeld and Gibson (1977)
Aphanocapsa PCC 6308	n.r.[c]	A	P	Allen (1984)
Aphanocapsa PCC 6714	CA-UV	A	A	Ihlenfeld and Gibson (1977)
Aphanocapsa PCC 6806	CA-UV	A	A	Ihlenfeld and Gibson (1977)
Aphanothece sp.	GLC	P (4.1)	P (4.7)	Steinbüchel and Tran (1996)[d]
Chroococcus PCC 7946	EM	P	n.t.[e]	Tomaselli (1997)[f]
Cyanothece PCC 7424	EM	P	n.t.	Tomaselli (1997)[f]
Gloeobacter PCC 7421	EM	P	n.t.	Tomaselli (1997)[f]
Gloeobacter violaceus ATCC 29082	GLC	A	P (2.8)	Steinbüchel and Tran (1996)[d]
Gloeocapsa PCC 6501	EM	P	n.t.	Rippka et al. (1971)
Gloeocapsa PCC 73106	EM	P	n.t.	Tomaselli (1997)[f]
Gloeocapsa ATCC 27928	GLC	P (1.7)	P (1.2)	Steinbüchel and Tran (1996)[d]
Gloeothece PCC 6909	GLC	P (2.6–6.4)[g]	P (3.3–4.7)[h]	Stal (1992)
Mycrocystis aeruginosa	EM	P[i]	n.t.	Jensen and Baxter (1981)
Mycrocystis aeruginosa (2 strains)	GLC	P	n.t.	Stal (1992)
Synechococcus PCC 6301	EM	A	n.t.	Tomaselli (1997)[f]
Synechococcus PCC 6307	EM	A	n.t.	Tomaselli (1997)[f]
Synechococcus PCC 7002	EM	A	n.t.	Tomaselli (1997)[f]
Synechococcus PCC 7942	GLC	A	n.t.	Suzuki et al. (1996)
Synechococcus sp. MA 19	GLC	P (21–27)[g]	n.t.	Miyake et al. (1996)
Synechococcus sp.	CA-UV	A	A	Ihlenfeld and Gibson (1977)
Synechococcus sp.	GLC	A	n.t.	Stal (1992)
Synechocystis PCC 6308	EM	A	n.t.	Tomaselli (1997)[f]

Table 13.3 Cyanobacterial strains tested for PHA presence (contd).

Species	PHA detection method[a]	PHA presence (P)[b] or absence (A)		References
		Photoautotrophy	Mixotrophy with acetate	
Synechocystis PCC 6803	EM	A	n.t.	Tomaselli (1997)[f]
Subsection II: Oscillatoriales				
Crinalium epipsammum SAG B22.89	GLC, CA-UV	A	n.t.	Stal *et al.* (1990)
Microcoleus chtonoplastes	GLC, CA-UV	A	n.t.	Stal *et al.* (1990)
Oscillatoria limnetica	CA-UV	A	A	Oren and Shilo (1979)
Oscillatoria limosa 23	GLC, CA-UV	P	P	Stal *et al.* (1990)
Phormidium sp. (5 strains)	GLC, CA-UV	A	n.t.	Stal (1992)
Plectonema borianum PCC 73110	GLC, CA-UV	A	n.t.	Stal *et al.* (1990)
Spirulina jenneri NK1	GLC, CA-UV, CA-HPLC	P (0.38)	n.t.	Vincenzini *et al.* (1990)
Spirulina laxissima MG5	GLC, CA-UV, CA-HPLC	P (0.30)	n.t.	Vincenzini *et al.* (1990)
Spirulina maxima (3 strains)	GLC, CA-UV, CA-HPLC	P (0.04–0.53)	P (2.4–3.2)[j,k]	Vincenzini *et al.* (1990); De Philippis *et al.* (1992a, 1992b)
Spirulina platensis	EM, CA-UV, IR	P (6.0)	P (6.0)	Campbell *et al.* (1982)
Spirulina platensis (4 strains)	GLC, CA-UV, CA-HPLC	P (0.05–0.73)	P (2.9)[k,l]	Vincenzini *et al.* (1990); De Philippis *et al.* (1992b)
Spirulina platensis SAG B85.79	GLC	P (0.6)	P (0.8)	Steinbüchel and Tran (1996)[d]
Spirulina subsalsa 85	GLC, CA-UV	P	n.t.	Stal *et al.* (1990)
Subsection III: Nostocales				
Anabaena cylindrica	n.r.	n.t.	P	Carr and Bradley (1973)
Anabaena cylindrica 10 C	CA-UV	A	P (2.0)[m]	Lama *et al.* (1996)
Anabaena cylindrica ATCC 27899	GLC	P (0.7)	P (0.5)	Steinbüchel and Tran (1996)[d]

Table 13.3 Cyanobacterial strains tested for PHA presence (contd.).

Species	PHA detection method [a]	PHA presence (P)[b] or absence (A)		References
		Photoautotrophy	Mixotrophy with acetate	
Anabaena doliolum HN-085	GLC	A	P (0.1)	Steinbüchel and Tran (1996)[d]
Anabaena hallensis HN-15	GLC	P (0.9)	P (2.6)	Steinbüchel and Tran (1996)[d]
Anabaena sp.	GLC	A	n.t.	Stal et al. (1990)
Anabaena torulosa SAG B26.79	CA-UV	A	n.t.	Lama et al. (1996)
Anabaena variabilis	CA-UV	A	A	Carr (1966)
Anabaena variabilis ATCC 29211	GLC	P (0.7)	P (2.6)	Steinbüchel and Tran (1996)[d]
Anabaenopsis siamensis	GLC	P (1.2)	P (5.0)	Steinbüchel and Tran (1996)[d]
Aphanizomenon gracile SAG B31.79	GLC	P (0.5)	P (5.5)	Steinbüchel and Tran (1996)[d]
Calothrix thermalis SAG B37.79	GLC	P (0.6)	P (1.1)	Steinbüchel and Tran (1996)[d]
Cyanospira rippkae ATCC 43194	CA-HPLC	A	n.t.	Our unpublished results
Geeotrichia raciborskii HN-03	GLC	P (0.7)	P (1.4)	Steinbüchel and Tran (1996)[d]
Nodularia haveyana SAG B50.79	GLC	P (0.6)	P (0.7)	Steinbüchel and Tran (1996)[d]
Nostoc commune HN-120	GLC	A	P (0.9)	Steinbüchel and Tran (1996)[d]
Nostoc commune SAG B1453–5	GLC	P (0.2)	P (0.6)	Steinbüchel and Tran (1996)[d]
Nostoc linckia	CA-HPLC	A	n.t.	Our unpublished results
Nostoc muscorum (2 strains)	GLC	P (0.3–0.9)	P (0.4–1.6)	Steinbüchel and Tran (1996)[d]
Nostoc sp.	GLC	A	n.t.	Stal (1992)
Nostoc sp. (symbiont of Cycas circinalis)	CA-HPLC	A	n.t.	Our unpublished results
Nostoc spp. (13 strains)	EM	P[(i)]	n.t.	Jensen (1980)
Scytonema sp. HN-16	GLC	A	P (3.9)	Steinbüchel and Tran (1996)[d]

302

Table 13.3 Cyanobacterial strains tested for PHA presence (contd.).

Species	PHA detection method [a]	PHA presence (P)[b] or absence (A)		References
		Photoautotrophy	Mixotrophy with acetate	
Tolypothrix tenuis (2 strains)	GLC	P (0.8–1.0)	P (2.7–3.4)	Steinbüchel and Tran (1996)[d]
Subsection IV: Stigonematales				
Chlorogloea fritschii	CA-UV	A	P (10.0)	Carr (1966)
Chlorogloea fritschii Mitra	EM, CA-UV	P	P	Jensen and Sicko (1971, 1973)
Fischerella SAG B.46.79	GLC	P (0.3)	P (1.2)	Steinbüchel and Tran (1996)[d]
Mastigocladus laminosus	GLC	A	P (2.5)	Steinbüchel and Tran (1996)[d]

Cyanobacteria are reported with the names which appeared in the original papers, but in many cases new generic and specific assignments have been introduced by the last edition of *Bergey's Manual of Systematic Bacteriology* (Vol. 3 1989) and by subsequent papers of taxonomic relevance.
[a] CA-UV = P(3HB) conversion to crotonic acid, detection and quantification by UV spectroscopy (Law and Slepecky (1961); CA-HPLC = P(3HB) conversion to crotonic acid, detection and quantification by reverse-phase high performance liquid chromatography (De Philippis *et al.* (1992a); EM = electron microscopy; GLC = gas liquid chromatography (Braunegg *et al.* (1978); IR = infrared spectroscopy; [b] in brackets are reported, if available, the intracellular PHA concentrations expressed as per cent on cell dry weight; [c] n.r. = not reported; [d] Steinbüchel, A. and Tran, H. (1996) personal communication; [e] n.t. = not tested; [f] Tomaselli, L. (1997) personal communication; [g] tested under various culture conditions; [h] tested after 48h of incubation with nitrate or under nitrogen-fixing conditions; PHA content of control cultures was 2.2 and 2.5 per cent of cell dry weight, respectively; [i] described as polymorphic bodies; [j] only two strains tested; [k] tested also mixotrophy with pyruvate; [l] only one strain tested; [m] tested also mixotrophy with propionate, valerate, acetate or glucose.

303

to autotrophic growth conditions, but for those strains that were unaffected by mixotrophic growth conditions, the fact cannot be excluded that the organic compound has not entered the cells. Other organic compounds such as pyruvate (Vincenzini *et al.*, 1990, De Philippis *et al.*, 1992b), propionate, valerate, glucose or citrate (Lama *et al.*, 1996) have shown a much lower stimulatory effect in comparison with acetate.

Table 13.3 does not include the few reports on PHA occurrence in cyanobacteria from natural samples because in these cases it is hard to define the nutritional conditions in which the polyesters have been synthesized by the cyanobacteria. The presence of P(3HB) granules has been observed in *Anacystis cyanea* collected from an eutrophic area of Lake Huron (Sicko-Goad, 1982) and in *Trichodesmium thiebautii* collected from a depth of 15 m in the northwestern Sargasso Sea (Siddiqui *et al.*, 1992b). Oval structures resembling P(3HB) granules were also detected, by means of ultrastructural analysis, in non-heterocystous cyanobacteria associated with *Trichodesmium* colonies in the southwestern Sargasso Sea (Siddiqui *et al.*, 1992a). Furthermore, a study on various natural samples taken from fresh and hypersaline waters and containing cyanobacteria as the dominant photoautotrophic population revealed the presence, at concentrations of about 0.02–0.03 per cent (on freeze dried weight of sample), of various kinds of (P3HB-*co*-P3HV) copolymers (Capon *et al.*, 1983).

As a concluding remark, ultramorphological evidence as well as the frequent presence of PHAs other than P(3HB) seem to exclude the fact that PHAs in cyanobacteria exist in the form of low molecular weight polyesters located in the cytoplasmic membrane structures in which only P(3HB) has been observed (Reusch and Sadoff, 1988).

Functions of PHAs in cyanobacteria

If the major functional roles claimed for P(3HB) in most of the bacteria so far investigated could be extended to all the other types of polyesters, as is reasonable, then PHAs should serve primarily as a carbon and/or energy reserve material, being intracellularly accumulated under certain unfavourable growth conditions such as limitation of nutrients other than carbon (Lafferty *et al.*, 1988; Anderson and Dawes, 1990, Dawes, 1992; Lee, 1996). Under certain circumstances, some additional roles for PHA biosynthesis and degradation have been proposed. Indeed, although less frequent, PHAs seem also to serve the bacterial cells to accomplish particularly stringent needs: for instance, the removal of superfluous reducing equivalents without dissipating useful resources, in the case of PHA biosynthesis, and the protection of sensitive enzymes from molecular oxygen by respiration of an oxidable substrate, in the case of PHA degradation (Senior and Dawes, 1971; Senior *et al.*, 1972; Stam *et al.*, 1986; McDermott *et al.*, 1989, De Philippis *et al.*, 1992a).

The role played by PHAs in cyanobacteria is still obscure in spite of the fact that evidence for the capability of cyanobacterial strains to synthesize the polyesters has greatly increased in recent years (Table 13.3). Indeed, no detailed information is yet available about factors controlling PHA accumulation and/or degradation in cyanobacteria during their photoautotrophic or mixotrophic growth so that no physiologycal function can be ascribed to cyanobacterial PHAs with confidence. On the other hand, the distribution of the PHA presence among cyanobacterial strains

(Table 13.3) demonstrates that any correlation between PHA occurrence and either morphological characteristics or metabolic peculiarities, such as nitrogen fixing capacity, should most probably be excluded. However, it is worth mentioning that Carr and Bradley (1973) reported on the occurrence of P(3HB) in *Anabaena cylindrica* grown in the presence of acetate pointing out that the polymer was located mainly, and possibly exclusively, in heterocysts where it would provide a reductant store of 2-carbon units, thus playing a certain role in the process of nitrogen fixation.

A role as an energy store seems quite unlikely under photoautotrophy as well as mixotrophy because cyanobacteria are unable to oxidize efficiently the acetyl-CoA originated from PHA degradation to yield significant ATP through tricarboxylic acid cycle-mediated reactions. Indeed, in these microorganisms, the cycle is generally interrupted at the level of α-ketoglutarate oxidation (Smith, 1982) and the enzymes of the glyoxylate shunt usually exhibit very low levels of activity (Smith and Hoare, 1977; Fay, 1983). Consequently, acetyl-CoA can only contribute to the synthesis of very few cell constituents, as stated by Smith (1982). On the other hand, the primary storage of energy for cyanobacteria is generally represented by granules of glycogen (Smith, 1982; Allen, 1984; Jensen, 1985; Shively, 1988): they are accumulated under virtually all culture conditions, provided that CO_2 fixation is not limited, and used as carbon and energy source under conditions of energy shortage, such as in the dark. A study on the effects induced by different growth conditions on the glycogen and P(3HB) content of *Spirulina maxima* (De Philippis *et al.*, 1992a) agrees with this statement, having demonstrated that glycogen, and not P(3HB), is the storage compound almost exclusively involved in the response of the cyanobacterial cells to nutrient imbalance known to stimulate the synthesis of PHAs in chemoheterotrophic bacteria. Finally, the possibility of a conversion of glycogen into PHAs, like that performed by the purple sulphur bacterium *Chromatium vinosum* – which in the dark oxidizes to P(3HB) the polyglucose accumulated during light periods while reducing to sulphide the intracellular globules of sulphur – was specifically investigated by Stal *et al.* (1990) using *Oscillatoria limosa*, a cyanobacterium capable of reducing elemental sulphur during anaerobic dark metabolism (Stal and Krumbein, 1986; Stal, 1991). The results demonstrated that the cellular PHA content was unaffected by anaerobic dark incubation of the cyanobacterial cells, although glycogen was metabolized.

All these findings strengthen the suggestion first advanced by Smith (1982) that PHAs must play only a minor role as storage compounds in cyanobacteria.

In *Oscillatoria limosa* (Stal *et al.*, 1990), the intracellular amount of PHA granules increased during the batch photoautotrophic growth, reaching a maximum when the culture entered the stationary phase. When *Oscillatoria limosa* cells were transferred from a culture in the stationary phase into a fresh culture medium, a dramatic decrease of PHA concentration was observed so that a role for PHA inclusions as a specific C-storage to be used for the synthesis of cell components necessary to enter into a new logarithmic phase of growth was hypothesized. However, the same putative role was not ascribed by Stal (1992) to PHA granules in *Gloeothece* sp. in spite of a pattern of PHA accumulation and mobilization quite similar to that previously observed in *Oscillatoria limosa*.

In *Spirulina maxima* (De Philippis *et al.*, 1992b), a significant accumulation of P(3HB) under photoautotrophic growth conditions was obtained by imposing phosphorus starvation on cells or by adding carbonylcyanide-*m*-chlorophenylhydrazone (CCCP), an uncoupler of photophosphorylation (Izawa and Good, 1972), to the culture medium. Since such conditions are known to generate an excess of reducing power by means of an imbalance in the formation of ATP and NADPH from photosynthesis (Bottomley and Stewart, 1976; Konopka and Schnur, 1981), P(3HB) occurrence in *Spirulina maxima* was explained by hypothesizing that P(3HB) acts as a buffer system capable of regulating the intracellular redox balance. On the other hand, in *Synechococcus* MA 19, Miyake and Asada (1995) suggested a role of carbon storage compound for PHB and energy storage for glycogen.

On the basis of these scant findings, it cannot be excluded that the role of PHAs in cyanobacteria during photoautotrophic growth might vary depending on the bacterial strain.

The presence of acetate in the culture medium seems to favour the accumulation of PHAs in most of the cyanobacterial strains so far tested (Table 13.3). This positive effect could be the consequence of the intracellular activation of acetate to acetyl-CoA, which can be directly utilized for PHA biosynthesis through the usual pathway (Figure 13.2). The recent experimental evidence of the incorporation of [1-^{13}C] acetate in P(3HB) by *Anabaena cylindrica* (Lama *et al.*, 1996) is consistent with this picture. However, on the basis of the results obtained with *Spirulina maxima* (De Philippis *et al.*, 1992a) and *Anabaena cylindrica* (Lama *et al.*, 1996), it should be emphasized that most of the acetate taken up by the cyanobacterial cells is utilized in the synthesis of cell constituents other than P(3HB).

From this brief overview it can be seen that information on the physiological function of PHAs in cyanobacteria is still incomplete but it also appears that the PHA production by these microorganisms cannot be as easily stimulated as in other bacteria. Evidently, the importance of PHAs for cyanobacteria and their growth cycle in laboratory cultures as well as in natural environments is not so high as for other bacteria.

Future prospects

The available data on the intracellular concentration of PHAs in cyanobacteria surely do not excite optimistic perspectives for an exploitation of these microorganisms as PHA producers. Notwithstanding this statement, it should be emphasized that the actual potential of cyanobacteria for PHA accumulation is still to be established and much basic research work must be carried out in order to achieve solid information on the factors which can trigger PHA synthesis. The dependence of PHA metabolism on variations in the intensity and spectral quality of incident light, for instance, is quite unexplored, and thus unknown, in spite of the considerable progress made in the last 20 years towards understanding the complex influence that light exerts upon the growth of these essentially photoautotrophic microorganisms, in which a close coordination is required between light-driven photosynthetic electron flow and other metabolic processes.

On the other hand, the possibility of utilizing photoautotrophic microorganisms in PHA production should not be discarded before assessing their actual potential because the use of sunlight and carbon dioxide for the production of PHAs could substantially reduce their manufactoring costs. Indeed, the present high price of PHAs, one of the key impediments to a greater use of the bacterial polyesters, is stimulating the development of biological processes which make use of substrates cheaper than the expensive organic acids presently employed by the industry for

PHA production with bacterial strains. The generation of transgenic plants (Poirier *et al.*, 1995) and recombinant cyanobacteria (Suzuki *et al.*, 1996; Miyasaka *et al.*, 1997) able to express PHA-biosynthetic genes of bacterial origin and to accumulate significant amounts of PHAs constitutes an enlightening example of the current research strategy.

Acknowledgements

The authors are indebted to Prof. A. Steinbüchel, Dr H. Tran and Dr L. Tomaselli for the communication of unpublished results. The results presented as L. Tomaselli's personal communication have been obtained in the framework of the EC Project 'BASIC – Biodiversity: Applied and Systematic Investigations of Cyanobacteria', contract no. BIO4-CT96-0256.

References

ALLEN, M.M., 1984, Cyanobacterial cell inclusions, *Annual Reviews Microbiology*, **38**, 1–25.

ANDERSON, A.J. and DAWES, E.A., 1990, Occurrence, metabolism, metabolic role and industrial uses of bacterial polyhydroxyalkanoates, *Microbiological Reviews*, **54**, 450–472.

ANDERSON, A.J., HAYWOOD, G.W., WILLIAMS, D.R. and DAWES, E.A., 1990, The production of polyhydroxyalkanoates from unrelated carbon sources, in Dawes, E.A. (Ed.), *Novel Biodegradable Microbial Polymers*, NATO ASI series, Vol. E 186, pp. 119–129, Dordrecht: Kluwer Academic Publishers.

ANDERSON, A.J., WILLIAMS, D.R., DAWES, E.A. and EWING, D.F., 1995, Biosynthesis of poly(3-hydroxybutyrate-*co*-3-hydroxyvalerate) in *Rhodococcus ruber*, *Canadian Journal of Microbiology*, **41** (Suppl. 1), 4–13.

BALLARD, D.G.H., HOLMES, P.A. and SENIOR, P.J., 1987, Formation of polymers of β-hydroxybutyric acid in bacterial cells and a comparison of the morphology of growth with the formation of polyethylene in the solid state, in Fontanille, M. and Guyot, A. (Eds), *Recent Advances in Mechanistic and Synthetic Aspects of Polymerization*, NATO ASI series, Vol. C 215, pp. 293–314, Lancaster: Reidel Publishing.

BARHAM, P.J., BARKER, P. and ORGAN, S.J., 1992, Physical properties of poly(hydroxybutyrate) and copolymers of hydroxybutyrate and hydroxyvalerate, *FEMS Microbiology Reviews*, **103**, 289–298.

BERGER, E., RAMSAY, B.A., RAMSAY, J.A. and CHAVARIE, C., 1989, PHB recovery by hypochlorite digestion of non-PHB biomass, *Biotechnology Technology*, **3**, 227–232.

Bergey's Manual of Systematic Bacteriology, 1989, Vol. 3, Baltimore: Williams and Wilkins.

BOTTOMLEY, P.J. and STEWART, W.D.P., 1976, ATP pools and transients in the blue-green alga, *Anabaena cylindrica*, *Archiv für Mikrobiologie*, **108**, 249–258.

BRADEL, R., KLEINKE, A. and REICHERT, K.H., 1991, Molar mass distribution of microbial poly(D(−)-3-hydroxybutyrate) in the course of intracellular synthesis and degradation, *Makromolecular Chemistry Rapid Communications*, **12**, 583–590.

BRANDL, H. and PÜCHNER, P., 1992, Biodegradation of plastic bottles made from 'Biopol' in an aquatic ecosystem under *in situ* conditions, *Biodegradation*, **2**, 237–243.

BRANDL, H., GROSS, R.A., LENZ, R.W. and FULLER, R.C., 1988, *Pseudomonas oleovorans* as a source of poly(β-hydroxyalkanoates) for potential applications as biodegradable polyesters, *Applied Environmental Microbiology*, **54**, 1977–1982.

BRANDL, H., GROSS, R.A., LENZ, R.W. and FULLER, R.C., 1990, Plastics from bacteria and for bacteria: poly(β-hydroxyalkanoates) as natural, biocompatible and biodegradable polyesters, *Advances in Biochemical Engineering/ Biotechnology*, **41**, 77–93.

BRANDL, H., BACHOFEN, R., MAYER, J. and WINTERMANTEL, E., 1995, Degradation and applications of polyhydroxyalkanoates, *Canadian Journal of Microbiology*, **41** (Suppl. 1), 143–153.

BRAUNEGG, G., SONNELEITNER, B. and LAFFERTY, R.M., 1978, A rapid gas chromatographic method for the determination of poly-β-hydroxybutyric acid in microbial biomass, *European Journal of Applied Microbiology Biotechnology*, **6**, 29–37.

BYROM, D., 1987, Polymer synthesis by microorganisms: technology and economics, *TIBTECH*, **5**, 246–250.

CAMPBELL, J., STEVENS, S.E. and BALKWILL, D.L., 1982, Accumulation of poly-β-hydroxy-butyrate in *Spirulina platensis*, *Journal of Bacteriology*, **149**, 361–363.

CAPON, B.J., DUNLOP, R.W., GHISALBERTI, E.M. and JEFFERIES, P.R., 1983, Poly-3-hydro-xyalkanoates from marine and freshwater cyanobacteria, *Phytochemistry*, **22**, 1181–1184.

CARR, N.G., 1966, The occurrence of poly-β-hydroxybutyrate in the blue-green alga, *Chloroglea fritschii*, *Biochimica Biophysica Acta*, **120**, 308–310.

CARR, N.G., 1988, Nitrogen reserves and dynamic reservoirs in cyanobacteria, in Rogers, L.J. and Gallon, J.R. (Eds), *Biochemistry of the Algae and Cyanobacteria*, pp. 13–21, Oxford: Clarendon Press.

CARR, N.G. and BRADLEY, S., 1973, Aspects of development in blue–green algae, in *Symposia of the Society for General Microbiology*, Vol. 23, pp. 161–188, Cambridge: Cambridge University Press.

DANIEL, M., CHOI, J.H., KIM, J.H. and LEBAULT, J.M., 1992, Effect of nutrient deficiency on accumulation and relative molecular weight of poly-β-hydroxybutyric acid by methylo-trophic bacterium, *Pseudomonas* 135, *Applied Microbiology Biotechnology*, **37**, 702–706.

DAWES, E.A., 1992, Storage polymers in prokaryotes, in Mohan, S., Dow, C. and Coles, J.A. (Eds), *Prokaryotic Structure and Function: A New Perspective*, pp. 81–122, Cambridge: Cambridge University Press.

DE KONING, G., 1995, Physical properties of bacterial poly((R)-3-hydroxyalkanoates), *Canadian Journal of Microbiology*, **41** (Suppl.1), 303–309.

DE KONING, G.J.M. and LEMSTRA, P.J., 1993, Crystallization phenomena in bacterial poly[(R)-3-hydroxybutyrate]. 2. Embrittlement and rejuvenation, *Polymer*, **34**, 4089–4094.

DE PHILIPPIS, R., SILI, C. and VINCENZINI, M., 1992a, Glycogen and poly-β-hydroxybutyrate synthesis in *Spirulina maxima*, *Journal of General Microbiology*, **138**, 1623–1628.

DE PHILIPPIS, R., ENA, A., GUASTINI, M., SILI, C. and VINCENZINI, M., 1992b, Factors affecting poly-β-hydroxybutyrate accumulation in cyanobacteria and in purple non-sulfur bacteria, *FEMS Microbiology Reviews*, **103**, 187–194.

DOI, Y., 1990, *Microbial Polyesters*, New York: VCH Publishers.

DOI, Y., SEGAWA, S., NAKAMURA, S. and KUNIOKA, M., 1990, Production of biodegradable copolyesters by *Alcaligenes eutrophus*, in Dawes, E.A. (Ed.), *Novel Biodegradable Microbial Polymers*, NATO ASI series, Vol. E 186, pp. 37–48, Dordrecht: Kluwer Academic Publishers.

DOI, Y., MUKAI, K., KASUYA, K. and YAMADA, K., 1994, Biodegradation of biosynthetic and chemosynthetic polyhydroxyalkanoates, in Doi, Y. and Fukuda, K. (Eds) *Biodegradable Plastics and Polymers*, pp. 39–51, Amsterdam: Elsevier.

EGGINK, G., VAN DER WAL, H., HUIJBERTS, G.N.M. and DE WAARD, P., 1993, Oleic acid as a substrate for poly-3-hydroxybutyrate formation in *Alcaligenes eutrophus* and *Pseudomonas putida*, *Industrial Crops Production*, **1**, 157–163.

FAY, P., 1983, *The Blue–Greens (Cyanophyta–Cyanobacteria)*, London: Edward Arnold.

FULLER, R.C., 1995, Polyesters and photosynthetic bacteria; from lipid cellular inclusions to microbial thermoplastics, in Blankeship, R.E., Madigan, M.T. and Bauer, C.E. (Eds),

Anoxygenic Photosynthetic Bacteria, pp. 1245–1256, Dordrecht: Kluwer Academic Publishers.

HAYWOOD, G.W., ANDERSON, A.J., CHU, L. and DAWES, E.A., 1988a, Characterization of two 3-ketothiolases possessing differing substrate specificities in the polyhydroxyalkanoate synthesizing organism *Alcaligenes eutrophus*, *FEMS Microbiology Letters*, **52**, 91–96.

HAYWOOD, G.W., ANDERSON, A.J., CHU, L. and DAWES, E.A., 1988b, The role of NADH- and NADPH-linked acetoacetyl-CoA reductases in the poly-3-hydroxybutyrate synthesizing organism *Alcaligenes eutrophus*, *FEMS Microbiology Letters*, **52**, 259–264.

HAYWOOD, G.W., ANDERSON, A.J. and DAWES, E.A., 1989, The importance of PHB-synthase specificity in polyhydroxyalkanoate synthesis by *Alcaligenes eutrophus*, *FEMS Microbiology Letters*, **57**, 1–6.

HAYWOOD, G.W., ANDERSON, A.J., EWING D.F. and DAWES, E.A., 1990, Accumulation of polyhydroxyalkanoates containing primarily 3-hydroxydecanoate from simple carbohydrate substrates by *Pseudomonas* sp. strain NCIMB 40135, *Applied Environmental Microbiology*, **56**, 3354–3359.

HAYWOOD, G.W., ANDERSON, A.J., WILLIAMS, D.R., DAWES, E.A. and EWING, D.F., 1991, Accumulation of poly(hydroxyalkanoate) copolymer containing primarily 3-hydroxyvalerate from simple carbohydrate substrates by *Rhodococcus* sp. strain NCIMB 40126, *International Journal of Biological Macromolecules*, **13**, 83–88.

HOLMES, P.A., 1988, Biologically produced (R)-3-hydroxyalkanoate polymers and copolymers, in Basset, D.C. (Ed.), *Developments in Crystalline Polymers*, Vol. 2, pp. 1–65, London: Elsevier.

IHLENFELD, M.J.A. and GIBSON, J., 1977, Acetate uptake by the unicellular cyanobacteria *Synechococcus* and *Aphanocapsa*, *Archives of Microbiology*, **113**, 231–241.

IZAWA, S. and GOOD, N.E., 1972, Inhibition of photosynthetic electron transport and photophosphorylation, *Methods in Enzymology*, **24**, 355–377.

JENSEN, T.E., 1980, Polymorphic bodies of various isolates of *Nostoc* (Cyanophyceae), *Microbios Letters*, **11**, 117–125.

JENSEN, T.E., 1985, Cell inclusions in the cyanobacteria, *Archiv für Hydrobiologie/Algological Studies*, **71**(38/39), 33–73.

JENSEN, T.E. and BAXTER, M., 1981, Vesicles and a vesicular crystalloid in *Microcystis aeruginosa* (Cyanophyceae), *Cytobios*, **32**, 129–137.

JENSEN, T.E. and SICKO, L.M., 1971, Fine structure of poly-β-hydroxybutyric acid granules in a blue-green alga, *Chlorogloea fritschii*, *Journal of Bacteriology*, **106**, 683–686.

JENSEN, T.E. and SICKO, L.M., 1973, The fine structure of *Chlorogloea fritschii* cultured in sodium acetate enriched medium, *Cytologia*, **38**, 381–391.

KAPLAN, D.L., MAYER, J.M., GREENBERGER, M., GROSS, R.A. and McCARTHY, S., 1994, Degradation methods and degradation kinetics of polymer films, *Polymer Degradation Stabilization*, **45**, 165–172.

KATO, M., BAO, H.J., KANG, C.K., FUKUI, T. and DOI, Y., 1996, Production of a novel copolyester of 3-hydroxybutyric acid and medium-chain-length 3-hydroxyalkanoic acids by *Pseudomonas* sp. 61-3 from sugars, *Applied Microbiology Biotechnology*, **45**, 363–370.

KAWAGUCHI, Y. and DOI, Y., 1992, Kinetics and mechanism of synthesis and degradation of poly(3-hydroxybutyrate) in *Alcaligenes eutrophus*, *Macromolecules*, **25**, 2324–2329.

KONOPKA, A. and SCHNUR, M., 1981, Biochemical composition and photosynthetic carbon metabolism of nutrient limited cultures of *Merismopedia tenuissima* (Cyanophyceae), *Journal of Phycology*, **17**, 118–122.

LAFFERTY, R.M., KORSATKO, B. and KORSATKO, W., 1988, Microbial production of poly-β-hydroxybutyric acid, in Rehm, H.J. and Reed, G. (Eds), *Biotechnology*, Vol. 6b, pp. 135–176, Weinheim: VCH Publishers.

LAMA, L., NICOLAUS, B., CALANDRELLI, V., MANCA, M.C., ROMANO, I. and GAMBACORTA, A., 1996, Effect of growth conditions on endo- and exopolymer biosynthesis in *Anabaena cylindrica* 10 C, *Phytochemistry*, **42**, 655–659.

LAW, J.H. and SLEPECKY, R.A., 1961, Assay of poly-β-hydroxybutyric acid, *Journal of Bacteriology*, **82**, 33–36.

LEE, S.Y., 1996, Bacterial polyhydroxyalkanoates, *Biotechnology and Bioengineering*, **49**, 1–14.

LEE, S.Y. and CHANG, H.N., 1995, Production of poly(hydroxyalkanoic acid), *Advances in Biochemical Engineering/Biotechnology*, **52**, 27–58.

LEMOIGNE, M., 1923, Production d'un acide β-oxybutyrique par certaines bactéries du groupe du *Bacillus subtilis*, *CR Hebdomadaires Seances Academie des Sciences*, **176**, 1761–1765.

LENZ, R.W., KIM, Y.B. and FULLER, R.C., 1992, Production of unusual bacterial polyesters by *Pseudomonas oleovorans* through cometabolism, *FEMS Microbiology Reviews*, **103**, 207–214.

MCDERMOTT, T.R., GRIFFITH, S.M., VANCE, C.P. and GRAHAM, P.H., 1989, Carbon metabolism in *Bradyrhizobium japonicum* bacteroids, *FEMS Microbiology Reviews*, **63**, 327–340.

MERGAERT, J., ANDERSON, C., WOUTERS, A., SWINGS, J. and KERSTERS, K., 1992, Biodegradation of polyhydroxyalkanoates, *FEMS Microbiology Reviews*, **103**, 317–322.

MIYAKE, M. and ASADA, Y., 1995, Biological role of polyhydroxybutyrate in cyanobacteria, in Mathis, P. (Ed.), *Photosynthesis: From Light to Biosphere*, pp. 403–406, Dordrecht: Kluwer Academic Publishers.

MIYAKE, M., ERATA, M. and ASADA, Y., 1996, A thermophilic cyanobacterium, *Synechococcus* sp. MA19, capable of accumulating poly-ß-hydroxybutyrate, *Journal of Fermentation and Bioengineering*, **82**, 512–514.

MIYASAKA, H., NAKANO, H., AKIYAMA, H., KANAI, S. and HIRANO, M., 1997, Production of PHA (polyhydroxyalkanoate) by a genetically engineered marine cyanobacterium, in *Abstracts of ICCDU IV (Fourth International Conference on Carbon Dioxide Utilization)*, p. 019.

MOSKOWITZ, G.J. and MERRICK, J.M., 1969, Metabolism of poly-β-hydroxybutyrate. II.Enzymatic synthesis of D(-)-β-hydroxybutyryl coenzyme A by an enoylhydrase from *Rhodospirillum rubrum*, *Biochemistry*, **8**, 2748–2755.

MUKAI, K. and DOI, Y., 1995, Microbial degradation of polyesters, in Singh, V.P. (Ed.), *Biotransformations: Microbial Degradation of Health Risk Compounds*, pp. 189–204, London: Elsevier.

OEDING, V. and SCHLEGEL, H.G., 1973, β-Ketothiolase from *Hydrogenomonas eutropha* H16 and its significance in the regulation of poly-β-hydroxybutyrate metabolism, *Biochemical Journal*, **134**, 239–248.

OREN, A. and SHILO, M., 1979, Anaerobic heterotrophic dark metabolism in the cyanobacterium *Oscillatoria limnetica*: sulfur respiration and lactate fermentation, *Archives of Microbiology*, **122**, 77–84.

POIRIER, Y., NAWRATH, C. and SOMERVILLE, C., 1995, Production of polyhydroxyalkanoates, a family of biodegradable plastics and elastomers, in bacteria and plants, *Bio/Technology*, **13**, 142–150.

REUSCH, R.N., 1989, Poly-β-hydroxybutyrate/calcium polyphosphate complexes in eukaryotic membranes, *Proc. Soc. Experimental Biology and Medicine*, **191**, 377–381.

REUSCH, R.N. and SADOFF, H.L., 1983, D-(-)-Poly-β-hydroxybutyrate in membranes of genetically competent bacteria, *Journal of Bacteriology*, **156**, 778–788.

REUSCH, R.N. and SADOFF, H.L., 1988, Putative structure and functions of poly-β-hydroxybutyrate/calcium polyphosphate channel in bacterial plasma membranes, *Proceedings of the National Academy of Sciences USA*, **85**, 4176–4180.

REUSCH, R.N., HISKE, T.W. and SADOFF, H.L., 1986, Poly-β-hydroxybutyrate membrane structure and its relationship to genetic transformability in *Escherichia coli*, *Journal of Bacteriology*, **168**, 553–562.

REUSCH, R.N., HISKE, T.W., SADOFF, H.L., HARRIS, R. and BEVERIDGE, T., 1987, Cellular incorporation of poly-β-hydroxybutyrate into plasma membranes of *Escherichia coli* and *Azotobacter vinelandii* alters native membrane structure, *Canadian Journal of Microbiology*, **33**, 435–444.

RIPPKA, R., NEILSON, A., KUNISAWA, R. and COHEN-BAZIRE, G., 1971, Nitrogen fixation by unicellular blue–green algae, *Archif für Mikrobiologie*, **76**, 341–348.

SASIKALA, C. and RAMANA, C.V., 1995, Biotechnological potentials of anoxygenic phototrophic bacteria. II. Biopolyesters, biopesticide, biofuel and biofertilizer, *Advances in Applied Microbiology*, **41**, 227–278.

SAWADA, H., 1994, Field testing of biodegradable plastics, in Doi, Y. and Fukada, K. (Eds), *Biodegradable Plastics and Polymers*, pp. 298–312, London: Elsevier.

SENIOR, P.J. and DAWES, E.A., 1971, Poly-β-hydroxybutyrate biosynthesis and the regulation of glucose metabolism in *Azotobacter beijerinckii*, *Biochemical Journal*, **125**, 55–66.

SENIOR, P.J. and DAWES, E.A., 1973, The regulation of poly-β-hydroxybutyrate synthesis in *Azotobacter beijerinckii*, *Biochemical Journal*, **134**, 225–238.

SENIOR, P.J., BEECH, G.A., RITCHIE, G.A. and DAWES, E.A., 1972, The role of oxygen limitation in the formation of poly-β-hydroxybutyrate during batch and continuous culture of *Azotobacter beijerinckii*, *Biochemical Journal*, **128**, 1193–1201.

SHIVELY, J.M., 1988, Inclusions: granules of polyglucose, polyphosphate and poly-β-hydroxybutyrate, *Methods in Enzymology*, **167**, 195–203.

SICKO-GOAD, L., 1982, A morphometric analysis of algal response to low dose, short term heavy metal exposure, *Protoplasma*, **110**, 75–86.

SIDDIQUI, P.J.A., BERGMAN, B. and CARPENTER, E.J., 1992a, Filamentous cyanobacterial associates of the marine planktonic cyanobacterium *Trichodesmium*, *Phycologia*, **31**, 326–337.

SIDDIQUI, P.J.A., BERGMAN, B., BJÖRKMAN, P.O. and CARPENTER, E.J., 1992b, Ultrastructural and chemical assessment of poly-β-hydroxybutyric acid in the marine cyanobacterium *Trichodesmium thiebautii*, *FEMS Microbiology Letters*, **94**, 143–148.

SMITH, A.J., 1982, Modes of cyanobacterial carbon metabolism, in Carr, N.G. and Whitton, B.A. (Eds), *The Biology of Cyanobacteria*, pp. 47–85, Oxford: Blackwell Scientific Publications.

SMITH, A.J. and HOARE, D.S., 1977, Specialist phototrophs, lithotrophs and methylotrophs: a unity among a diversity in prokaryotes? *Bacteriological Reviews*, **41**, 419–448

STAL, L.J., 1991, The sulphur metabolism of mat-building cyanobacteria in anoxic marine sediments, *Kieler Meeresforschstung*, **8**, 152–157.

STAL, L.J., 1992, Poly(hydroxyalkanoate) in cyanobacteria: an overview, *FEMS Microbiology Reviews*, **103**, 169–180.

STAL, L.J. and KRUMBEIN, W.E., 1986, Metabolism of cyanobacteria in anaerobic marine sediments, in *Actes du deuxième colloque international de bacteriologie marine*, Vol. 3, pp. 301–309, Brest: Gerbam Ifremer.

STAL, L.J., HEYER, H. and JACOBS, G., 1990, Occurrence and role of poly-hydroxyalkanoate in the cyanobacterium *Oscillatoria limosa*, in Dawes, E.A. (Ed.), *Novel Biodegradable Microbial Polymers*, NATO ASI series, Vol. E 186, pp. 435–438, Dordrecht: Kluwer Academic Publishers.

STAM, H., VAN VERSEVELD, H.W., D.VRIES, W. and STOUTHAMER, A.H., 1986, Utilization of poly-β-hydroxybutyrate in free-living cultures of *Rhizobium* ORS571, *FEMS Microbiology Letters*, **35**, 215–220.

STEINBÜCHEL, A., 1991, Polyhydroxyalkanoic acids, in Byrom, D. (Ed.), *Biomaterials: Novel Materials from Biological Sources*, pp. 124–213, New York: Stockton.

STEINBÜCHEL, A. and SCHLEGEL, H.G., 1991, Physiology and molecular genetics of poly(β-hydroxyalkanoic acid) synthesis in *Alcaligenes eutrophus*, *Molecular Microbiology*, **5**, 535–542.

STEINBÜCHEL, A. and VALENTIN, H.E., 1995, Diversity of bacterial polyhydroxyalkanoic acids, *FEMS Microbiology Letters*, **128**, 219–228.

STEINBÜCHEL, A., HUSTEDE, E., LIEBERGESELL, M., PIEPER, U., TIMM, A. and VALENTIN, H., 1992, Molecular basis for biosynthesis and accumulation of polyhydroxyalkanoic acids in bacteria, *FEMS Microbiology Reviews*, **103**, 217–230.

SUZUKI, T., DEGUCHI, H., YAMANE, T. and GEKKO, K., 1988, Control of molecular weight of poly-β-hydroxybutyric acid produced in fed-batch culture of *Protomonas extorquens*, *Applied Microbiology Biotechnology*, **27**, 487–491.

SUZUKI, T., MIYAKE, M., TOKIWA, Y., SAEGUSA, H., SAITO, T. and ASADA, Y., 1996, A recombinant cyanobacterium that accumulates poly-(hydroxybutyrate), *Biotechnology Letters*, **18**, 1047–1050.

TAIDI, B., ANDERSON, A.J., DAWES, E.A. and BYROM, D., 1994 Effect of carbon source and concentration on the molecular mass of poly(3-hydroxybutyrate) produced by *Methylobacterium extorquens* and *Alcaligenes eutrophus*, *Applied Microbiology and Biotechnology*, **40**, 786–790.

TAIDI, B., MANSFIELD, D.A. and ANDERSON, A.J., 1995, Turnover of poly(3-hydroxybutyrate) (PHB) and its influence on the molecular mass of the polymer accumulated by *Alcaligenes eutrophus* during batch culture, *FEMS Microbiology Letters*, **129**, 201–206.

TIMM, A. and STEINBÜCHEL, A., 1990, Formation of polyesters consisting of medium-chain-length 3-hydroxyalkanoic acids from gluconate by *Pseudomonas aeruginosa* and other fluorescent pseudomonads, *Applied Environmental Microbiology*, **56**, 3360–3367.

UTTLEY, N.L., 1986, Properties and possible applications of some biopolyesters, *Chimicaoggi*, **4** (7), 71–72.

VINCENZINI, M., SILI, C., DE PHILIPPIS, R., ENA, A. and MATERASSI, R., 1990, Occurrence of poly-β-hydroxybutyrate in *Spirulina* species, *Journal of Bacteriology*, **172**, 2791–2792.

WALLEN, L.L. and ROHWEDDER, W.K., 1974, Poly-β-hydroxyalkanoate from activated sludge, *Environmental Sciences and Technology*, **8**, 576–579.

YEOM, S.H. and YOO, Y.J., 1995, Effect of pH on the molecular weight of poly-3-hydroxybutyric acids produced by *Alcaligenes* sp., *Biotechnology Letters*, **17**, 389–394.

14

Pharmaceuticals and agrochemicals from microalgae

MICHAEL A. BOROWITZKA

Introduction

Living organisms are major sources of new biologically active molecules for the pharmaceutical, animal health and agrochemical industries. The screening of microbes, plants and animals for biologically active compounds has been a critical part of new drug discovery and development throughout this century. For example, about 13 500 naturally occurring antibiotics are known. Of these, 5500 are produced by actinomycetes, while about 3300 are produced by higher plants and about 90 are in current medical use (Berdy, 1989). These natural products serve not only as drugs in their own right, but they may also serve as structural models for the creation of synthetic analogues, and as models in structure–activity studies (e.g. Quinn *et al.*, 1993). There is an urgent need to discover new biologically active molecules as the rate of discovery of new 'lead' compounds has decreased steadily over the past few decades (cf. Omura, 1992). The microalgae are one exciting source of such new leads.

Algae have long been used for therapeutic purposes and the systematic examination of macro-algae for biologically active substances, especially antibiotics, began in the 1950s (Hoppe, 1979; Glombitza and Koch, 1989). In the 1970s the approach to screening organisms for biologically active compounds shifted from the use of pure compounds in primary screens, to using 'crude' extracts in preliminary screening and then using the bioactivity in the screen to direct the isolation and identification of the active compounds (Baker *et al.*, 1985; Reinehart, 1988). This approach was first applied on a large scale for the screening of marine algal and invertebrate extracts at the Roche Research Institute of Marine Pharmacology (RRIMP) in Australia (see von Berlepsch, 1980; Baker, 1984; Reichelt and Borowitzka, 1984 for reviews). More intensive investigations of microalgae, especially cyanobacteria, began in the 1980s (e.g. Cannell *et al.*, 1988a; Moore *et al.*, 1988; Kellam and Walker, 1989) and over the last decade the microalgae have become the focus for extensive screening. The development of new screens based on cell cultures or enzyme activities has also meant that much smaller amounts of material are required for the screens and that many more extracts and compounds can be screened (Reinehart, 1988; Suffness *et al.*, 1989; Yamaguchi *et al.*, 1989; Cardellina *et al.*, 1993). This means that the algal biomass required for primary screens can be easily cultured at the laboratory scale.

The microalgae are proving to be a valuable source of novel compounds, many with interesting biological activities which may lead to therapeutically useful agents. The diversity of chemical structures found in the microalgae is not surprising as the microalgae are a taxonomically and phylogenetically diverse group of microorganisms. They also have the advantage over many other organisms in that they can be cultured, thus providing an alternative to chemical synthesis for the production of these compounds. At present the cyanobacteria (blue–green algae) feature most prominently as sources of bioactive molecules. It is not clear whether this reflects an intrinsic property of this phylum or only the fact that many more of these algae have been screened in detail compared to the other algal phyla. The dinoflagellates are also providing many novel bioactive compounds, but they are much more difficult to isolate and culture and have not been studied as intensively. Other phyla which may be of interest, but where much further study is needed include the Haptophyta, the Cryptophyta and the Heterokontophyta/Chrysophyta.

This chapter provides an overview of the range of biologically active compounds found in algae and of the great diversity of chemical structures found to date. It does not attempt to be a complete compilation of all compounds discovered, nor of all activities found.

Antibiotics

The fact that microalgae may produce antibiotics has been known for some time (e.g. Harder and Opperman, 1953; Jorgensen and Steemann Nielsen, 1961). Since these early reports a large number of microalgal extracts and/or extracellular products have been shown to have antimicrobial (antibacterial, antifungal, antialgal, antiprotozoal) activity. For example, Cannell *et al.* (1988b) reported *in vitro* antibacterial activity in 10 per cent of the 300 freshwater algae screened, and Kellam *et al.* (1988) observed antifungal activity in about 5 per cent of the 532 marine and freshwater species screened. Despite these, and many other studies, the structure and identity of most of the active constituents is not yet known (see review by Pesando, 1990).

Those active compounds which have been identified belong to a diverse range of chemical classes. Table 14.1 summarizes some of the antimicrobial compounds isolated from microalgae. Acrylic acid [3] was the first antibiotically active compound to be isolated and unambiguously identified from algae (Sieburth, 1959, 1961) and since then has been found to be widespread (Glombitza, 1979). This is not surprising, as it may occur partially free or partially bound as the inactive dimethylpropiothetine [1] which can be split enzymatically or chemically to dimethylsulphide [2] and acrylic acid (Figure 14.1). Dimethylsulphide is a common product of the Chrysophyta (Malin *et al.*, 1994). Similarly, the antibiotically-active fatty acids such as γ-linolenic acid and eicosapentaenoic acid are very widespread among the algae (cf. Borowitzka, 1988a;

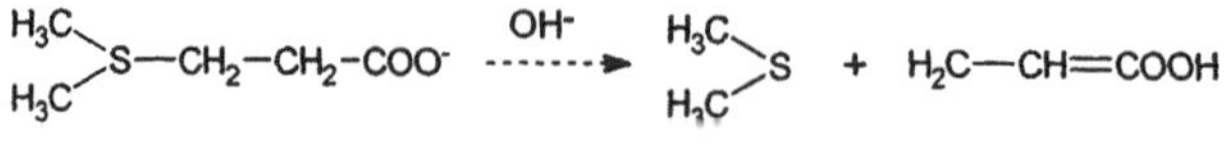

Figure 14.1 Formation of dimethylsulphide [2] and acrylic acid [3] from dimethylpropiothetine [1].

Viso and Marty, 1993; Dunstan *et al.*, 1994). The presence of γ-linolenic acid may also explain the activity of a methanolic extract of the cyanobacterium, *Spirulina platensis* against *Candida albicans* (Demule *et al.*, 1996).

Although antibacterial activity seems widespread amongst algal phyla, it is the antifungal activities which are of greatest interest and are most likely to find application in the near future. Moore's group in Hawaii have screened over 1000 strains of laboratory-cultured cyanobacteria for fungicidal activity using several species of fungi and yeasts and have observed activity in more than 10 per cent of the samples against one or more of the test organisms (Moon *et al.*, 1992). Some of the fungicides from cyanobacteria which have been characterized include the tjipanazoles **[4–7]** (Figure 14.2), isolated from the cyanobacterium, *Tolypothrix tjipanensis*, which are indolo[2,3-a]carbazoles, similar to those found in actinomycetes and slime moulds, but without a pyrrolo[3,4-c] ring (Bonjouklian *et al.*, 1991). They have little cytotoxicity and no *in vivo* activity against *Candida albicans*; however tjipanazole A1 **[4]** and A2 show appreciable fungicidal activity against rice blast and leaf rust wheat infections. The hapalindoles, such as hapalindole A **[8]** and related alkaloids, the ambiguines, were originally isolated from *Hapalosiphon fontinalis* and later also from *H. hibernicus, Fischerella ambigua* and *Westelliopsis prolifica*, and also show antibacterial and significant antimycotic activity (Moore *et al.*, 1987; Bonjouklian *et al.*,

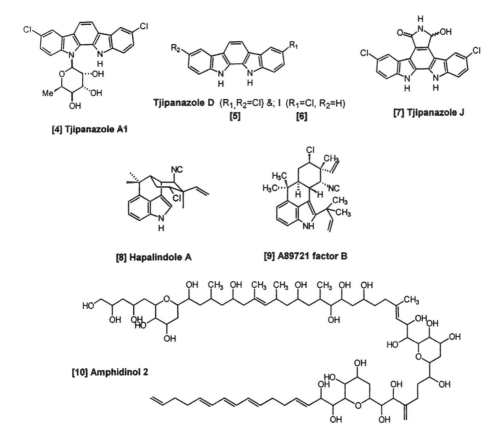

Figure 14.2 Antibiotically active compounds.

Table 14.1 Some of the antibiotically active compounds isolated from microalgae.

Type	Taxon	Activity	Reference
Chlorophyta			
Fatty acids	*Stichococcus bacillaris*	Bact	Harder and Opperman (1953)
	Protosiphon botryoides		
γ-Enolenic acid	*Dunaliella primolecta*	Bact	Ohta *et al.* (1995)
	Chlorococcum sp.		
Dinophyta			
Polyethers (gambieric acids A and B) [11, 12]	*Gambierdiscus toxicus*	Fung	Nagai *et al.* (1993, 1995)
Ciguatoxin [15]	*Gambierdiscus toxicus*	Fung	Nagai *et al.* (1990)
Okadaic acid [16]	*Prorocentrum lima*	Fung	Nagai *et al.* (1990)
Dinophysistoxin 1 [17]			
1,7-Deoxyyokadaic acid [18]			
Yessotoxin [20]	*Dinophysis fortii*	Fung	Nagai *et al.* (1990)
Goniodomin A [22]	*Alexandrium hiranoi* (= *Goniodoma pseudogonyaulax*)	Fung, yeast	Murakami *et al.* (1988); Murakami and Yamaguchi, 1989)
Amphidinol 2 [10]	*Amphidinium klebsii*	Fung	Satake *et al.* (1991; Paul *et al.* (1995)
β-ciketone	*Prorocentrum minutum*	Bact	Trick *et al.* (1984)
Accolyl-choline	*Amphidinium carteri*	Bact	Taylor *et al.* (1974)
Bacillariophyta			
Fatty acids	*Navicula delognei* f. *elliptica*	Bact	Findlay and Patil, 1984)
	Chaetoceros lauderi	Bact	Gauthier *et al.* (1978)
	Asterionella japonica	Bact	Pesando (1972)
Polysaccharides	*Chaetoceros lauderi*	Bact, fung	Pesando *et al.* (1980)
Asterionellins [23]	*Asterionella* sp	Bact	R. Wang (1992) cited in Shimizu (1993)
Chrysophyta			
	Stichochrysis immobilis	Bact	Berland *et al.* (1972)

316

Table 14.1 Some of the antibiotically active compounds isolated from microalgae (contd.).

Type	Taxon	Activity	Reference
Haptophyta			
Acrylic acid	*Phaeocystis*	Bact	Sieburth (1959)
Cyanobacteria			
Phenolics	*Nostoc muscorum*	Bact, yeast	De Cano *et al.* (1990)
Halogenated aromatics	*Calothrix brevissima*		Pedersen and DaSilva (1973)
Cyanobacterin [24]	*Scytonema hofmanni*	Bact, algae	Mason *et al.* (1981)
C13 chlorinated fatty acid [25]	*Schizothrix calcicola/*	Bact	Mynderse and Moore (1978)
	Oscillatoria nigriviridis		
Ambigol A [26]	*Fischerella ambigua*	Bact, fung	Falch *et al.* (1993)
Malingolide	*Lyngbya majuscula*	Bact	Cardinella *et al.* (1979)
Majusculamide C	*Lyngbya majuscula*	Fung	Carter *et al.* (1984)
Indolo[2,3-a]carbazoles	*Tolypothrix tjipanasensis*	Fung	Bonjouklian *et al.* (1991)
(Tjipanazoles) [4–7]			
Tjipanazole D [5]	*Fischerella ambigua*	Bact	Falch *et al.* (1993)
Cryptophycin 1 [27]	*Nostoc* sp. GSV 224	Fung	Schwartz *et al.* (1990)
Nostocyclamide	*Nostoc* sp. 31	Algae	Todorova and Juttner (1996)
Welwitindolineone A isonitrile [28]	*Hapalosiphon welwitschii* and *Westiella intricata*	Fung	Stratmann *et al.* (1994c)
Fischerellin A [39]	*Fischerella muscicola*	Fung	Li *et al.* (1989) Hagmann and Jüttner (1996)
Hormothamin	*Hormothamnion enteromorphoides*	Bact	Gerwick *et al.* (1988)
Calophycin [31]	*Calothrix fusca*	Fung	Moon *et al.* (1992)
Laxaphycin A [32] and B	*Anabaena laxa*	Fung	Frankmölle *et al.* (1992a, 1992b)
Scytonemin A [30]	*Scytonema* sp. (U-3-3)	Bact	Helms *et al.* (1988)
Scytophycins [34]	*Scytonema pseudohofmanni*	Fung	Ishibashi *et al.* (1986)

317

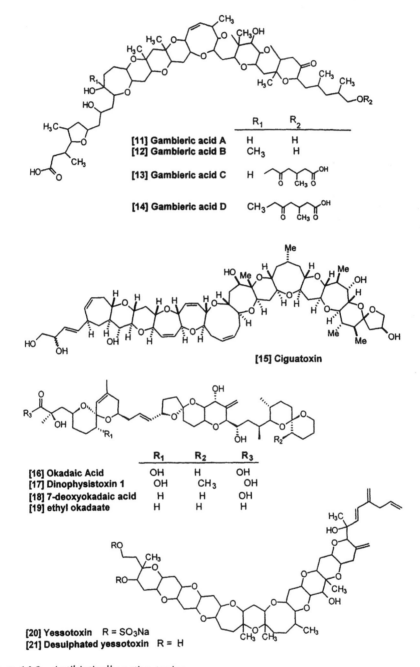

	R₁	R₂
[11] Gambieric acid A	H	H
[12] Gambieric acid B	CH₃	H
[13] Gambieric acid C	H	
[14] Gambieric acid D	CH₃	

[15] Ciguatoxin

	R₁	R₂	R₃
[16] Okadaic Acid	OH	H	OH
[17] Dinophysistoxin 1	OH	CH₃	OH
[18] 7-deoxyokadaic acid	H	H	OH
[19] ethyl okadaate	H	H	H

[20] Yessotoxin   R = SO₃Na
[21] Desulphated yessotoxin   R = H

Figure 14.3 Antibiotically active toxins.

1994). Several of the hapalindoles, including the A89271 factors [9], have been patented as antibacterial and antifungal agents (Moore and Patterson, 1988; Bonjouklian *et al.*, 1989, 1994).

The polyene-hydroxy compound, amphidinol 2 [10], produced by the marine dinoflagellate, *Amphidinium klebsii*, has some antifungal activity against

Aspergillus niger and, interestingly, several of the congeners of amphidinol found in the growth medium also have activity against the diatom, *Nitzschia* sp. (Paul *et al.*, 1995). Several of the dinoflagellate toxins such as the gambieric acids [11-14], ciguatoxin [15], okadaic acid and related compounds [16-18] and yessotoxin [19] also have *in vitro* anti-fungal activity.

Unfortunately, almost all of the studies have used only *in vitro* assays and, in analogy with macroalgae, it is likely that most of these compounds will have little or no application in medicine as they are either too toxic or are inactive *in vivo* (see Reichelt and Borowitzka, 1984; Borowitzka, 1995). They may, however serve as useful leads to new synthetic antibiotics or may find application in agriculture. Other algae may also find application as probiotics in aquaculture. For example, *Tetraselmis suecica* has been shown to inhibit strains of *Vibrio* sp. (Austin and Day, 1990), which are pathogens of cultured prawns.

Some algae also produce algaecides and these may find application in the control of algal blooms. Examples of such algaecides are the γ-lactone, cyanobacterin [24], produced by *Scytonema hofmanni* (Gleason *et al.*, 1986), fischerellin from *Fischerella muscicola* (Gross *et al.*, 1991), and an unidentified extracellular product of an *Oscillatoria* sp. (Bagchi *et al.*, 1990; Chauhan *et al.*, 1992). Cyanobacterin has been patented as a herbicide (Gleason, 1986).

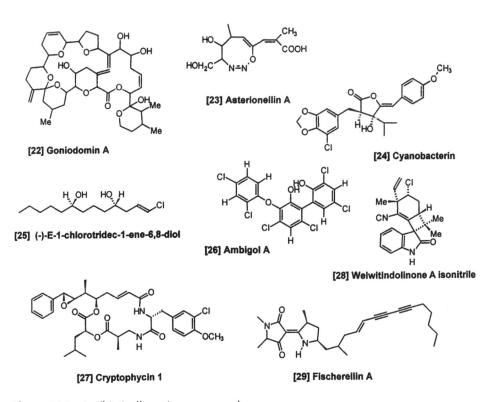

Figure 14.4 Antibiotically active compounds.

Antiviral activity

Microalgae, especially the cyanobacteria, are also a rich source of antiviral compounds. Rinehart *et al.* (1981) found that over 5 per cent of the extracts of cultured cyanobacteria screened by them showed antiviral activity against *Herpes simplex* virus type 2 (HSV-2), and more than 5 per cent had activity against respiratory syncytial virus. Similarly, Patterson *et al.* (1993a) screened some 600 strains of cyanobacteria and found antiviral activity in about 10 per cent of the extracts tested using live virus testing systems for inhibition of HSV-2 and human immunodeficiency virus type 1 (HIV-1), whereas about 2.5 per cent of the extracts were active against respiratory syncytial virus. Lau *et al.* (1993) have also screened extracts of over 900 strains of cyanobacteria for inhibition of reverse transcriptases of avian myeloblastosis virus and HIV-1, and they found that over 2 per cent of these algae showed promising activities.

Few of the active compounds have, however, been identified so far (Table 14.2). One compound whose structure is known is an anti-HIV sulpholipid [33] from *Lyngbya lagerheimii* (Gustafson *et al.*, 1989). Sulphated sterols have also been shown to have anti-HIV activity, but to date most of the studies have been done on invertebrates (eg. McKee *et al.*, 1994). Other compounds include anti-HSV-2 indolocarbazoles from *Nostoc sphaericum* (Knübel *et al.*, 1990) and a group of β-carbolines from *Dichothrix baueriana* (Larsen *et al.* in Patterson *et al.*, 1994). Ambiguol A [25] from *Fischerella ambigua* has been shown to inhibit HIV-1 reverse transcriptase and cyclooxigenase (Falch *et al.*, 1993).

A sulphated polysaccharide isolated from *Spirulina platensis* (Hayashi *et al.*, 1993; Hayashi, T. *et al.*, 1996) and named calcium spirulan, inhibits replication in enveloped viruses such as HSV-1, HIV-1, human cytomegalovirus, etc. This polysaccharide is composed of rhamnose, ribose, mannose, fructose, galactose, xylose, glucose, glucuronic acid, sulphate and calcium, and the calcium appears to be essential for maintaining the replication-inhibiting activity. Sulphated polysaccharides with antiviral activity are also known from several red algae and marine invertebrates (e.g. Beutler *et al.*, 1993; Bourgougnon *et al.*, 1993; Damonte *et al.*, 1994; Witvrouw *et al.*, 1994).

Cytotoxic, antitumour and antineoplastic metabolites

Cytotoxicity is defined as the inhibition of the proliferation of tumour cells *in vitro* (Suffness and Pezzuto, 1991) and is distinct from antitumour or antineoplastic activity which refer to *in vivo* activity in animal systems. Screening for cytotoxicity has

Table 14.2 Anti-viral compounds isolated from cyanobacteria

Alga	Compound	Virus	Reference
Dichothrix baueriana	β-Carbolines	HSV-2	Larsen *et al.* (1994)
Nostoc sphaericum	Indolocarbazole	HSV-2	Knübel *et al.* (1990)
Lyngbya lagerheimii and *Phormidium tenue*	Sulphonic acid containing glycolipids [33]	HIV-1	Gustafson *et al.* (1989)

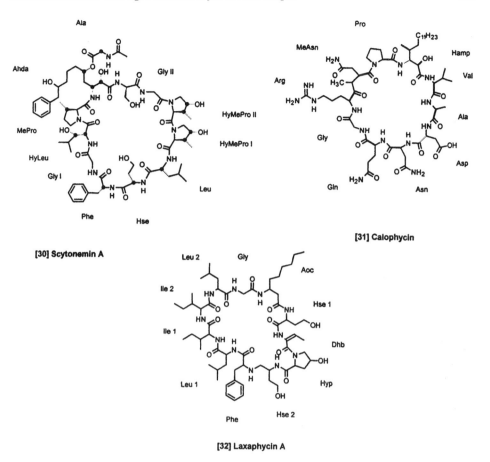

Figure 14.5   Antibiotically active peptides from cyanobacteria.

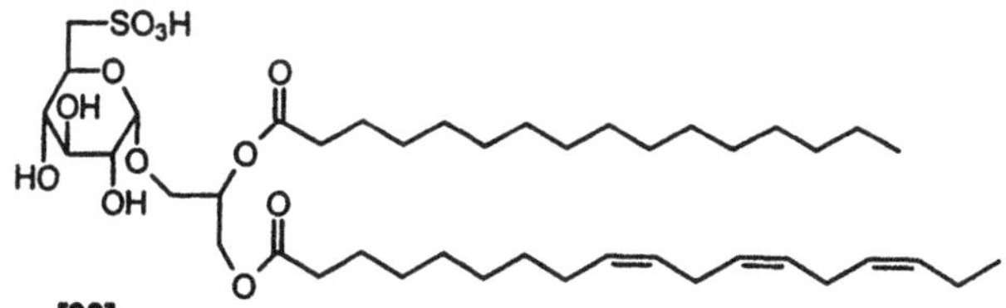

Figure 14.6   HIV-1 inhibiting sulphonic acid containing glycolipid from the cyanobacterium, *Lyngbya lagerheimii.*

focused mainly on the cyanobacteria. In their large-scale screening programme, Patterson *et al.* (1991) have found that approximately 7 per cent of the extracts tested inhibited proliferation of the human epidermoid carcinoma (KB) cell line, the primary screen used by them. The Scytonemataceae and Stigeonemataceae show the highest incidence of activity. A number of compounds responsible for this cytotoxicity have been characterized (Table 14.3).

The scytophycins (e.g. scytophycin B **[34]**) are polyketide-derived macrolides isolated from the scytonematacean genera *Scytonema* and *Tolypothrix* (Ishibashi *et al.*,

Table 14.3 Cytotoxic metabolites isolated from microalgae.

Alga	Compound	Cell line	Reference
Cyanobacteria			
Hormothamnion enteromorphoides	Hormothamnin A	Cytotoxic to several cancer cell lines	Gerwick et al. (1988)
	Hormithamnione [41]	P388	Gerwick et al. (1986)
		HL-60	
Westelliopsis prolifica	Westiellamide [40]	KB and LoVo	Prinsep et al. (1992b)
Scytonema mirabile	Scytophycin-type macrolides [34]	KB	Carmeli et al. (1990b)
	Isonitriles [44]	KB	Carmeli et al. (1990a)
	Mirabimides A–E [42, 43]	KB	Carmeli et al. (1991b);
		Solid tumour	Paik et al. (1994)
	Tantazole B [45]	KB	Carmeli et al. (1990c)
	Mirabazoles	KB	Carmeli et al. (1991a)
Tolypothrix byssoides	Tubercidin		Barchi et al. (1983)
Oscillatoria acutissima	Acutiphycins	KB, murine Lewis lung carcionoma *(in vivo)*	Barchi et al. (1984)
Dichthrix baueriana	Bauerines [50–52]	LoVo	Larsen et al. (1994)
		HSV-2	
Nostoc linckia	Borophycin [53]	LoVo	Hemscheidt et al. (1994)
Nostoc sp. GSV 224	Cryptophycins	Mammary, pancreatic, colon	Trimurtulu et al. (1994);
			Golakoti et al. (1995)
Phormidium tenue	Glycolipids	EBC-carrying human lymphoblastoid cells	Shirahashi et al. (1993)
Lyngbya majuscula	Microcolin A and B	P-338 leukaemia	Koehn et al. (1992)
	Malyngamide D [48]	KB	Gerwick et al. (1987)
	Curacin A [35]	Several cell lines	Gerwick et al. (1994a, 1995);
			Nagle et al. (1995)
	Majusculamide D	CCRF-CEM	Moore and Entzeroth, 1988

322

Table 14.3 Cytotoxic metabolites isolated from microalgae (contd).

Alga	Compound	Cell Line	Reference
Lyngbya	Debromoaplysiatoxin aplysiatoxin [54]	P338 lymphocytic mouse leukemia	Mynderse *et al.* (1977)
Aulosira fertilissima	Aulosirazole [55]	Solid tumour	Stratmann *et al.* (1994a)
Dinoflagellates			
Amphidinium sp.	Amphidinolides [36–38]	Antitumour, leukaemia	Kobayashi *et al.* (1986, 1989); Ishibashi *et al.* (1987)
Alexandrium hiranoi	Goniodomin A [22]	KB	Murakami and Yamaguchi (1989)
Prorocentrum lima	Prorocentrolide [39]	L1210 leukemia	Torigoe *et al.* (1988)
Gymnodinium breve	Hemibrevetoxin B [56]	Several tumour cell lines	Prasad and Shimizu (1989)
Cryptophytes			
Chrysophaeum taylori	6-Desmethoxy hormothamnione	KB	Gerwick, 1989

1986; Carmeli *et al.*, 1990b; Furusawa *et al.*, 1994) and the nostocacean genus *Cylindrospermum* (Jung *et al.*, 1991). The scytophycins inhibit the proliferation of a wide variety of mammalian cells including the KB cell line, human epidermoid carcinoma (Hep-2) cells, murine, intraperitoneally implanted P388 lymphocytic leukemia, Lewis lung carcinoma, human promyelocytic leukaemia (HL-60) cells and human colorectal adenocarcinoma (LoVo) cells (Ishibashi *et al.*, 1986; Moore *et al.*, 1986; Carmeli *et al.*, 1990b; Furusawa *et al.*, 1994). The scytophycins from *Scytonema pseudohofmanni* show cytotoxicity against KB cells at $1\,\mathrm{ng\,ml}^{-1}$. Scytophycins from *Scytonema ocellatum* and *S. burmanicum* have been patented as anticancer and antifungal agents (Patterson *et al.*, 1995). The acutiphycins from *Oscillatoria acutissima* are another group of macrolides with cytotoxicity against KB cells at $<1\,\mu\mathrm{g\,ml}^{-1}$, as well as having activity against murine, intraperitoneally implanted, Lewis lung carcinoma (Barchi *et al.*, 1984). One of the scytophycins, tolytoxin, has been shown to act by disrupting microfilament organization, resulting in a loss of the ability to form the contractile ring needed for cytokinesis in eukaryotic cells (Patterson *et al.*, 1993b). Curacin A [35], a thiazolene-containing lipid from *Lyngbya majuscula* with antiproliferative activity, acts by binding at the colcicine site on microtubules, thus inhibiting the polymerization of tubulin and microtubule formation (Gerwick *et al.*, 1995; Nagle *et al.*, 1995). The cryptophycins (e.g. cryptophycin 1 [27]), isolated from *Nostoc* sp. strain GSV 224 also act by inhibiting microtubule assembly and Golakoti *et al.* (1995) have carried out extensive structure–activity studies on these compounds.

Macrolides with antitumour action have also been isolated from dinoflagellates; e.g. amphidinolide-A [36] from an *Amphidinium* sp. (Kobayashi *et al.*, 1986) which is toxic to two kinds of leukaemia cells; L1210 (IC_{50} $2.4\,\mu\mathrm{g\,ml}^{-1}$) and L5178Y ($\mathrm{IC}_{50}$ $3.1\,\mu\mathrm{g\,ml}^{-1}$). Amphidinolides B and D [37, 38] differ from each other in stereochemistry and have markedly different activities: amphidinolide B is >100 times more active agains L1210 leukaemia cells than amphidinolide D (IC_{50} $0.00014\,\mu\mathrm{g\,ml}^{-1}$ versus $0.019\,\mu\mathrm{g\,ml}^{-1}$ respectively) (Ishibashi *et al.*, 1987; Kobayashi *et al.*, 1989). Further screening of dinoflagellates on the same scale as the screening of cyanobacteria is likely to prove very rewarding.

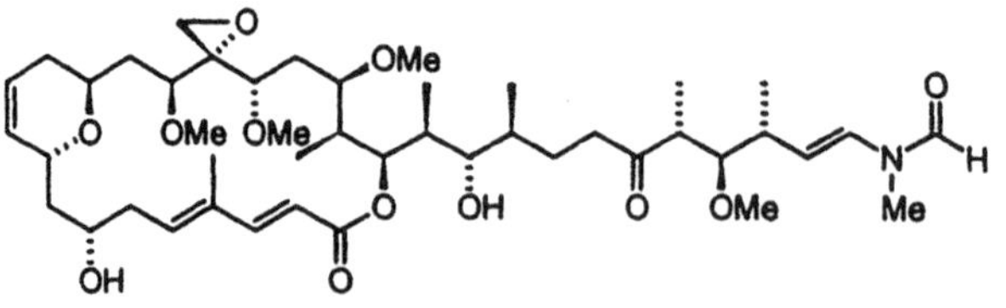

[34] Scytophycin B

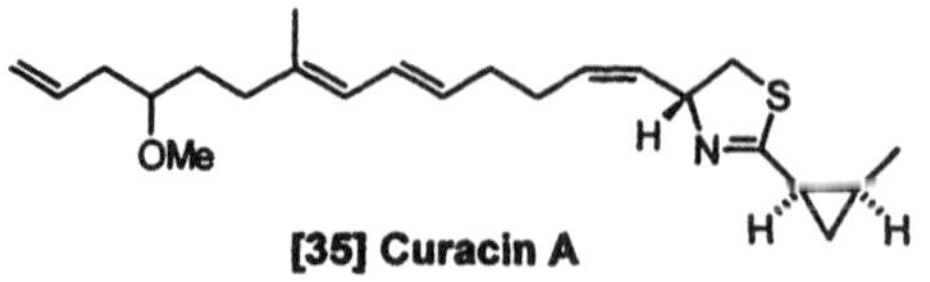

[35] Curacin A

Figure 14.7 Cyanobacterial cytotoxic compounds.

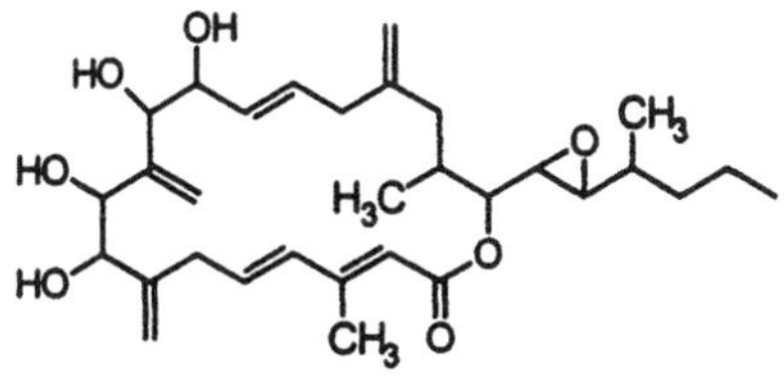

[36] Amphidinolide A

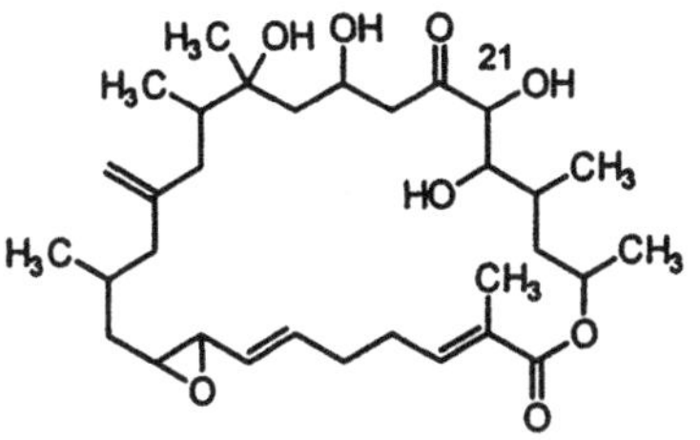

[39] Prorocentrolide

Amphidinolide B [37] & D [38]
(stereoisomeric at C21)

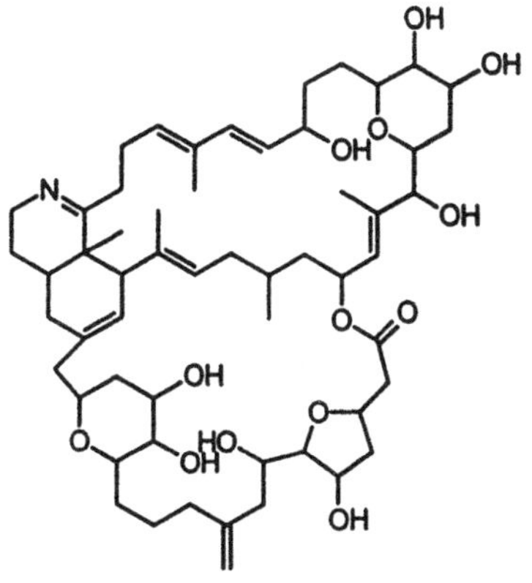

Figure 14.8 Cytotoxic macrolides from dinoflagellates.

The cytotoxic compounds, tubericidin and toyocamycin, both of which have been found in *Streptomyces,* have also been discovered in cyanobacteria (Barchi *et al.,* 1983; Patterson *et al.,* 1991). Another cytotoxic compound from cyanobacteria is westiellamide [40], a bistratamide-related cyclic peptide isolated from *Westelliopsis prolifica* (Stigeonemataceae) which is active at $2 \mu g \, ml^{-1}$ against the KB and LoVo cell lines (Prinsep *et al.,* 1992b). An apparently identical compound has been isolated from the ascidian *Lissoclonium bistratum* (Hambley *et al.,* 1992) and it is possible that endosymbiotic cyanobacteria or prochlorophytes are the sources of this compound.

The cytotoxic styrylchrochrome, hormothamnione [41], isolated from *Hormothamnium enteromorphoides,* is active against both P388 cells (IC_{50} $0.0046 \mu g \, ml^{-1}$) and HL-60 cells (IC_{50} $0.0001 \mu g \, ml^{-1}$) and appears to act by inhibiting RNA synthesis (Gerwick *et al.,* 1986).

Mirabimide-E [43], an unusual *N*-acylpyrrolidone with murine solid tumour selective cytotoxicity, has been isolated from the cyanobacterium, *Scytonema mirabile* (Paik *et al.,* 1994). Other *N*-acylpyrrolidones have also been found in the sponge *Dysidea herbacea* which contains large quantities of the symbiotic cyanobacterium *Oscillatoria spongeliae* (Hofheinz and Oberhänsli, 1977; Berthold *et al.,* 1982), as well as from the marine cyanobacterium *Lyngbya majuscula.*

Like the actinomycetes, some cyanobacteria produce a number of bioactive compounds. For example, the cyanobacterium, *Scytonema mirabile,* produces some 38 cytotoxins, representing four distinct classes of compounds. They are: three scytophycin-type macrolides, six isonitriles (mirabilene isonitriles A–F; [44]), five imides (mirabimides A–E; [42, 43]), 24 modified peptides including the tantazoles [45] and the mirabazoles (Carmeli *et al.,* 1990a, 1990b, 1990c, 1991a, 1991b). One strain of the cyanobacterium *Lyngbya majuscula* has also been shown to produce the mollusco-

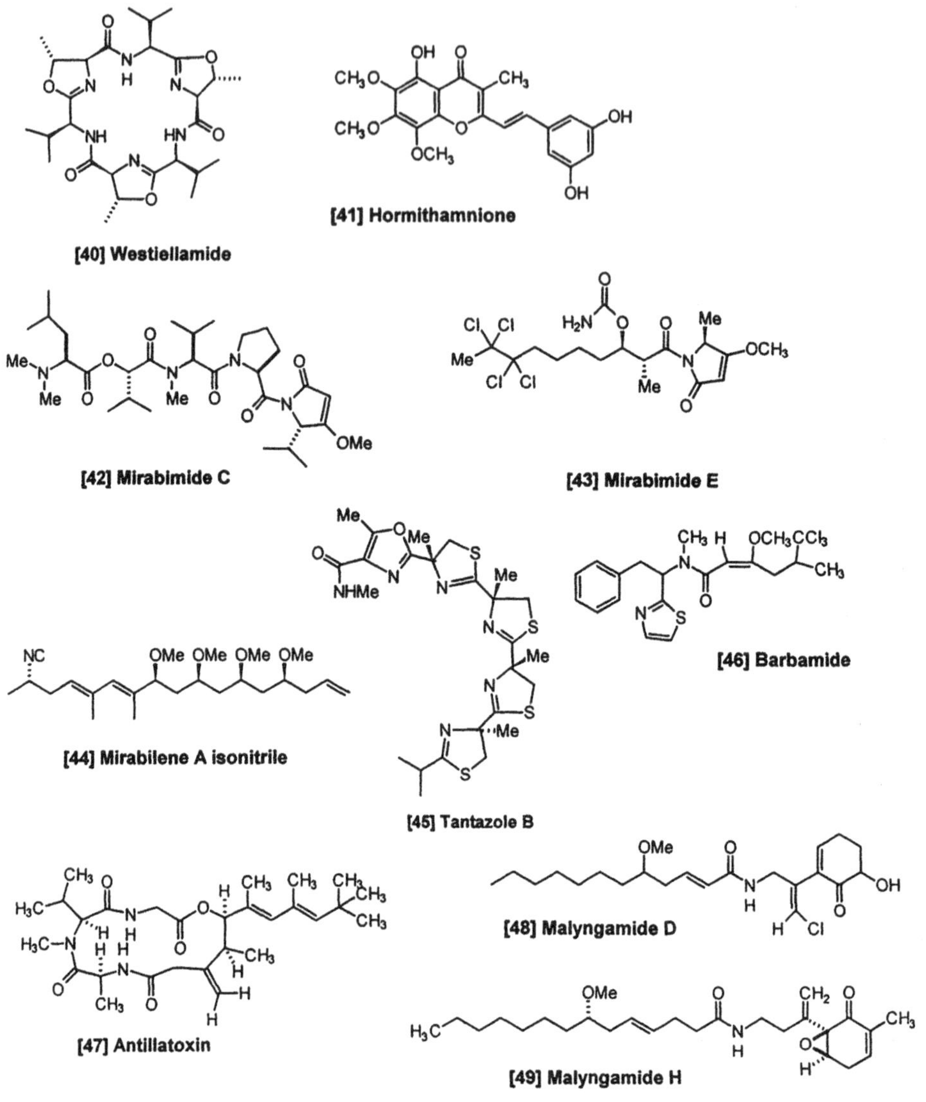

Figure 14.9 Cyanobacterial cytotoxins and other toxins.

cidal compound, barbamide [46], the ichthyotoxic compounds antillatoxin [47] and malyngamide H [49], and the brine shrimp toxin, curacin [35] (Orjala and Gerwick, 1996). A number of other bioactive compounds have been isolated from different strains of this alga and it would be interesting to know whether the differences observed are due to taxonomic confusion or whether genetically different races of this alga exist. Laingolide has been isolated from *Lyngbya bouillonii* and is structurally similar to antillatoxin, but no studies on possible bioactivities have been reported (Klein *et al.*, 1996).

An activity related to cytotoxicity is the ability to potentiate the cytotoxicity of other drugs. Tolyporphyrin [57], isolated from the cyanobacterium, *Tolypothrix nodosa*, potentiates the cytotoxicity of adriamycin or vinblastine in a vinblastine

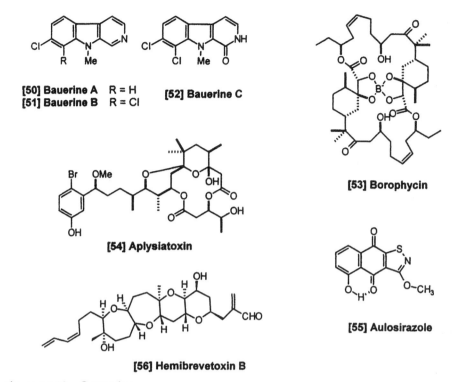

[50] Bauerine A R = H
[51] Bauerine B R = Cl

[52] Bauerine C

[53] Borophycin

[54] Aplysiatoxin

[55] Aulosirazole

[56] Hemibrevetoxin B

Figure 14.10 Cytotoxins.

resistant tumour cell line (SK-VLB) at $1\,\mu g\,ml^{-1}$ and may have an important role in cancer chemotherapy (Prinsep *et al.*, 1992a). *N*-methylwelwitindolinone C isothianate **[28]** from *Hapalosiphon welwitschii* and *Westiella intricata*, also reverses P-glycoprotein-mediated multiple drug resistance (MDR) in a vinblastine resistant subline (SK-VLB) of a human ovarian adenocarcinoma cell line (SK-OV-3) (Stratmann *et al.*, 1994c). Similar MDR activity has been reported in other cyanobacterial compounds such as hapalosin, welwistatin and dendroamide A (Stratmann *et al.*, 1994b; Ogino *et al.*, 1996; Zhang and Smith, 1996).

An indirect antitumour effect has also been observed in a glycoprotein-rich hot water extract of *Chlorella vulgaris* (Konishi *et al.*, 1985). The same researchers have

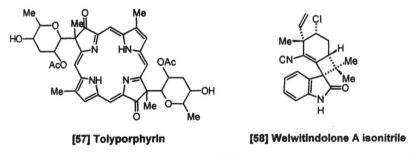

[57] Tolyporphyrin

[58] Welwitindolone A isonitrile

Figure 14.11 Cytotoxin potentiating and multiple drug resistance reversing compounds from cyanobacteria.

also characterized a glycoprotein with antitumour activity which is released into the culture supernatant by this alga (Noda *et al.*, 1996).

An alternative screen for potential anticancer activity, is screening for protein kinase inhibition, especially inhibition of protein kinase C, protein tyrosine kinase, and inosine monophosphate dehydrogenase. Reversible phosphorylation of proteins on serine, threonine and tyrosine residues by protein kinases and phosphatases is a principal mechanism by which eukaryotic cells respond to extracellular signals (Holmes and Borland, 1993) and has also been implicated as potential underlying biochemical cause of different kinds of cancer. Gerwick *et al.* (1994b) have found protein kinase inhibition in the cyanobacteria, Cryptophyta, Chrysophyta and Euglenophyta. One active compound with protein tyrosine kinase activity identified so far is malhamensilipin A [59] from the chrysophyte, *Poterioochromonas malhamensis.*

Toxins

Microalgae, especially the cyanobacteria and the dinoflagellates, produce a wide range of toxins, some of which have gained importance as pharmacological tools, although they have generally not found use in therapy (see review by Ikawa and Sasner, 1990). Of particular interest is saxitoxin [60] found in a number of dinoflagellates and the cyanobacterium *Anabaena flos-aquae*. Its effect is similar to tetrodotoxin. These toxins have been shown to block the influx of sodium through excitable nerve membranes, thus preventing the formation of action potentials (Hille, 1975). They have played a major role in developing the concept of Na^+ channels in particular and membrane channels in general (Hall *et al.*, 1990). Saxitoxin has been used as an aid in certain microsurgical procedures and as an experimental treatment for shortsightedness. The gonyautoxins show a similar mode of action. On the other hand, Maitotoxin, isolated from the dinoflagellate *Gambierdiscus toxicus*, seems to affect calcium transport and is also a useful tool in pharmacological studies (Gusofsky and Daly, 1990; Soergel *et al.*, 1992). Ciguatoxin [15] also affects cell membrane calcium transport.

Anatoxin-a [61], produced by *Anabaena flos-aquae* and several other cyanobacteria is a potent post-synaptic depolarizing neuromuscular blocking agent (Carmichael, 1992) and is a potent stereospecific agonist at nicotinic synapses. Anatoxin-a(s) [62] is a phosphate ester of a cyclic N-hydroxyguanine and acts as an irreversible peripheral anticholinesterase (Mahmood and Carmichael, 1987). Another structural type of toxin is the ichthyotoxic phosphorous substance, PB-1 [63] which is produced by dinoflagellates (DiNovi *et al.*, 1983). Although these toxins

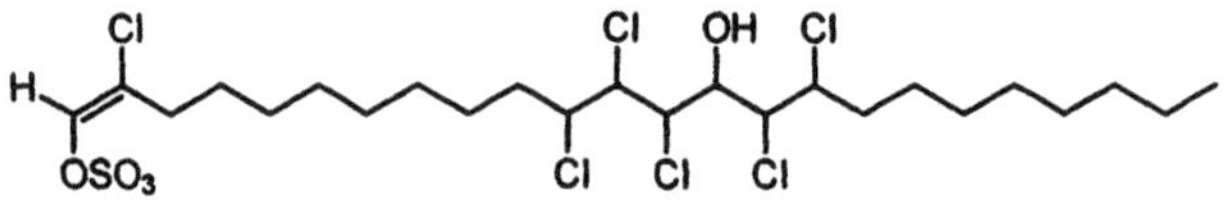

[59] Malhamensilipin A

Figure 14.12 Protein tyrosine kinase inhibitor from the chrysophyte, *Poterioochromonas malhamensis.*

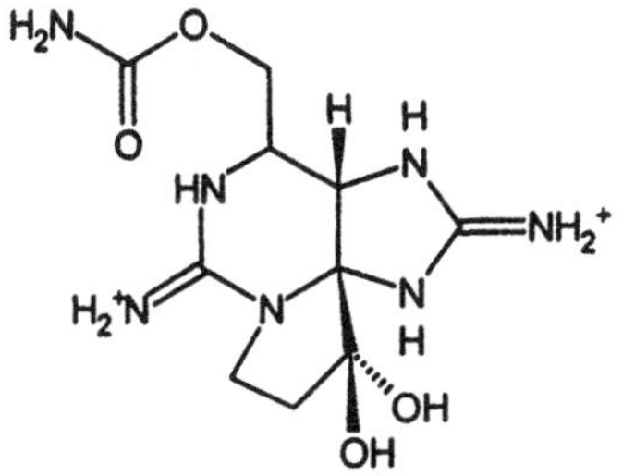

[60] Saxitoxin

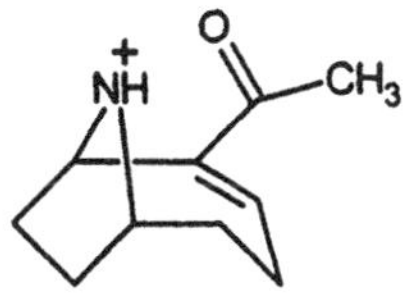

[61] Anatoxin a

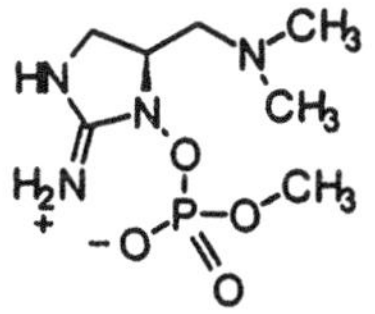

[62] Anatoxin a(S)

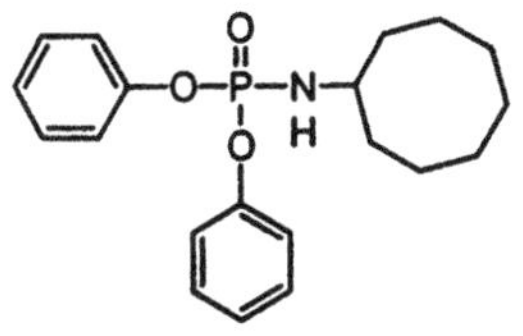

[63] PB-1 toxin

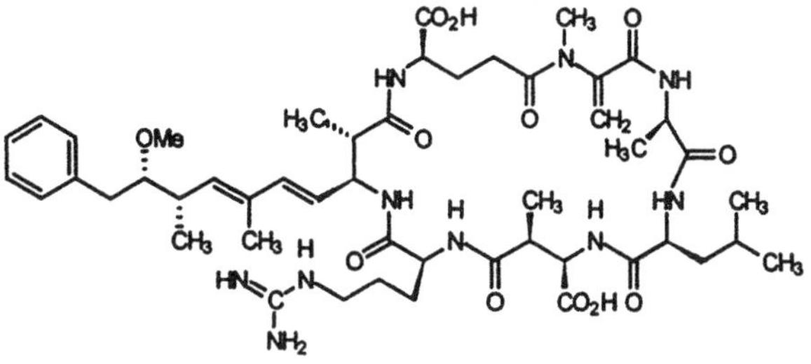

[64] Microcystin LR

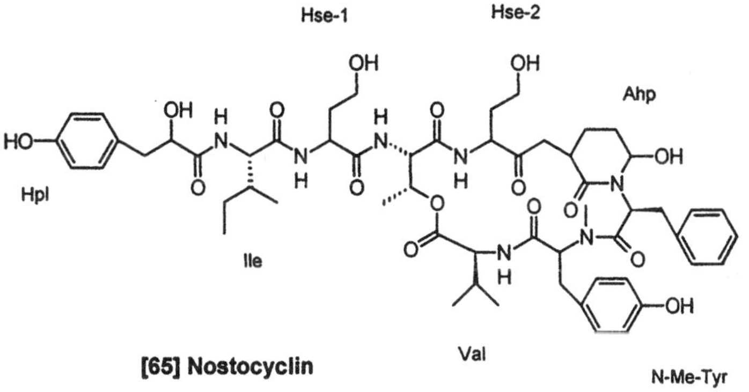

[65] Nostocyclin

Figure 14.13 Algal toxins.

themselves are not likely to be useful as pharmaceuticals, they can serve as models for the rational design of useful compounds.

Another interesting class of toxins are the neurotoxic brevetoxins, isolated from the dinoflagellate *Ptychodiscus brevis* and the raphidophyte *Chatonella marina* (Shimizu *et al.*, 1986; Haung and Wu, 1989; Khan *et al.*, 1995).

Okadaic acid **[16]** and 35-methyl okadaic acid (dinophysistoxin-1 **[17]**), produced by dinoflagellates, and the peptide hepatotoxins, microcystin and nodularin, produced by cyanobacteria, are also of interest in the study of cellular regulation. These toxins are protein phosphatase inhibitors (Mackintosh *et al.*, 1990; Chen *et al.*, 1993; Luu *et al.*, 1993). These algal toxins are therefore useful tools in studies of cellular regulation. A very large number of microcystins have now been characterized and these provide an excellent opportunity for structure–activity studies (Rinehart *et al.*, 1994). The cyanobacterium *Nostoc* not only produces the hepatotoxic cyclic heptapeptides, the microcystins, but also a non-toxic protein phosphatase I inhibitor, nostocyclin [65] (Kaya *et al.*, 1996).

The microcystins, especially microcystin-LR [64], also inhibit plant metabolism, especially photosynthesis (Kós *et al.*, 1995; Abe *et al.*, 1996) and may possibly find application as non- specific herbicides. Fischerellin A [29] from *Fischerella muscicola*, also shows herbicidal activity and is a potent inhibitor of photosystem-II activity (Hagmann and Jüttner, 1996).

Several of the microalgal toxins such as okadaic acid, microcystin LR and nodularin are tumour promotors (Falconer and Humpage, 1996), apparently inhibiting protein phosphatases leading to hyperphosphorylation of key proteins such as vimentin and cytokeratin (Falconer and Yeung, 1992; Yasunami *et al.*, 1993).

Other pharmacologically active compounds

Several hydrophilic extracts of cyanobacteria have also shown cardiotonic activity in isolated mouse atria. Although, in many cases, this activity can be attributed to tyramine, positive ionotropic activity in several species of scytonematacean algae has been attributed to compounds called the tolypophycins (Moore *et al.*, 1988) and, in the case of an *Anabaena* sp., to an unusual chlorine- containing cyclic peptide, puwainaphycin C (Gregson, 1986). The cyclic peptide, scytonemin A [30] , from a *Scytonema* sp., has also been shown to be a moderately strong calcium agonist (Helms *et al.*, 1988).

The diatom, *Nitzschia pungens* var. *multiseries,* is a potential source of the glutamate agonist, domoic acid (Laycock *et al.*, 1989). Domoic acid is structurally similar to glutamic and aspartic acid and acts as a potent excitant (agonist) of glutamate receptors; as such it could be an important tool in the study of neurodegenerative disease. *Nitzschia pungens* has been shown to contain more than 1 per cent of dry weight of domoic acid in culture and could therefore be an excellent source of this compound.

The cyanobacteria also produce a wide range of protease inhibitors and these may find application in medicine for treatment of strokes, coronary artery occlusions and pulmonary emphysema. For example, inhibitors of the serine protease, thrombin, could be used to control blood clot formation in these diseases. Thrombin acts by cleaving a peptide fragment from fibrinogen which then leads to the formation of fibrin, a major component of blood clots. Inhibition of this protease would thus also inhibit clot formation. Similarly, angiotensin-converting enzyme inhibitors are being developed as anti-hypertensive agents. Protease inhibitors are also being used in the treatment of HIV infections (Richman, 1996). Table 14.4 summarizes the various peptide protease inhibitors which have been isolated and identified from cyanobacteria so far. These inhibitors include linear and cyclic peptides as well as depsipeptides and have been isolated mainly from *Microcystis, Microchaete* and *Oscillatoria*. The structures of some of these are shown in Figure 14.14.

The brominated bi-indoles isolated from field-collected material of the intertidal cyanobacterium, *Rivularia firma* (Norton and Wells, 1982) illustrate some of the problems encountered in trying to develop a pharmacologically usable drug. The major compound isolated from this alga, (+)-1'-methoxy-2,3,5,5'-tetrabromo-3,4'-bi-1H-indole [70] showed both anti-inflammatory activity and anti-amphetamine activity. However, it also produced tremors in rats and mice. Two other less abun-

Table 14.4 Peptide protease inhibitors from cyanobacteria (adapted from Patterson, 1996).

Compound	Alga	Class	Enzyme inhibited	IC$_{50}$	Reference
A90720A	*Microchaete loktakensis*	DP	Thrombin	270 ng ml^{-1}	Bonjouklian *et al.* (1996)
			Trypsin	10 ng ml^{-1}	
			Plasmin	30 ng ml^{-1}	
Aeruginosin 298-A	*Microcystis aeruginosa*	LP	Thrombin	0.3 µg ml^{-1}	Murakami *et al.* (1994)
			Trypsin	1.0 µg ml^{-1}	
Aeruginosin 98-A	*Microcystis aeruginosa*	LP	Trypsin	0.6 µg ml^{-1}	Murakami *et al.* (1995)
			Plasmin	6.0 µg ml^{-1}	
			Thrombin	7.0 µg ml^{-1}	
Microginin	*Microcystis aeruginosa*	LP	Angiotensin-converting enzyme	7.0 µM	Okino *et al.* (1993a)
Micropeptin A	*Microcystis aeruginosa*	DP	Plasmin	26 nM	Okino *et al.* (1993b)
			Trypsin	72 nM	
Micropeptin 90	*Microcystis aeruginosa*	DP	Plasmin	0.1 µg ml^{-1}	Ishida *et al.* (1995)
			Trypsin	2 µg ml^{-1}	
Microviridin B	*Microcystis aeruginosa*	CP	Elastase	25 nM	Okino *et al.* (1995)
			Chymotrypsin	1.4 µM	
			Trypsin	34 µM	
Cyanopeptolin SS [69]	*Microcystis aeruginosa*	DP	Trypsin	0.2 µg ml^{-1}	Jacobi *et al.* (1996)
			Plasmin	1 µg ml^{-1}	
			Thrombin	5 µg ml^{-1}	
Oscillamide Y [66]	*Oscillatoria agardhii*	CP	Chymotrypsin	10 µM	Sano and Kaya (1995)
Oscillapeptin [67]		DP	Elastase	0.2 µM	
			Chymotrypsin	1.8 µM	
Oscillatorin [68]	*Oscillatoria agardhii*	CP	Chymotrypsin	13 µM	Sano and Kaya (1996)

DP = depsipeptide; LP = linear peptide; CP = cyclic peptide.

dant bi-indoles [71, 72] isolated from *R. firma* retained the anti-inflammatory activity without showing significant tremorigenic effects, and the remaining two [73, 74] showed no activity (Baker, 1984). The synthetic diacetate [75] of [70], had only a very slight tremorigenic effect while maintaining the anti-amphetamine activity as well as being active in the acetic acid writhing test. This study shows that the preparation of analogues can be useful in producing compounds with desired activities without undesirable side-effects. However, in the case of these compounds synthesis was not considered a viable way to produce the compounds and, as this species has not been culturable, these promising activities were not pursued.

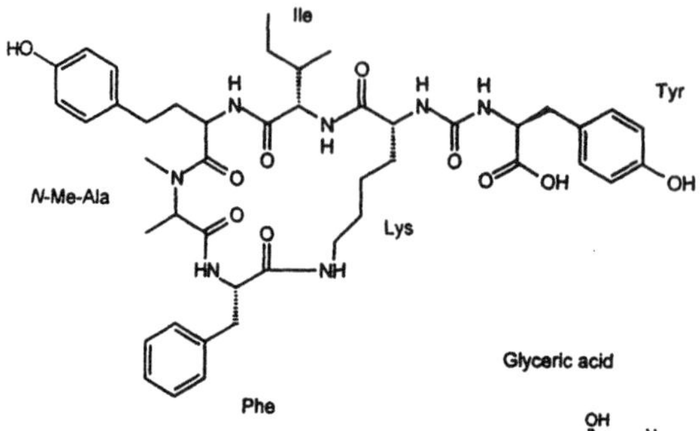

[66] Oscillamide Y

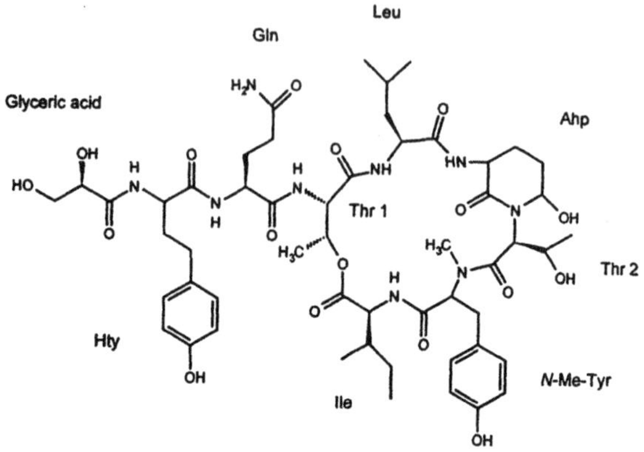

[67] Oscillapeptin G

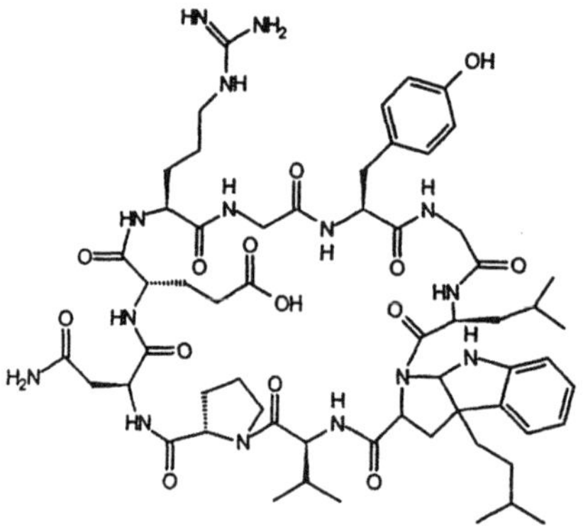

[68] Oscillatorin

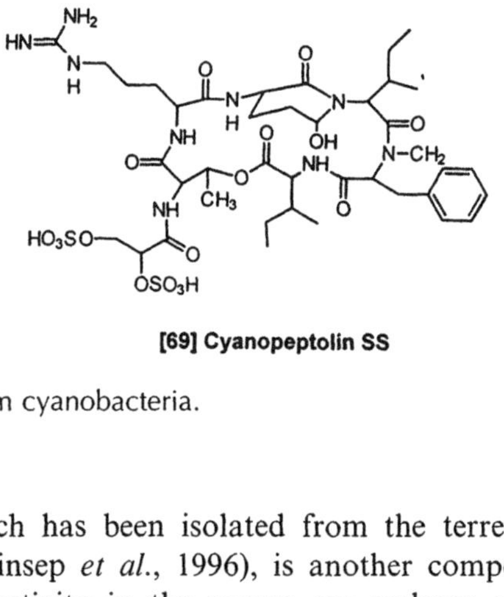

[69] Cyanopeptolin SS

Figure 14.14 Peptide protease inhibitors from cyanobacteria.

The diterpenoid, tolypodiol **[76]**, which has been isolated from the terrestrial cyanobacterium, *Tolypothrix nodosa* (Prinsep *et al.*, 1996), is another compound which shows strong anti-inflammitory activity in the mouse ear oedema assay, with an ED_{50} of $30\,\mu g\,ear^{-1}$. This compares favourably with the ED_{50} for hydrocortisone ($20\,\mu g\,ear^{-1}$).

There are many other interesting leads to pharmacologically active compounds in cultured microalgae. These include angiotensin-converting enzyme inhibitory activity in several species of Cyanophyceae, Chlorophyceae, Dinophyceae and Raphidophyceae (Yamaguchi *et al.*, 1989); these may lead to new antihypertensive agents. Similarly, Lincoln *et al.* (1990) found that extracts from a wide range of cultured microalgae affected electrically evoked muscle contraction and resting muscle tone in isolated guinea-pig ileum. Vasoconstrictive polyoxygenated long-chain

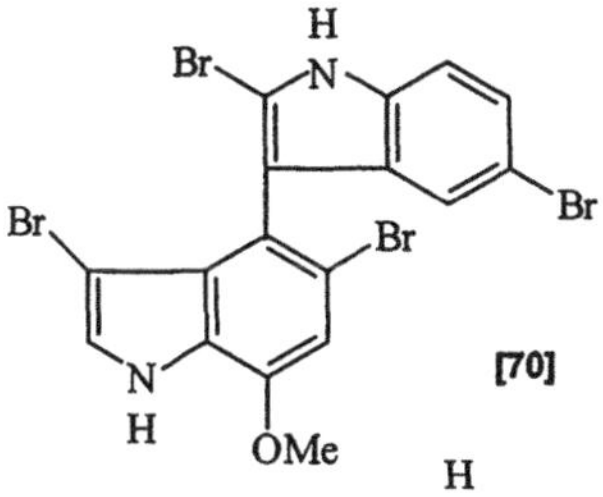

Figure 14.15 Brominated biuindoles isolated from the cyanobacterium, *Rivularia firma*. The major compound was [70] and [71–74] were minor constituents. Compound [75] is a synthetic analogue of [70].

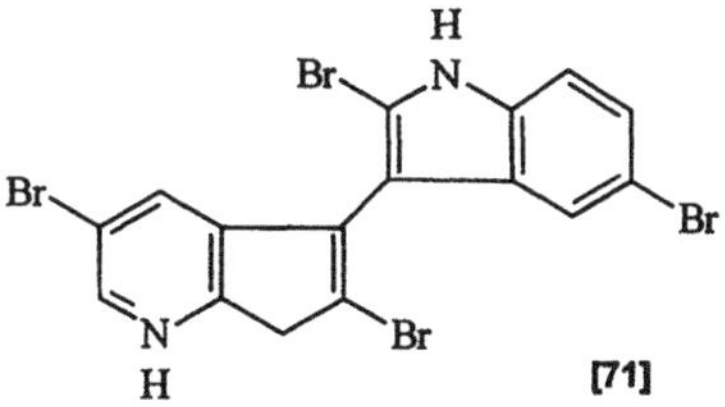

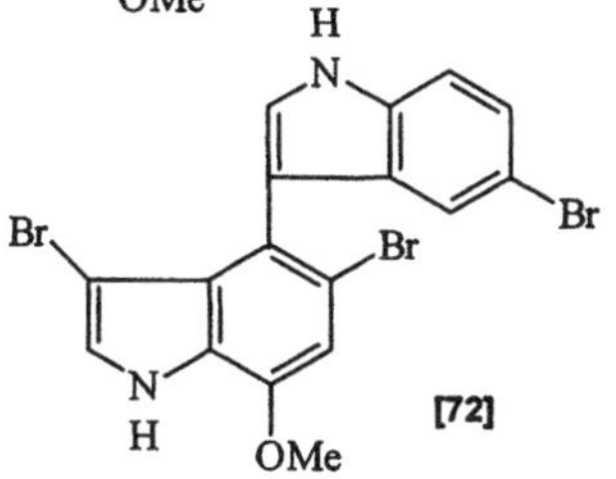

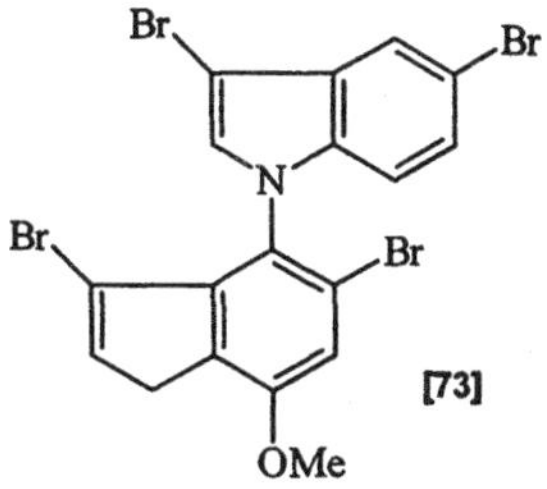

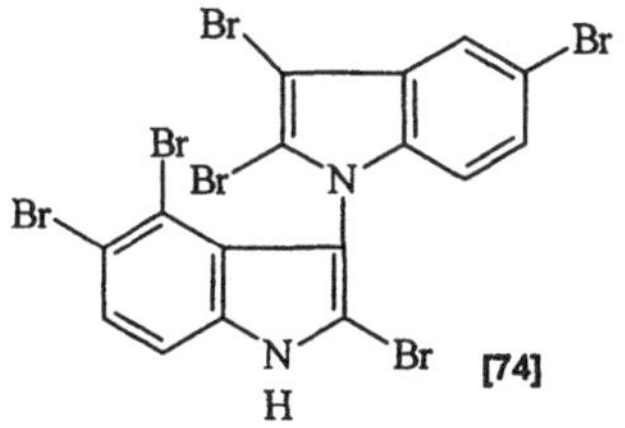

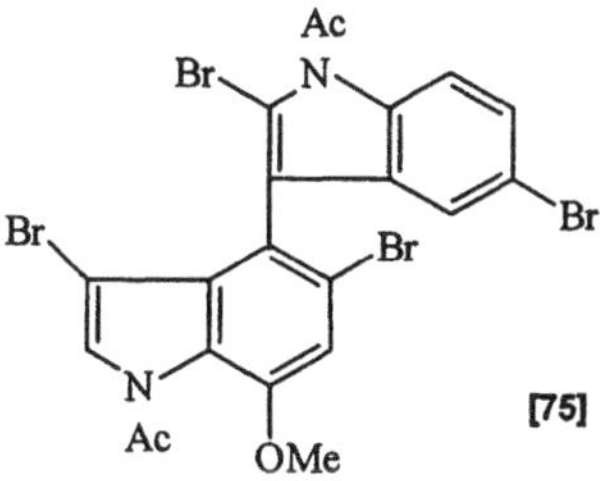

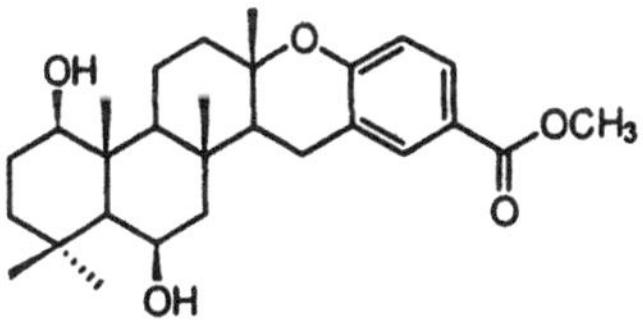

[76] Tolypodiol

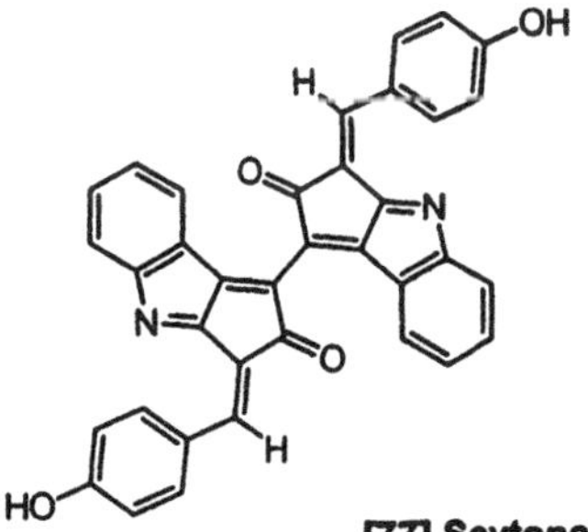

[77] Scytonemin

Figure 14.16 Bioactive compounds from cyanobacteria.

compounds with a similar activity to palytoxin and maitotoxin have also been isolated from the zooxanthella, *Symbiodinium* sp. (Nakamura *et al.*, 1993). Haploindolone A [8] and B, isolated from the cyanobacterium *Fischerella* (ATCC 53558), have been patented as vassopressin antagonists with possible application in the treatment of congestive heart failure, hypertension, oedema and hyponatraemia (Schwartz *et al.*, 1989). Our studies (unpublished) at RRIMP also observed promising pharmacological activities in several algae including the chlorophyte, *Dunaliella tertiolecta* (antihypertensive, bronchodilator, antiserotonin, polysynaptic block, analgesic, muscle relaxant, anti-oedema), and the diatoms *Phaeodactylum tricornutum* (nicotinic block) and *Navicula* sp. (neuromuscular block). Similarly, aqueous extracts of *P. tricornutum* and *D. tertiolecta* showed activity as a central nervous system (CNS) depressant and a potential muscle relaxant (Villar *et al.*, 1992) and *Chlorella stigmatophora* extracts showed anti-dopaminergic activity (Laguna *et al.*, 1993b). Aqueous extracts of *Tetraselmis suecica* and *Isochrysis galbana* also showed activities in CNS screens; however these activities have been difficult to characterize by standard pharmacological criteria (Laguna *et al.*, 1993a).

There are also several reports of algal extracts acting on the immune system. For example, extracts of *Spirulina platensis* have been shown to enhance chicken macrophage function *in vitro* as well as cell-mediated immune function in chickens and cats (Qureshi *et al.*, 1995, 1996; Qureshi and Ali, 1996). A glycoprotein-rich extract of *Chlorella vulgaris* seems to activate mature leukocytes and stimulate homeopoeitic stem cells in the bone marrow of mice and rats when administered subcutaneously or orally (Hasegawa *et al.*, 1990; Konishi *et al.*, 1990, 1996). This extract may find use in ameliorating the side-effects of cancer chemotherapy where leukopaenia often occurs.

The active principles in these extracts have not been characterized further in many of these studies, nor have the active compounds been isolated. However, the high proportion of pharmacologically active extracts indicates that microalgae are promising sources of bioactive molecules, some of which may eventually find application in medicine or veterinary science.

Other activities

Microalgae also produce other bioactive molecules of interest. For example, a number of cyanobacteria produce the pigment, scytonemin [77], in their sheath. This pigment acts as a ultraviolet sun screen, with the greatest absorption in the spectral range of UV-A (Proteau *et al.*, 1993) and therefore may have application in sun screens. Cyanobacteria also produce mycosporine-like compounds which also appear to act as UV-screens (Garcia-Pichel and Castenholz, 1993; Garcia-Pichel *et al.*, 1993) and at least one of these has been characterized (Böhm *et al.*, 1995). Recently bis-cyclo-penteno-indole alkaloids prepared by etherification or acylation of scytonemin have been patented for application as sun screens and/or anti-inflammatory compounds (Castenholz *et al.*, 1996). Matsunaga *et al.* (1993) have isolated another UV-A absorbing compound, a biopterin glucoside, from the marine planktonic cyanobacterium *Oscillatoria* sp.

Gibberellin and cytokinin-like activities have also been detected in the medium after cultivation of several green algae and in cyanobacterial extracts (Burkiewicz, 1987; Ördög and Pulz, 1996). The plant hormone, jasmonic acid, has also been found in *Chlorella, Euglena* and *Spirulina* (Ueda *et al.*, 1991a, 1991b).

Vitamins, carotenoids and fatty acids

Microalgae are also sources of known bioactive compounds such as vitamins and fatty acids. Many microalgae synthesize vitamins such as pro-vitamin A (β-carotene), vitamin B_{12}, vitamin B_6, biotin etc. (Borowitzka, 1988b). Of these, the green halophilic alga *Dunaliella salina* is the best natural source of β-carotene and is grown commercially as a source of β-carotene for use as a dietary supplement and a natural food colouring in Australia, USA and Israel (Borowitzka and Borowitzka, 1989b). Carotenoids such as β-carotene and fucoxanthin also have anti-tumour and cancer preventative activity (Okuzumi *et al.*, 1990; Gerster, 1993).

Microalgae, especially marine diatoms, dinoflagellates and chrysophytes, are also excellent sources of polyunsaturated fatty acids such as γ-linolenic acid, arachidonic acid, eicosapentaenoic acid and docosahexaenoic acid (Borowitzka, 1988a; Cohen and Cohen, 1991). These essential fatty acids are important for the treatment and prevention of a range of diseases, and are also important in human nutrition (Okuyama, 1992; Zevenbergen and Rudrum, 1993). Algal species can be selected for the preponderance of a particular fatty acid, and the content of these fatty acids can be manipulated by changing culture conditions (e.g. Sukenik, 1991; Chrismadha and Borowitzka, 1994). At present, heterotrophic production appears the most promising method for commercial production of these compounds from algae (Kyle and Gladue, 1991; Kyle *et al.*, 1991; Barclay *et al.*, 1994).

Production of biologically active substances by algal culture

A directed search for biologically active compounds as well as their production by algal culture requires some understanding of the culture conditions favouring their production. Many of the biologically active compounds of interest are secondary metabolites and thus are often most abundant in stationary phase or in slow-growing cultures. This is no problem for the chemist screening the algae, but can be a problem when large-scale algal culture is attempted in order to produce the useful metabolite commercially. For example, the β-carotene content in *Dunaliella salina* is greatest under conditions of high salinity, high light and nutrient limitation; i.e. under growth limiting conditions. Thus, when growth is best, the β-carotene content is least, a common feature of secondary metabolites. However, careful manipulation and optimisation of growth conditions has resulted in the reliable high productivities of β-carotene which make this alga a viable commercial source of β-carotene (Borowitzka and Borowitzka, 1989a). Apart from optimizing growth conditions some assessment should be made of the variation between clones of the same species. Some species show a great degree of variation in the production of specific metabolites between clones. For example, the dinoflagellate, *Prorocentrum lima* shows a high degree of interclonal variation in okadaic acid production (Lee *et al.*, 1989; Morton and Tindall, 1995).

A number of studies have shown that the production of the active compounds depended on the growth phase and/or culture conditions (e.g. Debro and Ward, 1979; van der Westhuizen and Eloff, 1983; Armstrong *et al.*, 1991; Patterson *et al.*, 1991; Morton and Bomber, 1994) and this means that culture conditions for the production of bioactive compounds must be optimized. For example, Patterson and Bolis (1993) have shown that tolytoxin accumulates throughout the growth cycle of *Scytonema ocellatum*. The optimum temperature and pH for production of tolytoxin were 25°C and pH 8.0 to 8.5 respectively. Continuous light of at least $25 \, \mu\text{mol photons m}^{-2} \, \text{sec}^{-1}$ was also necessary. They have also found that increasing the Ca^{2+} concentration in BG-11 medium increases the tolytoxin yield. Further improvements in yield to $3.5 \, \mathrm{mg \, l^{-1}}$ have been obtained by isolation of a higher yielding axenic clone of *S. ocellatum* from the parent culture (Patterson and Bolis, 1994).

The possibility of diurnal variation must also be considered and, if present, may be used to advantage. For example diurnal variation in saxitoxin content in *Alexandrium catenella* and *A. tamarense* has been observed by Kim *et al.* (1993) and in our laboratory in *Gymnodinium catenatum* (Soames, Hallegraeff and Borowitzka; unpublished results). Over a short period of about 1–3 hours the saxitoxin content of these algae can increase up to threefold before returning to the 'normal' basal level.

Nutrient limitation, especially N and P limitation, has been shown to be necessary for production of high levels of acutiphycin in *Oscillatoria acutissima* and maximum concentrations are achieved in early stationary-phase cultures. Addition of N, P or organic carbon greatly reduces formation of acutiphycin (Moore *et al.*, 1988). Carmichael (1986) and Rapala *et al.* (1993) have made similar observations on biotoxin production in planktonic cyanobacteria. On other hand, antibiotic production in *Nostoc muscorum* and *Scytonema* sp. was most affected by the nitrogen and iron content of the medium and was enhanced in actively growing cultures (Bloor and England, 1991; Chetsumon *et al.*, 1993a, 1993b). Species of *Microcystis* and *Oscillatoria* also produce more toxins under high N conditions, whereas P has no effect on anatoxin-a production (Sivonen, 1990; Rapala *et al.*, 1993).

Culture studies with the domoic acid-producing *Nitzschia pungens* f. *multiseries* showed that domoic acid was produced at a rate of 1 pg domoic acid $\mathrm{cell^{-1} \, d^{-1}}$ only during to the lag and stationary growth phases when cell division was low or zero, and not during exponential growth (Douglas and Bates, 1992). Domoic acid production also requires nitrogen (preferably as ammonium) and light (Bates *et al.*, 1991, 1993).

Other environmental factors may also be important; for example, the antibiotic cyanobacterin LU1 from *Nostoc linckia* is synthesized throughout the growth cycle, but is favoured by low temperatures (Gromov *et al.*, 1991). Similarly lipid production in diatoms is enhanced by silicon starvation, and low temperatures generally enhance the content of long-chain polyunsaturated fatty acids such as eicosapentaenoic acid in many algae. Microcystin production in *Microcystis aeruginosa* is also enhanced under red or green light compared to white light (Utkilen and Gjølme, 1992).

Several studies, and our own experience, have also shown that the desired activity may decline or be lost in culture. For example, Carmichael (1986) reports that anatoxin-a toxicity of *Anabaena flos-aquae* NRC-44-1 disappeared when the medium was changed from ASM-1 to BG-11. Gallon *et al.* (1994) have found that the size of a plasmid was reduced in the non-toxic culture. The reasons for this loss need to be understood and controlled before a reliable culture system for production can be developed.

Our understanding of the physiological and genetic control of secondary metabolite formation in algae is extremely limited, nor do we know much about the biosynthetic pathways. Such studies are necessary if we are to exploit fully the potential of the microalgae and if we wish to use genetic engineering to improve yields and/or to produce modified products. Of interest here is the work being done on engineering the biosynthesis of novel polyketides (cf. Hopwood, 1993; McDaniel *et al.*, 1993). Such work however requires the development of suitable probes and methods for transforming microalgae. Such systems are reasonably well developed for the cyanobacteria (cf. Elhai, 1994; Kindle and Sodeinde, 1994; Matsunaga and Takeyama, 1995); however in the eukaryotic algae the opportunities for genetic engineering are still limited to a few genera such as *Chlorella* and *Chlamydomonas* (e.g. Kindle and Sodeinde, 1994; Miller *et al.*, 1994). Classical mutagenic methods can also be applied and have already been used to produce β-carotene overproducing strains (Shaish *et al.*, 1991).

It has also been suggested that some of the genes for bioactive molecules may be carried on plasmids, as for example, the toxin-producing gene in *Microcystis* (Hauman, 1981). However, work by Schwabe *et al.* (1988) showed that different toxic strains had the same or different plasmids, and in one toxic and several non-toxic strains no plasmids could be found. More molecular biological studies are clearly needed to resolve this question as plasmid-borne genes may be easier to manipulate and could possibly be introduced into other species which are more easily cultured.

Culture versus 'wild'

One potentially confounding factor in some of the results of screens of microalgae and other organisms is the fact that often field collected material or non-axenic cultures are used. The assumption in these cases is that the activity derives from the most abundant organisms in the sample. There is some evidence however that this is not always the case (Faulkner *et al.*, 1993; Fenical, 1993; Numata *et al.*, 1993). For example, it is now clear that 13-demethylisodysidenin [78] and the chlorinated diketopiperazines [79] found in the sponge *Dysidea herbacea* are synthesized by the endosymbiotic cyanobacterium, *Oscillatoria spongeliae* (Berthold *et al.*, 1982; Unson and Faulkner, 1993). Similarly, [L-Val^2]nodularin isolated from the sponge *Theonella swinhoei* (deSilva *et al.*, 1992) is also likely to be synthesized by a cyanobacterial endosymbiont. Since symbiotic associations of marine invertebrates with microalgae and/or bacteria are very common it is quite likely that a number of the activities reported for marine invertebrates (e.g. Rinehart *et al.*, 1981; Schmitz *et al.*, 1993) are actually due to compounds produced by their microbial endosymbionts (see review by Kobayashi and Ishibashi, 1993). For example, the didemnins, a family of cyclic depsipeptides isolated from the tunicate, *Trididemnium* (Rinehart *et al.*, 1993) may be produced by the endosymbiotic prokaryotic prochlorophytes found in these organisms (Davidson, 1993). This is likely since cyanobacteria produce a wide range of bioactive depsipetides. However, since the symbiotic prochlorophyte, *Prochloron*, has not yet been cultured the role of this organism in the synthesis of these compounds has not been determined. It has also been proposed that at least some of the bioactive peptides found in sponges have a microalgal or bacterial origin (Fusetani and Matsunaga, 1993). Similarly, some of the activities found in field collected ('wild') microalgae actually may be produced by associated bacteria or fungi (e.g. Bewley *et al.*, 1996). It is therefore important that axenic cultures of microalgae are used in the screens to ensure that it is the algae which produce the active molecule of interest.

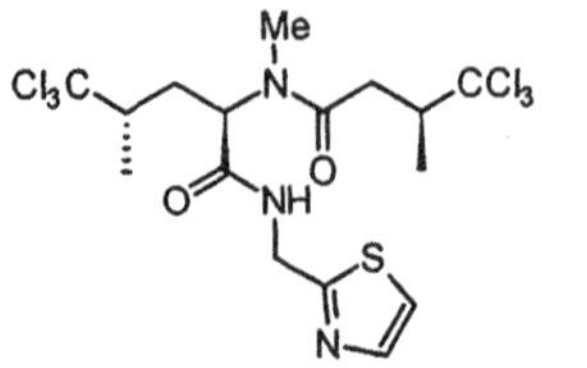

[78] 13-demethylisodysidenin

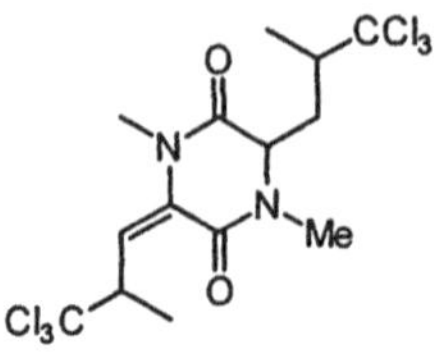

[79] diketopiperazine

Figure 14.17 Two chlorinated metabolites of *Oscillatoria spongeliae*, the endosymbiotic alga of the sponge, *Dysidea herbacea*.

Conclusion

During the last decade a large number of novel bioactive molecules have been isolated from microalgae, especially the cyanobacteria, and continued screening will certainly produce many more. Most microalgal phyla are still largely unexplored but continued isolation into culture of new species will make them available for screening. Several microalgal metabolites are now undergoing clinical trials and in the next few years we are likely to see some of them added to the list of compounds avialable to medicine. Screening and chemistry alone are however not the only activities. There are also several studies of the physiology of the production of these metabolites with the aim of optimizing their production. Ongoing work on new and improved large-scale culture systems will provide the system for producing the algae commercially.

References

ABE T., LAWSON, T., WEYERS, J.D.B. and CODD, G.A., 1996, Microcystin-LR inhibits photosynthesis of *Phaseolus vulgaris* primary leaves – implications for current spray irrigation practice. *New Phytologist*, **133**, 651–658.

ARMSTRONG, J.E., JANDA, K.E., ALVARADO, B. and WRIGHT, A.E., 1991, Cytotoxin production by a marine *Lyngbya* strain (cyanobacterium) in a large-scale laboratory bioreactor, *Journal of Applied Phycology*, **3**, 277–282.

AUSTIN, B. and DAY, J.G., 1990, Inhibition of prawn pathogenic *Vibrio* spp. by a commercial spray dried preparation of *Tetraselmis suecica*, *Aquaculture*, **90**, 389–392.

BAGCHI, S.N., PALOD, A. and CHAUHAN, V.S., 1990, Algicidal properties of a bloom-forming blue-green alga, *Oscillatoria* sp., *Journal of Basic Microbiology*, **30**, 21–29.

BAKER, J.T., 1984, Seaweeds in pharmaceutical studies and applications, *Hydrobiologia*, **116/117**, 29–40.

BAKER, J.T., REICHELT, J.L. and SUTTON, D.C., 1985, Pharmacologically active and related marine microbial products, in Blanch H.W., Drew, S. and Wang, D.I.C. (Eds.), *Comprehensive Biotechnology*, Vol. 3, pp. 225–230, Oxford: Pergamon.

BARCHI, J.J., MOORE, R.E., FURUSAWA, E. and PATTERSON, G.M.L., 1983, Identification of cytotoxin from *Tolypothrix byssoidea* as tubercidin, *Phytochemistry*, **22**, 2851–2852.

BARCHI, J.J., MOORE, R.E. and PATTERSON, G.M.L., 1984, Acutiphycin and 20,21-didehydroacutiphycin, new antineoplastic agents from the cyanophyte *Oscillatoria acutissima*, *Journal of the American Chemical Society*, **106**, 8193–8197.

BARCLAY, W.R., MEAGER, K.M. and ABRIL, J.R., 1994, Heterotrophic production of long chain omega-3 fatty acids utilizing algae and algae-like microorganisms, *Journal of Applied Phycology*, **6**, 123–129.

BATES, S.S., DEFREITAS, A.S.W., MILLEY, J.E., POCKLINGTON, R., QUILLIAM, M.A., SMITH, J.C. and WORMS, J., 1991, Controls on domoic acid production by the diatom *Nitzschia pungens* f *multiseries* in culture – nutrients and irradiance, *Canadian Journal of Fisheries and Aquatic Sciences*, **48**, 1136–1144.

BATES, S.S., WORMS, J. and SMITH, J.C., 1993, Effects of ammonium and nitrate on growth and domoic acid production by *Nitzschia pungens* in batch culture, *Canadian Journal of Fisheries and Aquatic Sciences*, **50**, 1248–1254.

BERDY, J., 1989, The discovery of new bioactive microbial metabolites: screening and identification, in Bushell, M.E. (Ed.), *Bioactive Metabolites from Microorganisms*, pp. 3–25, Amsterdam: Elsevier.

BERLAND, B.R., BONIN, D.J. and CORNU, A.L., 1972, The antibacterial substances of the marine alga *Stichochrysis immobilis* (Chrysophyta), *Journal of Phycology*, **8**, 383–392.

BERTHOLD, R.J., BOROWITZKA, M.A. and MACKAY, M.A., 1982, The ultrastructure of *Oscillatoria spongeliae*, the blue-green algal endosymbiont of the sponge *Dysidea herbacea*, *Phycologia*, **21**, 327–335.

BEUTLER, J.A., MCKEE, T.C., FULLER, R.W., TISCHLER, M., CARDELLINA, J.H., SNADER, K.M., *et al.*, 1993, Frequent occurrence of HIV-inhibitory sulphated polysaccharides in marine invertebrates, *Antiviral Chemistry and Chemotherapy*, **4**, 167–172.

BEWLEY, C.A., HOLLAND, N.D. and FAULKNER, D.J., 1996, Two classes of metabolites from *Theonella swinhoei* are localized in distinct populations of bacterial symbionts, *Experientia*, **52**, 716–722.

BLOOR, S. and ENGLAND, R.R., 1991, Elucidation and optimization of the medium constituents controlling antibiotic production by the cyanobacterium *Nostoc muscorum*, *Enzyme and Microbial Technology*, **13**, 76–81.

BÖHM, G.A., PFLEIDERER, W., BOGER, P. and SCHERER, S., 1995, Structure of a novel oligosaccharide-mycosporine-amino acid ultraviolet A/B sunscreen pigment from the terrestrial cyanobacterium *Nostoc commune*, *Journal of Biological Chemistry*, **270**, 8536–8539.

BONJOUKLIAN, R., MOORE, R.E., MYNDERSE, J.S., PATTERSON, G.M.L. and SMITKA, T.A., 1989, Hapalindoles, USA Patent Number 4870185.

BONJOUKLIAN, R., SMITKA, T.A., DOOLIN, L.E., MOLLOY, R.M., DEBONO, M., SHAFFER, S.A., *et al.*, 1991, Tjipanazoles, new antifungal agents from the blue-green alga *Tolypothrix tjipanasensis*, *Tetrahedron*, **47**, 7739—7750.

BONJOUKLIAN, R., MOORE, R.E., PATTERSON, G.M.L. and SMITKA, T.A., 1994, Preparation of hapalindole-related alkaloids from blue-green algae, USA Patent Number 5292650.

BONJOUKLIAN, R., SMITKA, T.A., HUNT, A.H., OCCLOWITZ, J.L., PERUN, T.J., DOOLIN, L., *et al.*, 1996, A90720A, a serine protease inhibitor isolated from a terrestrial blue-green alga *Microchaete loktakensis*, *Tetrahedron*, **52**, 395–404.

BOROWITZKA, M.A., 1988a, Microalgae as sources of essential fatty acids, *Australian Journal of Biotechnology*, **1**, 58–62.

BOROWITZKA, M.A., 1988b, Vitamins and fine chemicals, in Borowitzka, M.A. and Borowitzka, L.J. (Eds), *Micro-algal Biotechnology*, pp. 153–196, Cambridge: Cambridge University Press.

BOROWITZKA, M.A., 1995, Microalgae as sources of pharmaceuticals and other biologically active compounds, *Journal of Applied Phycology*, **7**, 3–15.

BOROWITZKA, L.J. and BOROWITZKA, M.A., 1989a, Industrial production: methods and economics, in Cresswell, R.C., Rees, T.A.V. and Shah, N. (Eds.), *Algal and Cyanobacterial Biotechnology*, pp. 294–316, London: Longman Scientific.

BOROWITZKA, L.J. and BOROWITZKA, M.A., 1989b, β-Carotene (provitamin A) production with algae, in Vandamme, E.J. (Ed.), *Biotechnology of Vitamins, Pigments and Growth Factors*, pp. 15–26, London: Elsevier Applied Science.

BOURGOUGNON, N., LAHAYE, M., CHERMANN, J.C. and KORNPROBST, J.M., 1993, Composition and antiviral activities of a sulfated polysaccharide from *Schizymenia dubyi* (Rhodophyta, Gigartinales), *Bioorganic and Medicinal Chemistry Letters*, **3**, 1141–1146.

BURKIEWICZ, K., 1987, Active substances in the media after algae cultivation, *Acta Physiologiae Plantarum*, **9**, 211–217.

CANNELL, R.J.P., KELLAM, S.J., OWASIANKA, A.M. and WALKER, J.M., 1988a, Results of large scale screen of microalgae for the production of protease inhibitors, *Planta Medica*, **54**, 10–14.

CANNELL, R.J.P., OWISANKA, A.M. and WALKER, J.M., 1988b, Results of a large-scale screening programme to detect antibacterial activity from freshwater algae, *British Phycological Journal*, **23**, 41–44.

CARDELLINA, J.H., MUNRO, M.H.G., FULLER, R.W., MANFREDI, K.P., McKEE, T.C., TISCHLER, M., *et al.*, 1993, A chemical screening strategy for the dereplication and prioritisation of HIV-inhibitory aqueous natural products extracts, *Journal of Natural Products*, **56**, 1123–1129.

CARDINELLA, H.J.H., MORE, R.E., ARNOLD, E.V. and CLARDY, J., 1979, Structure and absolute configuration of malygolide, an antibiotic from the marine blue-green alga *Lyngbya majuscula* Gomont, *Journal of Organic Chemistry*, **44**, 4039–4042.

CARMELI, S., MOORE, R.E. and PATTERSON, G.M.L., 1990a, Isonitriles from the blue-green alga *Sctyonema mirabile*, *Journal of Organic Chemistry*, **55**, 4431–4438.

CARMELI, S., MOORE, R.E. and PATTERSON, G.M.L., 1990b, Tolytoxin and new scytophycins from three species of *Scytonema*, *Journal of Natural Products – Lloydia*, **55**, 1533–1542.

CARMELI, S., MOORE, R.E., PATTERSON, G.M.L., CORBETT, T.H. and VALERIOTE, F.A., 1990c, Tantazoles, unusual cytotoxic alkaloids from the blue-green alga *Scytonema mirabile*, *Journal of the American Chemical Society*, **112**, 8195–8197.

CARMELI, S., MOORE, R.E. and PATTERSON, G.M.L., 1991a, Mirabazoles, minor tantazole-related cytotoxins from the terrestrial blue-green alga *Scytonema mirabile*, *Tetrahedron Letters*, **32**, 2593–2596.

CARMELI, S., MOORE, R.E. and PATTERSON, G.M.L., 1991b, Mirabimides A-D, new N-acyl-pyrrolidones from the blue-green alga *Scytonema mirabile*, *Tetrahedron*, **47**, 2087–2096.

CARMICHAEL, W.W., 1986, Algal toxins, *Advances in Botanical Research*, **6**, 47–101.

CARMICHAEL, W.W., 1992, A review: Cyanobacteria secondary metabolites – the cyanotoxins, *Journal of Applied Bacteriology*, **72**, 445–459.

CARTER, D.C., MOORE, R.E., MYNDERSE, J.S., NIEMCZURA, W.P. and TODD, J.S., 1984, Structure of majusculamide C, a cyclic depsipeptide from *Lyngbya majuscula*, *Journal of Organic Chemistry*, **49**, 236–241.

CASTENHOLZ, R., GARCIA-PICHEL, F., GERWICK, W.H., GRACE, K.J.S., JACOBS, R.S., PROTEAU, P. and ROSSI, J., 1996, New bis-cyclo-penteno-indole alkaloid compounds, US Patent Number 5,498,405.

CHAUHAN, V.S., MARWAH, J.B. and BAGCHI, S.N., 1992, Effect of an antibiotic from *Oscillatoria* sp. on phytoplankters, higher plants and mice, *New Phytologist*, **120**, 251–257.

CHEN, D.Z.X., BOLAND, M.P., SMILLIE, M.A., KLIX, H., PTAK, C., ANDERSEN, R.J. and HOLMES, C.F.B., 1993, Identification of protein phosphatase inhibitors of the microcystin class in the marine environment, *Toxicon*, **31**, 1407–1414.

CHETSUMON, A., FUJIEDA, K., HIRATA, K., YAGI, K. and MIURA, Y., 1993a, Optimization of antibiotic production by the cyanobacterium *Scytonema* sp. TISTR 8208 immobilized on polyurethane foam, *Journal of Applied Phycology*, 5, 615–622.

CHETSUMON, A., MIYAMOTO, K., HIRATA, K., MIURA, Y., IKUTA, Y. and HAMASAKI, A., 1993b, Factors affecting antibiotic production in bioreactors with immobilized algal cells, *Applied Biochemistry and Biotechnology*, 39, 573–586.

CHRISMADHA, T. and BOROWITZKA, M.A., 1994, Effect of cell density and irradiance on growth, proximate composition and eicosapentaenoic acid production of *Phaeodactylum tricornutum* grown in a tubular photobioreactor, *Journal of Applied Phycology*, 6, 67–74.

COHEN, Z. and COHEN, S., 1991, Preparation of eicosapentaenoic acid (EPA) concentrate from *Porphyridium cruentum*, *Journal of the American Oil Chemists Society*, 68, 16–19.

DAMONTE, E., NEYTS, J., PUJOL, C.A., SNOECK, R., ANDREI, G., IKEDA, S., *et al.*, 1994, Antiviral activity of a sulphated polysaccharide from the red seaweed *Nothogenia fastigiata*, *Biochemical Pharmacology*, 47, 2187–2192.

DAVIDSON, B.S., 1993, Ascideans: Producers of amino acid derived metabolites, *Chemical Reviews*, 93, 1771–1791.

DEBRO, L.H. and WARD, H.B., 1979, Antibacterial activity of freshwater green algae, *Planta Medica*, 36, 375–378.

DE CANO, M.M.S., DE MULÉ M.C.Z., DE CAIRE, G.Z. and DE HALPERIN, D.R., 1990, Inhibition of *Candida albicans* and *Staphylococcus aureus* by phenolic compounds from the terrestrial cyanobacterium *Nostoc muscorum*, *Journal of Applied Phycology*, 2, 79–81.

DEMULE, M.C.Z., DECAIRE, G.Z. and DECANO, M.S., 1996, Bioactive substances from *Spirulina platensis* (cyanobacteria), *Phyton-International Journal of Experimental Botany*, 58, 93–96.

deSILVA, E.D., WILLIAMS, D.E., ANDERSEN, R.J., KLIX, H., HOLMES, C.F.B. and ALLEN, T.M., 1992, Mutoporin, a potent protein phosphatase inhibitor isolated from the Papua New Guinea sponge *Theonella swinhoei* Gray, *Tetrahedron Letters*, 33, 1561–1564.

DiNOVI, M., TRAINOR, D.A., NAKANISHI, K., SANDUJA, R. and ALAM, M., 1983, The structure of PB-1, an unusual toxin isolated from the red tide dinoflagellate *Ptychodiscus brevis*, *Tetrahedron Letters*, 24, 855–858.

DOUGLAS, D.J. and BATES, S.S., 1992, Production of domoic acid, a neurotoxic amino acid, by an axenic culture of the marine diatom *Nitzschia pungens* f *multiseries* Hasle, *Canadian Journal of Fisheries and Aquatic Sciences*, 49, 85–90.

DUNSTAN, G.A., VOLKMAN, J.K., BARRETT, S.M., LEROI, J.M. and JEFFREY, S.W., 1994, Essential polyunsaturated fatty acids from 14 species of diatom (Bacillariophyceae), *Phytochemistry*, 35, 155–161.

ELHAI, J., 1994, Genetic techniques appropriate for the biotechnological exploitation of cyanobacteria, *Journal of Applied Phycology*, 6, 177–186.

FALCH, B.S., KÖNIG, G.M., WRIGHT, A.D., STICHER, O., RÜEGGER, H. and BERNARDINELLI, G., 1993, Ambigol A and B – new biologically active polychlorinated aromatic compounds from the terrestrial blue-green alga *Fischerella ambigua*, *Journal of Organic Chemistry*, 58, 6570–6575.

FALCONER, I.R. and HUMPAGE, A.R., 1996, Tumor promotion by cyanobacterial toxins, *Journal of Phycology*, 35, 74–79.

FALCONER, I.R. and YEUNG, D.S.K., 1992, Cytoskeletal changes in hepatocytes induced by *Microcystis* toxins and their relation to hyperphosphorylation of cell proteins, *Chemical Biological Interactions*, 81, 181–196.

FAULKNER, D.J., HE, H.H., UNSON, M.D. and BEWLEY, C.A., 1993, New metabolites from marine sponges: Are symbionts important? *Gazzetta Chimica Italiana*, 123, 301–307.

FENICAL, W., 1993, Chemical studies of marine bacteria: developing a new resource, *Chemical Reviews*, 93, 1673–1683.

FINDLAY, J.A. and PATIL, A.D., 1984, Antibacterial constituents of the diatom *Navicula delognei*, *Journal of Natural Products – Lloydia*, **47**, 815–818.

FRANKMÖLLE, W.P., KNÜBEL, G., MOORE, R.E. and PATTERSON, G.M.L., 1992a, Antifungal cyclic peptides from the terrestrial blue-green alga *Anabaena laxa*. 2. Structures of lax-aphycin-A laxaphycin-B, laxaphycin-D and laxaphycin-E, *Journal of Antibiotics*, **45**, 1458–1466.

FRANKMÖLLE, W.P., LARSEN, L.K., CAPLAN, F.R., PATTERSON, G.M.L., KNÜBEL, G., LEVINE, I.A. and MOORE, R.E., 1992b, Antifungal cyclic peptides from the terrestrial blue-green alga *Anabaena laxa*. 1. Isolation and biological properties, *Journal of Antibiotics*, **45**, 1451–1457.

FURUSAWA, E., MOORE, R.E., MYNDERSE, J.S., NORTON, T.R. and PATTERSON, G.M.L., 1994, New purified culture of *Scytonema pseudohofmanni* ATCC 53141 is used to produce scytophycins A,B,C,D and E, which are potent cytotoxins and anti-neoplastic agents, USA Patent Number 5281533.

FUSETANI, N. and MATSUNAGA, S., 1993, Bioactive sponge peptides, *Chemical Reviews*, **93**, 1793–1806.

GALLON, J.R., KITTAKOOP, P. and BROWN, E.G., 1994, Biosynthesis of anatoxin-a by *Anabaena flos-aquae* – examination of primary enzymic steps, *Phytochemistry*, **35**, 1195-1203.

GARCIA-PICHEL, F. and CASTENHOLZ, R.W., 1993, Occurrence of UV-absorbing, mycospor-ine-like compounds among cyanobacterial isolates and an estimate of their screening capacity, *Applied and Environmental Microbiology*, **59**, 163–169.

GARCIA-PICHEL, F., WINGARD, C.E. and CASTENHOLZ, R.W., 1993, Evidence regarding the UV sunscreen role of a mycosporine-like compound in the cyanobacterium *Gloeocapsa* sp., *Applied and Environmental Microbiology*, **59**, 170–176.

GAUTHIER, M.J., BERNARD, P. and AUBERT, M., 1978, Production d'un antibiotique lipidique photo-sensible pa la diatomée marine *Chaetoceros lauderi* (Ralfs), *Ann. Microbiol.*, **129B**, 63–70.

GERSTER, H., 1993, Anticarcinogenic effect of common carotenoids, *International Journal for Vitamin and Nutrition Research*, **63**, 93–121.

GERWICK, W.H., 1989, 6-Desmethoxyhormothamnione, a new cytotoxic strylchromone from the marine cryptophyte *Chrysophaeum taylori*, *Journal of Natural Products – Lloydia*, **52**, 252–256.

GERWICK, W.H., LOPEZ, A., VAN DUYNE, G.D., CLARDY, J., ORTIZ, W. and BAEZ, A., 1986, Hormothamnione, a novel cytotoxic strylchromone from the marine cyanophyte *Hormothamnion enteromorphoides* Grunow, *Tetrahedron Letters*, **27**, 1979–1982.

GERWICK, W.H., REYES, S. and ALVARADO, B., 1987, Two malyngamides from the Caribbean cyanobacterium *Lyngbya majuscula*, *Phytochemistry*, **26**, 1701–1704.

GERWICK, W.H., MROZEK, C., MOGHADDAM, M.F. and AGARWAL, S.K., 1988, Novel cyto-toxic decapeptides from the tropical marine cyanobacterium *Hormothamnion enteromor-phoides* Grunow. 1. Discovery, isolation and initial chemical and biological characterisation of the hormothamnins from wild and cultured material, *Experientia*, **45**, 115–121.

GERWICK, W.H., PROTEAU, P.J., NAGLE, D.G., HAMEL, E., BLOKHIN, A. and SLATE, D.L., 1994a, Structure of curacin-A, a novel antimitotic, antiproliferative, and brine shrimp toxic natural product from the marine cyanobacterium *Lyngbya majuscula*, *Journal of Organic Chemistry*, **59**, 1243–1245.

GERWICK, W.H., ROBERTS, M.A., PROTEAU, P.J. and CHEN, J.L., 1994b, Screening cultured marine microalgae for anticancer-type activity, *Journal of Applied Phycology*, **6**, 143–149.

GERWICK, W.H., NAGLE, D.G. and PROTEAU, P.J., 1995, New thiazoline-type compounds isolated from marine algae, US Patent Number 9,509,630.

GLEASON, F.K., 1986, Cyanobacterin herbicide, US Patent Number 4,626,271.

GLEASON, F.K., CASE, D.E., SIPRELL, K.D. and MAGNUSON, T.S., 1986, Effect of the natural algicide, cyanobacterin, on a herbicide-resistant mutant of *Anacystis nidulans* R2, *Plant Science*, **46**, 5–10.

GLOMBITZA, K.-W., 1979, Antibiotics from Algae, in Hoppe, H.A., Levring, T. and Tanaka, Y. (Eds), *Marine Algae in Pharmaceutical Science*, pp. 303–342, Berlin, New York: Walter de Gruyter.

GLOMBITZA, K.W. and KOCH, M., 1989, Secondary metabolites of pharmaceutical potential, in Cresswell, R.C., Rees, T.A.V. and Shah, M. (Eds), *Algal and Cyanobacterial Biotechnology*, pp. 161–238, Harlow: Longman Scientific and Technical.

GOLAKOTI, T., OGINO, J., HELTZEL, C.E., LE HUSEBO, T., JENSEN, C.M., LARSEN, L.K., *et al.*, 1995, Structure determination, conformational analysis, chemical stability studies, and antitumor evaluation of the cryptophycins. Isolation of 18 new analogs from *Nostoc* sp. strain GSV 224, *Journal of the American Chemical Society*, **117**, 12030–12049.

GREGSON, J.M., 1986, Isolation and structure determination of the puwainaphycins A-D, MSc University of Hawaii, Honolulu.

GROMOV, B.V., VEPRITSKIY, A.A., TITOVA, N.N., MAMKAYEVA, K.A. and ALEXANDROVA, O.V., 1991, Production of the antibiotic cyanobacterin LU-1 by *Nostoc linckia* CALU 892 (Cyanobacterium), *Journal of Applied Phycology*, **3**, 55–59.

GROSS, E.M., WOLK, C.P. and JUTTNER, F., 1991, Fischerellin, a new allelochemical from the freshwater cyanobacterium *Fischerella muscicola*, *Journal of Phycology*, **27**, 686–692.

GUSOFSKY, F. and DALY, J.W., 1990, Maitotoxin: a unique pharmacological tool for research on calcium-dependent mechanisms, *Biochemical Pharmacology*, **39**, 1633–1639.

GUSTAFSON, K.R., CARDELLINA, J.H., FULLER, R.W., WEISLOW, O.S., KISER, R.F., SNADER, K.M., *et al.*, 1989, AIDS-antiviral sulfolipids from cyanobacteria (blue-green algae), *Journal of the National Cancer Institute*, **81**, 1254–1258.

HAGMANN, L. and JÜTTNER, F., 1996, Fischerellin A, a novel photosystem-II-inhibiting allelochemical of the cyanobacterium *Fischerella muscicola* with antifungal and herbicidal activity, *Tetrahedron Letters*, **37**, 6539–6542.

HALL, S., STRICHARTZ, G., MOCZYDLOWSKI, E., RAVINDRAN, A. and REICHARDT, P.B., 1990, The saxitoxins: Sources, chemistry and pharmacology, in Hall, S. and Strichartz, G. (Eds), *Marine Toxins*, pp. 29–65, Washington, DC: American Chemical Society.

HAMBLEY, T.W., HAWKINS, C.J., LAVIN, M.F., VAN DER BRENK, A. and WATTERS, D.J., 1992, Cycloxasoline – a cytotoxic hexapeptide from the asidean *Lissoclonium bistratum*, *Tetrahedron*, **48**, 341–348.

HARDER, R. and OPPERMAN, A., 1953, Über antibiotische Stoffe bei den Grünalgen *Stichococcus bacillaris* und *Protosiphon botryoides*, *Archiv für Mikrobiologie*, **19**, 398–401.

HASEGAWA, T., YOSHIKAI, Y., OKUDA, M. and MONOTO, K., 1990, Accelerated restoration of the leukocyte number and augmented resistance against *Escherichia coli* in cyclophosphamide-treated rats orally administered with a hot water extract of *Chlorella vulgaris*, *International Journal of Immunopharmacology*, **12**, 833–840.

HAUMAN, J.H., 1981, Is a plasmid(s) involved in the toxicity of *Microcystis aeruginosa*?, in Carmichael, W.W. (Ed.), *The Water Environment: Algal Toxins and Health*, pp. 97–102, New York: Plenum Press.

HAUNG, J.M.C. and WU, C.H., 1989, Pharmacological actions of berevetoxin from *Ptychodiscus brevis* on nerve membranes, in Okaichi, T., Anderson, D.M. and Nemoto, T. (Eds), *Red Tides: Biology, Environmental Science and Toxicology*, pp. 379–382, New York: Elsevier.

HAYASHI, K., HAYASHI, T., MORITA, N. and KOJIMA, I., 1993, An extract from *Spirulina platensis* is a selective inhibitor of Herpes simplex virus type-1 penetration into HeLa cells, *Phytotherapy Research*, **7**, 76–80.

HAYASHI, T., HAYASHI, K., MAEDA, M. and KOJIMA, I., 1996, Calcium spirulan, an inhibitor of enveloped virus replication, from a blue-green alga *Spirulina platensis*, *Journal of Natural Products*, **59**, 83–87.

HELMS, G.L., MOORE, R.E., NIEMCZURA, W.P., PATTERSON, G.M.L., TOMER, K.B. and

GROSS, M.L., 1988, Scytonemin A, a novel calcium antagonist from a blue-green alga, *Journal of Organic Chemistry*, **53**, 1298–1307.

HEMSCHEIDT, T., PUGLISI, M.P., LARSEN, L.K., PATTERSON, G.M.L., MOORE, R.E., RIOS, J.L. and CLARDY, J., 1994, Structure and biosynthesis of borophycin, a new boeseken complex of boric acid from a marine strain of the blue-green alga *Nostoc linckia*, *Journal of Organic Chemistry*, **59**, 3467–3471.

HILLE, B., 1975, The receptor for tetrodotoxin and saxitoxin: a structural hypothesis, *Biophysics Journal*, **15**, 615–619.

HOFHEINZ, W. and OBERHÄNSLI, W.E., 1977, Dysidenin, a novel chlorine containing natural product from the sponge *Dysidea herbacea*, *Helvetica Chimica Acta*, **60**, 660–669.

HOLMES, C.F.B. and BORLAND, M.P., 1993, Inhibitors of protein phosphatase-1 and -2A; two of the major serine/threonone protein phosphatases involved in cellular regulation, *Current Opinion in Structural Biology*, **3**, 934–943.

HOPPE, H.A., 1979, Marine algae and their products and constituents in pharmacy, in Hoppe, H.A., Levring, T. and Tanaka, Y. (Eds), *Marine Algae in Pharmaceutical Science*, pp. 25–119, Berlin, New York: Walter de Gruyter.

HOPWOOD, D.A., 1993, Genetic engineering of *Streptomyces* to create hybrid antibiotics, *Current Opinion in Biotechnology*, **4**, 531–537.

IKAWA, M. and SASNER, J.J., 1990, The chemistry and physiology of algal toxins, in Akatsuka, I. (Ed.), *Introduction to Applied Phycology*, pp. 27–65, The Hague: SPB Academic Publishing.

ISHIBASHI, M., MOORE, R.E., PATTERSON, G.M.L., XU, C. and CLARDY, J., 1986, Scytophycins, cytotoxic and antimycotic agents from the cyanophyte *Scytonema pseudo-hofmanni*, *Journal of Organic Chemistry*, **51**, 5300–5306.

ISHIBASHI, M., OHIZUMI, Y., HAMASHIMATA, M., NAKAMURA, H., HIRATA, Y., SASAKI, T. and KOBATASHI, J., 1987, Amphidinolide B, a novel macrolide with potent antineoplastic activity from the marine dinoflagellate *Amphidinium* sp., *Journal of the Chemical Society. Chemical Communications*, **1987**, 1127–1129.

ISHIDA, K., MAURAKAMI, M., MATSUDA, H. and YAMAGICHI, K., 1995, Micropeptin 90, a plasmin and trypsin inhibitor from the blue-green alga *Microcystis aeruginosa* (NIES-90), *Tetrahedron Letters*, **36**, 3535–3538.

JACOBI, C., RINEHART, K.L., NEUBER, R., MEZ, K. and WECKESSER, J., 1996, Cyanopeptolin SS, a disulphated depsipeptide from a water bloom: structural elucidation and biological activities, *Journal of Phycology*, **35**, 111–116.

JORGENSEN, E.G. and STEEMANN NIELSEN, E., 1961, Effect of filtrates from cultures of unicellular algae on the growth of *Staphylococcus aureus*, *Physiologia Plantarum*, **14**, 896–908.

JUNG, J.H., MOORE, R.E. and PATTERSON, G.M.L., 1991, Scytophycins from a blue-green alga belonging to the Nostocaceae, *Phytochemistry*, **30**, 3615–3616.

KAYA, K., SANO, T., BEATTIE, K.A. and CODD, G.A., 1996, Nostocyclin, a novel 3-amino-6-hydroxy-2-piperidone-containing cyclic depsipeptide from the cyanobacterium *Nostoc* sp., *Tetrahedron Letters*, **37**, 6725–6728.

KELLAM, S.J. and WALKER, J.M., 1989, Antibacterial activity from marine microalgae, *British Phycological Journal*, **24**, 191–194.

KELLAM, S.J., CANNELL, R.J.P., OWSIANKA, A.M. and WALKER, J.M., 1988, Results of a large-scale screening programme to detect antifungal activity from marine and freshwater microalgae in laboratory culture, *British Phycological Journal*, **23**, 45–47.

KHAN, S., AHMED, M.S., ARAKAWA, O. and ONOUE, Y., 1995, Properties of neurotoxins separated from a harmful red tide organism *Chattonella marina*, *Israeli Journal of Aquaculture – Bamidgeh*, **47**, 137–141.

KIM, C.II., SAKO, Y. and ISHIDA, Y, 1993, Variation of toxin production and composition in axenic cultures of *Alexandrium catenella* and *A. tamarense*, *Nippon Suisan Gakkaishi – Bulletin of the Japanese Society of Scientific Fisheries*, **59**, 633–639.

KINDLE, K.L. and SODEINDE, O.A., 1994, Nuclear and chloroplast transformation in *Chlamydomonas reinhardtii* – strategies for genetic manipulation and gene expression, *Journal of Applied Phycology*, **6**, 231–238.

KLEIN, D., BRACKMAN, J., DALOZE, D., HOFFMANN, L. and DEMULIN, V., 1996, Laingolide, a novel 15-membered macrolide from *Lyngbya bouillonii* (Cyanophyceae), *Tetrahedron Letters*, **37**, 7519–7520.

KNÜBEL, G., LARSEN, L.K., MOORE, R.E., LEVINE, I.A. and PATTERSON, G.M.L., 1990, Cytotoxic, antiviral indolocarbazoles from a blue-green alga belonging to the Nostocaceae, *Journal of Antibiotics*, **43**, 1236–1239.

KOBAYASHI, J. and ISHIBASHI, M., 1993, Bioactive metabolites of symbiotic marine organisms, *Chemical Reviews*, **93**, 1753–1789.

KOBAYASHI, A., ISHIBASHI, M., NAKAMURA, H., OZUMI, Y., YAMASU, T., SASAKI, T. and HIRATA, Y., 1986, Amphidinolide-A, a novel antineoplastic macrolide from the marine dinoflagellate *Amphidinium* sp., *Tetrahedron Letters*, **27**, 5755–5758.

KOBAYASHI, J., ISHIBASHI, M., NAKAMURA, H., OHIZUMI, Y., YAMASU, T., HIRATA, Y., *et al.*, 1989, Cytotoxic macrolides from a cultured marine dinoflagellate of the genus *Amphidinium*, *Journal of Natural Products – Lloydia*, **52**, 1036–1041.

KOEHN, F.E., LONGLEY, R.E. and REED, J.K., 1992, Microcolins A and B, new immunosuppressive peptides from the blue-green alga *Lyngbya majuscula*, *Journal of Natural Products – Lloydia*, **55**, 613–619.

KONISHI, F., TANAKA, K., HIMEMO, K., TANIGUCHI, K. and NOMOTO, K., 1985, Antitumor effect inducer by a hot water extract of *Chlorella vulgaris* (CE): resistance to MethA tumor growth mediated by CE-induced polymorphonuclear leukocytes, *Cancer Immunol Immunother*, **19**, 73–80.

KONISHI, F., TANAKA, K., KUMAMOTO, S., HASEGAWA, T., OKUDA, M., YANO, I., *et al.*, 1990, Enhanced resistance against *Escherichia coli* infection by subcutaneous administration of the hot-water extract of *Chlorella vulgaris* in cyclophosphamide-treated mice, *Cancer Immunol Immunother*, **32**, 1–7.

KONISHI, F., MITSUYAMA, M., OKUDA, M., TANAKA, K., HASEGAWA, T. and NOMOTO, K., 1996, Protective effect of an acidic glycoprotein obtained from culture of *Chlorella vulgaris* against myelosuppression by 5-fluorouracil, *Cancer Immunol Immunother*, **42**, 268–274.

KÓS, P., GORZÓ G., SURÁNYI, G. and BORBÉLI, G., 1995, Simple and efficient method for isolation and measurement of cyanobacterial hepatotoxins by plant tests (*Sinapsis alba* L.), *Analytical Biochemistry*, **225**, 49–53.

KYLE, D.J. and GLADUE, R.M., 1991, Eicosapentaenoic acids and methods for their production, World Patent Number 9,114,427.

KYLE, D.J., REEB, S.E. and SICOTTE, V.J., 1991, Docosahexaenoic acid, methods for its production and compounds containing the same, World Patent Number 9,111,918.

LAGUNA, M.R., VILLAR, R., CADAVID, I. and CALLEJA, J.M., 1993a, Effects of extracts of *Tetraselmis suecica* and *Isochrysis galbana* on the central nervous system, *Planta Medica*, **59**, 207–214.

LAGUNA, M.R., VILLAR, R., CALLEJA, J.M. and CADAVID, I., 1993b, Effects of *Chlorella stigmatophora* extract on the central nervous system, *Planta Medica*, **59**, 125–130.

LARSEN, L.K., MOORE, R.E. and PATTERSON, G.M.L., 1994, β-Carbolines from the blue-green alga *Dichothrix baueriana*, *Journal of Natural Products – Lloydia*, **57**, 419—421.

LAU, A.F., SIEDLECKI, J., ANLEITNER, J., PATTERSON, G.M.L., CAPLAN, F.R. and MOORE, R.E., 1993, Inhibition of reverse transcriptase activity by extracts of cultured blue-green algae (Cyanophyta), *Planta Medica*, **59**, 148–151.

LAYCOCK, M.V., DE FREITAS, A.S.W. and WRIGHT, J.L.C., 1989, Glutamate agonists from marine algae, *Journal of Applied Phycology*, **1**, 113–122.

LEE, J.S., IGARASHI, T., FRAGA, S., DAHL, E., HOVGAARD, P. and YASUMOTO, T., 1989,

Determination of diarrhetic shellfish toxins in various dinoflagellate species, *Journal of Applied Phycology*, **1**, 147–162.

LI, C.W., CHU, S. and LEE, M., 1989, Characterizing the silica deposition vesicle of diatoms, *Protoplasma*, **151**, 158–163.

LINCOLN, R.A., STRUPINSKI, K. and WALKER, J.M., 1990, Use of an isolated guinea-pig ileum assay to detect bioactive compounds in microalgal cultures, *Journal of Applied Phycology*, **2**, 83–88.

LUU, H.A., CHEN, D.Z.X., MAGOON, J., WORMS, J., SMITH, J. and HOLMES, C.F.B., 1993, Quantification of diarrhetic shellfish toxins and identification of novel protein phosphatase inhibitors in marine phytoplankton and mussels, *Toxicon*, **31**, 75–83.

MACKINTOSH, C., BEATTIE, K.A., KLUMPP, S., COHEN, P. and CODD, G.A., 1990, Cyanobacterial microcystin-LR is a potent and specific inhibitor of protein phosphatase-1 and phosphatase-2A from both mammals and higher plants, *FEBS Letters*, **264**, 187–192.

MAHMOOD, N.A. and CARMICHAEL, W.W., 1987, Anatoxin-a(S), an anticholinesterase from the cyanobacterium *Anabaena flos-aquae* NCR-525–17, *Toxicon*, **25**, 1221–1227.

MALIN, G., LISS, P.S. and TURNER, S.M., 1994, Dimethyl sulfide: production and atmospheric consequences, in Green, J.C. and Leadbeater, B.S.C. (Eds), *The Haptophyte Algae*, pp. 303–320, Oxford: Clarendon Press.

MASON, C.P., EDWARDS, K.R., CARLSON, R.E., PIGNATELLO, J., GLEASON, F.K. and WOOD, J.M., 1981, Isolation of chlorine-containing antibiotic from the freshwater cyanobacterium *Scytonema hofmanni*, *Science*, **215**, 400–402.

MATSUNAGA, T. and TAKEYAMA, H., 1995, Genetic engineering in marine cyanobacteria, *Journal of Applied Phycology*, **7**, 77–84.

MATSUNAGA, T., BURGESS, J.G., YAMADA, N., KOMATSU, K., YOSHIDA, S. and WACHI, Y., 1993, An ultraviolet (UV-A) absorbing biopterin glucoside from the marine planktonic cyanobacterium *Oscillatoria* sp., *Applied Microbiology and Biotechnology*, **39**, 250–253.

McDANIEL, R., EBERT-KHOSLA, S., HOPWOOD, D.A. and KHOSLA, C., 1993, Engineered biosynthesis of novel polyketides, *Science*, **262**, 1546–1550.

McKEE, T.C., CARDELLINA, J.H., RICCIO, R., D'AURIA, M.V., IORIZZI, M., MINALE, L., *et al.*, 1994, HIV-inhibitory natural products. 11. Comparative studies of sulfated sterols from marine invertebrates, *Journal of Medicinal Chemistry*, **37**, 793–797.

MILLER, P.W., RUSSELL, B.L. and SCHMIDT, R.R., 1994, Transcription initiation site of a NADP-specific glutamate dehydrogenase gene and potential use of its promoter region to express foreign genes in ammonium-cultured *Chlorella sorokiniana* cells, *Journal of Applied Phycology*, **6**, 211–223.

MOON, S.S., CHEN, J.L., MOORE, R.E. and PATTERSON, G.M.L., 1992, Calophycin, a fungicidal cyclic decapeptide from the terrestrial blue-green alga *Calothrix fusca*, *Journal of Organic Chemistry*, **57**, 1097–1103.

MOORE, R.E. and ENTZEROTH, M., 1988, Majusculamide D and dehydroxymajusculamide D, two cytotoxins from *Lyngbya majuscula*, *Phytochemistry*, **27**, 3101–3103.

MOORE, R.E. and PATTERSON, G.M.L., 1988, Hapalindoles, USA Patent Number 4755610.

MOORE, R.E., PATTERSON, G.M.L., MYNDERSE, J.S., BARCHI, J., NORTON, T.R., FURUSAWA, E. and FURUSAWA, S., 1986, Toxins from cyanophytes belonging to the Scytonemataceae, *Pure and Applied Chemistry*, **58**, 263–271.

MOORE, R.E., CHEUK, C., YANG, X.Q.G., PATTERSON, G.M.L., BONJOUKLIAN, R., SMITKA, T.A., *et al.*, 1987, Hapalindoles, antibacterial and antimycotic alkaloids from the cyanophyte *Hapalosiphon fontinalis*, *Journal of Organic Chemistry*, **52**, 1036–1043.

MOORE, R.E., PATTERSON, M.L. and CARMICHAEL, W.W., 1988, New pharmaceuticals from cultured blue-green algae, in Fautin, D.G. (Ed.), *Biomedical Importance of Marine Organisms*, pp. 143–150, San Francisco: California Academy of Sciences.

MORTON, S.L. and BOMBER, J.W., 1994, Maximizing okadaic acid content from *Prorocentrum hoffmannianum* Faust, *Journal of Applied Phycology*, **6**, 41–44.

MORTON, S.L. and TINDALL, D.R., 1995, Morphological and biochemical variability of the toxic dinoflagellate *Prorocentrum lima* isolated from three locations at Heron Island, Australia, *Journal of Phycology*, **31**, 914–921.

MURAKAMI, M. and YAMAGUCHI, K., 1989, Biologically active compounds of microalgae, with special reference to a novel antifungal polyether macrolide from the dinoflagellate, *Alexandrium hiranoi*, in Miyachi, S., Karube, I. and Ishida, Y. (Eds), *Current Topics in Marine Biotechnology*, pp. 199–202, Tokyo: Japanese Society for Marine Biotechnology.

MURAKAMI, M., MAKABE, K., YAMAGUCHI, K., KONOSU, S. and WÄLCHLI, M.R., 1988, Goniodomin A, a novel polyether macrolide from the dinoflagellate *Goniodoma pseudogonyaulax*, *Tetrahedron Letters*, **29**, 1149–1152.

MURAKAMI, M., OKITA, Y., MATSUDA, H., OKINO, T. and YAMAGUCHI, K., 1994, Aeruginosin 298-A, a thrombin and trypsin inhibitor from the blue-green alga *Microcystis aeruginosa* (NIES-298), *Tetrahedron Letters*, **35**, 3129–3132.

MURAKAMI, M., ISHIDA, K., OKINO, T., OKITA, Y., MATSUDA, H. and YAMAGUCHI, K., 1995, Aeruginosins 98-A and B, trypsin inhibitors from the blue-green alga *Microcystis aeruginosa* (NIES-98), *Tetrahedron Letters*, **16**, 2785–2788.

MYNDERSE, J.S. and MOORE, R.E., 1978, The isolation of (-)-E-1-chlorotridec-1-ene-6,8-diol from a marine cyanophyte, *Phytochemistry*, **17**, 1325–1326.

MYNDERSE, J.S., MOORE, R.E., KASHIWAGI, M. and NORTON, T.R., 1977, Antileukemia activity in the Oscillatoriaceae: Isolation of debromoaplysiatoxin from *Lyngbya*, *Science*, **196**, 538–540.

NAGAI, H., SATAKE, M. and YASUMOTO, T., 1990, Antimicrobial activities of polyether compounds of dinoflagellate origins, *Journal of Applied Phycology*, **2**, 305–308.

NAGAI, H., MIKAMI, Y., YAZAWA, K., GONOI, T. and YASUMOTO, T., 1993, Biological activities of novel polyether antifungals, gambieric acids A and B from a marine dinoflagellate *Gambierdiscus toxicus*, *Journal of Antibiotics*, **46**, 520–522.

NAGAI, H., MIKAMI, Y. and YASUMOTO, T., 1995, Antifungals produced by dinoflagellates – gambieric acids, *Pesticide Science*, **44**, 92–94.

NAGLE, D.G., GERALDS, R.S., YOO, H.D., GERWICK, W.H., KIM, T.S., NAMBU, M. and WHITE, J.D., 1995, Absolute configuration of curacin A, a novel antimitotic agent from the tropical marine cyanobacterium *Lyngbya majuscula*, *Tetrahedron Letters*, **36**, 1189–1192.

NAKAMURA, H., ASARI, T., OHIZUMI, Y., KOBAYASHI, J., YAMASU, T. and MURAI, A., 1993, Isolation of zooxanthellatoxins, novel vasoconstrictive substances from the zooxanthella *Symbiodinium* sp, *Toxicon*, **31**, 371–376.

NODA, K., OHNO, N., TANAKA, K., KAMIYA, N., OKUDA, M., TADOAME, T., *et al.*, 1996, A water soluble antitumor glycoprotein from *Chlorella vulgaris*, *Planta Medica*, (in press).

NORTON, R.S. and WELLS, R.J., 1982, A series of chiral polybrominated biindoles from the blue-green marine algae *Rivularia firma*. Application of 13C NMR spin-lattice relaxation data and 13C-1H coupling constants to structure elucidation, *Journal of the American Chemical Society*, **104**, 3628–3635.

NUMATA, A., TAKAHASHI, C., ITO, Y., TAKADA, T., KAWAI, K., USAMI, Y., *et al.*, 1993, Communesins, cytotoxic metabolites of a fungus isolated from a marine alga, *Tetrahedron Letters*, **34**, 2355–2358.

OGINO, J., MOORE, R.E., PATTERSON, G.M.L. and SMITH, C.D., 1996, Dendroamides, new cyclic hexapeptides from a blue-green alga – multidrug-resistance reversing activity of dendroamide A, *Journal of Natural Products*, **59**, 581–586.

OHTA, S., SHIOMI, Y., KAWASHIMA, A., AOZASA, O., NAKAO, T., NAGATE, T., *et al.*, 1995, Antibiotic effect of linolenic acid from *Chlorococcum* strain HS-101 and *Dunaliella primolecta* on methicillin-resistant *Staphylococcus aureus*, *Journal of Applied Phycology*, **7**, 121–127.

OKINO, T., MATSUDA, H., MURAKAMI, M. and YAMAGUCHI, K., 1993a, Microginin, an angiotensin-converting enzyme inhibitor from the blue-green alga *Microcystis aeruginosa*, *Tetrahedron Letters*, **34**, 501–504.

OKINO, T., MURAKAMI, M., HARAGUCHI, R., MUNEKATA, H., MATSUDA, H. and YAMAGUCHI, K., 1993b, Micropeptins A and B, plasmin and trypsin inhibitors from the blue-green alga *Microcystis aeruginosa*, *Tetrahedron Letters*, **34**, 8131–8134.

OKINO, T., MATSUDA, H., MURAKAMI, M. and YAMAGUCHI, K., 1995, New microviridins, elastase inhibitors from the blue-gren alga *Microcystis aeruginosa*, *Tetrahedron*, **51**, 10679–10686.

OKUYAMA, H., 1992, Minimum requirements of n-3 and n-6 essential fatty acids for the function of the central nervous system and for the prevention of chronic disease, *Proceedings of the Society for Experimental Biology and Medicine*, **200**, 174–176.

OKUZUMI, J., NISHINO, H., MURAKOSHI, M., IWASHIMA, A., TANAKA, Y., YAMANE, T., *et al.*, 1990, Inhibitory effects of fucoxanthin, a natural carotenoid, on N-myc expression and cell cycle progression in human malignant tumor cells, *Cancer Letters*, **55**, 75–81.

OMURA, S., 1992, *The Search for Bioactive Compounds from Microorganisms*, New York: Springer Verlag.

ÖRDÖG, V. and PULZ, O., 1996, Diurnal changes of cytokinin-like activity in a strain of *Arthronema africanum* (cyanobacteria), determined by bioassays, *Archiv für Hydrobiologie, Suppl. Algological Studies*, **82** (in press).

ORJALA, J. and GERWICK, W.H., 1996, Barbamide, a chlorinated metabolite with molluscicidal activity from the Caribbean cyanobacterium *Lyngbya majuscula*, *Journal of Natural Products*, **59**, 427–430.

PAIK, S.G., CARMELI, S., CULLINGHAM, J., MOORE, R.E., PATTERSON, G.M.L. and TIUS, M.A., 1994, Mirabimide E, an unusual N-acylpyrrolinone from the blue-green alga *Scytonema mirabile*: Structure determination and synthesis, *Journal of the American Chemical Society*, **116**, 8116–8125.

PATTERSON, G.M.L., 1996, Biotechnological applications of cyanobacteria, *Journal of Scientific and Industrial Research*, **55**, 669–684.

PATTERSON, G.M.L. and BOLIS, C.M., 1993, Regulation of scytophycin accumulation in cultures of *Scytonema ocellatum*. 1. Physical factors, *Applied Microbiology and Biotechnology*, **40**, 375–381.

PATTERSON, G.M.L. and BOLIS, C.M., 1994, Scytophycin production by axenic cultures of the cyanobacterium *Scytonema ocellatum*, *Natural Toxins*, **2**, 280–285.

PATTERSON, G.M.L., BALDWIN, C.L., BOLIS, C.M., CAPLAN, F.R., KARUSO, H., LARSEN, L.K., *et al.*, 1991, Antineoplastic activity of cultured blue-green algae (Cyanophyta), *Journal of Phycology*, **27**, 530–536.

PATTERSON, G.M.L., BAKER, K.K., BALDWIN, C.L., BOLIS, C.M., CAPLAN, F.R., LARSEN, L.K., *et al.*, 1993a, Antiviral activity of cultured blue-green algae (Cyanophyta), *Journal of Phycology*, **29**, 125–130.

PATTERSON, G.M.L., SMITH, C.D., KIMURA, L.H., BRITTON, B.A. and CARMELI, S., 1993b, Action of tolytoxin on cell morphology, cytoskeletal organization, and actin polymerization, *Cell Motility and the Cytoskeleton*, **24**, 39–48.

PATTERSON, G.M.L., LARSEN, L.K. and MOORE, R.E., 1994, Bioactive natural products from blue-green algae, *Journal of Applied Phycology*, **6**, 151–157.

PATTERSON, G.M.L., MOORE, R.E., CARMELI, S., SMITH, C.D. and KIMURA, L.H., 1995, Scytophycin compounds, compositions and methods for their production and use, USA Patent Number 5439933.

PAUL, G.K., MATSUMORI, N., MURATA, M. and TACHIBANA, K., 1995, Isolation and chemical structure of amphidinol 2, a potent hemolytic compound from marine dinoflagellate *Amphidinium klebsii*, *Tetrahedron Letters*, **36**, 6279–6282.

PEDERSEN, M. and DASILVA, E.J., 1973, Simple brominated phenols in the bluegreen alga *Calothrix brevissima* West, *Planta*, **115**, 83–96.

PESANDO, D., 1972, Etude chimique et structurale d'une substance lipidique antibiotique prodiute par une diatomée marine: *Asterionella japonica*, *Rev. Intern. Oceanogr. Med.*, **25**, 49–69.

PESANDO, D., 1990, Antibacterial and antifungal activities of marine algae, in Akatsuka, I. (ed.), *Introduction to Applied Phycology*, pp. 3–26, The Hague: SPB Academic Publishing.

PESANDO, D., GNASSIA-BARELLI, M., GUEHO, E., RINAUDO, M. and DEFAYE, J., 1980, Isolement, étude structurale et propriétés antibiotiques et antifongiques d'un comosant polysaccaridique de la diatomée marine *Chaetoceras lauderi* Ralfs, *Oceanis. Fasc. Hors.-Sér.*, 561–568.

PRASAD, A.V.K. and SHIMIZU, Y., 1989, The structure of hemibrevetoxin-B: A new type of toxin in the Gulf of Mexico red tide organism, *Journal of the American Chemical Society*, **111**, 6476–6477.

PRINSEP, M.R., CAPLAN, F.R., MOORE, R.E., PATTERSON, G.M.L. and SMITH, C.D., 1992a, Tolyporphin, a novel multidrug resistance reversing agent from the blue-green alga *Tolypothrix nodosa*, *Journal of the American Chemical Society*, **114**, 385–387.

PRINSEP, M.R., MOORE, R.E., LEVINE, I.A. and PATTERSON, G.M.L., 1992b, Westiellamide, a bistratamide-related cyclic peptide from the blue-green algae *Westelliopsis prolifica*, *Journal of Natural Products – Lloydia*, **55**, 140–142.

PRINSEP, M.R., THOMSON, R.A., WEST, M.L. and WYLIE, B.L., 1996, Tolypodiol, an antiinflammatory diterpenoid from the cyanobacterium *Tolypothrix nodosa*, *Journal of Natural Products*, **59**, 786–788.

PROTEAU, P.J., GERWICK, W.H., GARCIAPICHEL, F. and CASTENHOLZ, R., 1993, The structure of scytonemin, an ultraviolet sunscreen pigment from the sheaths of cyanobacteria, *Experientia*, **49**, 825–829.

QUINN, R.J., TAYLOR, C., SUGANUMA, M. and FUJIKI, H., 1993, The conserved acid binding domain model of inhibitors of protein phosphatases 1 and 2A: molecular modelling aspects, *Bioorganic and Medicinal Chemistry Letters*, **3**, 1029–1034.

QURESHI, M.A. and ALI, R.A., 1996, *Spirulina platensis* exposure enhances macrophage phagocytic function in cats, *Immunopharmacology and Immunotoxicology*, **18**, 457–463.

QURESHI, M.A., KIDD, M.T. and ALI, R.S., 1995, *Spirulina platensis* extract enhances chicken macrophage functions after *in vitro* exposure, *Journal of Nutritional Immunology*, **3**, 35–45.

QURESHI, M.A., GARLICH, J.D. and KIDD, M.T., 1996, Dietary *Spirulina platensis* enhances humoral and cell-mediated immune functions in chickens, *Immunopharmacology and Immunotoxicology*, **18**, 465–476.

RAPALA, J., SIVONEN, K., LUUKKAINEN, R. and NIEMELÄ S.I., 1993, Anatoxin-A concentration in *Anabaena* and *Aphanizomenon* under different environmental conditions and comparison of growth by toxic and non-toxic *Anabaena*-strains – a laboratory study, *Journal of Applied Phycology*, **5**, 581–591.

REICHELT, J.L. and BOROWITZKA, M.A., 1984, Antibiotics from algae: results of a large scale screening programme, *Hydrobiologia*, **116/117**, 158–168.

RICHMAN, D.D., 1996, HIV therapeutics, *Science*, **272**, 1886–1888.

RINEHART, K.L., 1988, Screening to detect biological activity, in Fautin, D.G. (Ed.), *Biomedical Importance of Marine Organisms*, pp. 13–22, San Francisco: California Academy of Sciences.

RINEHART, K.L., SHAW, P.D., SHIELD, L.S., GLOER, J.B., HARBOUR, G.C., KOKER, M.E.S., et al., 1981, Marine natural products as sources of antiviral, antimicrobial, and antineoplastic agents, *Pure and Applied Chemistry*, **53**, 795–817.

RINEHART, K.L., SHIELD, L.S. and COHENPARSONS, M., 1993, Antiviral substances, in Attaway, D.H. and Zaborsky, O.R. (Eds), *Marine Biotechnology*, Vol. 1, pp. 309–342, New York: Plenum Publishing Corp.

RINEHART, K.L., NAMIKOSHI, M. and CHOI, B.W., 1994, Structure and biosynthesis of toxins from blue-green algae (cyanobacteria), *Journal of Applied Phycology*, **6**, 159–176.

SANO, T. and KAYA, K., 1995, Oscillamide Y, a chymotrypsin inhibitor from toxic *Oscillatoria agardhii*, *Tetrahedron Letters*, **36**, 5933–5936.

SANO, T. and KAYA, K., 1996, Oscillatorin – a chymotrypsin inhibitor from toxic *Oscillatoria agardhii*, *Tetrahedron Letters*, **37**, 6873–6876.

SATAKE, S., MURATA, M., YASUMOTO, T., FUJITA, T. and NAOKI, H., 1991, Amphidinol, a polyhydroxypolyene antifungal agent with an unprecedented structure from a marine dinoflagellate, *Amphidinium klebsii*, *Journal of the American Chemical Society*, **113**, 9859–9861.

SCHMITZ, F.J., BOWDEN, B.F. and TOTH, S.I., 1993, Antitumor and cytotoxic compounds from marine organisms, in Attaway, D.H. and Zaborsky, O.R. (Eds), *Marine Biotechnology, Vol 1*, Vol. 1, pp. 197–308, New York: Plenum Publishing Corp.

SCHWABE, W., WEIHE, A., BÖRNER, T., HENNING, M. and KOHL, J.G., 1988, Plasmids in toxic and nontoxic strains of the cyanobacterium *Microcystis aeruginosa*, *Current Microbiology*, **17**, 133–137.

SCHWARTZ, R.E., HIRSCH, C.F., SIGMUND, J.M. and PETTIBONE, D.J., 1989, Haploindolone compounds as vassopressin antagonists, USA Patent Number 4803217.

SCHWARTZ, R.E., HIRSCH, C.F., SESIN, D.F., FLOR, J.E., CHARTRAIN, M., FROMTLING, R.E., *et al.*, 1990, Pharmaceuticals from cultured algae, *Journal of Indian Microbiology*, **5**, 113–123.

SHAISH, A., BEN-AMOTZ, A. and AVRON, M., 1991, Production and selection of high β-carotene mutants of *Dunaliella bardawil* (Chlorophyta), *Journal of Phycology*, **27**, 652–656.

SHIMIZU, Y., 1993, Microalgal metabolites, *Chemical Reviews*, **93**, 1685–1698.

SHIMIZU, Y., CHOU, H.N., BANDO, H., VAN DUYNE, G. and CLARDAY, J., 1986, Structure of brevetoxin A (GB-1 toxin), the most potent toxin in the Florida red tide organism *Gymnodinium breve* (*Ptychodiscus breve*), *Journal of the American Chemical Society*, **198**, 514–515.

SHIRAHASHI, H., MURAKAMI, N., WATANABE, M., NAGATSU, A., SAKAKIBARA, J., TOKUDA, H., *et al.*, 1993, Isolation and identification of anti-tumor-promoting principles from the fresh-water cyanobacterium *Phormidium tenue*, *Chemical and Pharmaceutical Bulletin*, **41**, 1664–1666.

SIEBURTH, J.M., 1959, Acrylic acid, an 'antibiotic' principle in *Phaeocystis* blooms in Antarctic waters, *Science*, **132**, 676–677.

SIEBURTH, J.M., 1961, Antibiotic properties of acrylic acid, a factor in the gastrointestinal antibiosis of polar marine animals, *Journal of Bacteriology*, **82**, 72–79.

SIVONEN, K., 1990, Effects of light, temperature, nitrate, orthophosphate, and bacteria on growth of and hepatotoxin production by *Oscillatoria agardhii* strains, *Applied and Environmental Microbiology*, **56**, 2658–2666.

SOERGEL, D.G., YASUMOTO, T., DALY, J.W. and GISOVSKY, F., 1992, Maitotoxin effects blocked by SK&F 96365, an inhibitor of receptor mediated calcium entry, *Molecular Pharmacology*, **41**, 487–493.

STRATMANN, K., BELLI, J., JENSEN, C.M., MOORE, R.E. and PATTERSON, G.M.L., 1994a, Aulosirazole, a novel solid tumor selective cytotoxin from the blue-green alga *Aulosira fertilissima*, *Journal of Organic Chemistry*, **59**, 6279–6281.

STRATMANN, K., BURGOYNE, D.L., MOORE, R.E., PATTERSON, G.M.L. and SMITH, C.D., 1994b, Hapalosin, a cyanobacterial cyclic depsipeptide with multidrug-resistance reversing activity, *Journal of Organic Chemistry*, **59**, 7219–7226.

STRATMANN, K., MOORE, R.E., BONJOUKLIAN, R., DEETER, J.B., PATTERSON, G.M.L., SHAFFER, S., *et al.*, 1994c, Welwitindolinones, unusual alkaloids from the blue-green algae *Hapalosiphon welwitschii* and *Westiella intricata*. Relationship to fischerindoles and hapalindoles, *Journal of the American Chemical Society*, **116**, 9935–9942.

SUFFNESS, M. and PEZZUTO, J.M., 1991, Assays related to cancer drug discovery, in Hostettmann, K. (Ed.), *Methods in Plant Biochemistry*, Vol. 6, pp. 71–133, London: Academic Press.

SUFFNESS, M., NEWMAN, D.J. and SNADER, K., 1989, Discovery and development of antineoplastic agents from natural sources, in Scheuer, P.J. (Ed.), *Bioorganic Marine Chemistry*, Vol. 3, pp. 131–168, Berlin: Springer Verlag.

SUKENIK, A., 1991, Ecophysiological considerations in the optimization of eicosapentaenoic acid production by *Nannochloropsis* sp. (Eustigmatophyceae), *Bioresource Technology*, **35**, 263–269.

TAYLOR, R.F., IKAWA, M., SASNER, J.J., THURBERG, F.P. and ANDERSEN, K.K., 1974, Occurrence of choline esters in the marine dinoflagellate *Amphidinium carteri*, *Journal of Phycology*, **10**, 279–283.

TODOROVA, A. and JUTTNER, F., 1996, Ecotoxicological analysis of nostocyclamide, a modified cyclic hexapeptide from *Nostoc*, *Journal of Phycology*, **35**, 183–188.

TORIGOE, K., MURATA, M., YASUMOTO, T. and IWASHITA, T., 1988, Prorocentrolide, a toxic nitrogenous macrocycle from the marine dinoflagellate, *Prorocentrum lima*, *Journal of the American Chemical Society*, **110**, 7876–7877.

TRICK, C.G., ANDERSEN, R.J. and HARRISON, P.J., 1984, Environmental factors influencing the production of an antibacterial metabolite from the marine dinoflagellate, *Prorocentrum minimum*, *Canadian Journal of Fisheries and Aquatic Sciences*, **41**, 423–432.

TRIMURTULU, G., OHTANI, I., PATTERSON, G.M.L., MOORE, R.E., CORBETT, T.H., VALERIOTE, F.A. and DEMCHIK, L., 1994, Total structure of cryptophycins, potent antitumor depsipeptides from the blue-green alga *Nostoc* sp. strain GSV 224, *Journal of the American Chemical Society*, **116**, 4729–4737.

UEDA, J., MIYAMOTO, K., AOKI, M., HIRATA, T., SATO, T. and MOMOTANI, Y., 1991a, Identification of jasmonic acid in *Chlorella* and *Spirulina*, *Bulletin University of Osaka*, *Ser. B.*, **43**, 103–108.

UEDA, J., MIYAMOTO, K., SATO, T. and MOMOTANI, Y., 1991b, Identification of jasmonic acid from *Euglena gracilis* Z as a plant growth regulator, *Agricultural and Biological Chemistry*, **55**, 275–276.

UNSON, M.D. and FAULKNER, D.J., 1993, Cyanobacterial symbiont biosynthesis of chlorinated metabolites from *Dysidea herbacea* (Porifera), *Experientia*, **49**, 349–353.

UTKILEN, H. and GJØLME, N., 1992, Toxin production by *Microcystis aeruginosa* as a function of light in continuous cultures and its ecological significance, *Applied and Environmental Microbiology*, **58**, 1321–1325.

VAN DER WESTHUIZEN, A.J. and ELOFF, J.N., 1983, Effect of culture age and pH of culture medium on the growth and toxicity of the blue-green algae *Microcystis aeruginosa*, *Zeitschrift für Pflanzenphysiologie*, **110**, 157–163.

VILLAR, R., LAGUNA, M.R., CALLEJA, J.M. and CADAVID, I., 1992, Effects of *Phaeodactylum tricornutum* and *Dunaliella tertiolecta* extracts on the central nervous system, *Planta Medica*, **58**, 405–409.

VISO, A.C. and MARTY, J.C., 1993, Fatty acids from 28 marine microalgae, *Phytochemistry*, **34**, 1521–1533.

VON BERLEPSCH, K., 1980, Drugs from marine organisms. The target of the Roche Research Institute for Marine Pharmacology in Australia, *Naturwissenschaften*, **67**, 338–342.

WITVROUW, M., ESTE, J.A., MATEU, M.Q., REYMEN, D., ANDREI, G., SNOECK, R., et al., 1994, Activity of a sulfated polysaccharide extracted from the red seaweed *Aghardhiella tenera* against human immunodeficiency virus and other enveloped viruses, *Antiviral Chemistry and Chemotherapy*, **5**, 297–303.

YAMAGUCHI, K., MURAKAMI, M. and OKINO, T., 1989, Screening of angiotensin-converting enzyme inhibitory activities in microalgae, *Journal of Applied Phycology*, **1**, 271–275.

YASUNAMI, J., KOMORI, A., OHTA, T., SUGANUMA, M. and FUJIKI, H., 1993, Hyperphosphorylation of cytokeratins by okadaic acid class tumor promoters in primary human keratinocytes, *Cancer Research*, **53**, 992–996.

ZEVENBERGEN, J.L. and RUDRUM, M., 1993, The role of polyunsaturated fatty acids in the prevention of chronic diseases, *Fett Wissenschaft Technologie – Fat Science Technology*, **95**, 456–460.

ZHANG, X.Q. and SMITH, C.D., 1996, Microtubule effects of welwistatin, a cyanobacterial indolinone that circumvents multiple drug resistance, *Molecular Pharmacology*, **49**, 288–294.

15

Physiological principles and modes of cultivation in mass production of photoautotrophic microalgae

AMOS RICHMOND

Introduction

The idea to grow microalgae as a source of biomass for various economic purposes originated some 70 years ago, with the realization that certain algal species such as *Chlorella* proliferate vary rapidly (Warburg, 1919). Development of the biotechnology involved in mass production of microalgae gathered impetus in Germany, the USA and Japan after the Second World War. Support to this research was provided (in the early 1970s) by reports of the United Nations Advisory Committee on International Action to Avert an Impending Protein Crisis (1967) which foresaw a sharp shortage in food commodities, particularly protein, towards the end of the century and recommended that non-conventional crops such as photoautotrophic microorganisms should be considered as a source of feed and food for a rapidly expanding humanity.

There is indeed a very substantial biodiversity of microalgal species from which many organic products for human needs may be produced. Most staple food and animal feed may be thus provided by different species of microalgae in which it is possible to enhance production or storage capacity of many useful cell constituents (e.g. proteins, carbohydrates, fatty acids, pigments, vitamins, alkaloids, antibiotics and other drugs), by modifying the cells' environment and genetics.

Many species of microalgae (especially for aquaculture) can be cultivated in sea water. This is of particular importance in many parts of the world, rich in sunshine, where a growing shortage of sweet water for agriculture is becoming acute. As humanity in the arid and semi-arid regions will greatly expand in coming generations, it will become of utmost importance to utilize saline water, presently useless for conventional agriculture. Also, water use efficiency, i.e. the amount of water required to produce a given quantity of cell product or cell mass, is much higher in microalgae grown in closed systems as compared with field crops. The concept of producing microalgae was originally based on their high protein content, particularly impressive in *Spirulina* (up to 70 per cent of the dry weight), but perhaps the most important consideration in developing microalgae for agricultural purposes concerns the very high areal yield of desirable products which may be theoretically expected from algae, significantly higher than that obtainable from conventional crops.

What the limits are of productivity of photoautotrophic microorganisms is controversial. Raven (1988) calculated the maximal output rate of photosynthetic activity in a hypothetical culture exposed to 2000 μmol photons $m^{-2} s^{-1}$ as being ca. $11 g l^{-1} h^{-1}$, assuming the quantum demand was 16 (mol photons per mol CO_2 fixed in cell mass), cell content was 50 per cent carbon and all incident photons are absorbed by the algal suspension. This would correspond to 110 g dry cell mass per m^2 per 10 hours of strong daylight or ca. 400 tons $ha^{-1} yr^{-1}$. Pirt et al. (1983) however assumed that the quantum demand may be much lower than 16, suggesting a yield potential which exceeds even Raven's purely theoretical projection.

Actual production rates achieved to date with strictly photoautotrophic microorganisms such as *Spirulina* sp., fall far short of the theoretical limits, yielding at most 45 tons $ha^{-1} yr^{-1}$ (for Spirulina, by the Dainippon Inc. Corp. in Bangkok, Thailand). The theoretical yield potential serves nevertheless as an essential beacon, because economic considerations indicate that a substantial increase in production rates alone would significantly curtail production costs. A major issue in commercial cultivation of photoautotrophic algae therefore concerns sustained trapping of high irradiance at as high an efficiency as possible throughout the year; the more effectively this task is accomplished, the sounder would become the economic basis for algal biotechnology.

Some 50 years have elapsed since the inception of commercial microalgaculture in Japan. Two major outlets for microalgae have since become well established commercially. The first concerns supplementation of human food, particularly 'health food' products, suggesting a rich array of proteins, vitamins, as well as essential fatty acids. The second concerns products in demand for aquaculture, a rapidly expanding venture. Future trends point to a surge in both demand for fish and aquatic animals as well as in the portion of aquatic products originating in aquaculture. Suitable species of microalgae, therefore, should enjoy a substantial increase in commercial demand. In what follows, the principles for mass production of microalgae will be elucidated.

Microalgal nutrition

Microalgae may exhibit several nutritional modes (Kaplan *et al.*, 1986), generally grouped into two major classes, i.e. autotrophy and heterotrophy.

Autotrophs (or lithotrophs) obtain all their nutritional needs from inorganic compounds, the energy for their metabolism being derived from light or oxidation of inorganic compounds or ions. Autotrophs therefore are subdivided according to their energy source; those using light energy are phototrophs.

Heterotrophs obtain their material and energy needs from organic compounds synthesized by other organisms. A common form of heterotrophy is auxotrophy, in which organisms must complete their phototrophic nutrition with small quantities of essential organic compounds, e.g. vitamins.

This chapter relates mainly to production of phototrophic microalgae which depends, at optimal temperature, on two basic inputs, i.e. nutrition and light.

Identifying nutritional requirements

The optimal mineral composition required by several algal species has been researched extensively (Richmond, 1986). Culture media differ greatly from what is usually available for algae in their natural habitats, commercial production of algae mandating careful optimization of mineral nutrients concentration and balance to ascertain that nutritional factors do not limit growth and productivity.

A growth medium used for culturing a specific microorganism should rest on tests aimed to identify the most effective combination of minerals that make it up. This may not be a simple task, since nutritional requirements are affected by the specific growth conditions to which the culture is exposed, such as population density, light intensity, temperature, and pH. When conditions permit fast growth rates, the optimal mineral concentrations (particularly nitrogen) often shifts towards a substantially higher value. Also, once uptake of nutrients takes place with the advent of cell growth and proliferation, specific deficiencies may develop. As long as the deficiency (or diversion from the optimal nutrient balance) is not of major proportions, decline in growth may not be significant, remaining therefore unnoticed. The nutritional balance however is modified continuously along the production course and is thus most probably diverted from optimal, causing in extreme cases culture breakdown. Comparison of growth media reported in the literature for a given microalga, e.g. *Nannochloropsis* sp., is illuminating, revealing a significant disagreement between the culture media used by different workers, both in respect to mineral concentration as well as the ratio between major macronutrients. According to Boussiba *et al.* (1987) the optimal growth medium for *Nannochloropsis salina* contained ca. ten times more nitrate and ca. five times more phosphate than that used by Renaud *et al.* (1991), claimed to be optimal for outdoor cultures under tropical conditions. Also, whereas Boussiba *et al.* (1987) applied phosphorous and nitrate in a ratio of 1:20, Renaud *et al.* (1991) used an approximate 1:10 ratio (and applied one-third as much $FeCl_3$). Finally, whereas Yu *et al.* (1987) add biotin, vitamin B_{12}, thiamin and soil extract to the growth medium, Boussiba *et al.* (1987) did not find any organic substrate mandatory. Such differences in growth media could be attributed to several factors, e.g. diversity in subspecies and cultivars.

The complexity involved in optimizing the growth medium is well depicted in the work of Soeder *et al.* (1967), demonstrating the influence of growth conditions on optimum nutrient levels: *Chlorella fusca* was first grown under suboptimal conditions, varying the macroelement concentrations over a wide range but keeping their proportions and the total microelement concentration constant. Repeating the experiments under improved conditions for the growth of *Chlorella* yielded a significantly different relationship between growth versus mineral concentration. Also, the combined improvements in temperature, CO_2 supply, and light substantially reduced salt tolerance in fresh water *Chlorella*, thereby illuminating a basic physiological principle: any extreme limitation to growth and productivity, e.g. low temperature, deficient nitrogen, low light, etc., masks the effect of potential secondary limitations (e.g. salinity) which unfold as soon as the extreme limitation is removed or lessened.

In conclusion, mass cultivation of microalgae, particularly when grown commercially under the changing outdoor environment, requires nutrient optimization tests throughout the year to ascertain that minerals, carbon and growth factors are never growth-limiting.

Inorganic carbon

The kinetics of inorganic carbon transformations as related to algal growth have been elucidated by Goldman *et al.* (1974). All algae growing chemo- or photoauto-trophically use dissolved CO_2 or one of its hydrated forms to synthesize organic compounds. In water, CO_2 may appear as H_2CO_3, HCO_3^- or CO_3^{2-}, depending on the pH. In most natural freshwater the major pH buffer is the CO_2-H_2CO_3-HCO_3^--CO_3^{2-} system, which is also a very useful buffer system for mass algal cultures maintained at an alkaline pH. The total available molar concentration of dissolved inorganic carbon (C_r) is as follows:

$$C_r = CO_2(aq) + H_2CO_3 + HCO_3^- + CO_3^{2-} \tag{15.1}$$

$CO_2(aq)$ represents the aqueous carbon dioxide concentration.

The relative concentrations of the inorganic carbon species determine the pH and in turn are determined by the pH. At equilibrium the $CO_2(aq)$ concentration is much higher than that of H_2CO_3; thus $CO_2(aq)$ concentration may be considered to present the sum of the dehydrated as well as hydrated forms of CO_2.

$$HCO_3^- \rightarrow CO_2 + OH^- \tag{15.2}$$

$$CO_3^{2-} + H_2O \rightarrow CO_2 + 2OH^- \tag{15.3}$$

The pH of natural waters typically ranges between 8.0 and 8.5 at or near equilibrium with atmospheric CO_2. At this pH, HCO_3^- is the major carbon species, CO_2 being the form taken up by growing algae (Goldman *et al.*, 1974).

As CO_2 is taken up by photosynthesis, the pH rises, CO_3^{2-} becoming the major inorganic carbon species. It can be converted to CO_2 by hydration which in turn increases the pH (reaction 15.3). Indeed, because of OH^- production by these reactions, it is not uncommon to observe, in late afternoon, pH values as high as 10 or 11 in active algal systems such as sewage treatment ponds, fish ponds, or *Spirulina* production ponds. Since CO_2 is not fixed at night (at which period only respiration which releases CO_2 into the medium takes place), the pH of such systems may descend to ca. 5.0 by dawn, demonstrating the advantage in buffering the pH of the growth medium.

pH must be maintained at all times at the optimum range for the cultivated species to prevent depletion of available carbon. This is accomplished by introducing into the medium either CO_2 or $NaHCO_3$. If carbon is not added to an actively growing culture, severe carbon limitation will rapidly develop, resulting in reduced growth and productivity; extreme deficiency leading to photooxidation and death. Carbon supplementation represents the most costly nutritional input in commercial algaculture.

Organic carbon

The modes of organic carbon nutrition vary greatly in algae, even within a single species. A good example is *Euglena gracilis* grown under different environmental conditions. Strain L incorporates ^{14}C acetate efficiently in light but not in dark. Other *E. gracilis* strains use a variety of organic compounds for growth in the dark, o.g. var. *bacilaris* which may use several sugars (sucrose, glucose, etc.) or acids (pyruvate, succinate, aspartate, glutamate, etc.). Another strain, vischer, uses only acetate and butyrate (Cook, 1968).

Organic carbon nutrition in a number of species of the genus *Chlorella* has been extensively studied with respect to production of cell-mass. These species lie near the midpoint between obligate autotrophy and obligate heterotrophy, shifting rapidly and reversibly between growth on organic substrates in the dark and with carbon dioxide in light. In *Chlorella pyrenoidosa*, heterotrophic cellular synthesis proceeds with high efficiency albeit with a very limited number of substrates.

Mixotrophic cultures carry an obvious advantage. If inorganic carbon can be taken up efficiently and metabolized during daylight and organic carbon taken up during the night, a very significant surge in productivity along the diurnal cycle may be expected. Indeed, Lee *et al.* (1996) claimed extraordinarily high yields of *Chlorella sorokiniana* cell mass cultured outdoors in a tubular photobioreactor using mixotrophic as well as heterotrophic modes of nutrition. 0.1 M glucose at night was completely consumed, yielding 0.35 g of cell mass per 1.0 g glucose. During daylight, the growth yield of glucose was as high as 0.49. Such experiments provide hope that productivity of microalgae per reactor volume may be greatly increased, matching the output rates of commercial fermentors based on bacteria, fungi or yeast.

Temperature and light: the major growth limitations outdoors

Response to temperature

Algal cells continuously respond to the temperature which affects the nutritional requirements, the rates and nature of cellular metabolism and finally, cellular composition.

Temperature requirements of algae vary over a wide range, from cryophilic (cold loving), to mesophilic (which includes all the species presently used in commercial algaculture), to thermophilic (heat loving), i.e., species for which the optimal temperature for growth is well over 40°C. In general, temperature dependence of algal growth displays an exponential increase of yields until the optimum temperature is reached, illuminating the importance of maintaining optimal temperature along daylight. Important for large-scale production of microalgae outdoors is the fact that growth essentially ceases when temperature declines by 15 to 20°C below the optimal for the species.

The majority of algae exhibit a temperature growth response curve with a rather wide plateau, culminating in a sharp decline once the optimum is passed. Indeed, exceeding the optimum by only 2 to 4°C may result in a pronounced decline in growth and culture death. This response to above optimal temperature may be accentuated when light intensity is high, although light intensity in itself modifies the optimal temperature for growth.

A most prominent effect of temperature on cell metabolism is reflected in its influence on dark respiration. The respiration rate, as a rule (particularly in warm temperature algae) increases exponentially with the rise in temperature. This carries far-reaching ramifications on the production of algal cell mass, for when night temperature is high, the loss in biomass due to intensive respiration may be very damaging to the overall output rate.

McCombie (1960) studied the effect of nutrients on cell response to temperature. Doubling the nutrient salts concentration in the growth medium of *Chlamydomonas reinhardi* caused appreciable changes in the growth–temperature relationship, e.g. a higher growth maximum, compared with that attained in standard medium. A similar effect was revealed by Spoehr and Milner (1949): cultures of *Chlorella* grown on a given nutrient medium produced the best yield at 20°C, whereas in an improved nutrient medium, the highest yield was obtained at 25°C.

In open systems, the relationship between pond and ambient temperature is of practical significance. Culture temperature in an open pond is determined by the ambient temperature, the extent and duration of solar irradiance (most of which is converted to heat while being absorbed in the algal mass), and the relative humidity which governs the extent of evaporative cooling. In addition, the depth and the overall surface to volume ratio, as well as the materials from which the pond was constructed, represent important factors in stabilizing pond temperature. Relative humidity has a most profound effect on the difference between ambient and pond temperature: In tropical Bangkok, culture temperatures may exceed air temperatures by as much as 13°C on sunny days, ambient and culture temperatures being similar on cloudy days. In arid lands in contrast, pond daytime temperature may be lower by over 10 degrees than ambient temperature. Even in ponds covered with 0.2-mm thick transparent polyethylene sheets to elevate temperature in winter, culture temperature lags behind air temperature by 6 to 8°C throughout the diurnal cycle (Richmond, 1986).

It is well to note that temperature may also have a significant influence on the chemical composition of algae. Sato and Murata (1980) concluded that temperature was one of the most important environmental factors influencing fatty acid composition in *Anabaena variabilis*. Lipid synthesis was considerably stimulated in the first five hours following a growth–temperature shift from 22 to 38°C. During this period, the relative content of palmitic acid increased and that of palmitoleic acid decreased in the diacylmonogalactosylglycerol, becoming restored to the original values after 20 hours.

Temperature was also reported to exert a strong influence on the chemical composition of several species of marine phytoplankton. A basic condition for successful commercial production is that cultures are maintained monoalgal, and the decisive effect that temperature exerts on species competition must therefore be always considered. *Chlorella* sp. as an example, quickly proliferates in *Spirulina* culture, to become the dominant species within two or three weeks as soon as the summer heat recedes and the average day temperature reaches approximately 20°C, i.e. approximately 15°C below the optimum for *Spirulina* (Richmond, 1986).

Clearly, daylight temperature not far from optimal represents a requirement without which sound microalgal industry cannot develop outdoors. Most suitable are unique environmental niches around the world, in which fluctuations in ambient temperature during the year are relatively small.

Response to light

When the nutritional requirements are satisfied and the temperature is not far from optimal, maximal culture productivity may be obtained when light represents the sole limitation to productivity. Optimally, phototrophic cultures should therefore be limited by light only. Therein rests the major challenge, both scientific and economic, involved in microalgal biotechnology, i.e. to devise the methodology by which to utilize best high level irradiance such as solar energy for phototrophic production.

Efficient utilization of light by the individual cells in a culture is associated with many constrains. One relates to the fact that light energy attenuates exponentially in passing through the culture column. The higher the cell density, the lesser light penetration into the culture, thereby affecting two light zones in the culture: the illuminated volume, in which enough light is present to support photosynthesis (i.e., the photic zone); and the dark volume, in which net photosynthetic productivity cannot take place, since light intensity is below the compensation point. The higher the population density (and the longer the light path), the more complex becomes the basic requirement for efficient utilization of light, i.e., an even distribution of light to all cells.

The other difficulty in harnessing solar, or any other strong light source, for photosynthesis rests with the fact that the higher the light intensity, the lower, as a rule, the efficiency by which the light impinging on a culture surface may be utilized to produce chemical energy. The obscurity concerning the effect of light is particularly evident outdoors where the photon flux density (PFD) changes within a few hours from zero at dawn to over $2000 \, \mu E \, m^{-2} \, s^{-1}$ at noon, i.e. a photon flux that may be much above saturation of the single cell. Within a few additional hours thereafter PFD declines to zero as night sets in. This very large and fast change in irradiance along the day may shift the culture from being light-limited in the morning to being light-inhibited at summer noon and thereafter light-limited once again as light intensity diminishes in the afternoon and early evening.

The complexity of the effect of light has been augmented over the years by erroneous application of the so called 'light curve' (Goldman, 1980) to interpret the growth response of mass cultures to light. This model provides a generalized shape of the light response curve, evidenced by the rates of growth or photosynthesis. As elucidated by Goldman (1980), the main features of this curve are as follows. At some very low light intensity, the resulting low growth-rate is balanced by decay (compensation point), net growth being zero. As light becomes more intense growth is accelerated, the initial slope of the curve representing maximal efficiency of growth in response to light. As light intensity is further increased, the light saturation function is reached, at which point the growth rate is the maximal attainable. A further increase in light intensity above this point would not result in further increase in growth rate. Rather, light intensity becomes injurious as manifested initially by decreased growth rate and culminating, upon further increase of light, in photodamage and culture death.

In optically thin cultures and when all growth conditions are optimal, the intensity of the light source is indeed the sole determinant of light availability to the cells. Under these conditions, the effect of the intensity of the light source on algal photosynthesis and growth is faithfully portrayed by the 'light curve'. The very limited use of this model, particularly for mass cultures, rests with the fact that it is correct only for very thin cultures, i.e. of such low population density that mutual shading by the cells is essentially absent. In reality, however, mass cultures should not be maintained in low cell concentrations with minimal mutual shading, for two basic reasons. First, the photon flux density (PFD) outdoors is much greater, up to an order of magnitude greater than the photosynthetic saturating light intensity, and the most practical approach to cope with this is to increase cell density to the point in which mutual shading causes cells to receive the strong solar light intermittently. The high PFD prevailing outdoors is thereby diminished for the individual cells, being distributed to an increased number of cells. The light each cell receives at any one time is thus not only a function of the intensity of the light source but is also, and often more so, dependent on cell density. Unlike the prediction imbued in the 'light curve', any growth rate may indeed be manifested under a given light source, depending on the population density (Figure 15.1). Second, the major parameter in mass culture is the yield of cell mass or some specific product, per unit reactor-volume or reactor-area. Yield is not solely a function of the growth rate but also of cell concentration. Optically very thin or extremely dense cultures yield below maximal output rates because, as in any biological phenomenon related to optimal exploitation of resources per unit area, there is a certain optimal stand which results in the highest areal harvest. This is the 'optimal cell density' (OPD) which permits the most efficient exploitation of irradiance reaching the culture. It is worth noting that at optimal cell density, light penetrates to only a small fraction of the water column or the light path. At any given instant therefore, most of the cells at OPD are in darkness. Growth rate is highest when mutual shading and thus light-limitation are at the permissible minimum, i.e. that cell mass which in effect protects from photoinhibition that may take place in mid day. Conversely, in cultures of extremely high cell densities, e.g. three or four times the OPD, the growth rate – even in a highly illuminated culture outdoors – would be much reduced, due to severe light limitation to the individual cell. The OPD falls somewhere in the middle of these extremes, the growth rate at this point being about one half the maximal attainable under the given conditions.

The importance of maintaining cell density in relation to the intensity of the light source is seen in Figure 15.2D; when cell density is too low in relation to the light source, photodamage may cause culture collapse. The extent of damage by excess light depends also on the state of photoacclimatization of the cells. Cells acclimatized to relatively low light (shade adapted) prior to exposure to high intensity solar radiation (as in Figure 15.2) will become photodamaged at a lower irradiation dose than cells which are high light-acclimatized. No matter what the physiological state of the cells however, maintaining the optimal cell density in continuous cultures represents a practical approach by which to eliminate or control photodamage, as indicated by the Fv/Fm ratio (a parameter delineating the photochemical quantum yield of photosystem II which is lower in cells exposed to excess radiation) (Figure

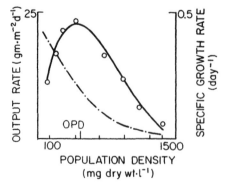

Figure 15.1   The effect of population density on the specific growth rate (dashed line) and the output rate (continuous line) in outdoor *Spirulina* cultures. Nutrients and temperature are not growth-limiting. Peak solar irradiance – above 2000 μE m^{-2} sec^{-1} (after Richmond, 1988).

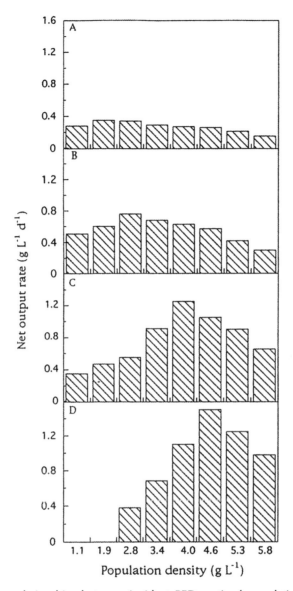

Figure 15.2 Interrelationships between incident PFD, optimal population density and net output rate. **A** = 90% shade; **B** = 60% shade; **C** = 30% shade; **D** = no shade, full sunlight (from Hu and Richmond, 1994).

15.3). In batch cultures grown outdoors, it is advisable to have the inoculum acclimatized to high light (Falkowski *et al.*, 1985).

Since the individual cells in mass cultures are illuminated intermittently, light availability is best described in terms of the light–dark (L–D) cycle, a function of the duration of cell-exposure to the photic and dark zones in the reactor, as well as the relative length of such exposure in each zone. Indeed, the parameter delineating the light available per cell is only meaningful in terms of the entire 'light regime' to which the average cell in the culture is exposed. In species or strains sensitive to

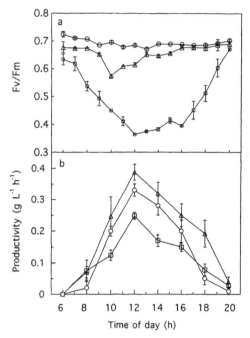

Figure 15.3 Fv/Fm (a) and hourly output rate (b) throughout the day as affected by the population density in outdoor cultures of *S. platensis* during June and July. Population density (g l^{-1}) = 1.8; $\triangle$ = 8.5; O = 15.2 (from Hu *et al.*, 1996b).

photoinhibition, a pertinent subparameter in the light regime would be the frequency of exposure to the highest photon flux densities present at the very outer layer of the photic zone. The average duration as well as the overall PFD of each such damaging exposure to high light would affect the overall extent of photodamage.

Thus, the light regime to which the average cell in the culture is exposed when all other conditions for growth are optimal, is governed by four major factors. One is the intensity of the light source (erroneously interpreted to be the sole factor affecting rates of photosynthesis and growth). Another important factor concerns the population density. At optimal density, cells are exposed to relatively short flashes of light followed by relatively long periods of darkness; the higher the frequency, the more efficiently may light (particularly high PFD) be utilized for photosynthesis (Grobbelaar *et al.*, 1996; Hu *et al.*, 1998). Another factor governing the light regime therefore is the frequency of the L–D cycle which is strongly dependent on the length of light-path across which axis the algal cells are moved back and forth from the illuminated to the dark volume in the reactor (Hu *et al.*, 1998). Finally, the extent of turbulent streaming, which affects the rate of cell movement and thus the frequency of the L–D cycle, exerts a strong influence on the light regime, elucidated below.

The duration of the light and the dark phase in the L–D cycle is related to the relative width of the photic in relation to the dark zone of the reactor. Gitelson *et al.* (1996) have measured the penetration depth of photosynthetic active radiation (PAR) in relation to algal density. Light penetration depth (defined as the depth at which down-welling irradiance is decreased to 10 per cent of incident irradiance), differed for blue and red lights, compared with green light, highlighting the complexity of what may be regarded as the 'photic volume', particularly when cell density is ultra high. The general relationship between light penetration depth and population density is nevertheless such that the former (and hence the photic volume) is strictly dependent on the population density. In ultra-high population densities (i.e. over 10 g dry wt of cell mass per litre culture; Hu *et al.*, 1996a), the photic zone may be only ca. 1 to 4 mm deep in flat plate reactors, placing obvious advantage on high frequency L–D cycles.

The interrelationships between the light path and optimal culture density in affecting the output rate of cell mass in *Spirulina platensis* is shown in Table 15.1, adapted from Hu *et al.* (1996b). Both volume and areal yield increased significantly when the light path of a *Spirulina platensis* culture was reduced ca. eightfold, from 10.4 to 1.3 cm. This increase in output reflects improved utilization of irradiance due to the increase in OPD as well as the frequency of the L–D cycles stemming from the reduction in light path.

It must be borne in mind however, that an as yet unidentified growth-inhibitory activity was found to exert severe limitation to growth in cultures of ultra-high cell densities, mandating continuous removal from the culture medium. Such inhibitory activity carries a far reaching impact on the practical use of ultra-high cell densities for mass cultivation of microalgae. Clearly, to effect an increased output rate, the reduction in areal volume associated with the reduction in the light-path described so far must be compensated by a surge in the optimal population density. This, however, cannot take place unless the inhibitory activity is eliminated (Javanmardian and Palsson, 1991). If the removal rate of the inhibitory activity associated with high cell density is insufficient, cell density could not be maintained at the high level facilitated by the narrowed light-path, eliminating thereby the advantages associated with reduction in light-path. In conclusion, the length of the light-path of a reactor should be such as to yield the highest areal output rate in cell mass or cell products, mandating a careful tailor made optimization for each species.

Effect of mixing

Mixing represents a major input in phototrophic cultures. It is meant to induce turbulent flow of the algal suspension, for two different, yet correlated functions.

Table 15.1 The effect of the light path on basic parameters related to growth and productivity of cell mass of *Spirulina platensis*.

Parameter	Light path (cm)		Effect
	10.4	1.3	
Culture volume (l m^{-2})	104	13.0	8-fold decrease
Optimal culture density (g l^{-1})	1.7	15.8	9-fold increase
Specific growth rate (μ h^{-1})	0.014	0.021	50% increase
Estimate of frequency of L–D cycle (msec)	<700	<100	ca. 7-fold decrease
Volume yield (g l^{-1} day^{-1})	0.3	4.3	14-fold increase
Areal yield (g m^{-2} day^{-1})	33	51	50% increase

One concerns mass transfer of materials into the cell and out, regarding which mixing affects reduction of the cells' boundary layers. The other important function of mixing relates to the light regime.

When light is the sole limiting factor for growth and the culture maintained at OPD, mixing represents the most practical means by which to attempt to distribute radiation evenly to all cells in the culture, improving thereby the light regime, a prerequisite for efficient utilization of light. In open raceways, when stirring is insufficient, the efficiency of solar energy utilization declines as the pattern of flow becomes laminar. In addition, dissolved oxygen (DO) builds up as a result of slow mixing, causing inhibition of photosynthesis.

The positive effect that stirring exerts on the output of biomass is accentuated as the population density and light limitation increase. Richmond and Grobbelaar (1986) showed that relatively slow stirring in an open raceway, e.g., considerably less than ca. $30\,\text{cm sec}^{-1}$, plays havoc on the output rate of cell mass at a population density far above optimal – a situation which may readily occur in large-scale *Spirulina* reactors due to practical pressures to improve harvesting efficiency. The finding that an increase in the rate of stirring shifted the OPD to a higher value illuminates the mode of action involved (Richmond and Vonshak, 1978). Since the output rate per given volume at steady state is a function of both the specific growth rate and cell density (Pirt, 1975), an increase in OPD should result in an increase in the output rate.

A basic interrelationship exists between the intensity of the light source, the OPD, the mixing rate and the output rate (Table 15.2). Varying the rate of mixing in cultures exposed to 'low' and 'high' incident light showed that at 'low' light, a low mixing rate (e.g. $0.6\,\text{l air min}^{-1}\,\text{l}^{-1}$ algal suspension) resulted in an OPD of ca. $2\,\text{g l}^{-1}$. As the rate of mixing was increased (all other conditions kept constant), the OPD shifted up (to ca. $5\,\text{g l}^{-1}$), and as appropriate, the output rate increased from 70 (at the minimal mixing rate) to $100\,\text{mg dry wt l}^{-1}\,\text{h}^{-1}$. Two aspects concerning the relationship between the output rate and cell density at this low light intensity deserve attention: when cell density was below $2\,\text{g l}^{-1}$, there was no response to

Table 15.2 Interrelationships between the light source, optimal cell density, rate of mixing and output rate of cell mass in *Spirulina platensis* (from Hu and Richmond, 1996).

Light μmol photon m^{-2} s^{-1}	Mixing rate l(air) l^{-1} min^{-1}	Optimal cell density g dry wt l^{-1}	Maximal output rate mg dry wt l^{-1} hr^{-1}
500	0.6	2.0	70
	2.0	5.0	100
	4.0	5.0	100
900	0.6	4.0	110
	2.0	8.0	160
	4.0	10.0	220
	6.0	7.0	150
1800	0.6	6.0	200
	2.0	9.0	300
	4.0	17.0	400
	6.0	14.0	300

increased mixing rates (i.e. 0.6–4.2 l air min^{-1} to 1 culture), indicating that the rate of mixing was not limiting productivity under these conditions. Indeed, the magnitude of the mixing effect was strictly dependent on the strength of the light source which in turn affected culture density. Thus, at 'high' PFD, 1800 µmol m^{-2} s^{-1} (the order of magnitude of solar energy outdoors at noon), the output rates of cell mass reflected strong dependence on the rate of mixing at this energy flux, an increase in mixing (from 0.6 to 4.2 l min^{-1} l^{-1}), resulted in doubling the output rate. A further increase in mixing rate was harmful. Clearly, the higher the intensity of the light source, the higher the optimal population density would become and the more significant would be the degree to which the rate of mixing affects the output rate (Hu and Richmond, 1996). It should be stressed however, that such a dominant effect of the mixing rate was affected in a culture of *Spirulina platensis*, a very large filamentous cell-aggregate requiring a high amount of stirring energy. Small, single-cell species, e.g. *Nannochloropsis salina*, require a much smaller energy for optimal stirring. In such species, the effect of the mixing rate on the output rate would be considerably smaller.

Reactors for mass production of microalgae

Reactors may be roughly classified in two groups – open and enclosed. The open systems are by far the most prevalent in commercial production at present, but extensive research is going into developing a cost-effective enclosed reactor in which growth conditions are better controlled. A general description of the various reactor types presently used or experimented with, follows.

Open reactors

Most industrial reactors (some exceptions being the *Dunaliella* producing lagoons in Australia and mixotrophic cultures of *Chlorella*) are open raceways, i.e., shallow ponds (water level ca. 15–20 cm high), each covering an area of 1000 to 5000 m^2, constructed as a loop in which the culture is circulated by a paddle wheel. The pond may be lined in various fashions and would have some device by which to disperse carbon dioxide into the culture. Harvesting, i.e. removing cell mass from the culture, is done mostly by screening, centrifugation and air-flotation. Drying is accomplished using either drum- or spray-drying or by extrusion of the algal paste derived by filtration. This production mode has the great advantage of being relatively simple, but it has many disadvantages which relate to the factors governing outdoor production of photoautotrophic microorganisms.

The inherent photosynthetic potential cannot be expressed when day temperature is below some 15 to 20°C of the optimal for growth of a microorganism such as *Spirulina platensis*. The lack of temperature control, therefore, represents a major drawback of the open raceway, particularly in winter. Another disadvantage concerns the long light-path, stemming from practical requirements which mandate that the water level in large basins would not be much lower than 15 cm. The long light-path which thus ensues facilitates maintenance of very dilute (e.g., a few hundred mg of dry wt per litre) algal suspensions, resulting in a large culture volume containing little biomass (ca. 0.1 per cent w/v). The temperature of this large body of water has a long response lag to the ambient temperature, often reaching optimal temperature only at midday in winter. Of even greater significance is the fact that dilute cultures become readily contaminated by other microorganisms and are costly in harvesting.

Open reactors may be divided into two basic classes: one, in which the reactor is perfectly levelled; and another, in which the reactor rests on an inclined plane (Richmond and Becker, 1986). In the former, the culture is usually circulated by a paddle wheel or by a rotating arm. In the latter, the culture is circulated as it flows downstream, and is returned thereafter to the head by an appropriate pump (Richmond, 1986). Inclining systems, particularly the inclined surface developed in Trebon (Setlik *et al.*, 1970) should have received more attention because they permit, in principle, good turbulent-streaming and a relatively short light path, conducive to high OPD and output rates.

Levelled open systems require a rather simple technology to construct and to maintain, but do not meet the major requirements for efficient utilization of sunlight, a strong light source. Also, the paddle wheel cannot practically affect in large areas the extent of turbulent streaming required to create a high frequency D–L cycles. As a result of all its drawbacks, the open raceway yields low outputs of algal mass, i.e., ca. 10 to 15 g dry weight of cell mass $m^{-2} d^{-1}$, or ca. $100 mg l^{-1} d^{-1}$.

There is little experience with inclined systems, theoretically superior to raceways. Nevertheless, these systems also lack temperature control, and would undergo excessive evaporative loss, particularly in hot dry climates. Finally, the unstirred lagoons which seem to be well used in Australia for the production of *Dunaliella*, do not seem suitable for industrial production of any other microalgal species.

The inadequacy of the open systems has been prompting development of better controlled, enclosed reactors aimed to yield much greater productivity by being more effective in use of the high photon flux density (PFD) existing outdoors. In continuous cultures, productivity (P) at steady state is the product of $\mu x v$, μ representing the specific growth rate (h^{-1}), x the concentration of algal mass ($g l^{-1}$), and v, culture volume (litres). The most appropriate practical method therefore by which to increase P per unit area significantly is to greatly increase cell concentration (and thus mutual shading) yet maintain, in a light-limited system, a high growth rate. This goal represents the major challenge in photobioreactor research. An example for these relationships may be seen in Table 15.3.

Enclosed reactors

Various types of enclosed reactors designed to provide an improved alternative to the open raceway have been researched. In the UK, Pirt *et al.* (1983), in France,

Table 15.3 Steady state productivity as a function of the specific growth rate (μ), cell concentration (x) and reactor volume (v) in *Spirulina platensis* (data from Hu et al., 1996b).

Light-path	(v) l		(x) $g l^{-1}$		(μ) h^{-1}		Productivity $m^{-2} h^{-1}$
10.4 cm	104	×	1.7	×	0.014	=	2.5 g
1.3 cm	12	×	15.8	×	0.021	=	4.0 g

Gudin and Chaumont (1978) and in Italy, Torzillo *et al.* (1986) and Tredici *et al.* (1991) pioneered the concept of growing photoautotrophic microorganisms in tubular or flat plate reactors, and many variations on the original themes have since been suggested as shall be elucidated.

The materials for construction of photobioreactors represent a significant practical issue both from standpoint of biological performance and investment cost. Reactors for photoautotrophs must be fully translucent and should not lose this property overtime. None of the materials commonly available to date, except glass, seems to be fully translucent without any loss in transparency, particularly outdoors, over time. Materials used for the reactor may be rigid, e.g. glass, Plexiglas or polycarbonate, or collapsible, e.g. Teflon, PVC and the most prevalent, polyethylene. There is obviously an interplay between cost of materials and their rate of amortization and there is at present no clear choice for any material, an impediment in development of this field.

Of all suggested designs for photobioreactors, two major types may be discerned, i.e. tubular and flat plates. Tubular reactors are made of three components: the tube arrangement which makes up the main reactor volume in which the culture circulates, a degasser to scrub excess supersaturating oxygen, and a pump to affect circulation. The latter represents an important component of photobioreactors: Pirt *et al.* (1983) grew *Chlorella* cultures in a tubular loop reactor recycled either by centrifugal pump, a rotary positive displacement pump, a peristaltic pump or an air-lift. The maximum specific growth rate of cultures recycled either by centrifugal or rotary positive displacement pump was less than that of cultures recycled by a peristaltic pump. The adverse effects on the cells caused by centrifugal and rotary positive displacement pump were proportional to the rotation speed of the blades and could be avoided by using air-lift for circulation. Likewise, Gudin and Chaumont (1991) worked for several years with a tubular system in which circulation was affected by a screw-pump which was finally found to cause cell damage in *Porphyridium*.

Pirt *et al.* (1983) have provided a thorough analysis of the tubular reactor they devised, portraying the design elements and characteristics of this reactor type. The reactor was made of 52 m, 1 cm bore glass tubing, enclosing 4.6 l and covering an area of ca. 0.5 m^2, the total surface area (at the inner radius) being about three times higher. The tubes were interconnected with U bends and were positioned horizontally, one above the other. The loop outlet was connected via a 5 cm bore tube to a 30 cm, vertical gas riser, to which the inlet tube was also connected. The ratio of liquid flow rate to gas flow rate was about 0.75. As typical for a narrow light-path reactor, virtually all the light was absorbed in a thin layer next to the tube walls; only ca. 1 per cent of the light impinging on the reactor surface penetrated deeper than 1 mm. Turbulent streaming to ensure that all cells in the culture are equally exposed to light was shown to be of particular significance. The Reynolds number (N_r) was directly proportional to the gas flow rate in the airlift and a 52 m^2 reactor with 3 m (1 cm bore) riser required 3.0 l min^{-1} of compressed air to establish an N_r of 2000, the minimal value assumed to be required for turbulent streaming. Under conditions favourable to growth, the species used by Pirt *et al.* (1983) i.e. 'Consortium MA 003', did not adhere to the glass tubing in either light-limited or nitrogen-limited cultures. Wall growth has been usually observed as the first sign of adverse growth conditions, such as trace element deficiency, pH or temperature limitation. Effective methods for removing wall-growth *in situ* consisted of increasing incident light intensity and/or increased turbulence as well as increasing trace element concentration. The only steady state cell concentrations experimented with, i.e. 5 and $8 \, \mathrm{g} \, l^{-1}$ yielded 38.6 per cent and 41.6 per cent photosynthetic efficiency, respectively. These figures were obtained by relating the entire area of the tube (considering the inner radius) to the mean total light energy obtained by combining the inputs of direct beam irradiance as well as diffuse and scattered light.

Biomass yield in response to light increased with increasing Reynolds numbers. When cell concentration was doubled (from 5 to 10 g dry weight l^{-1}, a similar increase in N_r (from 2000 to 3500) was required to achieve the maximum growth yield. Nevertheless, total light energy uptake was limited to ca. $38 \, \mathrm{W} \, \mathrm{m}^{-2}$ i.e., some 20 per cent of direct beam solar energy at noon. Indeed, Pirt *et al.* (1983) reached the surprising conclusion that their tubular photobioreactor yielded approximately the same areal output of biomass as that reported for an open raceway, in which culture temperature is not controlled, the flow rate less than optimal and the overall light yield small.

Other pioneers in developing tubular reactors were the researchers at the CNR and the University of Firenze (Torzillo *et al.*, 1986). They initially used tubes of a large bore (14 cm) made from flexible, transparent, 0.3 mm thick polyethylene, to simulate a volume-ground area ratio comparable to that of the open raceway, i.e. ca. 100 l of culture suspension per m^2 of illuminated surface. The tubes were deemed inadequate in mechanical strength and were thus replaced by tubes made of 4 mm thick polyethyl methacrylate (Plexiglas), a much more costly material. Each reactor was made up of several tubes placed side by side on the ground, lined with white polyethylene sheets and joined together to form a loop by PVC connections, each incorporating a narrow port for scrubbing oxygen. At the exit of the tubular circuit, culture suspension fell into a receiving tank. A diaphragm pump raised the culture into a feeding tank containing a siphon which facilitated intermittent discharge into the reactor. At intervals of 4 min, portions of culture suspension were discharged into the reactor, moving the culture in the tubes at $26 \, \mathrm{cm} \, \mathrm{sec}^{-1}$ obtained by adjusting the flow rate of the pump to $4000 \, l \, h^{-1}$. Production of biomass in this manner was higher than that obtained by continuous circulation at the same pump rate. The maximal length of the circuit was 500 m, containing a volume of 8000 l and a surface of $80 \, m^2$ which was correctly calculated on the basis of the total ground area effectively occupied by the tubes, including the 3 cm interspace between them. Seasonal productivity of *Spirulina* was higher than in open raceways, mainly due to increasing the temperature closer to optimal, which also expanded the growing season of *Spirulina* well into the autumn.

Dissolved oxygen build-up represents an acute problem associated with tubular reactors designed in the serpentine-like formation described so far. Another difficulty rests with the fact that the longer the serpentine and the smaller the tube bore, the higher the pressure necessary to maintain turbulent streaming in the reactor. The duration of the run between aeration (degassing) stages is dependent on the distance between stages and the linear flow velocity. 'Run duration' rather than 'run distance' or cycle should be considered because CO_2 depletion and O_2 formation are time-dependent under given environmental conditions and turbulence. The linear velocity of culture suspension determines the maximum permissible length of tubing between aeration or deoxygenation points. Manifolds connecting several loops in parallel have been introduced by Richmond *et al.* (1993) in order to reduce run duration and O_2 build-up as well as decrease the overall pressure required to recycle the culture in a long serpentine tubing. Their reactor consists of 0.2 mm thick transparent polycarbonate tubing, 3 cm internal diameter, co-extruded with UV resistant outer coating, using an air-lift to circulate culture suspension. The suspension flows down from a gas separator into a manifold which distributes the culture among four pipes spaced 3 cm apart. At the far end, the pipes connect to a similar manifold with eight openings: four for incoming pipes and four for the outgoing pipes which run for another 20 m (parallel to the outgoing pipes) back to the airlift riser and gas-separator. The long meandering serpentine formation, in which the required turbulence is difficult to maintain was thereby cut down to one-quarter, as was photosynthetic oxygen pressure which may accumulate to harmful levels in long run distances.

Control of temperature represents an acute problem in hot, sunny climates, for enclosed reactors represent, in effect, solar collectors which readily become overheated. An effective method for cooling is required to prevent temperature from rising above optimal. Evaporative cooling, i.e. spraying water on the reactor tube surface represents a satisfactory solution in dry climates. Gudin and his colleagues (Chaumont *et al.*, 1988) devised a 100 m^2 reactor (700 l) consisting of polyethylene tubes arranged in five identical units, in which temperature control was achieved by manipulating the degree to which the entire reactor was submerged in a pool of water. In general, thermotolerant strains have a clear advantage in closed systems, reducing the cooling requirements. Torzillo (1997) used *Spirulina platensis* M2 strain which grew at 42°C and could tolerate 46°C for a few hours, i.e., some 10°C above the common optimal for the species.

In an attempt to maximize capture of radiation incident on the plane of the photobioreactor and achieve full utilization of the land area it occupied, Torzillo *et al.* (1993) came up with a unique solution: a two-plane photobioreactor consisting of a 245 cm long loop made of 2.6 cm bore Plexiglas tubes arranged in two planes. Each plane was made of 4.6 m long tubes connected by glass U bends. The tubes in the lower plane were arranged in the vacant space between two tubes in the upper plane. In this fashion the two planes received all the incident radiation falling on the land area occupied by the reactor. The culture flowed from one plane to the other, the two planes forming a single loop. The tubes were placed in north–south orientation on a horizontal table covered with a white polyethylene sheet which reflected radiation escaping the two tubular planes. The photobioreactor contained 145 l of culture and covered an overall area of 7.8 m^2 corresponding to the sum of all the tubes' external diameters, including the gas risers and return tubes, i.e. ca. 18 l m^{-2}.

Circulation was affected by either two air-lifts (one for each plane of tubes) or by two peristaltic pumps. The height of the gas riser was correctly adapted to the biomass concentration tested, since the higher the cell concentration, the higher becomes the optimal flow rate. Thus the gas risers were 2.65 m and 3.65 m high with biomass concentration of 3.5 g l^{-1} and 6.3 g l^{-1}, respectively. The degassers consisted of a 40 cm long piece of Plexiglas tubing, 14 cm bore, the air flow rate for each gas riser being ca. 18 l min^{-1}, i.e. ca. 1 l air for every 4 l culture.

A disadvantage of tubular reactors arranged horizontally on the ground surface is evidenced in that total energy received by such tubes may be significantly lower, year round, than the energy received by the same tubes arranged vertically at an angle with the ground, ensuring thereby full exposure of the entire reactor surface to irradiance throughout the day. Lee *et al.* (1995) arranged a tubular reactor into an α-shape configuration, in which the culture was lifted up by an air-lift to a receiver tank 5 m above ground. It then flowed down the tubular photobioreactor to reach the opposite set of air-lift risers in which the culture was lifted 5 m to another receiving tank, flowing down to parallel tubes connected to the base of the first set of air-lift tubes. As the flow of culture in the α-photobioreactor does not change in direction except when it is lifted up in the air-lift system, high Reynolds numbers were achievable at relatively low air supply rates. A short cycle time of culture in the bioreactor did not allow dissolved oxygen to accumulate above 200 per cent air saturation due to the proximity of the opposite air-lift riser in which the O_2 is scrubbed. As the α-photobioreactor was placed at an angle with the horizontal, a fairly constant rate of solar irradiance was received throughout the day, allowing high biomass productivity (e.g. 72 g d wt m^{-2} ground area d^{-1}), obtained by maintaining a high optimal culture density of 10 g l^{-1} (Lee *et al.* 1995).

A common type of vertical reactor made of polyethylene sleeves is extensively used for production of food-chain components for aquaculture. Volume of each such bag ranges from a few litres to as high as hundreds of litres. Baynes *et al.* (1979) described a system in which each bag, containing hundreds of litres, is supported by a galvanized weldmesh frame placed on a concrete base. Cultures are illuminated externally by banks of fluorescent lamps. Such installations may be housed either in a special structure in which temperature is controlled and light is available for 24 hours or in a greenhouse, where temperature control would be somewhat limited but in which solar light would be available throughout the day. As the available light intensity for these large tubes is increased, the optimal population density and output rate of cell mass should rise.

Cohen and Arad (1989) looped polyethylene sleeves around iron rods connecting to a suitable frame so as to form a row of vertical column pairs, each 2 m high, all 20 such columns (totaling 500 l) being vertically suspended from an iron frame. Algal culture (25 l) was pumped into each column and mixed by means of compressed air stream containing 3–4 per cent CO_2. A later development of this idea saw a system of interconnected 10 cm diameter sleeves, covering a total area of 80 m^2 containing 8 m^3 of culture suspension or 100 l m^{-2}. When sleeve diameter was decreased by half (i.e. one quarter of the volume and only one half of the irradiated surface area) volume productivity was reported to increase eightfold.

A vertical glass column reactor for outdoor production with a narrow 2.6 cm bore was devised by Hu and Richmond (1994). The tubes were 150 cm long fixed on solid plastic-sheet panels facing the sun, at a 60° tilt. T-glass joints passing through the bottom stoppers served as air inlets as well as sampling ports. When exposed to full sunlight at the proper, optimal cell density (4.6 g cell mass l^{-1}), this reactor yielded record output rates for *Isochrysis galbana* outdoors i.e. 1.6 g dry wt l^{-1} d^{-1}.

The 'biocoil', developed and patented by Robinson *et al.* (1993) represents a 'cross' between the vertical and horizontal designs for tubular photobioreactors, in that the horizontally lying tubes are stacked to form an uprising coil. It consists of an upstanding helically wound tube (2.5 cm bore), in which either compressed air, peristaltic or centrifugal pumps may all be used to induce circulation. The tubes are made from food grade PVC, or other transparent, non toxic material, internal diameter ranging from 16 to 60 mm, wound onto a central core structure. Defined fluid flow patterns may be achieved in the biocoil reactor, and the hydrodynamic stress on cells may be varied by altering the flow rate. Optimal oxygen transfer rates may be established for all types of biocoil reactors operated at 10 and up to 14 000-l capacities. Good turbulence (Reynolds numbers 5000–8000) and oxygen transfer as well as low shear have been claimed to be effectively provided by the biocoil, in which dead zones were not formed even at very high culture densities. The versatility of the biocoil, and the absence of restrictions on scale-up, make it a seemingly promising reactor type for cultivation of photoautotrophic cells.

Chrismadha and Borowitzka (1994) described a helical photobioreactor, designed according to the Robinson patent: 2.5 cm bore polyethylene tubes, coiled around a sheet frame, illumination provided by fluorescent light placed in the centre of the core. The diatom *Phaeodactylum tricornutum* strain MUR 136 reached a maximum cell density of 10.24×10^6 cells ml^{-1} and a maximum specific growth rate of 0.62 d^{-1} when using a centrifugal pump for circulation. Significantly lower cell density and specific growth rates were obtained when a peristaltic pump was used, probably because of the lower flow rate induced by this pump.

Flat plate panels

Ramos de Ortega and Roux (1986) were first to test flat rigid panels as photobio-reactors, made of either PVC or polycarbonate alveoli. Tredici *et al.* (1991) and Tredici and Materassi (1992) developed a rigid panel with alveoli, as a promising culturing device. The alveolar panel was set vertically at an angle to the ground, and referred to as a 'vertical alveolar panel' (VAP). Several sizes of VAP reactors of 1.6 cm alveolar diameter were constructed, surface areas of the panels ranging from 0.5 to 2.2 m^2. Sheets made of polycarbonate are initially about 95 per cent transparent to PAR. Commercially available sheets may be cut to desired sizes and closed at one end, glued with a rectangular Plexiglas bar. The top opening may be capped with a proper material or simply left open to air. In the top and bottom section of the panel the inner walls (forming the alveoli) were removed for a length of ca. 15 cm to ensure free communication of the culture suspension from the entire panel. The panels were placed facing south with the proper tilt angle, the degree of inclination being changed monthly in order to maximize the amount of radiant energy received by direct beam radiation. Mixing and deoxygenation of the culture suspension were affected by bubbling air at the bottom of the reactor, the flow rate varying from 0.5 to 1.0 l per litre cell suspension per minute. When photon fluency was doubled (ca. 2600 µm photon m^{-2} s^{-1}) air supply more than doubled. In this fashion, a sort of a 'bubble column reactor' was realized which, unlike mechanically agitated reactors, is simple to construct and to operate.

A clear advantage of rigid flat panels is reflected in the option to set it at an optimal tilt angle, securing maximal exposure to direct beam. The relationships between orientation and tilt angle and biomass productivity was investigated by Lee and Low (1991) who devised a 10 l tubular loop photobioreactor similar to that of Pirt *et al.* (1983). It consisted of two panels each with 34 Pyrex glass tubes (1 m long, 1.2 cm bore), connected at one side by hinges and placed in an east–west direction. The bioreactor could be folded up and erected at any angle to the ground but since the experiments were carried out in Singapore i.e., practically on the equator, the overall biomass output rate of a fed-batch culture over an 8 h^{-1} d^{-1} period (26–30 g biomass m^{-2} bioreactor surface area d^{-1}) was comparable, regardless of the tilt angle. Recently, this relationship was theoretically elucidated and experimentally tested for *Spirulina platensis* cultures grown in flat plate bioreactors under the climatic conditions of south Israel (latitude approx. 31°; Hu *et al.*, 1998).

As already demonstrated by Tredici *et al.* (1997), the reactor tilt angle exerted a significant effect on the optimal population density and hence productivity of biomass, owing to its control over the amount of solar radiation entering the reactor. A direct relationship between solar energy and productivity was observed: the higher the solar energy flux admitted by varying the reactor tilt angle according to season (i.e. according to the solar angle on the horizon), the higher the sustainable productivity. Small tilt angles of 10° to 30° in summer and larger angles, ca. 60° in the winter resulted in maximal productivities for these seasons. Photosynthetic efficiency was calculated for the different tilt angles in all seasons. Efficiency was low in winter due to temperature limitations on growth of the test organism *Spirulina platensis*. In summer, efficiency was highest in the 90° reactors whereas productivity was highest at 10°, indicating that for optimal tilt angles in the season (as regards productivity) a significant amount of radiation could not be effectively used by the culture. Up to 35 per cent enhancement in annual output rate was estimated to be achievable by adjusting the tilt angle seasonally (Hu *et al.*, 1998).

The reactor used for these experiments was a modification of the flat panel developed by Tredici *et al.* (1991), made of fully translucent 6 mm glass plates (glued together with silicon rubber), eliminating the alveoli which interfere with turbulent streaming (Figure 15.4). Mixing was effectively established using two (0.5 cm diameter) compressed air tubes running horizontally along the entire length of the reactor, one on the bottom and the other at half its height. Temperature control was based on evaporative cooling, provided by a series of water sprinklers placed 6 cm apart, running across the upper side of the front panel and activated by a temperature control sensor set in the culture (Hu *et al.*, 1996b).

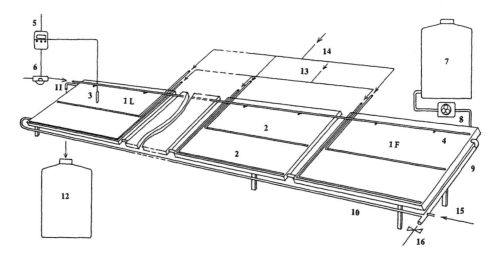

Figure 15.4 A schematic diagram of the flat inclined modular photobioreactor (FIMP) showing the first (1F) and last (1L) units in the cascade; 2, perforated tubes for bubbling compressed air; 3, thermo-sensor; 4, cooling water line and sprinklers; 5, thermostat; 6, solenoid valve; 7, culture medium reservoir; 8, pump; 9, air-lift pipe; 10, recycling tubing, connecting the first with the end reactor in the row; 11, harvesting outlet; 12, harvesting reservoir; 13, compressed air supply; 14, line supplying 2% CO₂ enriched air; 15, compresses air port for the air lift; 16, sampling port.

Special attention in designing this reactor was given to the effect of the length of the light-path, as well as that of the rate of mixing on the productivity of *Spirulina platensis*. The increase in volumetric productivity i.e., algal dry mass per given culture volume shown in Table 15.2 is self evident and is indeed well documented. Also, the growth rate has been known to increase with the decrease in tube diameter (or light path). The very significant increase in areal productivity of *Spirulina platensis* that took place as the light path decreased, however, represents a novel observation of important practical significance, relating to a unique advantage of flat plates over tubular reactors: Whereas a very narrow light path (e.g. 1.0 cm) in the flat plate does not impede the degree of turbulence, a decrease in bore in a tubular reactor extending over long distances, would mandate application of very significant light pressure, not practically feasible.

A novel configuration for a flat plate photobioreactor was designed by Pulz (1994) using plates with rectangular channels in which a meandering flow along the long axis of the channels takes place. The reactor is arranged in two parallel module groups, each containing several plate modules. Either one or both module groups may be operated together. A recirculating-piston type pump (probably not suitable for all microalgal species) was used, producing a Reynolds number of 3000. The highlight of this system is portrayed in that when, for example, the tubular modules are placed 20 cm apart, a ground area of $1\,m^2$ would have in effect a photosynthetic-active surface area of $10\,m^2$ with a 350 l volume. The short (ca. 3 cm) light path coupled with high turbulent flow permitted cell densities of $5\,g\,l^{-1}$, resulting in good areal productivity (e.g. $35\,g\,m^{-2}\,d^{-1}$, computed on the basis of the irradiated surface). If productivity is computed on the basis of the occupied ground area, however, an outstanding productivity of ca. $175\,g\,m^{-2}\,d^{-1}$ of dry cell-mass was reported, reflecting record high productivity and photosynthetic efficiency. The reason for this is that at 20 cm apart, the panels shade each other, being thus exposed mainly to diffuse and scattered light, i.e. relatively low energy radiation compared with direct beam. The great surge in photosynthetic efficiency (PE) obtained with this system is a result of very large reactor surface per ground area, holding a large thinly spread volume exposed to the photosynthetically more efficient low energy light. In essence, the record PE is achieved by a very extended and costly reactor hardware; another disadvantage being the relatively low OPD, stemming from the availability of low PFD per unit reactor surface.

Fibre optics and light emitting diodes

Mori (1985) introduced a revolutionary concept to address the issue of light attenuation and distribution in dense cultures by using fibre-optic light radiations which are inserted into the culture. Low-heat visible solar light is focused into the optical fibres with Fresnel lenses (which filter out infrared and ultraviolet wavelengths) with the aid of a device that tracks the sun and focuses its rays into quartz optical fibres, specially manufactured to be transparent to PAR. In a modified approach, the fibre-optics cables are passed into transparent acrylic rods, encased in air filled acrylic tubes which form the centres of the light radiators. The maximal light path through the 2 litre suspension, starting from the luminant surface on to the algal cell was 1.5 mm, providing a very high radiative surface. Since the light radiators are essentially heat-free, cooling is not necessary. Gas bubbles introduced in the base of the tank aerate and circulate the algal suspension.

Another interesting approach by which to achieve maximal photosynthetic rates in ultrahigh cell densities (i.e., very high extent of mutual shading) was described by Javanmardian and Palsson (1991) who designed a very efficient photobioreactor bent on achieving production rates of biomass at the theoretically achievable maximum. The reactor was made of acrylic with light radiators inside the bioreactors arranged as concentric vertical cylinders to provide 1 m^2 radiative surface per 3.3 l of algal suspension (i.e. a ratio 50 times higher than that typical to an open raceway and ca. 5–10 times greater than available in flat plate and tubular reactors). An on-line ultrafiltration unit is used to dialyze the culture medium at a relatively high flow rate, allowing selective separation of the waste and/or secreted inhibitors and exchanging them with fresh medium.

Lee and Palsson (1994) described an innovative design using high efficient light-emitting diodes (LED, gallium aluminium arsenide chips) for introduction of light into the culture suspension. This development permitted design of very small photobioreactors. The volumes of the illumination chamber were either 80 cm^3 or 52 cm^3, for a depth of 1.55 cm or 1.00 cm, respectively. At a gas flow rate of 0.1 l min^{-1} with a liquid circulation rate of 0.1 l min^{-1}, the calculated superficial velocities were between 0.42 cm sec^{-1} and 0.67 cm sec^{-1} (depending on the distance between the walls of the illumination chamber). Lee and Palsson (1994) have calculated theoretical values for gas mass transfer and light-intensity requirements to support very high-density algal cultures using LED providing 680 nm monochromatic red light. A cell concentration of more than 2 × 10^9 cells ml^{-1} (more than 6.6 v/v) with doubling times of 12 hours and an oxygen production rate as high as 10 μmol O$_2$ l^{-1} h^{-1} were achieved using on-line ultrafiltration to provide fresh medium and remove inhibitors. A further development of this system provides LED delivering as high as 500 W m^{-2} of light into the culture medium. Gas transfer by internal sparging had the capacity to transfer 250 μmol O$_2$/CO$_2$ l^{-1} h^{-1}. Continuous dilution prevented nutritional limitations, and eliminated build-up of inhibitory substances. When the photobioreactor operated at a dilution rate of 6 reactor volumes per 24 hours, it supported ultrahigh density cultures of 4 × 10^9 cells ml^{-1} and total cell volume fractions of 9.4 per cent v/v (about 25 g dry weight l^{-1}). Oxygen production rate at its peak was 13 to 15 μmol l^{-1} culture h^{-1}. Continuous perfusion allowed for long-term stable oxygen production, while oxygen production in batch mode ceased when the stationary phase was reached due apparently to inhibitory substances.

Evaluation of photobioreactor performance

Surprisingly perhaps, the strategy by which to utilize the high PFD existing outdoors at the highest efficiency is to 'dilute' it by exposing the cultures to diffuse and scattered light, rather than direct beam. As already suggested, this may be accomplished by placing flat plate panels outdoors in close proximity to one another, resulting in a large algal volume, spread, in effect, over a small ground area (Pulz, 1994). The optimal population density in such plates would necessarily be greatly reduced to about one-third or less of that maintained in flat plates exposed to direct beam radiation. Since, however, photosynthetic efficiency (PE) involved in utilizing low-energy light could be as much as five times higher than that obtained by exposure to direct beam radiation, the output rate of cell mass per ground area of such a system would be the highest attainable, yielding the highest efficiency possible (on ground basis) for strong light utilization. Record areal output rates and highest PE, however, do not impart economic significance on this system: the volumetric yield (i.e., the amount of cell mass per unit of reactor volume) is necessarily much reduced in such a reactor arrangement, and the capital investment involved in stacking expensive photobioreactor hardware to yield low volumetric yields represents a financial burden that seems to counteract the mere benefit of utilizing PAR in a most efficient manner. Indeed, economic constraints in microalgal biotechnology favour high volumetric yields, reflecting the advantage of perceiving photobioreactor efficiency in terms of output rate per unit reactor volume as well as its irradiated area.

Two major parameters should be used for evaluation and comparison of photobioreactors, providing together a reliable assessment of reactor performance. One is the volumetric yield (i.e. the output rate per unit reactor volume), the other is areal yield (i.e. the output rate per unit reactor area). The areal yield is more complex: yield may be related to the ground area the reactor occupies or to the area of the irradiated reactor surface. The difference between these modes of calculating areal yield is very significant and care should be taken to discern between these subparameters. Indeed, computation of the areal yield is fraught with pitfalls. In tubular reactors set above ground, the areal yield in relation to the irradiated surface of the tube must be very carefully measured to include all the irradiated surfaces, both the surface receiving direct beam radiation and the rest of the tube surface receiving mainly diffuse and scattered light. The same is true for flat plate reactors as well as for any reactor set in an angle above ground. To calculate the areal yield of such reactors on the basis of the occupied ground is much more complex than often realized because of peripheral effects. The correct approach is to perceive the reactor as placed in the middle of a very large reactor area. This would correctly focus the effect of shading by surrounding reactors. If such peripheral effects are not brought into account, the single reactor occupies, in effect, a much larger ground since it receives full radiation throughout the day, reflected from adjacent areas. If the reactors are set apart in such distances so as to eliminate shading, the ground area occupied would be much larger than if partial shading (e.g., early morning or the afternoon) is permitted. Reactor setting therefore is to a large extent dependent on economic considerations involved in the cost of land, infrastructure and reactor volume. In tubular reactors laid on the ground, the bare area left in between the tubes must be accounted for (since it affects the rate of irradiance received on the entire tube surface , both direct and diffuse, throughout the day). It is thus incorrect to calculate areal yield of tubular reactors set upright on the basis of the tube surface alone or to base calculations solely on the area of the tube cross section in horizontally laid tubes.

To sum up, the conditions mandatory to obtain highest photosynthetic productivity in photobioreactors have been identified as follows. First, the relationship between the intensity of the light source and the population density must be kept at its optimal. It exerts a major effect on a basic parameter in a light-limited culture i.e. the light regime for the individual cells.

Second, a relatively narrow light path which ensures light–dark cycles of highest possible frequency is mandatory for ultra-high cell densities. The high frequency has been shown to affect an exponential increase in output rate in *Spirulina platensis* (Hu *et al.*, 1998), reflecting a more efficient photosynthesis. This seems true when growth is solely light-limited. It is essential, however, to optimize the light path specifically for each species. Some species (e.g. the slow growing *Nannochlropsis salina*), require a much longer light path for maximal areal productivity compared with other species such as *Spirulina* sp.).

Indeed, an acute complication involved in optimization of the light path concerns inhibitory substances associated with high cell density, that may limit cell-growth and thereby decrease the optimal density required for maximal areal output. Such inhibitory activity (e.g. Lee and Palsson, 1994) may become growth limiting at a cell concentration in which light limitation is not the optimal to ensure highest areal output. Such inhibitor-induced reduction in optimal cell density erases the advantage of a narrow light path, imparting an advantage on a lower optimal cell density associated with a longer light path, but higher (or perhaps much higher) volume, to yield the maximal areal output in cell mass. Thus, inhibitory substances released into the medium as cell concentration rises above a certain (most probably species specific) threshold, should not be permitted to accumulate.

Third, a high rate of turbulent mixing must prevail to facilitate full expression of the photosynthetic potential in ultrahigh density cultures.

Using these guidelines should be useful for designing more efficient photobioreactors. Indeed, the microalgal industry faces at present the need to define a more effective mode than the open raceway for mass production of different algal species and products. In this vein, it could be instructive to define the general type of reactor most efficient for the production of cell mass, keeping in mind that requirements to produce cell products such as pigments or essential fatty acids most efficiently, may greatly differ.

Various parameters associated with the performance of different classes of photobioreactors are compared in Table 15.4, and a new parameter defined as 'reactor efficiency' is suggested. This quantity is obtained by measuring Vm, the volumetric output rate (which as a rule would increase with a decrease in the light-path), and modifying it by the overall irradiated photobioreactor area (A_I) required to produce a given cell mass. Although the larger the size of the irradiated area per ground area, the higher, as a rule, the photosynthetic efficiency, the large reactor hardware involved burdens investment cost. The most efficient reactor should thus be regarded as an economic entity, considering most effective a reactor which yields high volume output rates (requiring thereby relatively small culture volumes to produce a given quantity of cell mass), yet would require the smallest volume to area ratio for production of cell mass. Clearly, the higher the Vm and the smaller the A_I required for production of a given quantity of output, the more efficient the reactor.

Using this methodology, analysis of the reactors shown in Table 15.4 indicates that a fully exposed, flat plate reactor represents in principle a very efficient design. Since the actual cost of reactor volume in reference to its productivity has not been calculated, the practical conclusions drawn from Table 15.4 should be regarded as tentative, and may vary widely for different species.

Nevertheless, certain points can be treated with sufficient clarity. Whereas photosynthetic efficiency (PE) represents a valuable parameter in evaluating performance of mass cultures, it must be used with great caution. First, a high PE does not necessarily represent high productivity, simply because high PE is readily obtainable at low light intensity, which results in low productivity. This is one reason why the flat plates placed close (e.g. 20 cm apart) are so efficient in terms of photochemical

Table 15.4 Comparative analysis of different classes of photobioreactors, based on estimated reactor capacity required to produce 1 kg day^{-1} dry cell mass. *Spirulina platensis* production data serve for illustration (based on Richmond, 1996).

	Open raceway (15 cm deep)	Tubular reactor horizontal, 2.9 cm OD[c] (no distance between tubes)	Flat plate, vertical, fully exposed (2.5 cm light path)	Flat plates, vertical (20 cm apart, 2.5 cm light path)[e]
Approximate volume reactor parameters required to produce 1 kg *Spirulina* day^{-1}	5550 (l) (150 l m^{-2})	629 (l) (17 l m^{-2})	416 (l) (25 l m^{-2})	1138 (l) (125 l m^{-2})
Occupied ground area (m^2)	37.0	37.0	17.0	9.1
Total irradiated reactor area (m^2)	37.0	37.0	34.0[d]	91.0[d]
Approximate optimal culture density (g l^{-1})[a]	0.40	4.8	6.0	2.2
Assumed output rate				
Volumetric (g l^{-1} day^{-1})	0.18	1.60	2.4	0.90
Areal (ground) (g m^{-2} day^{-1})	27.0	27.0	60.0	110.0
Relative photosynthetic efficiency (open raceway =1)	1.0	1.0	2.2	4.0
Reactor efficiency[b] $\times 10^2$	0.5 (0.18 × 37^{-1})	4.3 (1.6 × 37^{-1})	7.0 (2.4 × 34^{-1})	0.9 (0.9 × 91^{-1})

[a] A rough estimation – 2.5 times the volumetric output rate.
[b] Reactor efficiency: obtained by dividing the volumetric output rate (g l^{-1} day^{-1}) by the irradiated photobioreactor area (m^2) required to produce a given weight of dry cell mass per day.
[c] Outer diameter; inside diameter (ID) is 2.5.
[d] Irradiated area includes both reactor surfaces.
[e] Based on Pulz (1994).

conversion of down welling irradiance: instead of the strong, less efficient direct beam, available irradiance is mostly scattered and diffuse, effecting a very large irradiated reactor area and volume per ground area. While this arrangement is photosynthetically most efficient, it requires a very extensive reactor size which produces a relatively low density output, thereby upscalling costs.

The open raceway represents another extreme – it produces the lowest volumetric output, requiring just as large an irradiated area to produce a given cell mass as the tubular reactor. It is thus least efficient. The fully exposed flat plate comes up as most efficient, producing the highest volumetric yield yet requiring a similar irradiated area as the tubular and the open raceway, which yield lower volumetric outputs. Also, it requires a much smaller irradiated area producing a much higher volume yield than flat plates placed close together. The properly tilted, fully exposed flat plate comes out therefore as most efficient, yielding the highest volumetric output per lowest irradiated area required to produce a given quantity of cell-mass.

Maintenance of mass cultures

Successful maintenance of microalgal cultures aimed at obtaining maximum productivity requires continuous assessment of culture performance. On-line information is required to facilitate a quick and reliable evaluation of the physiological state of the culture, detecting developments which if left unattended could culminate in significant losses. In essence, the basic requisite for correct maintenance concerns consistent evaluation of some basic parameters from which it is possible to infer the relative performance and state of the culture, in reference to the maximal performance possible under the given environmental circumstances (Richmond, 1986). In practical terms, the aim for management of mass cultures is to sustain the optimal state, i.e. in which growth and metabolic activity leading to growth and product formation are carried out at their genetically inherent maxima, limited only by irradiance.

Measurement of growth and productivity

Net growth may be estimated quickly by measuring changes in the overall turbidity of the culture. This, however, provides only a rough estimation of growth and should be followed routinely with other measurements such as dry weight or total organic carbon (TOC). The concentration of cell chlorophyll and protein are well suited to express growth in algae, but should be used with caution in outdoor cultures, being strongly affected by environmental conditions. Chromatic adaptation or even a slight nitrogen deficiency may, within days, affect considerable changes in the cells' chlorophyll and protein content, which do not necessarily correspond with changes in the concentration of cell mass. Cell protein content may also change greatly during the diurnal cycle (Torzillo *et al.*, 1984).

Measurements of growth should be accompanied by detailed microscopic evaluations, the most important reason for which is tracing possible development of foreign algal species as well as protozoa or fungi which may quickly proliferate and either take over the cultivated species or cause damage. Build-up in the number of other organisms in the culture should be regarded as a warning signal that the cultured species may have come under stress: the temperature may have deviated too much from optimal, bestowing a relative advantage on competing species, or it may signal that the concentration of a particular nutrient had declined below the minimal threshold, limiting culture growth.

Maintaining optimal population density (OPD)

The population density represents a major parameter in the production of photo-autotrophic mass, exerting far-reaching effects on the general performance and productivity of the culture. Optimal population density (OPD) is defined as that cell mass which results in the highest output rate of biomass or desired products, under given environmental conditions. Experimentation to elucidate the relationships between the specific growth- and output-rates and between the OPD indicated that the latter varied between the seasons (Hu *et al.*, 1998). Plotting the output rate of the reactor against population density in *Spirulina platensis* during the course of the year revealed that the higher the temperature (up to the optimal) and the availability of irradiance per cell, the more pronounced became the dependence of the output rate on the OPD, provided inhibitory activity is removed. The sharp decline in output obtained at very high population densities is explained by the decrease in growth rate resulting from increased light limitation and also by the relative cell expenditure on maintenance energy which also increases as irradiance per cell decreases. Likewise, reducing the population density much below the optimal range resulted in a significant decrease in the output rate, apparently due to increased photoinhibition as mutual shading decreases. Also, as the population density is reduced below optimal, the utilization of high solar irradiance, i.e. high above saturating PFD, becomes increasingly less efficient.

At the optimal population density, the specific growth rate in continuous cultures is about one-half its maximal, growth being strongly light-limited due to mutual shading. At this density, cultures outdoors are exposed to the optimal light regime comprised of dark–light cycles in which a period of intense light at the top of the photic zone is followed by an about ten times longer period of darkness or low irradiance in the dark volume of the reactor. Clearly, the performance of each individual cell in the culture is a function of both the PFD impinging on the surface, as well as culture density and the extent of turbulence which together determine the light regime for the average cell (Richmond, 1988; Hu *et al.* 1998). Stirring the culture therefore, represents one factor which elevates the optimal population density and also, as culture density increases, the extent of mixing carries ever increased importance in determining productivity. Special attention, therefore, should be taken to ensure that the culture is stirred at the optimal rate (i.e. inducing highest productivity; Hu and Richmond, 1996), which may be determined by repeated tests.

Preventing nutritional deficiencies

Correct culture maintenance requires routine tests to check any possible development of a deficiency in mineral nutrients. Nutrient elements in the growth medium are being constantly depleted in continuous cultures, having been absorbed in the biomass. In addition, the concentration of some elements, relative to others in the nutrient formula may be grossly altered. One practical method to supplement the nutritional status in the culture is to monitor the concentration of nitrogen, using it as a guideline for adding, in equivalent amounts, the entire set of elements making up the growth medium. Carbon and phosphorus however, should be monitored specifically and added to the culture separately. No doubt such a protocol is prone to error as the concentration of some elements (particularly minor), may with time either build up or become depleted, irrespective of the extent of algal growth and nitrogen utilization. In fact, a nutritional imbalance in the growth medium may, among other factors such as inhibitory activity, be at the root of the observation in one commercial outfit that *Spirulina* cultures grown continuously outdoors become 'old', requiring a complete replacement of the growth medium every few weeks.

Culture pH is rising continuously as a result of depletion of carbon through photosynthesis; the daily rise in pH could be rather small (0.1 or 0.2 units) in highly buffered (e.g., 0.2 M bicarbonate) *Spirulina* medium, or as high as a few pH units in lightly buffered growth media or natural water bodies. pH may be maintained by an inflow of CO_2 or by occasional addition of inorganic acid in a growth medium of high alkalinity. Sodium bicarbonate may be used in addition to CO_2 to adjust the pH for species such as *Spirulina*, the optimum pH of which is in the range of 8.5 to 10.0. It is worth noting that carbon nutrition represents a major component in the operating cost of commercial production of microalgae.

Assessing culture performance: on-line monitoring of photosynthetic activity

Measuring dissolved oxygen

Very high concentrations of dissolved oxygen (DO) may build up in actively growing cultures of photoautotrophs. Concentrations of 35 to 43 mg O_2 l^{-1}, representing maximal supersaturation values (with respect to air) of 400 to 500 per cent, are common in *Spirulina* cultures grown in large (e.g. 1000 m^2), open raceways in which mixing is insufficient, or in enclosed tubular systems in which the DO concentration builds up rapidly, in *Spirulina* cultures grown in summer in tubular reactors (Vonshak *et al.*, 1994). The full array of effects which very high concentrations of DO exert on commercially grown algal species has not been sufficiently studied. There is evidence that excessive DO results in decreased yields of cell mass as well as in pigment content in *Spirulina* (Tredici and Chini Zittelli, 1997). In addition, it would promote, under suitable conditions, photoinhibition and photo-oxidation followed by quick death of the culture (Abeliovich and Shilo, 1972; Richmond, 1986). For some species, DO inhibits CO_2 assimilation by interfering with the carboxylase activity involved in photosynthesis. In *Spirulina*, one effect of DO concentration relates to the protein content, 45 per cent of O_2 in the gas phase greatly reducing protein content from 48 per cent of dry weight to 22 per cent (Torzillo *et al.*, 1984). In an open raceway, vigorous stirring effects a significant decrease in O_2 concentration, bestowing an additional advantage on intense stirring.

The DO is readily measured with an oxygen electrode and is recognized as a reliable and sensitive indicator of the state of the culture, as related to growth and productivity. An early detection of an inexplicable decrease in dissolved oxygen or a decline in the normal rate of the daily increase under given environmental conditions, serves as a reliable warning signal that the culture is stressed and may quickly deteriorate if corrective measures are not taken (Richmond, 1986). Experience shows that as a quick 'first aid' remedy, a significant volume of the culture under such circumstances should be immediately replaced with fresh medium.

In situ *monitoring of chlorophyll fluorescence*

The chlorophyll fluorescence technique may be used to evaluate the photosynthetic activity of the culture and its relative performance under given environmental conditions, as manifested by the quantum yield of PS II, which relates the radiation absorbed by the photosynthetic apparatus and the CO_2 assimilated. A decline in the quantum yield may indicate stress conditions, on which basis Torzillo *et al.* (1996) suggest that measurements of chlorophyll fluorescence could serve as a useful tool to evaluate environmental effects on the physiological state of the culture. Measurements of chlorophyll fluorescence may well complement DO measurements facilitating a rapid and accurate assessment of the relative 'well being' of the culture.

Maintenance of monoalgal cultures

Contaminants (i.e. microorganisms different from the cultured species) often represent a major limitation to the overall productivity in microalgacultures, particularly in open systems outdoors. Maintaining a monoalgal culture under non-sterile conditions is obviously complex, requiring an imaginative approach based on extensive research and experience. The ecological niche that in effect is formed in non-sterile large-scale reactors is exposed to pressures of competition and succession. One good example is the work of Goldman and Ryther (1976), who found that competition among five tested species of marine phytoplanktons in outdoor cultures was highly dependent on temperature. The two most important commercially grown species, *Spirulina* and *Dunaliella*, thrive in growth media which represent extreme environments for most other species. The high bicarbonate (0.2 M) and pH (up to ca. 10.2) which are optimal to *Spirulina* and the high salt concentration (2.0 to 4.0 M) in which *Dunaliella* may be cultivated form effective barriers against most contaminants. Nevertheless, certain *Chlorella* and *Spirulina* species may play havoc in a *Spirulina platensis* culture (Richmond, 1988). Algal species that require growth media of a more general nature, such as *Haematococcus*, *Porphyridium* or *Nannochloropsis*, in which a large variety of other species could grow well, are much more susceptible to contamination.

It is important to note that in cultures growing vigorously (responding to the optimal environment for its species), contamination is not, as a rule, a great problem. It is when the culture is limited by conditions that impede cell growth and development – e.g. some nutritional limitation or temperature too far from optimal – that the culture may become quickly contaminated. A decline in temperature which coincides with an increase in the organic load in the medium imparts a clear advantage to mixotrophic *Chlorella* spp. which may rapidly take over a culture of the strictly photoautotrophic *Spirulina*. This process is accelerated if intensive harvesting by screening takes place, *Chlorella* being ca. 3–5 µm in diameter and its population, similar to that of the small *Spirulina minor*, is steadily enriched in the course of harvesting by screening (Vonshak *et al.*, 1982). In cultures maintained at low population densities (i.e. high light per cell) *Chlorella* was particularly successful in becoming rapidly dominant, whereas in relatively dense *Spirulina* cultures, *Chlorella* contamination has always been less severe (Richmond *et al.*, 1990). This is explained by the fact that cyanobacteria are as a rule sensitive to high light intensity but have extremely low maintenance energy requirements (Mur, 1983). Organisms such as *Chlorella* spp. which require relatively high maintenance energy are thus placed as a disadvantage at a low flux of light per cell.

Several strategies are available for control of contaminants; one example is the case of *Spirulina platensis* contaminated by *Chlorella*. First, high alkalinity and a high pH (10.3 and over), were shown to impede growth of *Chlorella* (Richmond *et al.*, 1982). Repeated pulses of 1–2 mM NH_3, followed by a 30 per cent dilution of the culture also comprise an effective treatment which is based on the differential sensitivity of *Spirulina* and *Chlorella* cells to NH_3. An increase in the organic load of the culture medium due to cell lysis or leakage of cell constituents (e.g. glycerol in the case of *Dunaliella*) are of particular significance in mass cultures. It is a common cause for the loss of entire cultures of strict photoautotrophs such as *Spirulina* or *Dunaliella*.

Grazers, mainly of the amoebae type, may afflict great damage in cultures of unicellular, colonial or filamentous species, and amoebae grazing on *Chlorella* and *Spirulina* were observed in some commercial ponds which were improperly maintained. Grobbelaar (1981) described a case in which a culture containing mainly *Chlorella* sp. was infected by a *Stylonichia* sp. Within five days, *Scenedesmus* sp. dominated the algal population since its colonies were too large to be taken in by the *Stylonichia*. Addition of ammonia (2 mM) arrested the development of these grazers, evidenced when ammonia was used as the main nitrogen source (Lincoln *et al.*, 1983). In one work the use of rotifers (*Brachionus plicatilis*), efficient unicellular grazers, has been suggested for keeping *Spirulina* cultures free of unicellular contaminants such as *Monoraphidium minutum* and *Chlorella vulgaris* (Mitchell and Richmond, 1987).

Similarly, certain types of zooplankton have been known to decimate, within days, cultures of *Haematococcus* and *Nannochloropsis*. In each case, treatment must be 'tailor fit' for the specific case, considering the unique characteristics of the algal species. In enclosed systems, it is of utmost importance to keep the reactor clean by preventing cells and debris from accumulating on the reactor walls ('wall growth') or in specific corners of the reactor where, due to reduced turbulence, cell debris tend to accumulate. Cleaning enclosed reactors periodically is thus mandatory, in spite of burden on the cost of production. Cultures of species such as *Nannochloropsis salina* known to withstand disinfectants (e.g. based on chlorine) may be very effectively treated with such chemicals, which are most effective against zooplankton and amoebae, provided that the cultured algae are not seriously affected.

Acknowledgements

It is a pleasure to thank Dr G. Torzillo for reading the manuscript, offering valuable criticism and suggestions.

References

ABELIOVICH, A. and SHILO, M., 1972, Photooxidative death in blue-green algae, *J. Bacteriol.*, **111**, 682.

ANONYMOUS, 1967, *International Action to Avert the Impending Protein Crisis*, UN Publications sale No. E. 68X1112.

BAYNES, S.M., EMERSON, L. and SCOTT, A.P., 1979, Production of algae for use in the rearing of larval fish, Res. Tech. Rep., MAFF Dir. Fish. *Res. Lowestoft*, **53**, 13–18.

BOUSSIBA, S. and RICHMOND, A., 1979, Isolation and characterization of phycocyanin from the blue-green alga *Spirulina platensis*, *Arch. Microbiol.*, **120**, 155–159.

BOUSSIBA, S., VONSHAK, A., COHEN, Z., AVISSAR, Y. and RICHMOND, A., 1987, Lipid and biomass production by the halotolerant microalga *Nannochloropsis salina*, *Biomass*, **12**, 37–47.

CHAUMONT, D., THEPENIER, C., GUDIN, C. and JUNJAS, C., 1988, Scaling up a tubular photobioreactor for continuous culture of *Porphyridium cruentum* from laboratory to pilot plant 1981–87, in Stadler, T., Mollion, J., Verdus, M.-C. *et al.* (Eds), *Algal Biotechnology*, pp. 199–217, London: Elsevier Applied Science.

CHRISMADA, T. and BOROWITZKA, M.A., 1994, Effect of cell density and irradiance on growth, proximate composition and eicosapentaenoic acid production of *Phaeodactylum tricornutum* grown in a tubular photobioreactor, *J. Appl. Phycol.*, **6**, 67–74.

COHEN, E. and (MALIS) ARAD, S., 1989, A closed system for outdoor cultivation of *Porphyridium*, *Biomass*, **18**, 59–67.

COOK, J.R., 1968, The cultivation and growth of *Euglena*, in Buetow, D.E. (Ed.), *The Biology of Euglena*, Vol. 1, p. 244, New York: Academic Press.

FALKOWSKI, P.G., DUBINSKY, Z. and WYMAN, K., 1985, Growth-irradiance relationships in phytoplankton, *Limnol. Oceanography*, **30**, 311–321.

GITELSON, A., HU, Q. and RICHMOND, A., 1996, The photic volume in photobioreactors supporting ultra-high population densities of the photoautotroph *Spirulina platensis*, *Applied & Environmental Microbiology*, **62**, 1570–1573.

GOLDMAN, J.C., 1980, Physiological aspects in algal mass cultures, in Shelef, G. and Soeder, C.J. (Eds), *Algal Biomass*, pp. 343–59, Amsterdam: Elsevier/North Holland Biomedical Press.

GOLDMAN, J.C. and RYTHER, J.H., 1976, Waste reclamation in an integrated food chain system, in Tourbier and Pierson (Eds), *Biological Control of Water Pollution*, pp. 197–214, Philadelphia: University of Pennsylvania Press.

GOLDMAN, J.C., OSWALD, W.J. and JENKINS, D., 1974, The kinetics of inorganic carbon limited algal growth, *J. Water Pollut. Cont. Fed.*, **46**, 553.

GROBBELAAR, J.U., 1981, Open semi-defined systems for outdoor mass culture of algae, in Grobbelaar, J.U., Soeder, C.J. and Toerien, D.F. (Eds), *Wastewater for Aquaculture*, Fr. Series C., No. 3, pp. 24, Bloemfontein, SA: UOFS Publications.

GROBBELAAR, J.U., NEDBAL, L. and TICHY, V., 1996, Influence of high frequency light/dark fluctuations on photosynthetic characteristics of microalgae photoacclimated to different light intensities and implications for mass algal cultivation, *J. Appl. Phycol.*, **8**, 335–343.

GUDIN, C. and CHAUMONT, D., 1978, Bioconversion de l'energie solaire dans un systeme double couche, in Martigues, J. (Ed.), *Heliosynthese et Aquaculture*, pp. 71–88, CNRS.

GUDIN, C. and CHAUMONT, D., 1980, A biotechnology of photosynthetic cells based on the use of solar energy, *Biochem. Soc. Trans.*, **8**, 4.

GUDIN, C. and CHAUMONT, D., 1991, Cell fragility – the key problem of microalgae mass production in closed photobioreactors, *Bioresources Technology*, **38**, 145–151.

HU, Q. and RICHMOND, A., 1994, Optimizing the population density in *Isochrysis galbana* grown outdoors in a glass column photobioreactor, *J. Appl. Phycol.*, **6**, 391–396.

HU, Q. and RICHMOND, A., 1996, Productivity and photosynthetic efficiency of *Spirulina platensis* as affected by light intensity, cell density and rate of mixing in a flat plate photobioreactor, *J. Appl. Phycol.*, **8**, 139–145.

Hu, Q., GUTTERMAN, H. and RICHMOND, A., 1996a, Physiological characteristics of *Spirulina platensis* cultured in ultra-high cell densities, *J. Phycol.*, **32**, 1066–1073.

Hu, Q., GUTTERMAN, H. and RICHMOND, A., 1996b, A flat inclined modular photobioreactor (FIMP) for outdoor mass cultivation of photoautotrophs, *Biotechnol. & Bioeng.*, **51**, 51–60.

Hu, Q., Hu, Z.Y., COHEN, Z. and RICHMOND, A., 1997, Enhancement of eicosapentaenoic acid (EPA and γ-linolenic acid (GLA) production by manipulating cell density in outdoor cultures of *Monodus subterraneus* (Eustigmatophyte) and *Spirulina platensis* (Cyanobacteria), *Eur. J. Phycol.*, **32**, 81–86.

Hu, Q., ZARMI, Y. and RICHMOND, A., 1998, Interactions between light intensity, light-path and culture density effecting the output rate of *Spirulina platensis*, *Eur. J. Phycol.* (in press).

JAVANMARDIAN, M. and PALSSON, B.O., 1991, High-density photoautotrophic algal cultures: design, construction, and open operation of a novel photobioreactor system, *Biotechnol. & Bioeng.*, **38**, 1182–1189.

KAPLAN, D., RICHMOND, A., DUBINSKY, Z. and AARONSON, S., 1986, Algal nutrition, *in* Richmond, A. (Ed.), *Handbook of Microalgal Mass Cultures*, pp. 147–198, Boca Raton, FL: CRC Press.

LEE, Y.K. and LOW, C.S., 1991, Effect of photobioreactor inclination on the biomass production of an outdoor algal culture, *Biotechnol. Bioeng.*, **38**, 995–1000.

LEE, C.-G. and PALSSON, B.O., 1994, High-density algal photobioreactors using light-emitting diodes, *Biotechnol. & Bioeng.*, **44**, 1161–1167.

LEE, Y.K., DING, S.Y., LOW, C.S., CHANG, Y.C., FORDAY, W.L. and CHEW, P.C., 1995, Design and performance of an α-type tubular photobioreactor for mass cultivation of microalgae, *J. Appl. Phycol.*, **7**, 47–51.

LEE, Y.K., DING, S.Y., HOE, C.H. and LAW, S.C., 1996, Mixotrophic growth of *Chlorella sorokiniana* in outdoor enclosed photobioreactor, *J. Appl. Phycol.*, **8**, 163–169.

LINCOLN, E.P., HALL, T.W. and KOOPMAN, B., 1983, Zooplankton control in mass algal cultures, *Aquaculture*, **32**, 331.

McCOMBIE, A., 1960, Actions and interactions of temperature, light intensity and nutrient concentration on the growth of the green algae *Chlamydomonas reinhardi* Dangeard, *J. Fish Res. Board Canada*, **17**, 871.

MITCHELL, S.A. and RICHMOND, A., 1987, The use of rotifers for the maintenance of monoalgal mass cultures of *Spirulina*, *Biotechnol. & Bioeng.*, **30**, 164–168.

MORI, K., 1985, Photoautotrophic bioreactor using visible solar rays condensed by fresnel lenses and transmitted through optical fibres, in Scott, C.D. (Ed.), *Proceedings of the 7th Symposium on Biotechnology for Fuels and Chemicals*, pp. 331–345, Gatlinburg, Tennessee: John Wiley & Sons.

MUR, L.R., 1983, Some aspects of the ecophysiology of cyanobacteria, *Ann. Microbiol.*, **134B**, 61–72.

PIRT, S.J., 1975, *Principles of Microbe and Cell Cultivation*, p. 274, Oxford: Blackwell Scientific.

PIRT, S.J., LEE, Y.K., WALACH, M.R., PIRT, M.W., BALYUZI, H.H.M. and BAZIN, M.J., 1983, A tubular bioreactor for photosynthetic production of biomass from carbon dioxide: design and performance, *J. Chem. Tech. Biotechnol.*, **33B**, 35–58.

PULZ, O., 1994, Open-air and semi-closed cultivation systems for the mass cultivation of microalgae, in Pheng *et al.* (Eds), *Algal Biotechnology in the Asia-Pacific Region*, pp. 113 117, Kuala Lumpur. University of Malaya.

RAMOS DE ORTEGA and ROUX, J.C., 1986, Production of *Chlorella* biomass in different types of flat bioreactors in temperate zones, *Biomass*, **10**, 141–156.

RAVEN, J.A., 1988, Limits to growth, in Borowitzka, M.A. and Borowitzka, L.J., (Eds), *Microalgal Biotechnology*, pp. 331–356, Cambridge: Cambridge University Press.

RENAUD, S.M., PARRY, D.L., THINH, L.V., KUO, C., PADOVAN, A. and SAMMY, N., 1991, Effect of light intensity on the proximate biochemical and fatty acid composition of *Isochrysis* sp. and *Nannochloropsis oculata* for use in tropical aquaculture, *J. Appl. Phycol.*, **3**, 43–53.

RICHMOND, A. (Ed.), 1986, *Handbook of Microalgal Mass Culture*, Boca Raton, FL: CRC Press.

RICHMOND, A., 1988, A prerequisite for industrial microalgaculture: efficient utilization of solar irradiance, in Stadler, T., Mollion, J., Verdus, M.-C. *et al.* (Eds), *Algal Biotechnology*, pp. 237–244, London: Elsevier Applied Science.

RICHMOND, A., 1992, Open systems for the mass production of photoautotrophic microalgae outdoors: physiological principles, *J. Appl. Phycol*, **4**, 281–286.

RICHMOND, A., 1996, Efficient utilization of high irradiance for production of photoautotrophic cell mass: a survey, *J. Appl. Phycol.*, **8**, 381–387.

RICHMOND, A. and BECKER, E.W., 1986, Technological aspects of mass cultivation – a general outline, in Richmond, A. (Ed.), *Handbook of Microalgal Mass Culture*, pp. 245–263, Boca Raton, FL: CRC Press.

RICHMOND, A. and GROBBELAAR, J.U., 1986, Factors affecting the output rate of *Spirulina platensis* with reference to mass cultivation, *Biomass*, **10**, 253–264.

RICHMOND, A. and VONSHAK, A., 1978, *Spirulina* culture in Israel, *Arch. Hydrobiol. (Beih. Ergeben Limnol.)* **11**, 274–280.

RICHMOND, A., KARG, S. and BOUSSIBA, S., 1982, The effect of the bicarbonate system on the competition between *Spirulina platensis* and *Chlorella* sp., *Plant and Cell Physiol.*, **23**, 1411–1417.

RICHMOND, A., LICHTENBERG, E., STAHL, B. and VONSHAK, A., 1990, Quantitative assessment of the major limitations on productivity of *Spirulina platensis* in open raceways, *J. Appl. Phycol.*, **2**, 195–206.

RICHMOND, A., BOUSSIBA, S., VONSHAK, A. and KOPEL, R., 1993, A new tubular reactor for mass production of microalgae outdoors, *J. Appl. Phycol.*, **5**, 327–332.

ROBINSON, L.F., MORRISON, A.W. and BAMFORTH, M.R., 1993, European Patent No. 261872.

SATO, N. and MURATA, N., 1980, Temperature shift-induced responses in lipids in the blue green alga *Anabaena variabilis*. The central role of diacylamonoalactosylglycerol in thermo adaptation, *Biochim. Biophys. Acta*, **619**, 353–362.

SETLIK, I., SUST, V. and MALEK, I., 1970, Dual purpose circulation units for considerations and scheme of pilot plant, *Algological Studies, Trebon*, **11**, 111–164.

SOEDER, C.J., SCHULZE, G. and THIELE, D., 1967, Einfluss verschiedener Kulturbedingungen auf das Wachstum in Synchronkulturen von *Chlorella fusca* SH et Kr., *Hydrobiol.*, Suppl. **33**, 127–138.

SPOEHR, H.A. and MILNER, H.W., 1949, The chemical composition of *Chlorella*: effect of environmental conditions, *Plant Physiol.*, **24**, 120–129.

TORZILLO, G., 1997, Tubular photobioreactors, in Vonshak, A. (Ed.), *Spirulina platensis (Arthrospira): Physiology, Cell-Biology and Biotechnology*, pp. 101–116, London: Taylor & Francis.

TORZILLO, G., ACCOLLA, P., PINZANI, E. and MASOJIDEK, J., 1966, In situ monitoring of chlorophyll fluorescence to assess the synergistic effect of low temperature and high irradiance stresses in *Spirulina* cultures grown outdoors in photobioreactors, *J. Appl. Phycol.*, **7**, 1–9

TORZILLO, G., GIOVANNETTI, L., BOCCI, F. and MATERASSI, R., 1984, Effect of oxygen concentration on the protein content of *Spirulina* biomass, *Biotechnol. Bioeng.*, **26**, 1134–1135.

TORZILLO, G., PUSHPARAJ, B., BOCCI, F., BALLONI, W., MATERASSI, R. and FLORENZANO, G., 1986, Production of *Spirulina* biomass in closed photobioreactors, *Biomass*, **11**, 61–64.

TORZILLO, G., CARLOZZI, P., PUSHPARAJ, B., MONTAINI, E. and MATERASSI, R., 1993, A two-plane photobioreactor for outdoor culture of *Spirulina*, *Biotechnol. Bioeng.*, **42**, 891–898.

TREDICI, M.R. and CHINI ZITTELLI, G., 1997, Cultivation of *Spirulina platensis* (*Arthrospira*) in flat plate reactors, in Vonshak, A. (Ed.), *Spirulina platensis* (*Arthrospira*): *Physiology, Cell-Biology and Biotechnology*, pp. 117–130, London: Taylor & Francis.

TREDICI, M.R. and MATERASSI, R., 1992, From open ponds to vertical alveolar panels: the Italian experience in the development of reactors for the mass cultivation of phototrophic microorganisms, *J. Appl. Phycol.*, **4**, 221.

TREDICI, M.R., CARLOZZI, P., CHINI ZITTELLI, G. and MATERASSI, R., 1991, A vertical alveolar panel (VAP) for outdoor mass cultivation of microalgae and cyanobacteria, *Bioresources Technology*, **38**, 153–159.

VONSHAK, A., ABELIOVICH, A., BOUSSIBA, S. and RICHMOND, A., 1982, On the production of *Spirulina* biomass: The maintenance of pure culture under outdoor conditions, *Biotech. and Bioeng.*, **25**, 341–351.

VONSHAK, A., TORZILLO, G. and TOMASELLI, L., 1994, Use of chlorophyll fluorescence to estimate the effect of photoinhibition in outdoor cultures of *Spirulina platensis*, *J. Appl. Phycol.*, **6**: 31–34.

WARBURG, O., 1919, Uber die Geschwindigkeit der photochemischen Kohlensäurezersetzung in lebenden Zellen, *Biochem. Z.*, **100**, 230–237.

YU, S., Y., HUBBARD, J.S., HOLZER, G. and TORNABENE, T.G., 1987, Total lipid production of the green alga *Nannochloropsis* sp. Q II under different nitrogen regimes, *J. Phycol.*, **23**, 289–296.

16

Economic evaluation of microalgal processes and products

MICHAEL A. BOROWITZKA

Introduction

The commercial use of algae has a long history and many species of algae are used as food, feed and as sources of valuable chemicals. For example, several species of the macroalgae *Caulerpa, Porphyra, Iridaea, Gigartina* and *Ulva* are harvested for human food or hydrocolloids, whilst species of *Laminaria, Macrocystis, Eucheuma, Gracilaria* and *Chondrus* are also farmed (Lipkin, 1985; Tseng and Fei, 1987; Avila and Seguel, 1993; Merrill, 1993; Trono, 1994). A few microalgae such as some species of *Nostoc* and *Aphanizomenon* are also harvested from the wild for human consumption (cf. Martinez, 1988; Martinez *et al.*, 1995), however most commercially used microalgae are cultured. These include *Dunaliella, Chlorella* and *Spirulina* which are cultured for food and for the extraction of high value chemicals (Borowitzka and Borowitzka, 1989b; Belay *et al.*, 1994), a wide range of species such as *Tetraselmis chuii, Isochrysis galbana, Chaetoceras muelleri, Nannochloropsis* spp., *Skeletonema costatum*, etc., which are used as feed in the aquaculture of fish, shellfish and crustaceans (Liao *et al.*, 1983; Fulks and Main, 1991; Benemann, 1992) and several cyanobacteria for use as biofertilizers (Subrahmanyan, 1972; Roger, 1989; Dubey and Rai, 1995). Many other species of microalgae are being considered for possible commercial production.

This chapter will only consider the economics of microalgae production. It will describe the processes and considerations which are essential in evaluating whether a particular alga, algal product and/or algae process may be commercially viable. The focus will be on the methods which a biologist/biotechnologist can use to determine whether the alga and the process are possible candidates for commercialization.

The algae and algal products

The greatest challenge remains the identification of suitable products and services.
There is still inadequate recognition of the fact that the science, however outstanding,

represents a relatively small element in the commercial success of biotechnology
(P. Dunnill, *Chem. and Ind.*, 2 July 1984).

This quote is a reminder that good science alone is no guarantee that a process or
product will be commercially successful. Commercial-scale success requires that science
and business interact, and success is more likely if this interaction takes place as early
as possible. Figure 16.1 is a flow chart outlining a strategy for developing an algal
product from the preliminary concept to a commercial process. A rational and
systematic approach helps to maximize the chance of success and minimizes expensive failures. This process has many checkpoints, mainly based on the economic

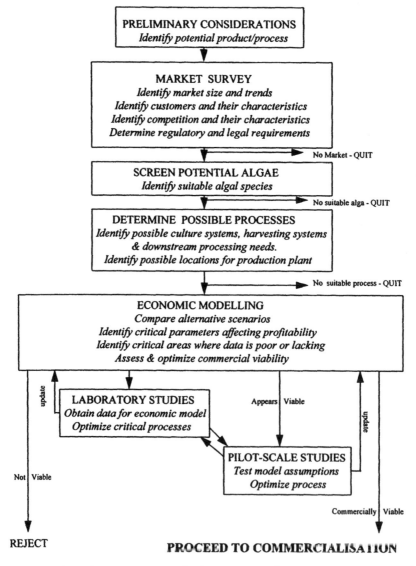

Figure 16.1 Flow chart summarizing the basic steps in the process of proceeding from the
basic concept to possible commercialization of a microalgal product.

viability of the product, where one can decide that further development is unnecessary as the likelihood of an economically successful product is very low. These checkpoints may however also provide independent confirmation that the proposed product can make money.

The first step in the process of selecting a particular microalga and/or algal product for commercial development is usually the preliminary identification of possible products. This may be done by a survey of the patent and scientific literature, by a preliminary market survey seeking opportunities for algal products, or by having an inspired idea.

The patent and literature survey establishes how novel the concept is, as well as whether the process may be patentable later on. The patent literature is also a good place to start to assess what alternative sources and processes may exist. Patent searches are quite easy and several on-line national and international patent databases exist. A survey of the scientific literature, of course, can provide much data on biology and engineering which will be required for preliminary modelling of the process.

Once a possible product or process has been identified, then an as detailed as possible market survey should be carried out. The purpose of this survey is to determine whether there will be customers for the proposed algal product. This avoids the development of a product which no one will buy.

This market survey needs to establish factors such as: who are the customers, what is the market size, the market growth potential and what limits this, and the price of the product. Possible threats from other, non-algal, sources also need consideration. An example of this type of analysis is shown in Table 16.1. It is also desirable to determine what factors affect the present price. Is it high demand and limited supply? Would a lower price increase the market size or is this largely independent of price? Is the price maintained at its present value because there are only a few suppliers who

Table 16.1 Sale price, market size, market value and alternative sources of selected algal products (based on Borowitzka, 1992a).

Product	Approx. price ($ kg^{-1})	Market size (t year^{-1})	Market value ($ ×10^6$)	Other sources	Number of producers
Protein-rich biomass	0.13–1			Grains, fishmeal, animal meal	Very many
Glycerol	1–2			Fats	Very many
Tocopherol (Vitamin E)	68	2500	170	Peanut oil, etc.	Many
Phycocyanin (food grade)	< 500	?	5	None	Many
β-Carotene	> 800	> 100	8—180	Synthetic, carrot extract, palm oil	Very few
Astaxanthin	3000	50–100	30–60	Synthetic, *Phaffia*	Very few
Canthaxanthin	2000	?	> 100	Synthetic	Very few

control the price? Answers to these questions are important in establishing whether there is a market and whether it is worthwhile to attempt to enter this market. Even at this early stage, possible regulatory and social barriers (or advantages) to market entry of a new algal product should be identified.

How does one go about a market analysis? For a product which is already on the market the first step is to determine the current wholesale price of the product, and what discounts there are available for large purchases. This can usually be found out by contacting a supplier. The wholesale price does not provide information on what the actual cost of production is, but a reasonable assumption is that production costs are at least 30–50 per cent less than the bulk wholesale price. If there are few suppliers of the product, then it is likely that the profit margin is higher. It can be more difficult to obtain reliable data on market size, but there are numerous commercial and government databases which may contain relevant information. These databases often also provide information on market trends; i.e. has the market been growing or is it declining. Discussions with potential customers, regulatory authorities, wholesalers, retailers, etc. are also very important to determine the characteristics of the market and whether an algal alternative may have some advantage over the existing products.

If the product is new, then market analysis is much more difficult. The obvious first question is: Why it is the product not being produced and/or sold at present? One needs to determine if there is a potential unmet demand or whether it will be necessary to create a demand. Developing a completely new market can be very expensive and may take some time. It is also critical do find out whether the proposed application or product has to meet any particular regulatory requirements. The costs of obtaining registration for new food products, food additives or pharmaceuticals can range from 10s to 100s of millions of dollars and the process will take several years; does the potential market justify such an investment?

In parallel with the market analysis, a detailed evaluation of the algal species which it is proposed to exploit must be undertaken. The simplest form of this type of analysis, which evaluates the theoretical maximum worth of the algal biomass, is illustrated in Table 16.2. This table clearly shows that an alga suitable for commercial development should have a high content of a very valuable product; preferably one where there are few, if any, alternative sources. Thus, even though *Dunaliella salina* can contain up to 40 per cent of its dry weight as glycerol (Borowitzka and Borowitzka, 1988), the low price of glycerol means that, on a glycerol basis, the alga is worth only about $0.80 per kg; abundant and very cheap alternative sources of

Table 16.2 Estimated value of algal biomass, based on product value and cell content.

Alga	Product	Cell content (% dry wt)	Value of product ($ kg^{-1})	Value of alga ($ kg^{-1})
Chlorella	Health food	100	25	25.00
Dunaliella salina	Glycerol	40	2	0.80
	β-Carotene	5	600	30.00
		10	600	60.00
Haematococcus pluvialis	Astaxanthin	2	6000	60.00
Spirulina platensis	Phycocyanin	2	500	10.00

glycerol thus make the algal product uncompetitive. On the other hand, *D. salina* contains more than 5 per cent dry weight of β-carotene (Borowitzka and Borowitzka, 1988), and this values the algal biomass at more than $30.00 per kg. Similarly, astaxanthin from *Haematococcus pluvialis,* at only 1 per cent of cell dry weight, still values this alga at about $30.00 per kg due to the high value of the astaxanthin. The best strains of *Haematococcus* have a significantly higher content of astaxanthin (e.g. Boussiba and Vonshak, 1991; Borowitzka, 1992b), making this an attractive potential candidate for development. Furthermore, the only natural alternative to astaxanthin is the yeast *Phaffia rhodozyma*, which has a much lower cell content of astaxanthin (Meyer *et al.*, 1993; Johnson and Schroeder, 1995).

Table 16.2 shows one of the reasons why microalgal biomass for use as health food or even as animal feed is attractive: 100 per cent of the biomass is sold. For example, *Chlorella* biomass sells for over $30 per kg wholesale, and algae used as feeds for many aquaculture species cost hatcheries from about $60 to over $600 per kg to produce (Fulks and Main, 1991; Benemann, 1992; Coutteau and Sorgeloos, 1992). Some algal pastes are available and these sell for over $500 per kg.

Alternatively there are also small and specialized, but profitable markets for certain algal products. These include the protein phosphatase inhibitors, the microcystins and nodularins (Carmichael, 1992) which sell for about $280 per g and have an annual market size of only several grams. This is a good sideline for a research laboratory, but not the basis of an industry. Similarly the production of isotopically-labelled metabolites from microalgae such as *Chlorella* and *Chlamydomonas,* is a market of about $500 000 per year. These compounds are used in research and also in medical diagnostic tests (Behrens *et al.*, 1994). The latter application is relatively new and it is likely that this will lead to an expanding demand for these compounds.

This kind of extremely simple analysis can often show whether it is even worth considering an alga or algal product for commercialization. Combined with consideration of the present market as well as future market trends, such an analysis can rapidly narrow down the possible options and saves a lot of futile research and development effort.

If these basic evaluations give encouraging results then the analysis must be refined further to include an assessment of possible large-scale culture systems, the harvesting process and any further downstream processing which may be required to produce a saleable product. Some consideration of the likely location of the production plant will also aid in selecting technically and biologically feasible systems. Climate, water sources, power sources, labour availability, air quality, environmental regulations, etc. all have some influence on the methods for algal culture, harvesting and downstream processing which may be possible.

It is essential, even at this stage, that the *whole* process is considered, and that all the costs are included in estimating the actual production cost of the algal biomass and/or product. A full engineering and economic analysis is not necessary at the beginning as sufficient information is available in the literature for some 'order of magnitude' estimations. These will allow the ranking of several algae, algal products and processes, all of which appear to have commercial potential to see which appears to have the greatest potential for commercial success.

For large-scale culture (in most cases large-scale culture means an annual production of at least several tons dry weight) it is desirable that an alga has one or more of the features listed in Table 16.3. Similarly, the algal products should meet at least some of the desirable characteristics listed in Table 16.4. In reality, no single alga or

Table 16.3 Desirable characteristics of algae for mass culture.

Desirable characteristic	Advantages	Disadvantages
Growth in extreme environment	Reduces problems with competing species and predators	Only limited numbers of species available and some extreme environments are difficult to maintain on a large scale (i.e. cold)
Rapid growth rate	Provides competitive advantage over competing species and predators; reduces pond area required	Growth rate is usually inversely related to cell size; i.e., fast growing cells are usually very small
Large cell size, colonial or filamentous morphology	Reduces harvesting costs	Large cells usually grow slower than smaller cells
Wide tolerance of environmental conditions	Less control of culture conditions required for reliable culture	–
Tolerance of shear force	Allows cheaper pumping and mixing methods to be used	–
High cell content of product	Higher value of biomass	Products are usually secondary metabolites, high concentrations mean slower growth

Table 16.4 Desirable product characteristics and their advantages and disadvantages.

Product characteristic	Advantages	Disadvantages
High value	Greater income	Attractive to competitors as well. Often high value products have only a small market
Large, increasing market	Greater potential income	Attractive to competitors
Cannot be chemically synthesized	Potentially less competition	May mean that product has no established market
Only occurs in algae	No competition from other sources	May mean that product has no established market
Stable	Less difficulty in extraction, purification and storage	
Little or no requirement for product registration	Rapid and lower cost entry into market	
Large production cost to sale price differential	More flexibility in pricing and higher profitability allow faster development of the technology and market	

algal product meets all of the above criteria. Unfortunately there are few high value products which are unique to algae and/or which are produced by algae in large amounts. Similarly, the number of algae which grow in very selective environments is limited and most algae require quite specific growth conditions.

Not all of the desirable characteristics listed in the above tables are of equal importance, and their relative weighting will change with factors such as the proposed culture system, the site of production, the type and size of market, etc. The culture system to be used has a significant effect on final costs; not just on the cost of the biomass, but also on the costs of downstream processing. Table 16.5 compares some of the features of the main large-scale algal culture systems available today. In some cases there are several alternative systems which can be considered for a particular alga, whereas in others the constraints of the alga's biology or the nature of the desired product mean that there are only one or two options which can be considered. For example, the freshwater alga *Haematococcus* cannot be grown reliably in open-air systems such as raceway ponds as it is quickly overgrown by other algal species. *Dunaliella salina*, on the other hand, grows in an extremely selective high salinity environment and can therefore be grown in any of the various open-air systems as long as the climate is suitable and an adequate cheap source of high salinity brine is available. The very extensive open pond systems used by South Australia and Western Betatene Ltd in Australia are possible because land costs are very low, site preparation costs are minimal, brine is available and only requires pumping, and the climate allows almost year-round cultivation. In Israel, on the other hand, the high land and site preparation costs, as well as the high cost of salt, means that Nature Beta Technologies Ltd at Eilat uses a two stage, highly controlled, culture system (Ben-Amotz, 1995). The location of the proposed production plant will affect the choice of preferred culture system to some extent; factors such as land costs, climate, water sources and pollution all need to be considered. Because of the complexity of this task some sort of economic modelling should be the centrepiece of the evaluation process.

Economic modelling

Computer models of the process with an emphasis on the economics of the process are a powerful tool for decision making. Even simple models allow the evaluation of the whole process and can be used to determine which aspects of the process have the greatest effect on the final outcome – the production cost of the product. The principal components of an algal production process which should be included in such models are shown diagrammatically in Figure 16.2. It is best to start with a simple model and then refine it progressively as new and better data become available. The initial model will probably require many assumptions, however as long as these are within realistic limits this does not present a problem. One very useful aspect of building a model is that this process highlights the areas of ignorance and uncertainty, and can be used to determine what data need to be collected. Judicious use of models can save a lot of time and money during the research and development process. Sensitivity analysis (an analysis of changes in assumptions and the resulting changes in profitability) can be used to determine which aspects of the process are most important to success and can be used to focus attention on those aspects which have the greatest effect on the final profitability of the process. The models also help

Table 16.5 Values used in the baseline model of the cost of large scale microalgal production (alga = *Chlorella*).

Total pond area	31.8 ha	
Number of ponds	50	
Pond depth	25 cm	
Pond width	12 m	
Mixing speed	30.0 cm s^{-1}	
Paddle wheel efficiency	50%	
Average cell density at harvest	0.8 g l^{-1}	
Harvest efficiency	80%	
Centrifuge capacity	15 m^3 h^{-1}	
Hours of centrifuge operation	18 h d^{-1}	
Proportion of pond harvested	80%	20% remains as inoculum for next cycle
Proportion of medium recycled	90%	10% of medium is lost in processing
Number of times ponds set up per year	1	
Proportion of down time	10%	
Rate of evaporation	3 cm d^{-1}	
NaNO$_3$ concentration in medium	1.0 g l^{-1}	
K$_2$HPO$_4$ concentration in medium	0.1 g l^{-1}	
Cell N	5.5% of dry weight	
Cell P	1.1% of dry weight	
Algal productivity	20 g m^{-1} d^{-1}	
Annual growth period	300 days	
Total annual production	395 t dry weight	
Land cost	$300 ha^{-1}	
Site preparation costs	$10 000 ha^{-1}	
Culture system costs	$150 000 ha^{-1}	Lined raceway ponds
Harvesting system costs	$33 000 ha^{-1}	Varies with harvesting efficiency
Engineering (15% of site, culture and harvesting system costs)	$28 950 ha^{-1}	
Contingency (5% of total construction costs)	$11 100 ha^{-1}	
Labour costs	$60 000 per labour unit	Average cost for personnel including all on-costs
Power costs	$0.15 kW h^{-1}	
Water costs	$0.05 m^{-3}	
N cost	$1.00 kg^{-1}	
P cost	$1.50 kg^{-1}	
Flocculant cost	$0.45 kg^{-1}	Used at 1 kg m^{-3} water
NaCl cost	$0.02 kg^{-1}	Not required for base model

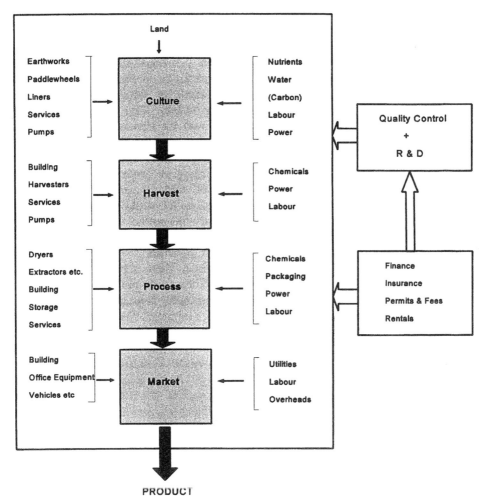

Figure 16.2 Schematic diagram showing the main components of an economic model for algal production. In the main box the four steps of the process (culturing, harvesting, processing and marketing) are shown centrally, with fixed cost components on the left and variable cost components on the right. Quality control and R&D are separate from the main process and are not considered in detail in this chapter. Insurance, interest costs etc. are also listed separately (from Borowitzka, 1992a).

in ensuring that the ultimate goal, a cost-effective and profitable process, is not forgotten.

The availability of powerful spreadsheet programs such as Excel, Lotus and Quattro Pro make it easy for almost anyone to develop meaningful models without the need of any specialized programming skills. A good example of such a model, and of what can be done with it, is the model described by Borowitzka (1992a). This model considers an algal plant with 50 raceway ponds, each of 6360 m^2 area (total area of 31.8 ha) with an operating depth of 25 cm, mixed with paddle wheels, and culturing an alga such as *Chlorella* or *Phaeodactylum*. The pond size and design was optimized for efficient mixing at a flow rate of 30 cm s^{-1} using the data of Oswald (1988a). Harvesting in this model is by flocculation and flotation as the first step, followed by centrifugation. Construction costs are based on local experience in Australia, and harvesting equipment costs are based on present prices. Energy requirements and efficiency of the harvesting equipment are based on data from the studies of Mohn (Mohn, 1988; Mohn and Cordero-Contreras, 1990), and other energy requirements are based on estimates found in the papers by Oswald (1988a, 1988b). Other cost relativities are based on data from Peters and Timmerhause (1980), Black (1983) and Reisman (1988). The capital costs of the plant were amortized over a period of 12 years and the harvesting equipment was amortized over six years. The baseline parameters and values of the model are listed in Table 16.5. The cost of the harvested biomass predicted by the model is $13.18 per kg. This cost does not yet include the cost of drying or further processing of the biomass. The cost of drying and packaging will add about 10 per cent to the product cost; i.e. the cost of the dry biomass in the above model would be about $14.50. If there is the need to extract the biomass and to purify the extract, as is the case with β-carotene from *D. salina* or phycocyanin from *Spirulina*, then the additional cost can be considerably higher. At present rates of interest, provision for interest costs would add at least another 3–5 per cent to the product cost. There is also no allowance in the model for ongoing research and development. Research and development costs of at least 5 per cent of the final market price are probably conservative and would be much greater if the company is a small start-up company. The production of the algal biomass is only the first step and, although a significant cost factor, may not be the only significant cost in many cases.

The model used here represents only one possible scenario and many of the assumptions and the final cost of the biomass will change greatly for other algae, other processes and at other sites. However, it serves to illustrate how such models can be used, especially in assessing the relative impact of particular variables on the final biomass production cost. The cost of the algal biomass is most sensitive to the productivity of the system; i.e. both the daily areal productivity and the annual productivity which is also a function of the number of days the plant is in operation per year. Thus, a 50 per cent increase in the productivity from 20 to $30 \, \mathrm{g \, m^{-2} \, d^{-1}}$ results in a 19 per cent decrease in the production cost per kg of biomass (Figure 16.3). The effects of increasing the number of days the plant can operate is shown clearly in Fig. 16.3. The high proportion of fixed costs (ponds, harvesting equipment, etc.) as well as labour costs are the reason for this. Similarly, if the harvesting efficiency is increased from 80 to 100 per cent the biomass cost declines by 20 per cent (Figure 16.5). The model also shows that the ability of recycling the medium is very important. If no recycling takes place then the cost of the biomass increases by about 42 per cent. This is not only the cost of water, but also the cost of the nutrients still contained in that water. Thus recycling is important not only to avoid problems and costs in the disposal of large amounts of nutrient-rich water, but also in maximizing the utilization of the expensive nutrients added.

Figure 16.6 shows the impact of a 10 per cent increase in the costs of consumables, labour and capital. It shows that labour costs are an important cost component. The base model assumes that 13 full-time persons are required and a 10 per cent increase in labour costs increases the cost of the biomass by about 2.5 per cent. The cost of labour also interacts strongly with the size of the plant. The increase in costs as the plant size is reduced (Figure 16.4) has a large labour component as the number of persons required to operate the plant does not increase directly as the plant size and capacity increases. After labour, power costs are most sensitive to price increases as

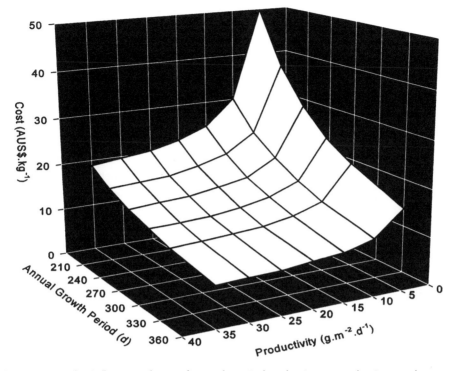

Figure 16.3 The influence of annual growth period and primary production on the cost of harvested biomass, calculated using the base model.

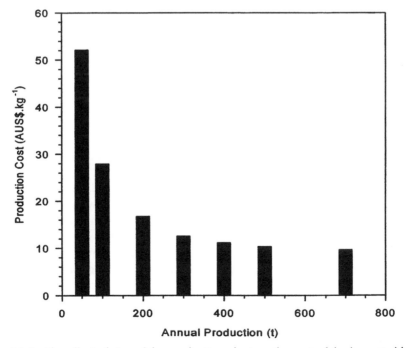

Figure 16.4 The effect of size of the production plant on the cost of the harvested biomass calculated using the base model.

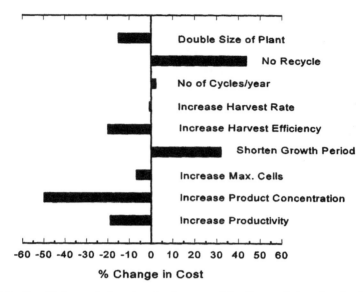

Figure 16.5 Sensitivity analysis using the base model: effects of changing basic operating conditions. Plant size – increase from 50 ponds to 100 ponds; water recycling – effect of no recycling after harvesting; setting up – set up twice a year instead of once; harvest rate – increase by 10%; harvest efficiency – increase by 10%; annual growth period – shorten from 300 days to 250 days; maximum cell concentration reached – increase by 10%; product concentration – increase by 50%; productivity – increase by 50% (the change in the cost of the harvested biomass is expressed as percentage change from the base model value).

there is a large power requirement for the paddle wheels and for harvesting. Increases in nutrient and water costs, on the other hand, have a much lower effect on the final cost of the product. The cost of flocculant is also important in this model. However, if harvesting can be by filtration as for *Spirulina*, or by autoflocculation as for some strains of *Chlorella* then significant cost savings can be achieved.

Figure 16.6 also includes the impact of the cost of NaCl. Although NaCl is not required in the base model, it can be a significant cost factor for algal systems such as the culture of *Dunaliella salina* which requires about $200–250\,\mathrm{g\,l^{-1}}$ of NaCl (Borowitzka and Borowitzka, 1988). This is the reason why most *Dunaliella* plants are located near salt lakes or salt works where brines are available as a free or low cost salt source.

Carbon dioxide can also be a large cost for an algal plant. In open systems at normal pH (pH 7 to 9) the transfer efficiency of CO_2 is low, and it is doubtful whether the increased productivity achieved by CO_2 addition in open systems offsets the high cost of the CO_2 required. CO_2 addition is used in only very few commercial-scale open-air algal plants such as in some *Spirulina* plants where a high pH and high bicarbonate concentration are important for species maintenance, and in the cascade system growing *Chlorella* in the Czech Republic where it is necessary to achieve the very high productivities and biomass obtained in this system (Livansky and Pilarski, 1993; Doucha and Livansky, 1995).

The above analysis clearly points to the major parameters requiring particular attention in the development of a microalgal process. These are detailed in the following subsections.

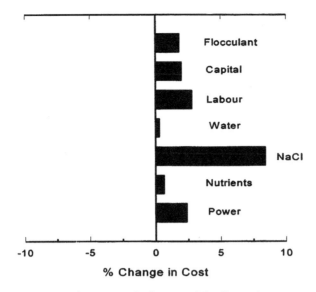

Figure 16.6 Sensitivity analysis using the base model: effects of a 10% increase in the cost of a range of 'consumables' on the cost of harvested algal biomass (expressed as percentage change from the base model value). The cost of addition of NaCl, as would be required in the culture of *Dunaliella salina*, is also shown.

Productivity

Any factor which can increase the net productivity of the plant (on an annual basis) at little incremental capital and/or operating cost, is of fundamental importance. The simplest and most important way to improve productivity and reduce costs is site selection. Most large-scale algal production systems require daylight as artificial lighting is very expensive. A microalgal plant therefore should be located at a site where climatic conditions provide optimal growth conditions for the longest possible period. If possible, the plant should be able to produce algae for the whole year. Figure 16.3 illustrates the impact of increasing the annual growth period from 210 days to 330 days. The impact of increasing productivity is also shown. Selection for faster growing, more productive strains optimized for the prevailing climatic conditions is also important. To do this it is important to have a good understanding of those factors which limit growth and productivity. An excellent example of such studies is given in the papers by Vonshak and coworkers (Richmond *et al.*, 1980; Vonshak *et al.*, 1982, 1983, 1996; Vonshak, 1987). Productivity can also be improved through better culture systems; however the incremental cost of these culture systems in relation to the improvement in productivity must be evaluated carefully.

Simulation modelling can also be used to optimize the design and operations of the plant to optimize biomass production. A good example of such a model is presented in the paper by Sukenik *et al.* (1991) for the outdoor pond production of *Isochrysis galbana*.

If the algal biomass will be extracted to produce the final product, then the cell concentration of the product is also very important. Increased product concentration not only decreases the effective unit cost of the raw biomass, but it also generally reduces the cost of extraction and purification. Research to improve product yield is therefore of paramount importance to the commercialization of algal products.

Harvesting

Harvesting represents a significant capital and operating cost component. It is therefore important to select an alga with properties which simplify harvesting (i.e. large cell size or filaments, high specific gravity in comparison to the medium, reliable autoflocculation etc; cf Table 16.3). In practice, this can be achieved only in part, and continued research and development of more efficient and cost-effective harvesting systems is required.

The cost of harvesting may be offset against the cost of achieving higher productivity. Thus, for example, the Australian *Dunaliella* producers use very extensive open pond culture systems with their associated lower productivity and offset this low productivity by having extremely efficient and cost effective harvesting and extraction methods (Curtain *et al.*, 1987; Borowitzka and Borowitzka, 1989a, 1990). The net effect of this is that their production costs for *D. salina* β-carotene are the lowest in the world. The Israeli producer is able to harvest by centrifugation as their cell densities are higher in their intensive culture system. Similarly, the thin-layer cascade system at Trebon, Czech Republic, achieves cell densities of up to $15 \, g \, l^{-1}$ without significantly reducing productivity, and this results in an up to 85 per cent drop in the costs of mixing and harvesting as compared with conventional raceway ponds (Doucha and Livansky, 1995).

The choice of harvesting system also depends on the final product desired. For example, if the algal biomass is to be used for food or feed then certain flocculants such as $AlOH_3$ and many polymers, may not be able to be used due to potential toxicity. On the other hand, if the biomass is to be extracted, it may be possible to use these flocculants as long as they do not contaminate the final product; the flocculants should however be non-toxic to the algae so that the medium can be recycled to reduce costs.

Some applications require no harvesting. For example, in aquaculture the algae are often fed directly to the rotifers, crustaceans, shellfish or fish (Fulks and Main, 1991; Benemann, 1992; Lee and Tamaru, 1993). This reduces the overall cost of algal production.

Extraction and purification

Although the cost of these is not considered in the economic model presented here because of the very great differences between various algae and products (e.g. Nonomura, 1987; Cohen *et al.*, 1993; Mishra *et al.*, 1993; Anon., 1994; Chauhan *et al.*, 1994; Mendes *et al.*, 1995), extraction and purification represent important cost components of the overall process. Algae present several problems to the chemical engineer; for example, the extraction of astaxanthin from *Haematococcus pluvialis* is complicated by the fact that the astaxanthin-containing aplanospores have a thick, resistant cell wall requiring either mechanical breakage or enzyme treatment. Alternatively, the delicate cells of *Dunaliella salina* are easily damaged, leading to rapid oxidation and loss of the β-carotene. Each alga, and each product presents its own unique problems and more emphasis on large-scale extraction and processing methods optimized for algae, is required.

Labour

Algal production is relatively labour intensive. Labour is required for pond and equipment maintenance, monitoring of the cultures, harvesting, extraction and further processing. Any improvements in the design of the process and automating the operations of the plant which decrease the labour requirement, without a disproportionate increase in capital costs, need to be considered carefully as a possible means of reducing production costs.

Reliability

For commercial production the whole process of culture, harvesting and subsequent downstream processing must be reliable. Unfortunately, many algal systems are notoriously unreliable and subject to sudden 'crashes' and similar problems. Furthermore, experience has shown that great difficulties exist in scale-up, and that extrapolations from small cultures are often very unreliable. This is particularly the case in the large open pond systems in use today. This means that extensive piloting must be undertaken before the final production plant is built (cf. Borowitzka and Borowitzka, 1989a, 1989b). Although large-scale microalgal culture has now been undertaken for over 40 years, our experience is still limited to only a few species and, even for these, our understanding of their biology and ecology is still very incomplete. Furthermore, open-air cultures, which are the preferred system for most existing commercial operations, are at the mercy of the elements and thus require good monitoring systems and effective rapid response strategies. These assume a good understanding of the algae and their interaction with the environment. Well trained personnel and clearly defined operation procedures help, but it is obvious that commercial-scale microalgal culture must be supported by good basic biology.

Closed systems, such as the tubular photobioreactors which are being developed in several parts of the world (Chaumont *et al.*, 1988; Hoshino *et al.*, 1991; Borowitzka, 1996), are another way to improve reliability since the growth environment can be much more closely monitored and controlled. At this time, however our understanding of the behaviour of algae in most of these systems is still very limited, and many algal species are not suitable for culture in the systems which are presently available. These systems are still very new, however, and it is reasonable to expect major advances in this field in the future. Several species of microalgae also can be grown heterotophically in fermenters (Gladue, 1991; Kyle *et al.*, 1991; Barclay *et al.*, 1994). These systems are very reliable and the high cell yields result in low biomass costs.

The cost of algal production – case studies

The profitability of algal processes is also very dependent on the scale of the operation. It is a truism that the larger the operation, the smaller the unit cost of algal

Table 16.6 Comparison of characteristics of different culture systems.

	Capital Cost	Running Cost	Cell Yield	Reliability
Unstirred shallow ponds	Medium[a]	Low	Low	Variable[b]
Raceways (paddle-wheel)	High	Higher	Higher	Good[b]
Cascade system	Medium–High	Medium	Higher–Very high	Good[b]
Tubular photobioreactor	High–Very high	Medium–High	Very high	Excellent
Fermenter	Very high	High–Very high[c]	Very high	Very good

[a]Depends on land and water cost as very large pond area is required; [b]the range of species which can be cultured in these 'open' systems is very limited; [c]heterotrophic culture is very much cheaper compared to cultures requiring light.

production. This is illustrated in Figure 16.4 which shows the cost per kg of harvested biomass as the size (capacity) of the plant increases from 50 to 700 t y^{-1}. It is clear that the main factor for the high cost of algal production for aquaculture (up to $600 per kg) is due to the small scale of the algal culture systems in hatcheries (Coutteau and Sorgeloos, 1992). This cost could be reduced significantly if algal production were centralized and a process were available for shipping a preserved algal biomass of high and reliable nutritional quality.

The actual cost of algal production is generally not disclosed by commercial producers. Some appreciation of the costs can however be obtained from the various models which have been published. Table 16.7 summarizes the main published data on the cost of algal production. The original costs have been adjusted to 1997 prices by assuming a 5 per cent per annum inflation factor. These costs vary from $0.30 to $600 per kg dry biomass. It should be noted that those cost estimates based on actual commercial operations are at the top of this range, with the small-scale aquaculture hatchery algal production units being the most expensive. The very low costs are from some extremely optimistic models for extremely large-scale plants (e.g. Benemann *et al.*, 1982; Regan and Gartside, 1983) and experience to date shows them to be unrealistic. The lowest actual cost is about $7.00 per kg for feed-grade *Spirulina* production in Thailand (Tanticharoen *et al.*, 1993) and the average cost for algal biomass for health food is between $15 and $30 per kg. The model described earlier is well within this range. The costs shown in Table 16.7 only consider the cost of the dry biomass and do not include further processing such as tableting, pasteurizing, packaging, etc. Other costs such as advertising, marketing and administration are also not included.

It is interesting to compare the different models and to examine the relative impact of various factors on this cost. Figure 16.7 shows the proportion of costs attributed to capital costs, operating costs and maintenance costs. With the exception of the Thai *Spirulina* plant, operating costs represent 50 per cent or more of the total costs, with capital costs making up the bulk of the remainder.

A breakdown of the capital components is shown in Figure 16.8. In all cases the culture system and associated site preparation costs make up more than 50 per cent of the total capital costs. The cost of site preparation is very variable. Surprisingly, the harvesting system represents only 10 to 30 per cent of the capital costs. Land costs are a small component, despite the large land area required. This reflects the

Table 16.7. Estimates of cost of algal biomass.

Alga	Culture system	Culture area/ volume	Productivity $(g\,m^{-1}\,d^{-1})$	Cost $(\$ \, kg^{-1})$	Reference
Scenedesmus	Raceway	4 ha	20	7	Becker and Venkataramaran (1980)
Microalgae	Tubular reactor	10 ha	16	7	Tapie and Bernard (1988)
Microalgae	Ponds	10000 ha	20	0.3	Regan and Gartside (1983)
Microalgae	Ponds	800 ha	17	0.4	Benemann et al. (1982)
Microalgae	Raceway	4 ha	20	1.0	Barclay et al. (1987)
Porphyridium cruentum	'Solar water heater'	34 ha	30	14	Anderson and Eakin (1984) – cited in Tapie and Bernard (1988)
Spirulina	Raceway	10 ha	14	6	Richmond (1983)
Spirulina	Raceway	5 ha	12	12	Rebeller (1982)
Spirulina	Ponds	20 ha	12	5	Rebeller (1982)
Estimates based on actual commercial operations					
Chlorella (autotrophic)	Raceway		25–30	22	Kawaguchi (1980)
Chlorella (mixotrophic)	Raceway		25–30	23	Kawaguchi (1980)
Spirulina	Raceway	5 ha	3.2	22	Jassby (1988)
Spirulina	Raceway	1.5 ha	7.3	7–9	Tanticharoen et al. (1993)
Dunaliella	Raceway	2 ha	4	10	Mohn and Cordero-Contreras (1990)
Chlorella	Cascade	1 ha	18	12	Personal observation
Marine microalgae	Bags, towers and tanks ('closed systems')			>600	Coutteau and Sorgeloos (1992); Fulks and Main (1991); personal observations
Microalgae	Raceway	32 ha	20	15	Baseline model

All costs are in Australian dollars and have been adjusted for inflation by using a 5 per year year inflation factor.

403

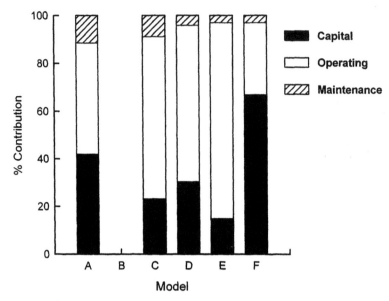

Figure 16.7 Relative proportions of capital, maintenance and operating costs in a number of models of algal production systems. The models are: **A** = Becker and Venkatamaran (1980) model for the production of *Scenedesmus* in India; 50 000 m² pond area, paddle-wheel race-way ponds, 20 cm deep, a growth period of 360 days y⁻¹ with a productivity of 20 g m⁻² d⁻¹, harvested by centrifugation followed by drum drying. **B** = Kawaguchi (1980) model for the production of *Chlorella* in Taiwan; no details are supplied but data are based on actual operations. **C** = Neenan *et al.* (1986) model for microalgae production for liquid fuels in the USA; 8 600 000 m² total pond area, 15 cm deep, a growth period of 360 days y⁻¹ with a productivity of 25 g m⁻² d⁻¹, harvested by microstraining followed by centrifugation. **D** = Jassby (1988) model for *Spirulina* production in California, USA.; 50 000 m² total area of paddle-wheel driven raceway ponds; 20 cm deep (?), a growth period of 360 days y⁻¹ with a productivity of 20 g m⁻² d⁻¹, harvested by vibrating filter followed by spray drying. **E** = the base model described in this chapter. **F** = Tanticharoen *et al.* (1993) model for feed-grade *Spirulina* production grown on treated starch wastewater in Thailand; 15 000 m² concrete raceway ponds, 30 cm deep, a growth period of 360 days y⁻¹ with an annual production of 40 t, harvested with a vibrating filter and spray dried (based on actual operations).

fact that most algal plants are located in marginal land areas where the land cost is low.

The breakdown of operating costs shown in Figure 16.9 shows much greater variation between the models than was seen in the Figures 16.7 and 16.8. This illustrates the difficulty in estimating actual operating costs as well as differences between different algal culture and harvesting systems and also local variations in cost relativities. Labour is generally a large cost component; only in the Becker and Venkatamaran model of an algal plant located in India (Model A), does labour represent less than 20 per cent of total operating costs. Nutrient costs are significant in two of the models (Models A and C), especially in the Neenan *et al.* model (Model C) for a very large plant for producing biomass for liquid fuels. The costs of utilities (electricity, oil, etc.) are also very variable and reflect the relative costs of power in various locations as well as the efficiencies of the harvesting system. The costs of water are generally low. The final category is 'other'; in the Kawaguchi case study

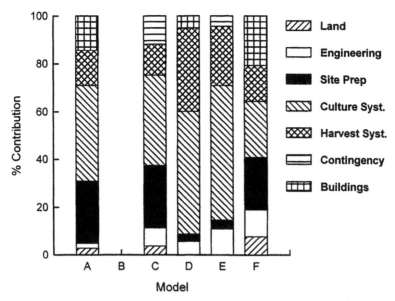

Figure 16.8 Breakdown of the capital costs as a proportion of total capital cost in the models shown in Figure 16.7.

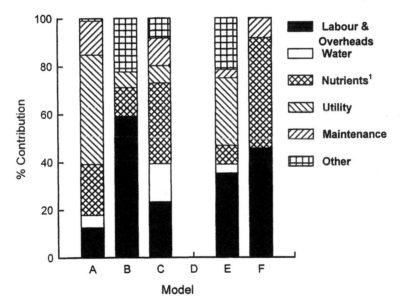

Figure 16.9 Breakdown of the operating costs as a proportion of total operating costs in the models shown in Figure 16.7. ([1] In the case of model **F**, the costs for water, nutrients and utilities are combined under 'Nutrients').

(Model B) this consists mainly of the cost of drying and packing, costs which are not included in the other case studies. In the base model (Model E), 'other' represents the cost of flocculant used in harvesting.

Conclusions

The main conclusion is that microalgae are generally expensive to produce. The reasons for this are mainly the need to supply light which results in either expensive culture systems and/or low productivities compared to heterotrophically produced cultures. Harvesting is also often a major cost factor. A critical task for the person developing a commercial process is to minimize these costs. To do this effectively it is always essential to consider the whole process from culture to harvesting to down-stream processing. Only when this is done can the work be directed to those aspects of the process where optimization will lead to the greatest cost savings. Where out-door culture is considered, either in open ponds or in closed photobioreactors, the selection of a site with the optimum climate for the algae being cultured will lead to significant cost savings. For example, the best use of the capital intensive plant and equipment requires that it be used all year round. This requires continuous sunshine and a good temperature regime.

Despite the high costs of algal production there are several commercially viable large algal plants in Australia, Israel, China, India, Thailand, Taiwan and the USA. Furthermore, the high cost of algal production for feed for aquaculture species is offset by the high value of the final product (fish, shellfish, crustaceans) produced. Furthermore, large-scale algal culture is still a very new activity and over the last few decades the costs of microalgal production have continued to decline due to improved technology and also due to the increasing size of the algal plants. The continued development of new culture systems as well as our improved understanding of algal biology and physiology is certain to lead to new products and processes being developed and new microalgal products reaching the marketplace.

References

Anon., 1994, Separation of colouring matter from *Spirulina*, Japan Patent Number 6,271,783.

Avila, M. and Seguel, M., 1993, An overview of seaweed resources in Chile, *Journal of Applied Phycology*, **5**, 133–9.

Barclay, W.R., Terry, K.L., Nagle, N.J., Weissman, J.C. and Goebel, R.P., 1987, Potential of new strains of marine and inland salt-adapted microalgae for aquaculture, *Journal of the World Aquaculture Society*, **18**, 218–228.

Barclay, W.R., Meager, K.M. and Abril, J.R., 1994, Heterotrophic production of long chain omega-3 fatty acids utilizing algae and algae-like microorganisms, *Journal of Applied Phycology*, **6**, 123–129.

Becker, E.W. and Venkatamaran, L.V., 1980, Production and processing of algae in pilot plant scale. Experiences of the Indo-German project, in Shelef, G. and Soeder, C.J. (Eds), *Algae Biomass*, pp. 35–50, Amsterdam: Elsevier/North Holland Biomedical Press.

Behrens, P.W., Sicotte, V.J. and Delente, J., 1994, Microalgae as a source of stable iso-topically labeled compounds, *Journal of Applied Phycology*, **6**, 113–121.

Belay, A., Ota, Y., Miyakawa, K. and Shimamatsu, H., 1994, Production of high quality *Spirulina* at Earthrise Farms, in Phang, S.M., Lee, K., Borowitzka, M.A. and Whitton, B. (Eds), *Algal Biotechnology in the Asia-Pacific Region*, pp. 92–102, Kuala Lumpur: Institute of Advanced Studies, University of Malaya.

Ben-Amotz, A., 1995, New mode of *Dunaliella* biotechnology: two-phase growth for β-car-otene production, *Journal of Applied Phycology*, **7**, 65–68.

BENEMANN, J.R., 1992, Microalgae aquaculture feeds, *Journal of Applied Phycology*, **4**, 233-245.

BENEMANN, J.R., GOEBEL, R.P., WEISSMAN, J.C. and AUGENSTEIN, D.C., 1982, Microalgae as a source of liquid fuels, *Report to DOE, Office of Energy Research*, May 1982, 1–17.

BLACK, J.H., 1983, Operating-cost estimation, in Jelen, F.C. and Black, J.H. (Eds), *Cost and Optimisation Engineering*, New York: McGraw-Hill.

BOROWITZKA, M.A., 1992a, Algal biotechnology products and processes: matching science and economics, *Journal of Applied Phycology*, **4**, 267–279.

BOROWITZKA, M.A., 1992b, Comparing carotenogenesis in *Dunaliella* and *Haematococcus*: implications for commercial production strategies, in Villa, T.G. and Abalde, J. (Eds), *Profiles on Biotechnology*, pp. 301–310, Santiago de Compostela: Universidade de Santiago de Compostela.

BOROWITZKA, M.A., 1996, Closed algal photobioreactors: design considerations for large-scale systems, *Journal of Marine Biotechnology* (in press).

BOROWITZKA, M.A. and BOROWITZKA, L.J., 1988, *Dunaliella*, in Borowitzka, M.A. and Borowitzka, L.J. (Eds), *Micro-algal Biotechnology*, pp. 27–58, Cambridge: Cambridge University Press.

BOROWITZKA, L.J. and BOROWITZKA, M.A., 1989a, Industrial production: methods and economics, in Cresswell, R.C., Rees, T.A.V. and Shah, N. (Eds), *Algal and Cyanobacterial Biotechnology*, pp. 294–316, London: Longman Scientific.

BOROWITZKA, L.J. and BOROWITZKA, M.A., 1989b, β-Carotene (provitamin A) production with algae, in Vandamme, E.J. (Ed.), *Biotechnology of Vitamins, Pigments and Growth Factors*, pp. 15–26, London: Elsevier Applied Science.

BOROWITZKA, L.J. and BOROWITZKA, M.A., 1990, Commercial production of β-carotene by *Dunaliella salina* in open ponds, *Bulletin of Marine Science*, **47**, 244–252.

BOUSSIBA, S. and VONSHAK, A., 1991, Astaxanthin accumulation in the green alga *Haematococcus pluvialis*, *Plant and Cell Physiology*, **32**, 1077–1082.

CARMICHAEL, W.W., 1992, A review: Cyanobacteria secondary metabolites – the cyanotoxins, *Journal of Applied Bacteriology*, **72**, 445–459.

CHAUHAN, V.S., KOTHARI, R.M. and RAMAMURTHY, V., 1994, Method for efficient extraction of phycocyanin from *Spirulina*, *Biotechnology Techniques*, **8**, 525–528.

CHAUMONT, D., THEPENIER, C., GUDIN, C. and JUNJAS, C., 1988, Scaling up a tubular photo-reactor for continuous culture of *Porphyridium cruentum* from laboratory to pilot plant (1981–1987), in Stadler, T., Mollion, J., Verdus, M.C., Karamanos, Y., Morvan, H. and Christiaen, D. (Eds), *Algal Biotechnology*, pp. 199–208, London: Elsevier Applied Science.

COHEN, Z., REUNGJITCHACHAWALI, M., SIANGDUNG, W. and TANTICHAROEN, M., 1993, Production and partial purification of gamma-linolenic acid and some pigments from *Spirulina platensis*, *Journal of Applied Phycology*, **5**, 109–115.

COUTTEAU, P. and SORGELOOS, P., 1992, The use of algal substitutes and the requirement for live algae in the hatchery and nursery of bivalve molluscs: an international survey, *Journal of Shellfish Research*, **11**, 467–476.

CURTAIN, C.C., WEST, S.M. and SCHLIPALIUS. L. 1987, Manufacture of β-carotene from the salt lake alga *Dunaliella salina*; the scientific and technical background, *Australian Journal of Biotechnology*, **1**, 51–57.

DOUCHA, J. and LIVANSKY, K., 1995, Novel outdoor thin-layer high density microalgal culture system: productivity and operational parameters, *Algological Studies*, **76**, 129–147.

DUBEY, A.K. and RAI, A.K., 1995, Application of algal biofertilizers (*Aulosira fertilissima* Tenuis and *Anabaena doliolum* Bhardawaja) for sustained paddy cultivation in northern India, *Israel Journal of Plant Sciences*, **43**, 41–51.

FULKS, W. and MAIN, K.L., 1991, The design and operation of commercial-scale live feeds production systems, in Fulks, W. and Main, K.L. (Eds), *Rotifer and Microalgae Culture Systems*, pp. 3–52, Honolulu: The Oceanic Institute.

GLADUE, R., 1991, Heterotrophic microalgae production: potential for application to aquaculture feeds, in Fulks, W. and Main, K.L. (Eds), *Rotifer and Microalgae Culture Systems*, pp. 275–286, Hololulu: The Oceanic Institute.

HOSHINO, K., HAMOCHI, M., MITSUHASHI, S. and TANISHITA, K., 1991, Measurements of oxygen production rate in flowing *Spirulina* suspension, *Applied Microbiology and Biotechnology*, **35**, 89–93.

JASSBY, A., 1988, *Spirulina*: a model for microalgae as human food, in Lembi, C.A. and Waaland, J.R. (Eds), *Algae and Human Affairs*, pp. 149–179, Cambridge: Cambridge University Press.

JOHNSON, E.A. and SCHROEDER, W., 1995, Astaxanthin from the yeast *Phaffia rhodozyma*, *Studies in Mycology*, , 81–90.

KAWAGUCHI, K., 1980, Microalgae production systems in Asia, in Shelef, G. and Soeder, C.J. (Eds), *Algae Biomass Production and Use*, pp. 25–33, Amsterdam: Elsevier/North Holland Biomedical Press.

KYLE, D.J., REEB, S.E. and SICOTTE, V.J., 1991, Docosahexaenoic acid, methods for its production and compounds containing the same, World Patent Number 9,111,918.

LEE, C.S. and TAMARU, C.S., 1993, Live larval food production at the Oceanic Institute, Hawaii, in McVey, J.P. (Ed.), *CRC Handbook of Mariculture. Crustacean Aquaculture*, Vol. 1, pp. 15–28, Boca Raton: CRC Press.

LIAO, I.C., SU, H.M. and LIN, H.H., 1983, Larval foods for penaeid prawns, in McVey, J.P. (Ed.), *CRC Handbook of Mariculture. Crustacean Aquaculture*, Vol. 1, pp. 43–69, Boca Raton, Florida: CRC Press.

LIPKIN, Y., 1985, Outdoor cultivation of sea vegetables, *Plant and Soil*, **89**, 159–183.

LIVANSKY, K. and PILARSKI, P.S., 1993, Carbon dioxide supply to algal cultures II. Efficiency of CO_2 absorption from a natural gas supplied to the recirculation pipe of a cultivation unit, *Archiv Fur Hydrobiologie*, **69**, 113–123.

MARTINEZ, M.R., 1988, *Nostoc comune* Vauch., a nitrogen-fixing blue-green alga, as source of food in the Philippines, *The Philippine Agriculturalist*, **71**, 295–307.

MARTINEZ, M.R., PALACPAC, N.Q., GUEVARRA, H.T. and BOUSSIBA, S., 1995, Production of indigenous nitrogen-fixing blue-green algae in paddy fields of the Philippines, in Thirakhupt, V. and Boonakijjinda, V. (Eds), *Mass Cultures of Microalgae. Proceedings of the Research Seminar and Workshop, Silpakorn University, Thailand. November 18–23, 1991*, pp. P51–60, Thailand: UNESCO.

MENDES, R.L., COELHO, J.P., FERNANDES, H.L., MARRUCHO, I.J., CABRAL, J.M.S., NOVAIS, J.M. and PALAVRA, A.F., 1995, Applications of supercritical CO_2 extraction to microalgae and plants, *Journal of Chemical Technology and Biotechnology*, **62**, 53–59.

MERRILL, J.E., 1993, Development of Nori markets in the Western World, *Journal of Applied Phycology*, **5**, 149–154.

MEYER, P.S., DUPREEZ, J.C. and KILIAN, S.G., 1993, Selection and evaluation of astaxanthin-overproducing mutants of *Phaffia rhodozyma*, *World Journal of Microbiology and Biotechnology*, **9**, 514–520.

MISHRA, V.K., TEMELLI, F. and OORAIKUL, B., 1993, Extraction and purification of omega-3 fatty acids with an emphasis on supercritical fluid extraction – a review, *Food Research International*, **26**, 217–226.

MOHN, F.H., 1988, Harvesting of micro-algal biomass, in Borowitzka, M.A. and Borowitzka, L.J. (Eds), *Micro-Algal Biotechnology*, pp. 395–414, Cambridge: Cambridge University Press.

MOHN, F.H. and CORDERO-CONTRERAS, O., 1990, Harvesting of the alga *Dunaliella* – some consideration concerning its cultivation and impact on the production costs of β-carotene, *Berichte des Forschungszentrums Jülich*, **2438**, 1–50.

NEENAN, B., FEINBERG, D., HILL, A., McINTOSH, R. and TERRY, K., 1986, Fuels from microalgae: technology, status, potential and research requirements, *Solar Energy Research Institute, Report*, SERI/SP-231-2550, pp. 1–158.

NONOMURA, A.M., 1987, Process for producing a naturally derived carotene-oil composition by direct extraction from algae, US Patent Number 4680314.

OSWALD, W.J., 1988a, Large-scale algal culture systems (engineering aspects), in Borowitzka, M.A. and Borowitzka, L.J. (Eds), *Micro-Algal Biotechnology*, pp. 357–394, Cambridge: Cambridge University Press.

OSWALD, W.J., 1988b, Micro-algae and waste-water treatment, in Borowitzka, M.A. and Borowitzka, L.J. (Eds), *Micro-algal Biotechnology*, pp. 305–328, Cambridge: Cambridge University Press.

PETERS, M.S. and TIMMERHAUSE, K.D., 1980, *Plant Design and Economics for Chemical Engineers*, 3rd edn, New York: McGraw-Hill, .

REBELLER, M., 1982, Techniques de culture et de récolte des algues spirulines, *Doc. IFP*, pp. 1–14.

REGAN, D.L. and GARTSIDE, G., 1983, *Liquid fuels from micro-algae in Australia*, Melbourne: CSIRO.

REISMAN, H.B., 1988, *Economic Analysis of Fermentation Processes*, Boca Raton: CRC Press.

RICHMOND, A., 1983, Phototrophic microalgae, in Rehm, H.-J. and Reed, G. (Eds), *Biotechnology*, Vol. 3, pp. 111–143, Weinheim: Verlag Chemie.

RICHMOND, A., VONSHAK, A. and ARAD, S., 1980, Environmental limitations in outdoor production of algal biomass, in Shelef, G. and Soeder, C.J. (Eds), *Algae Biomass*, pp. 65–72, Amsterdam: Elsevier/North Holland Biomedical Press.

ROGER, P.A., 1989, Cyanobactéries et rhiziculture, *Bulletin de La Societe Botanique de France-Actualites Botaniques*, **136**, 67–81.

SUBRAHMANYAN, R., 1972, Some observations on the utilization of blue-green algal mixtures in rice cultivation in India, in Desikachary, T.V. (Ed.), *Taxonomy and Biology of Blue-Green Algae*, pp. 281–293, Madras: University of Madras.

SUKENIK, A., LEVY, R.S., LEVY, Y., FALKOWSKI, P.G. and DUBINSKY, Z., 1991, Optimizing algal biomass production in an outdoor pond – a simulation model, *Journal of Applied Phycology*, **3**, 191–201.

TANTICHAROEN, M., BUNNAG, B. and VONSHAK, A., 1993, Cultivation of *Spirulina* using secondary treated starch wastewater, *Australasian Biotechnology*, **3**, 223–226.

TAPIE, P. and BERNARD, A., 1988, Microalgae production: technical and economic evaluations, *Biotechnology and Bioengineering*, **32**, 873–885.

TRONO, G.C., 1994, The mariculture of seaweeds in the tropical Asia-Pacific region, in Phang, S.M., Lee, Y.K., Borowitzka, M.A. and Whitton, B.A. (Eds), *Algal Biotechnology in the Asia-Pacific Region*, pp. 198–210, Kuala Lumpur: Institute of Advanced Studies, University of Malaya.

TSENG, C.K. and FEI, X.G., 1987, Macroalgal commercialisation in the Orient, *Hydrobiologia*, **151/152**, 167–172.

VONSHAK, A., 1987, Strain selection of *Spirulina* suitable for mass production, *Hydrobiologia*, **151/152**, 75–77.

VONSHAK, A., ABELIOVICH, A., BOUSSIBA, S., ARAD, S. and RICHMOND, A., 1982, Production of *Spirulina* biomass: effects of environmental factors and population density, *Biomass*, **2**, 175–185.

VONSHAK, A., BOUSSIBA, S., ABELOVICH, A. and RICHMOND, A., 1983, Production of *Spirulina* biomass: maintenance of monoalgal culture outdoors, *Biotechnology and Bioengineering*, **25**, 341–349.

VONSHAK, A., CHANAWONGSE, L., BUNNAG, B. and TANTICHAROEN, M., 1996, Light acclimation and photoinhibition in three *Spirulina platensis* (cyanobacteria) isolates, *Journal of Applied Phycology*, **8**, 35–40.

Index